Alexander Nimmo
Master Engineer
1783–1832
Public Works and Civil Surveys

NOËL P. WILKINS

Foreword by
Peter B. Heffernan

IRISH ACADEMIC PRESS
DUBLIN • PORTLAND, OR

For Molly

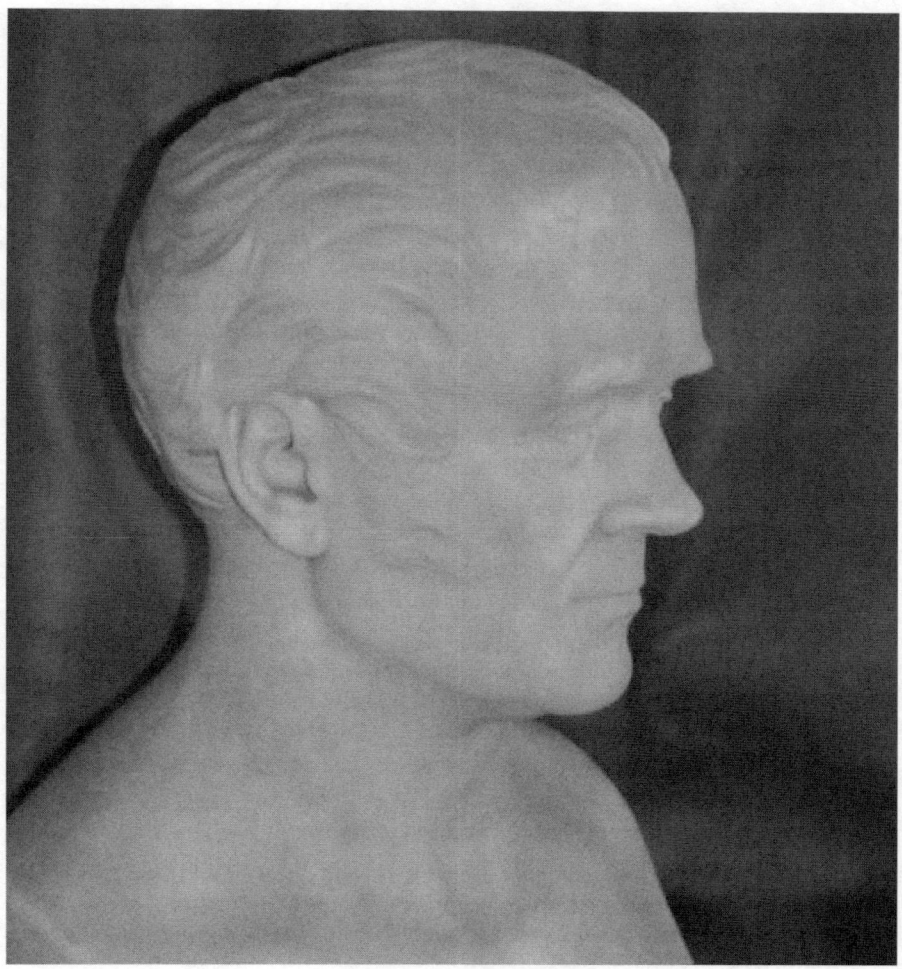

Alexander Nimmo (1783–1832)
Bust in the collection of the Royal Dublin Society, sculpted in 1845 by John E. Jones. Jones trained as an engineer under Nimmo and supervised the construction of the timber bridge at Youghal, 1829–1832. (Photo by the author with permission of the Royal Dublin Society.)

First published in 2009 by Irish Academic Press

2 Brookside,
Dundrum Road,
Dublin 14, Ireland

920 NE 58th Avenue, Suite 300
Portland, Oregon,
97213-3786, USA

Copyright © Noël P. Wilkins 2009

www.iap.ie

British Library Cataloguing-in-Publication Data
An entry can be found on request

978 0 7165 2995 8 (cloth)

Library of Congress Cataloging-in-Publication Data
An entry can be found on request

All rights reserved. Without limiting the rights under copyright reserved alone, no part of this publication may be reproduced, stored in or introduced into a retrieval system, or transmitted, in any form or by any means (electronic, mechanical, photocopying, recording or otherwise), without the prior written permission of both the copyright owner and the above publisher of this book.

Printed by Antony Rowe, Chippenham, Wiltshire

Contents

Brief Chronology	vii
Preface and Acknowledgements	ix
Foreword by Peter B. Heffernan	xii
1. A Young Man of Considerable Genius and Erudition	1
2. The Inverness Perambulation and the *Edinburgh Encyclopaedia*	23
3. The Bogs Commission, 1809–1813	43
4. A Kerry Apprenticeship	57
5. First Foray into Connemara	79
6. Major Harbour Engineering: Dunmore and Cork Harbours	101
7. Grand Juries, Nimmo and the Emergence of the Ordnance Survey of Ireland	127
8. A Labyrinth of Legislation: Famine, Relief Work and the Emergence of the Office of Public Works	148
9. The Fisheries Commission and Nimmo's Coastal Survey	163
10. Master of all he Surveyed? Nimmo in Connaught	187
11. Nimmo's Irish Bridges	214
12. Private Commissions in Ireland	253
13. The Wallasey Pool Plan and other Works in Britain	276
14. Making the Way Permanent: Nimmo and Railways	318
15. A Most Humane and Cultivated Man	339

Appendix 1. A Gazetteer of Nimmo's Piers and Harbours. 340

Appendix 2. A Listing of Nimmo's Reports, Maps, Plans
and Writings and their Locations. 398

Appendix 3. Account of all the structural works, and their state
of completion, carried on by Nimmo in the western
district as listed by him in June 1830. 417

Appendix 4. Statement of the Commissioners for Auditing
the Public Accounts in Ireland regarding the
Accounts of Alexander Nimmo. 420

Sources and Select Bibliography 422

Index 427

Brief Chronology

1783	Born in Fife, Scotland.
1797–98	Attended St Andrews University.
1799–1800	Attended University of Edinburgh.
1802–5	Second Master at Fortrose Academy. Commenced independent scientific observations; paper read at Royal Society of Edinburgh.
1804	Experimenting on hydrography of Loch Ness.
1805–11	Rector of Inverness Academy.
1806	Surveys the boundaries of Inverness-shire for the Commission for Highland Roads and Bridges.
1809	Surveys the High Bridge to Killin road. Elected member of the Geological Society of London.
1811	Elected Fellow of the Royal Society of Edinburgh. Joined the Commission for the Bogs of Ireland.
1811–13	With the Bogs Commission in Kerry and Connemara.
1812	Articles on 'Bridge' and 'Carpentry' published in the *Edinburgh Encyclopaedia*.
1814	Bogs Commission reports published. Surveying roads in Mayo and Leitrim with the Irish Post Office. Surveys Waterford harbour and presents first design for Dunmore harbour. Visits the Continent.
1815–32	Supervising engineer, Dunmore harbour.
1815	Surveys river and port of Cork.
1816	Designs Ballingaule village, County Waterford.
1817	Appointed engineer under Loan Act. Investigates steamship boilers and valves. Modifies diving bell and its air supply.
1818	Elected Member of the Royal Irish Academy.
1819	Evidence to Select Committee on Irish Poor. Appointed engineer under the Moiety Act. First design for Courtown harbour.
1820–24	Engineer with the Commissioners for Irish Fisheries. Carries out coastal survey.

1821	Survey of Sligo harbour and Killala harbour.
1822–32	Engineer of the western district.
1822–23	Most of his roads and piers in the western district started.
1823	Designs Wellesley Bridge, Limerick. Consulted on Newry Canal and proposed Lough Neagh drainage. Redesigns Courtown harbour. Advises on Mersey navigation with Telford.
1824	Dispute with the Admiralty. Evidence to Select Committee on Survey and Valuation. Evidence to House of Lords Enquiry into State of Ireland. The coastal survey terminated.
1824–25	Advising Hibernian Mining Company.
1824–27	Advising American and Colonial Steam Navigation Company.
1825	Plan and survey of Limerick to Waterford railway. Plan for Drogheda harbour.
1826	Evidence to Committee on Norwich and Lowestoft Navigation Bill. Evidence to the Select Committee on Emigration.
1827	Tours Ireland and England with Rowan Hamilton. Evidence to Select Committee on Milford Haven Communication. Designs pier at Hobbs' Point and ferries on River Severn. Designs bridge for Swansea. Advising on Mersey navigation with Telford. Proposes railway, Rathdrum to Wicklow harbour.
1827–28	Plans Wirral canal and Wallasey Pool docks with Stevenson and Telford. Consultant to Duchy of Lancaster.
1828	Elected Member of the Institute of Civil Engineers, London.
1830	Plan for Aberystwyth port. Tests new low-pressure locomotive boiler with Vignoles.
1830–31	Engineer of Liverpool and Leeds Railway Company. Consultant, St Helens Canal and Railway Company. Consultant, Preston and Wigan Railway Company.
1830–31	Designs Dublin to Kingstown railway.
1831	Plan for ship canal, Dublin to Kingstown.
1832	Dies, aged 49, in Dublin on 20 January. Book entitled *Piloting Directions* and chart of Irish coast published posthumously.

Preface and Acknowledgements

Throughout his life Alexander Nimmo, schoolmaster turned civil engineer, was rarely mentioned without the epithet 'eminent' attached to his name. Why was this? He was an engineer of acknowledged genius and erudition who came from an academic rather than a practical background and who made a lasting impression on the landscape and seascape of Ireland – and on his contemporaries in engineering, public and political life. He started his engineering career with Telford in Inverness, and came to Ireland in 1811 to join the Bogs Commission survey that had started in 1809. He spent the rest of his life in Ireland, with short periods spent in England, especially in his later years, and died at the age of 49 in 1832. A full account of his life and works has never been written.

In the first three decades after its enactment in 1801, some persons in Ireland, native Irish and British alike, saw potential in the Union for economic improvement, structural development and social advancement of the country and were willing to work towards its realisation. These forward-looking men needed technical experts to devise schemes and implement developments that they could promote through their influence in political and business circles. Nimmo more than any other engineer was the one on whom they and the government came to rely for this expertise. He had in abundance the necessary scientific and engineering skills, the broad vision for development, the appreciation of natural resources and an indomitable disposition that was not dismayed by even the most distressing conditions of famine that were a regular feature of Irish life in those decades.

Working as an official government engineer, he designed harbours and piers in almost every maritime county; he carried out some of the earliest hydrography ever done in Ireland; he designed and built bridges and roads in every province; he laid down most of the Ring of Kerry road; he designed railways and obtained Acts of Parliament for them, including Ireland's very first proposed line (the Limerick to Waterford) and the historic Dublin to

Kingstown line. He improved steam-engines, fixed windmills, used and improved diving bells. He made the first ever geological map of Connemara and, it is argued here, may even have pre-dated Griffith in making the first geological map of the whole country. He determined latitudes and longitudes, mountain heights and sea depths. He went down mines, up mountains and far out to sea, out to and beyond the hundred-fathom line. He was friendly with the powerful and famous, but empathic with the Irish poor, defending and praising them before many Select Committees of both Houses of Parliament. He played a seminal role in the emergence of the Ordnance Survey of Ireland, the Fisheries Commission, the Hydrographic Survey and the Office of Public Works, institutions that more than anything else signalled the dawning of a new, technological, post-Union Ireland.

Another epithet — elusive — must also attach to him. We do not know the exact date of his birth, or know if he ever married. His grave went unmarked and its location is now long forgotten. He left no private papers or testament that we know of, and his family had all died or left Ireland by 1850. Personal privacy in the midst of public works was the distinguishing feature of his life and the private man, in particular, has remained mostly hidden. This elusiveness, together with his influence on the landscape and seascape of Ireland make recording his life and works a difficult but worthy subject. Although the available public records alone cannot meaningfully reveal the full nature and depth of his character, and his own writings are almost entirely technical, I have tried to see through to the *persona* of this private man, to enliven our knowledge of him, and to describe, for the first time, all that he successfully achieved, along with those plans and schemes that he attempted without success. One can only hope that this effort, coming almost two hundred years late, brings some justice to the memory of a man who so enriched his adopted country.

The information could not have been gathered without the help of the staffs of many archives and libraries, including: the National Archives of Ireland, the National Archives of Scotland, the National Archives of the UK (Kew), the Public Records Office of Northern Ireland, Liverpool Records Office, Sheffield Archives and the Irish Architectural Archive; the National Library of Ireland, the National Library of Scotland, the British Library, Cambridge University Library, University College London Library, James Hardiman Library NUI Galway, St Andrews University Library, Durham University Library, Inverness Library (Highland Archives), Firestone Library, Princeton University NJ; the Highland Folk Museum Kingussie, the records of the Port of Cork, the Crawford Municipal Gallery Cork, the Institute of Civil Engineers London, the Royal Irish Academy, the Royal Society of Edinburgh, the records office of Edinburgh University, the Maritime Museum Liverpool and the Geological Society of London. To the staff of all these I express my sincere thanks for their patient and expert help.

Colleagues and friends Eoin McLoughlin, An tOllamh Gearóid Ó'Tuathaigh, Professor James Houghton, Paul Duffy (Galway County Council), Tom and Des Kenny, Tim Collins, Noel O'Neill, Dr Martin Feely and Siuvaun Comer all gave invaluable help in numerous ways. Robert Preece, archivist, Inverness Royal Academy, was a mine of information on Nimmo's tenure there. Many others too helped with comments, advice and encouragement in what was a long research task.

Particular thanks are due to Mícheal O'Cinnéide of the Marine Institute for his positive support and lively encouragement when the task seemed almost too daunting, and to Dr Peter Heffernan, Chief Executive of the Institute, who organised sponsorship, encouraged the work and contributed the Foreword. Without their confidence and help the book would never have come to fruition. To the sponsors, the Marine Institute, the Ordnance Survey of Ireland and the Heritage Council, I owe a particular debt of gratitude; I trust the book lives up to their high standards and high expectations of it. Where it fails to do that, the fault is entirely my own. The opinions expressed and analyses of events represent only my own views and may not necessarily be representative of those of the sponsors.

The Trustees of the various libraries, and document owners, have generously given their permission to quote from manuscript material in their possession which I gratefully acknowledge, especially the Trustees of the National Library of Scotland for permission to quote extensively from Nimmo's *Journal*. Reproductions of pictures are acknowledged in the picture legends.

I am happy to thank the National University of Ireland Galway for library, computing and considerable other support over many years, and my family for their boundless patience and understanding.

Financial Support of the Marine Institute, the Ordanance Survey of Ireland and the Heritage Council Publication Grant Scheme is gratefully acknowledged by the author and publisher.

Foreword

I am delighted to introduce this important book on the life of Alexander Nimmo (a full account of whose life and works has never been written) both in the west, but most especially in the field of maritime development. While he was Scottish by birth, Nimmo proved himself to be Irish by choice and dedication and he was our first indigenous (if not native) hydrographer/oceanographer. But his work in England and Wales was also noteworthy.

In the book about fifty-five piers and harbours in fourteen Irish counties are attributed to him. His work on and use of diving bells is revealed, as is his surveying out to the 100-fathom line and his analysis of seabed sediments off the coasts of Ireland, England and France. On land, the book associates him with the design or supervision of more than thirty stone bridges (until now only two were confidently attributed to him); he made innovative plans for what is now Birkenhead in England, and of course he built Dunmore Harbour and designed harbours at Kilmore Quay, Rossaveal (illustrated in the book), at Greencastle and Berehaven. In this way, he contributed to the development of harbours that are now major centres of Irish fisheries.

Nimmo's contribution was an early, important attempt in addressing the deficit of the physical and institutional infrastructure in Ireland with the new technology of the Industrial Revolution. Many of the structures that he left behind (built mostly in times of famine and disease and meant mostly to be 'temporary relief works') have been altered or modified since his time. But thankfully many of them have survived and we can still identify his works, distinguishing the original from the more recent modifications and additions, even now, two hundred years after him. The gazetteer at the end of the book does this most noticeably. He also designed Ireland's first railway and supported a steamship line to America from Valentia.

The persistence into our own time of so many 'temporary' works is a tribute to Nimmo and to the skills and workmanship of the starving men who participated in their construction. They are important human heritage items

FOREWORD

in our maritime culture and tradition and, indeed, in our wider heritage canvas. To have them recorded here, and to have their designer properly acknowledged, is a duty of care that this book has sought to fulfil.

The book has many valuable echoes and lessons for us in 'modern Ireland'. The energy, vision and the project management skills shown by these early commissions and engineers was remarkable; the speed and responsiveness of the parliamentary and committee system in Westminster in the early nineteenth century is worthy of note. The book brings out the essential role of the British administration in the technical development of Ireland and its infrastructure. It will be of great interest to students of engineering, history and public administration.

On behalf of my colleagues in the Marine Institute, the Ordnance Survey and the Heritage Council, I wish to express deep appreciation to Noël Wilkins for his dedication and scholarship in bringing us this insight into our heritage.

PETER B. HEFFERNAN
Chief Executive Officer
The Marine Institute

CHAPTER 1

A Young Man of Considerable Genius and Erudition

In 1783 a son, Alexander, was born in Cupar, Fife to John Nimmo a watchmaker in Kirkcaldy.[1] At least two more sons would be born to him in due course and all three would eventually work together in Ireland.

Young Alexander was raised in Kirkcaldy, probably over the shop. Watchmaking was hardly a very lucrative activity in a provincial town in the days before mass production and mass ownership of timepieces, and it is not surprising that Nimmo's business had changed to a general hardware shop by 1822.[2] Observing and helping his father during his youth must have introduced Alexander early on to the intricacies of time and motion, the necessity for precision and accuracy and the principles of physics and mechanics. High Street looks out on the Firth of Forth and on cold, clear days young Alexander could see the smoke rising above 'Auld Reekie', only ten miles away on the far side of the Firth. Edinburgh was then in the midst of the construction of the new town, and ordinary Kirkcaldy folk could not have been unaware of the thriving bustle across the water. How much more exciting it must have appeared than Kirkcaldy, which had only one street, causing it to be known as 'the lang toon'. By any measure, the construction of Edinburgh New Town was an event of Renaissance proportions in Scotland. Rennie, Telford, Stevenson and other renowned engineers were drawn there and worked in the planning, design and construction that were taking place even as the intellectual, commercial and urban élite of the Kingdom enlivened and graced the new environment.

Young Nimmo proved to be a highly intelligent, even gifted, youngster who came to the attention of William Ferguson of Raith, just west of Kirkcaldy. Ferguson had two sons, Robert (1770–1840) and Ronald (1773–1841), who were tutored privately at Raith from 1882 to 1887 by John Playfair. Playfair, a brilliant teacher of mathematics and an intimate of Adam Smith, was a graduate of St Andrews University who had resigned his church ministry to take up the tutoring post and to pursue his mathematical studies in greater earnest.[3] While employed at Raith he wintered with the

Fergusons in Edinburgh, where he moved easily through literary, scientific and intellectual society. Among his friends were the geologist James Hutton and other eminent professors. His move from the Church had proved a successful strategy, since he was soon (1785) appointed Professor of Mathematics at the University of Edinburgh, where he subsequently enjoyed a long, illustrious career.

Ferguson senior was the patron of three bursaries, called Bayne's Bursaries, in the University of St Andrews. Each paid a stipend lasting four years to enable a young man to attend the university. One of the bursaries fell vacant in 1796 and it is very likely that Ferguson sought the advice of his learned friend Playfair as to suitable young men for nomination. In the event, Ferguson nominated Alexander Nimmo to the bursary, with a stipend of ten pounds sterling (about 120 Scottish pounds) attaching to it. The money, which was to be paid in a moiety every term, was payable by the Magistrates and Town Council of the Burgh of Cupar 'at the terms and manner specified in their Bond and Act of Council'.[4] The Council consented to Nimmo's nomination in July 1796, recording the decision at their meeting on 2 November that year. Ferguson specified that Nimmo was to receive the benefit of the bursary 'commencing the term of Whitsunday last [June] notwithstanding the date hereof' and went on to request that the very reverend and learned principals and professors of the United Colleges of St Salvator and St Leonard 'admit and receive the said Alexander Nimmo to the said Bursary and the privileges and perquisites thereof'.' A recommendation backed by Professor Playfair of Edinburgh University would have greatly influenced the professors in Nimmo's favour. In February 1797 Alexander, who was then 14 years old, signed the Register of the University and was duly admitted to the College of St Salvator and St Leonard.[5]

Had Nimmo not been granted a bursary, it is doubtful that he could ever have aspired to education at the university. His father's trade was hardly sufficient to keep him in the college without financial support, and Ferguson's specification that Alexander was to receive payment from Whitsuntide 1796, although he did not register in the college until early 1797, may have been an effort to help in a practical way. The privileges and perquisites attaching to the bursary would also have been useful. For example, we know from the minutes of the college that other bright students not fortunate enough to receive bursaries received such perquisites as 'vacant seats at the Bursars' table', and one James Dick was granted 'his dinner and supper'.[6]

A little west from Kirkcaldy lies the town of Burntisland. It was here that William Bald was born in 1789. Living only six miles apart, the two boys may have known each other as youngsters, or later on as young men when they both lived in Edinburgh. Bald went to Edinburgh at 12 years of age, where, on soon leaving school, he became apprenticed to John Ainslie, civil engineer and cartographer. From there he would rise, through surveying and cartography, to become a well-renowned engineer, who, as we shall see, would subsequently weave in and out of Nimmo's professional life.

In St Andrews, Nimmo studied Greek, Latin and Mathematics for two years and logic, ethics and natural philosophy in his second, senior year. Coming home over the hills he cannot have failed to notice the progressive expansion of Edinburgh on the distant shore, and perhaps he longed to move further afield. St Andrews was a very fine university, but for mathematics and natural philosophy (nowadays called physics) there was nowhere in Scotland to beat Edinburgh in those days. Thus, after two years he changed universities and headed off across the Firth. He was still only 16 years old when he matriculated in Edinburgh University in 1799, but this was not an unusually tender age for Scottish university students at the time.[7]

Scientific, legal, financial and literary life was vibrant and the city's intellectual exuberance must have impressed any clever young man with a mathematical or scientific bent. John Playfair, mathematician, physicist, geologist and friend of Nimmo's patron, whose name perfectly reflected his 'mild and placid manner', was Professor of Mathematics;[8] John Robison, engineer, chemist and naval organiser, whom James Watt described as 'a man of the clearest head and the most science of anybody I have known', was Professor of Natural Philosophy.[9] Sir James Hall, another friend of the geologist Hutton, was advancing the latter's theories through laboratory experiments, while Playfair was advancing them by pedagogy. All of these scientists were deeply immersed in the emerging new science of geology. They were convinced Plutonists, that is, they subscribed to the theory that granite had an igneous origin. In vernacular, if not jocular, terms they were known as 'firemen'. On the other side, Robert Jameson, mineralogist and soon-to-be Professor of Natural History, was propagating his view of the origin of granite by precipitation from a primordial fluid that was thought to have covered the primitive earth. He and his followers were therefore known as Neptunists or 'watermen'. The heat and steam of the firemen–watermen debate attracted students to Edinburgh from all over the Old and the New Worlds and in this intellectual ferment the modern science of geology came into being.

The young man from Kirkcaldy, who had once watched the city growing on his youthful horizon, had at last arrived, and the intellectual milieu, as well as the physical fabric of the growing city, was all that he could ever have hoped for. He threw himself into the academic life of this Athens of the North, attending Playfair's lectures in mathematics and Robison's in natural philosophy. For his part, Playfair probably remembered Nimmo from the Bayne's Bursary and was happy to take the young man under his tutelage. Another student of Playfair's at the time was David Brewster (1781–1858), only two years older than Nimmo and destined to become one of Scotland's finest exponents of science. By 1802, when he was only 21 years old, he was editor of the *Edinburgh Magazine*, later called the *Edinburgh Philosophical Journal*, which he had

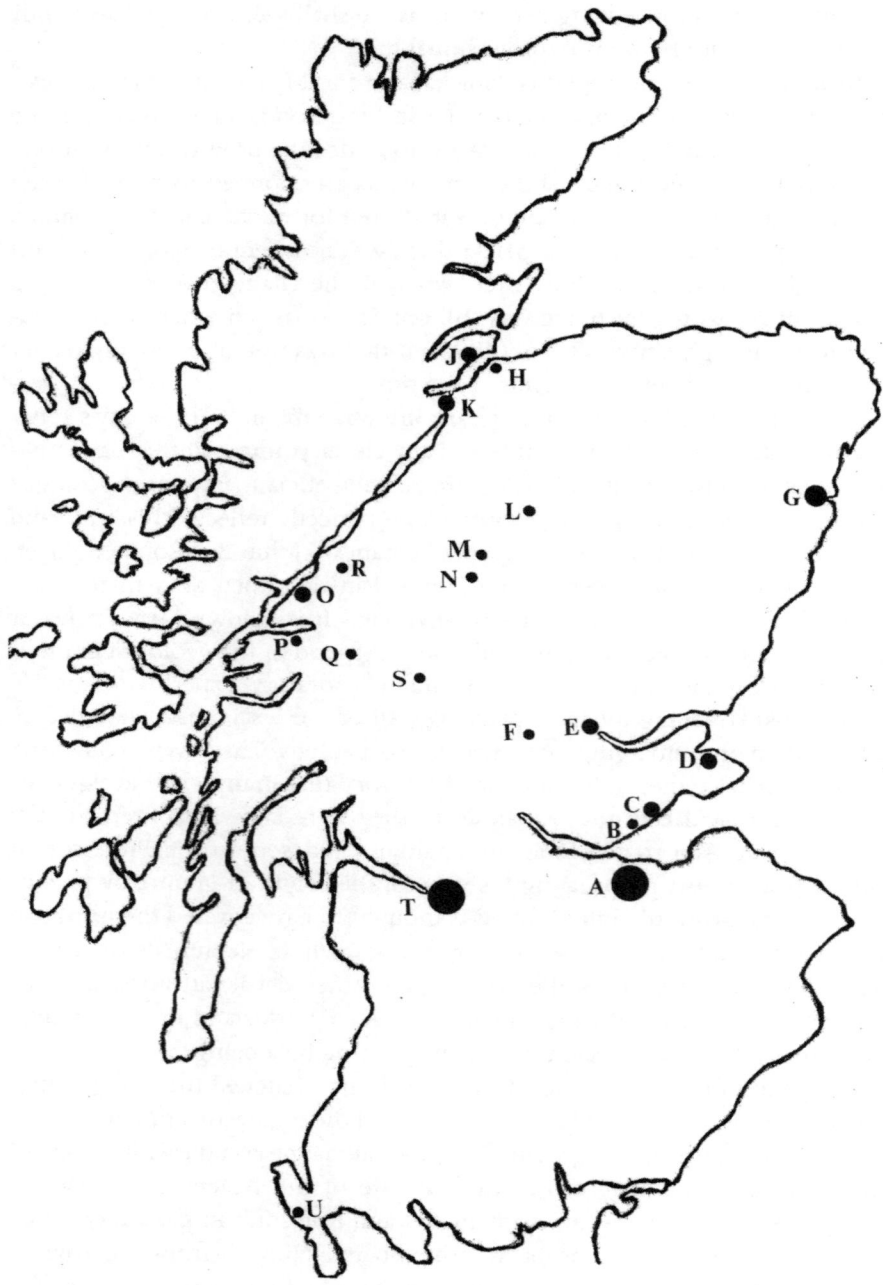

Figure 1.1
Map showing locations in Scotland of places mentioned in the text.
A Edinburgh; **B** Burntisland; **C** Kirkcaldy; **D** St Andrews; **E** Perth; **F** Crieff; **G** Aberdeen; **H** Fort George; **J** Fortrose; **K** Inverness; **L** Grantown; **M** Kingussie; **N** Dalwhinnie; **O** Fort William; **P** Ballachulish; **Q** Ard Feadh (Kings house); **R** Spean Bridge; **S** Killin; **T** Glasgow; **U** Port Patrick.

co-founded with Professor Jameson. Some years later (1807) he founded and edited the *Edinburgh Encyclopaedia*, to which Nimmo would contribute a number of distinguished articles. That the two were acquainted while they were students of Playfair's in 1799 and 1800 seems inevitable. Had they not, it is almost inexplicable that Nimmo, when just a schoolmaster in distant Inverness, would be the person selected by Brewster to contribute so much to the *Encyclopaedia*: there were certainly sufficient intellectual and academic lights illuminating Edinburgh and Glasgow at the time to make it unnecessary to call upon a provincial schoolmaster for his input.

After graduating (and we have no firm record of his degree or his graduation; once again, this was not unusual in universities of the time), Nimmo became a tutor in Edinburgh until he was appointed second master (of three) at the Academy of Fortrose, a town on the Moray Firth just north of Inverness (see figure 1.1). When and why he left Edinburgh to go to such an apparently isolated teaching post is not known, but slender financial means must surely have influenced his decision. If he had successfully carried his Bayne's Bursary with him on transfer from St Andrews, it would have finished in 1800, having run its full four years. Without financial support thereafter, he would have needed to take up paid employment where and when it was offered to him. Maybe Playfair advised him to go north for experience, and might even have recommended him to the Presbytery of Chanonry, which administered Fortrose Academy and was the body that appointed the masters. As a friend of Nimmo's wrote many years later, 'Mr Nimmo was one of those ... who were waiting in Edinburgh until Playfair should find situations in which they were likely to be useful both to the public and themselves.'[10] Whatever his reason, Nimmo had left Edinburgh and was at his new schoolmaster's post in Fortrose early in 1802. Although not a minister or preacher of the Gospel, he subscribed, in February 1802, to the 'Formula' – a statement of adherence to the precepts of the Scottish Presbyterian Church – as a condition of his new post.[11]

His teaching duties encompassed 'arithmetic, book-keeping and drawing; the elements of Euclid, navigation, land surveying and other mensurations; architecture, fortifications and gunnery. Also the elements of chemistry and natural philosophy.' For this, he received a salary of thirty-five pounds sterling, class fees, and 'the profit from boarding gentlemen's sons under his immediate inspection' together with 'private along with public tuition in that excellent and commodious house attached to his office in the Academy'. In 1802 he had fifteen students in his various classes whereas Mr MacKaine, the third master and teacher of English reading and grammar, had sixty. During term, Nimmo was expected to accompany his students to church every Sabbath.

This might have been the start of a scholarly, quiet, reasonably comfortable life in a regional backwater. The bucolic delights of the location were included in an advertisement for a new master in 1809 as follows: 'The fine and healthy situation of Fortrose, and its convenience for sea-bathing are well-known. Provisions, especially fish of the best kinds, are sold as reasonably here as in most towns of the North; and there is near and frequent access, by water, to several other good markets. The town's harbour, for delivering coals, etc. is at the foot of the Academy garden.'[12] But by April 1805 a new Rector, Mr James Tod, was making life difficult for the students and therefore for Nimmo. Complaints were made to the Trustees regarding the harshness and extreme severity of discipline practised by Tod towards the students, and how this very materially injured the reputation of the Academy.[13] A good number of students had left his classes in disgust and others had been deterred from entering. The Trustees wrote to him suggesting he resign his office forthwith. Understandably he refused to do this, insisting on a proper public enquiry into his conduct. An enquiry was duly held, and the charges against him were eventually dropped. It was held that he had always acted from the best of motives and, much more importantly, the Trustees were not prepared to admit to their own misjudgement in appointing him: 'The Presbytery', the enquiry found, 'having formerly expressed (by various advertisements) their public approbation of Mr Tod's conduct, believing that he has at all times acted by the best and purest motives ... unanimously express it as their opinion that Mr Tod should continue to act as Rector'.[14]

Two months later the students were examined by one of the trustees. Only five students of Mr Tod's class turned up and very thin numbers from the other classes. The minutes of the meeting of the Presbytery on 18 June record: 'Their teachers, particularly Mr Nimmo second master appeared quite discontented at the effects of Mr Tod's severity, so that being offered an office in the neighbouring Academy of Inverness, Mr Nimmo, although much attached to Fortrose, will probably soon resign his office here which, as he is a young man of considerable genius and erudition will be a detriment to the Fortrose Institution, it being very difficult to find a well qualified successor to his office.'[15] The outcome of the enquiry was mentioned again and created further division in the Presbytery. Eventually Tod agreed to resign at the end of the December 1805 session. By then Nimmo had already departed to a new post as Rector of Inverness Academy and that terminated his links with Fortrose.

Rector of Inverness Academy was without doubt a far better position than second master in Fortrose, and Nimmo's departure from the latter may not have been entirely a principled reaction to Tod's behaviour, as analysis of the chronology of events shows. The position of Rector, or Headmaster, had fallen vacant in March 1805 on the death of Alexander McGregor, the incumbent. At its meeting on 2 April 1805, the Board of Directors had before it a letter from Nimmo offering himself as a candidate for the vacancy, together with certificates of his ability and character. Leaving aside the unseemliness of his haste, he had submitted this letter more than two full weeks *before* the complaint against Tod was raised at the Presbytery on 17 April, but maybe Nimmo was aware of Tod's unsatisfactory behaviour long before then. The directors at Inverness decided to advertise the post in the Edinburgh papers.

At a general meeting held on 22 May to examine candidates, applications were considered from Nimmo, Mr Pollock of Glasgow, Mr Tulloch an assistant master at Inverness, and Mr Adie of Nielston near Glasgow. It was decided to ask some eminent professors of Edinburgh University to help with making the choice for the post by setting up a comparative trial of the candidates. Professors Finlayson, Dalziel, Hill, Playfair and Leslie, or any three of them, were suggested as possible examiners. Mr Pollock objected to a comparative trial, and although he was a highly regarded candidate, he effectively withdrew his application by absenting himself from it. Later on he applied for, and got, Nimmo's old job at Fortrose Academy, and when Tod eventually resigned he was made its Rector – it is an ill wind that blows no one some good!

When the professors (Joseph Hill, chair of Humanities, John Playfair, chair of Natural Philosophy, James Finlayson, chair of Logic and Thomas Charles Hope, chair of Chemistry) convened to interview the candidates on 27 June, Nimmo alone presented for examination. Messrs Tulloch and Adie did not appear, for whatever reasons. It was nine days earlier, on 18 June, that the Presbytery of Chanonry had let it be known that Nimmo had been offered a post at Inverness, so perhaps everyone knew in advance that he was in an unassailable position even before the comparative trial. Playfair's membership of the panel cannot have done Nimmo's candidacy any harm: as recently as the previous December, Playfair had read a paper written by Nimmo to the Royal Society of Edinburgh, so he was well up-to-date with the progress of his ex-student.[16]

The professors reported as follows:

> Agreeable to the request of the Directors of the Inverness Academy signified by James Grant Esq. Provost of Inverness in his letters to Professor Playfair of 22nd May last and the 5th curt., we the undersigned met this day at the College Library to examine such candidates for the place of Rector in the Academy of Inverness as might come forward to submit their qualifications to a comparative trial. There appeared before us only Mr Nimmo at present at the Academy of Fortrose who was accordingly examined by us on the subjects of Mathematics, Natural Philosophy, Chemistry, the Greek and Latin and French languages in all of which he acquitted himself very much to our satisfaction. In the sciences above named in which he underwent a long and particular examination he showed great accuracy and extent of knowledge as well as ingenuity and soundness of judgement. We can have no doubt therefore in recommending him, the said Mr Alexander Nimmo, to the Directors as possessing the qualifications which they require in the Rector of their Academy in an eminent degree and as likely to discharge the duties of that office much to his own credit and the satisfaction of all concerned.[17]

At their meeting on 13 July the directors thanked the professors for their help, noting especially that Playfair 'had taken particular trouble in the business', and they duly appointed Nimmo to the post.[18]

The post of Rector did not pay greatly, but was one of significant reputation and social status.[19] Since he was then only 22 years old, Nimmo's appointment was quite an achievement. But despite its kudos, and the advancement of his fortune that rectorship represented, Inverness Academy can hardly have been a 'plum' posting for a young, highly talented, ambitious lowland man of wide scientific interests and acknowledged mathematical and linguistic skills, who had already experienced the urbanities and luminaries of Edinburgh. Inverness was a far cry from the bustle of the capital. Communication with other Scottish towns was minimal: there was no regular coach service to Edinburgh, carriage roads being almost non-existent and requiring a difficult ferry crossing at Dunkeld. The great mountainous region of the Northern Highlands brooded over the north and west; to the south, Monadliath, the Grampian Highlands and the Moor of Rannoch cut off easy direct access to the lowlands. Post horses carried the mail from Aberdeen three days a week and it was 1811 before regular coaches ran between that city and Inverness. Inverness had a sailing-ship connection with Edinburgh (Leith) and London, but to all intents and purposes it was the isolated capital of a remote, rugged and unruly, if no longer rebellious, region.

What could have induced Nimmo to remain in this remote town far from the scientific and intellectual communities of the greater university cities? Why, after Fortrose, did he not return to Edinburgh with whatever money he may have saved by then? To answer these questions and understand his motivation we need to consider the social and economic circumstances that prevailed in northern Scotland at the time. These, more than his rector's responsibilities, would come to dominate his outlook and ultimately would shape his later life and career. But first we must consider his tenure of this new post.

A contemporary item in the National Archives of Scotland purports to give the number of students attending Inverness Academy in 1800/01.[20] This was stated to average 230 in 1800 and 245 in 1801. These are unbelievably large numbers, even for an academy servicing the educational needs of the local Highland gentry. English was said to be the largest class, with 108 pupils, then arithmetic, with 80, bookkeeping 49, French 22, Mathematics 16, Geography 12, drawing 7 and Gaelic a lowly 4. If these were the true numbers attending the various subjects then it seems that many pupils must have been counted a number of times, which would account for the large total number. There were, after all, only six classrooms in the whole establishment. Joseph Mitchell, writing in 1883, said there were about 300 pupils during Nimmo's time, which was before he himself had been a pupil there.[21] However, Mitchell was recalling from memory events that had occurred almost eighty years earlier and he went on to say of the Academy, 'still it was a poor place'. The figures, right or wrong, at least indicate the range of subjects taught just before Nimmo took over.

A further item in the Scottish archives, unsigned and undated but probably dating from the 1790s, is titled 'A Plan for Inverness Academy'.[22] It sets out the rules and regulations that were to govern the school. There were to be two sessions a year – 16 July to 16 December and 6 January to 1 June. Along with the rector, there were four masters, for English, Latin, Writing and Arithmetic. The rector was to supervise and direct the others, inform them of the proper mode of teaching and acquaint the directors with the progress of the school. He himself was required to teach Chemistry, Natural Philosophy, Astronomy, Civil and Natural History and his teaching hours were limited to three hours a day, from 10 to 11 and 1 to 3, with no teaching on Saturdays or Sundays. For this he was to receive 1½ guineas per pupil, per session plus one-seventh of the fees of all the other masters. One other of the written regulations is worthy of note: in the maths and philosophy classes 'order was to be preserved by fines; in the others, by the whip.'

By the time Nimmo departed the Academy in 1811 the curriculum had become immensely more sophisticated and leaned noticeably to the scientific. A final document in the archive entitled 'Relating to the Academy at Inverness, 1810' describes the curriculum in the year before he left.[23] First class was English and Grammar, second Latin and Greek, and third Arithmetic and Bookkeeping. These were the traditional subjects of the existing masters. Fourth class was new and astonishing: 'Elements of Euclid and their application to plane and spherical trigonometry; mensuration of solids and surfaces in all its parts; Geography with the use of Globes; Navigation and the most useful parts of practical Astronomy; Naval, Civil and Military Architecture; Practical Gunnery; Perspective and Drawing'. In the fifth or highest class, which was taken by the rector, were taught Civil and Natural History, Experimental Philosophy and Chemistry. There were still only four masters, so Nimmo must have been kept busy in those years.

Pupils generally remained at the academy for two years. In their first year they obtained a knowledge of 'Arithmetic and the different exchanges with foreign practice, bookkeeping, drawing, French, Euclid's elements, plane trigonometry, mensuration of surfaces, gauging, navigation and fortification'. In the second year they studied 'spherical trigonometry, natural and experimental philosophy in all their branches, Algebra, Fluxions etc.' According to this document, the number of pupils was 'about 80 and increases each year' – a much more likely number than that given earlier. The addition of 'fortifications' and 'gunnery' to the curriculum no doubt reflected a response to the Napoleonic Wars then ongoing on the Continent and the need for informed young recruits to join the Highland regiments.

There can be little doubt that Nimmo, in his relatively short tenure as rector, had profoundly influenced the nature of the curriculum at the academy. The

Board of Directors acclaimed his success at a meeting in 1807, expressing great satisfaction at the outcome of the latest student examinations.[24] Strangely, when Nimmo presented a proposal to the meeting making the case for a library in the school, the board was divided and referred the matter to a committee of visitors that they had set up earlier. Undaunted, Nimmo went on to request funds to build a wall around his garden, and this was also referred to the visitors, but with the rider that if they approved the proposal the funds could be allocated. However, in 1811 Nimmo was obliged to pay £19 that was outstanding for two years for work on his house.[25] Later still (1819), Mr Ettles, a bookseller, demanded the return of books that Nimmo had bought before he retired but had never paid for, so the visitors may not have been overly generous.

Nimmo's tenure was not entirely felicitous. A complaint had been made against one of the masters – Mr Robert Moodie, teacher of Gaelic and Drawing – that his teaching to a private evening class was disrupting his service to the academy and Nimmo, having enquired into the matter, satisfied the board that Moodie no longer taught the offending class. The board was happy that the problem seemed to have gone away, but in truth it had not. The following May (1808) Moodie resigned his teaching post, stating in his letter of resignation: 'The principal cause of dissatisfaction and that which involves all the others, is that suspicion and jealousy with which the teachers of Inverness Academy are watched ... It becomes, of course, unsafe for a man who considers teaching as his business for life to remain long in such a situation, although it may do very well for those who are travelling towards the Church through the office of a teacher.'[26] He went on: 'There are, besides, some circumstances which must make the particular office I have now resigned either a warfare or a slavery but these are at most beneath notice.' What his real grievances were we do not know. That Nimmo might have contributed to his bitterness is most unlikely, since both men had become friends in Fortrose and remained so for the rest of Nimmo's life. On 29 July, just two months after his resignation, Moodie changed his mind and reapplied for his old job, on the grounds that the working conditions had changed for the better. The directors were in no mood to accede and they rejected his application.[27] Having spent some years teaching in Dundee, he moved to London, became a journalist and later editor of the *Sunday Times*. He is remembered as a writer on miscellaneous topics, especially but not exclusively on natural history, spelling his name Mudie. An excess of officiousness such as Moodie's letter may hint at, does not at all seem to fit the picture of Nimmo that other records paint. Indeed, he may have been less than meticulous in some of his own school duties: in 1807, having delayed too long on his surveying work (chapter 2), the directors had had to call him back to the academy: 'I had already exceeded the term of absence allowed from my occupations here,' he wrote, 'and I was given to understand that my return was much wished for.'[28]

Moodie's resignation was to return to colour, or maybe grace, Nimmo's reputation. Very many years later Mitchell recounted a story allegedly about Nimmo, but which is probably a conflation of Moodie's case and other events. According to this, the students were wont to assemble in the school hall on Sundays to hear a prayer and, headed by their masters, march in procession to church. Nimmo, 'who was a social [sic] man, used sometimes to neglect this duty on Sundays. The Magistrates, austere men, particularly when their authority was questioned, censured him and he resigned his appointment to the great loss of the Institution.'[29] If this were true, and there is absolutely no contemporary evidence to back it up, it would suggest a certain contrariness about Nimmo: he had already resigned from Fortrose Academy allegedly on principled grounds: to resign one master's job on principle is understandable; to resign two would be positively contrary.

There was, however, one particular event that marred his tenure at Inverness and that resonates discordantly with his apparently principled departure from Fortrose. In February 1810 a letter was sent to the directors complaining about the severity of a beating administered by Nimmo to one of the pupils. The pupil's father wrote:

> On Monday 29th ult. [my son] suffered a most cruel beating from Mr Nimmo, who, not satisfied with using his fists, seized a heavy mahogany ruler and with it beat the boy upon the shoulders, back and thigh. The marks and bruises are evidences of the usage he received which might, and still may prove of very bad consequences. However, I shall not at present enter into a more detailed account of the injuries received as that will best appear from the evidence of the medical gentlemen and the masters who saw the boy afterwards as well as the testimony of the other boys who were present.[30]

He went on to say that he did not think the boy was perfectly innocent, but he wanted action from the Board. Nimmo explained the facts of the case as regards the manner and degree of chastisement he had inflicted and the directors unanimously agreed that he had erred extremely in both aspects. Being perfectly aware of the need for chastisement when necessary, they were convinced it ought invariably to be inflicted with solemnity and a severity proportional to the offence but in no case whatever with any other instrument than a tawse.[31] They therefore recommended Nimmo to beware in future of yielding to 'suddenness of provocation or hastiness of temper'. The matter went no further, but it certainly does not reflect well on Nimmo. It is ironic that he should have displayed something of the same behaviour that, when exhibited by Tod, had allegedly caused his resignation from Fortrose.

In July 1809 Mr Campbell, a teacher in the Academy, announced that he was to open a class in Gaelic, whether in the academy or independently is not clear.[32] This indicates that individual teachers of the Academy were enterprising in starting new courses, and Nimmo, too, was active in teaching out of hours. Mitchell tells us that Thomas Rhodes, a self-educated carpenter from Hull who had come with his brother to erect the wooden lock gates for the Caledonian Canal then under construction, had placed himself under Nimmo to learn Mathematics and Drawing. He must have been an excellent student: soon after Nimmo left in 1811, Rhodes switched roles from student to tutor and started coaching others when his day's work on the canal was over. One of his charges was Mitchell, who wrote this charming vignette of Rhodes's 'class':

> Every evening after working hours I used to change my mason's apron, dress and spend two hours with him in drawing and other instruction. We worked together, he at his problems and I drawing part of the works. About 9 o'clock his wife laid on the table a bottle of strong Edinburgh ale of which he gave his wife and me a champagne glass full and I then took my departure.[33]

Both Rhodes and Mitchell would go on to make famous careers as engineers, Rhodes being especially active in Ireland in connection with the Shannon navigation and the Port of Limerick. Mitchell became a right-hand man to Telford until the latter's death. They were Nimmo's first 'academic son and grandson' that we know about. He was to teach many more young engineers later on in his Irish and English engineering practices.

Nimmo resigned the rectorship in June 1811, when he took up a position with the 'Commission to Examine and Enquire into the Nature and Extent of the Bogs of Ireland and the Practicability of Draining and Cultivating them, and the best means of effecting the same'. This will be referred to hereafter as the Bogs Commission. According to the official report of the commissioners, he joined them in January of 1811.[34] This cannot be exactly true, since he received his rector's salary of twenty-five pounds for the January to June session as normal on 2 June 1811.[35] At the directors' meeting on 7 January 1811, Nimmo even proposed extensive alterations to the regulations of the school that were designed to improve its teaching[36] – hardly the actions of a man who had already committed himself to a completely new career abroad! In May, when the directors decided on the examination dates and intimated these to the rector and other teachers, there was still no hint of any impending departure.[37] When his letter of resignation was read at the meeting of 11 June 1811 the directors must have been quite stunned: there was no sentiment of regret or expression of thanks for a job well done, although they had many times previously expressed their satisfaction at the performance of the teachers and students under his direction.[38] The post of rector was advertised later in the month, and Mr Matthew Adam was appointed on 9 July, without any 'comparative trial' to slow the process down. The board 'was requested of Provost Grant to intimate to Mr Adam his appointment, and that by the regulations of the Academy six months previous notice is to be given by him in case of his giving up his charge'. This, surely, was a case of locking the door after the horse had bolted? It certainly suggests peevishness in respect of Nimmo's sudden departure. Adam continued in office for many years, his abiding interest being in increasing his salary at all costs. He even alleged in a letter to the board that Nimmo had been unable to live comfortably on what he had earned and had needed financial help from his father.[39] Since Nimmo had received at least £200 for his work with the Commissioners for Highland Roads and Bridges in addition to his academy emoluments (estimated by one of the directors to have been about £219 per annum; the rector was paid a fixed salary and a fee from each student registered, so that the exact sum is difficult to calculate), it seems unlikely that he was quite as impecunious as Adam suggested. Even more damning, Adam claimed that Nimmo would have been imprisoned for debt, had his brother not helped him out to the tune of £180 at the time of his departure for Ireland! We do not know whether there was any truth in these allegations. How could Nimmo have accumulated such debts? Where would his father and brother have got such funds? Throughout his life, irregularities with money were to dog Nimmo's career. However, his brother George would benefit later on from Nimmo's engineering works, a recompense that suggests there could have been at least a grain of truth in what Adam alleged.

Nimmo's departure from the Academy would not just signal a change of job; it was to be a complete career change from teacher to engineer, from academic to practical journeyman scientist. What mediated this dramatic change after almost a decade as a schoolmaster? His original post at Fortrose could be seen as a first step in a serious teaching career and his move to Inverness a further determined and significant ascent up the ladder of the teaching profession, suggesting that he was fully committed to that career. But if he really had felt called to the teaching life, how do we explain his abandonment of that calling while it was still full of prospects?   Mr Tulloch, the candidate who had failed to appear for the comparative trial when Nimmo was appointed rector, was elected conjunct Professor of Mathematics at Aberdeen University in July 1811, and Nimmo might realistically have anticipated similar or even greater advancement in due course, had he continued with his teaching career. His departure to Ireland changed all that. So, did the Bogs Commissioners just pluck a very clever schoolmaster from his teaching post and offer him a lucrative position as engineer simply because there was no one else available even half-trained for the job? Or was Nimmo glad to get away for his own reasons?

We can put these questions in a slightly different frame by briefly looking forward three short years. Nimmo joined the Bogs Commission in June 1811; by December 1813 he had completed extensive surveys and produced comprehensive reports, of immense scope and depth and including remarkable maps, on most of the County of Kerry with parts of Limerick and Cork, and on the Connemara region of County Galway. His written descriptions alone of the geology of these areas were remarkably accurate and, in the words of a modern reviewer 'so detailed that a crude geological map could be constructed from them' even now.[40] His sketch map of the geology of Connemara[41] is the first such map ever made of that area. That, and his geological knowledge of Ireland, impressed even George Bellas Greenough, President of the Geological Society of London. It beggars belief that any erstwhile schoolmaster, no matter how erudite, could have successfully concluded such complex surveys and comprehensive practical reports without having attained considerable skill in practical geology and extensive field surveying experience. If his original sojourn in Fortrose had been a way of living closer to the 'geological coalface' of the Scottish mountains with a view to developing his skills in mineralogy and applied geology, rather than a serious first step in a teaching career, then his move to, and ultimate departure from, Inverness must be viewed as part of that same canvas. What his move to Inverness had really achieved was to bring him closer to the engineering and surveying activities that were beginning in the Highlands in 1803, while retaining his access to the geological resources of the area. And, of course, the rectorship was a reasonably well-paid job for which he was eminently suited. Viewed in this way, his teaching posts may, in fact, have been steps along a different career path that had nothing to do with any real ambitions as a teacher. If this conclusion is to be credible, we will need to see evidence for his applied scientific development over the decade spent as a teacher, together with evidence of some catalyst that was of sufficient import to cause him to resign an esteemed teaching post and emigrate from Scotland.

As a student, Nimmo had thrown himself into the intellectual life of Edinburgh and espoused the Plutonist views of Playfair, a position that ultimately proved to be not only geologically correct but one that probably stood him in good stead later when he was seeking the post in Inverness. Nor had the Edinburgh milieu at the time been one of thought, talk and theory alone. Engineers and architects, surveyors and cartographers, instrument makers and inventors, explorers and mineralogists were all engaged in the city, side by side with the intelligentsia of the university. While Nimmo had been attending lectures, other young men, like his Burntisland neighbour William Bald, were serving their apprenticeships and honing their skills with John Ainslie, the leading Scottish cartographer of his day.

A young Irishman, Richard Griffith, who would soon become first engineer of the Bogs Commission, also came under the influence of these eminent men of science when he attended Edinburgh University in 1805 and 1806.[42] Griffith attended the Neptunist lectures of Jameson the 'waterman' but balanced his approach – or hedged his bets – by carrying out geological experiments with the Plutonist 'fireman' Sir James Hall. All his life Griffith would show himself to be a consummate political operator like this, shrewdly balancing one interest against another. Nimmo was never quite as deft at this, and for that reason would never enjoy the political preferment that his Irish counterpart did.

Griffith made a geological tour of the Hebrides at the same time as William Bald, and it would be strange indeed if the two men had not met or cooperated in that venture. Whether or not Griffith met Nimmo in the course of that tour, or subsequently, we do not know for certain: Griffith may well have passed through Inverness on his way to the Hebrides while Nimmo was at the Academy. Nimmo was in correspondence with Playfair at the time and was about to make his own survey of the boundaries of Inverness-shire on behalf of the Commission for the Highland Roads and Bridges in 1806. Some contact and exchange of information between the two, possibly mediated through Playfair, would have been natural. Whether or not it occurred, both men were fated to become colleagues in the Bogs Commission and in other engineering works in Ireland within a few years.

William Bald, who also was to make an eminent engineering reputation in Ireland starting with service to the Mayo Grand Jury and with the Bogs Commission, was actually surveying the Hebrides and the north of Scotland in those years.[43] From 1803 to 1809, aged from 14 to 20 and entirely alone, he surveyed estates in the northern highlands and the isles, developing an acknowledged and deserved reputation for precise and accurate work. In July 1806 he was in the Ardnamurchan peninsula in northern Argyll at the exact time that Nimmo was surveying its shared boundary with Inverness. It stretches credulity that Nimmo and Bald did not meet, or that all three men, Nimmo, Bald and Griffith, trained in Edinburgh and active in surveying the north of Scotland, two of whom came from neighbouring Scottish towns, did not know one another, or at least know of one another, in those early years of the century. All three would join the Bogs Commission and were amongst the most skilled and influential engineers to work in Ireland in the first half of the nineteenth century. Thomas Rhodes, whom Nimmo had taught, would form a fourth distinguished engineer with Scottish connections to make a significant engineering impact in Ireland, but he was never with the Bogs Commission, nor did he enjoy the Edinburgh experience of the others. Bald and Rhodes (among others) would eventually take up and complete many of Nimmo's unfinished Irish works after his death in 1832.

In those days, however, Nimmo was not yet an experienced practical scientist let alone an engineer. When he first took up his schoolmaster's job in Fortrose, he endeavoured to continue with the scientific interests that he had absorbed in the Scottish capital. What he particularly needed was practical experience in field geology and mineralogy, exactly what northern Scotland could give him. Once settled in Fortrose, he commenced scientific surveys and geological observations on land, on the seashore and in Loch Ness. The earliest evidence of his work still extant is his sketch of Fort George, a military fort situated near Ardersier on the southern shore of the Moray Firth opposite Chanonry Point.[44] (By coincidence, it was built by his co-townsmen, the Adam family of Kirkcaldy.) The sketch is merely a tiny, folded paper plan, about two inches square, of the famous fort and dated 1805 or 1806 (the date on it was altered, it seems from the later to the earlier year). It is accompanied in the archive by an explanatory hand-written note headed 'Farr [Inverness-shire] 23 November 1864' stating: 'The inclosed is a sketch of Fort George drawn by Mr Nimmo Rector of the Inverness Academy in May 1864 [sic; this was surely a slip, as Nimmo was long dead by 1864] which I had with me in India. It is worn out by having been carried many years in my pocket book. [Initialled] A.M.' The sketch and note were the possessions of Col. Alex Mackintosh of Farr. The sketch is too small and lacking in detail to have much practical use or value except as a memento. This rather trivial item, whose principal interest probably lies in its peculiar provenance, is nevertheless the earliest extant example of a drawing by Nimmo, and it indicates that he was already taking note of important structures during those early years in the north.

The Moray Firth narrows dramatically at Chanonry Point due to promontories on the north and the south shores. Both are comprised of material deposited as terraced raised beaches, an observation that did not go unnoticed by Nimmo, who was familiar with such topography along the Firth of Forth near Kirkcaldy. However, he observed that deposition was still ongoing, and he sent an account of his observations to Playfair. The latter communicated a report entitled 'An account of the removal of a large mass of stone to a considerable distance along the Murray Firth', under Nimmo's sole authorship, at the meeting of the Royal Society of Edinburgh held on 3 December 1804.[45] The report was never published in full so we do not know its detail. Nowadays we know that the displacement of material along the shore is an important feature of coastal evolution and its effects have been well described in modern times, particularly at Chanonry Point and Ardersier, exactly where Nimmo made his original observations.[46] Nimmo and Playfair would appear to be among the earliest to have drawn this process of active coastal evolution to the attention of the scientific community.

In the same year that he was making these observations on the coast (1804) he was also carrying out experiments on Loch Ness in company with his friend Simon Fraser of Foyers.[47] They were trying to discover the temperature of the water at the bottom of the Loch and whether the lower strata had 'any impregnation of salt agreeable to the theory of Count Rumford'. They lowered a vessel of half a gallon several times to a depth of 120 fathoms (about the maximum depth of the Loch) and were surprised to see that in the course of the operations they had drifted about a mile up the Loch although what little wind was there blew down the Loch. They accounted for their contrary movement by supposing a sub-aqueous eddy to be returning the water to one end that was being blown by the wind to the other. They obviously realised that the submerged vessel must have been under the influence of a compensating return current at depth and, like a deep sheet anchor, it was towing them in a direction contrary to the surface flow. This was a sophisticated deduction at the time, but the much more interesting methodology and outcome of the substantive experiment unfortunately were never reported. There were no reversing thermometers available at the time, so we can only wonder how they hoped to record accurately the temperature at such a depth? They may have possessed an instrument known as a self-registering (Six's) thermometer, which functions like today's common 'minimum–maximum' thermometer. That instrument would have recorded the extremes of temperature encountered, but would not have indicated the exact temperature at any given depth.

We do not know either how the lowered vessel was supposed to collect water from 120 fathoms, a depth far greater than any from which water samples were previously collected. Lowered to such a depth, a vessel containing air would require very heavy sinkers and would implode or burst its seams under the high hydrostatic pressure. What they required was a vessel open at both ends during the descent that could be closed by a remote messenger device dropped on a line from the surface once the vessel had attained the required depth. No such vessel existed at the time. Robert Stevenson designed one somewhat akin to this in 1812, which he called a 'hydrophore'. The version for use in shallow water – up to a couple of fathoms – was attached to a metal pole and had only one opening sealed by a peg that could be drawn by pulling on an attached line.[48] His deep-water version, for use in 30 to 40 fathoms, had a capacity of about one pint and weighed about half a hundredweight. With both versions, Stevenson lowered the vessel full of air; neither version, if filled with air on descent, would be likely to work at the depths encountered by Nimmo and Fraser. Whatever type of vessel they used, they succeeded in dropping it to the bottom a number of times, which suggests that it cannot have been air-filled during descent. Apart from that, we know nothing, unfortunately, of its construction. Despite the technical problems, their endeavour was a very early attempt at deep limnological or hydrographical investigation. Later in life, Nimmo would rekindle his interest in hydrographical topics and would even come to call himself a 'hydrographer'.

These sparse records are all that we know of his earliest scientific endeavours but they confirm that he was engaged in practical scientific experiments and observations way beyond the demands of an ordinary schoolmaster's job, and his life could reasonably have continued in this vein had circumstances been different or his inclinations less ambitious. But practical undertakings far grander than these isolated experiments were getting underway at Inverness, as Nimmo must certainly have been aware. When the rectorship had first become vacant in 1803 before these undertakings started, Nimmo had not applied for the position. But when the post had fallen vacant again in 1805, things were altogether different. By then, Telford and his men were busy in Inverness and the surrounding areas, and the bustle of major engineering projects was in the air. Nimmo so much wanted to be close to the new developments that he offered himself for the Inverness post even before it was advertised publicly. His decision to move to Inverness was to be a crucial one for him and the circumstances that precipitated it revolved around the profound social and political changes that were occurring in Highland life.

In contrast to the sophisticated life of Edinburgh and the general optimism in the central lowlands in the late eighteenth century, things were different in the Highlands, the *ultima Thule* of Britain. Notwithstanding the well-reported peregrinations of a small number of famous and intrepid literary travellers, the Highlands were largely unknown even to the Scottish lowlanders of Edinburgh and Glasgow, not to mention the powers-that-be of England. The only roads into and out of this fearful, mountainous wilderness were the drove roads used to bring cattle to the great fairs at Crieff and Falkirk.[49] The military roads made by General Wade earlier in the century had already fallen into disrepair and were never, in any case, suitable for wheeled carriages and civilian commerce. The disaster of 1745 had emptied many estates of their lairds, who chose exile to the open prison of their forfeited estates. As the century progressed the human population started to grow again as did the sheep population (to an even greater extent) and a profound change began to occur in attitudes to the Highlands and its people.

The year of Nimmo's birth, 1783, was an important one both in international affairs and in Scottish domestic politics. The Treaty of Versailles brought the American War of Independence to an end and Scottish soldiers of the highland regiments, having proved their loyalty to the Hanoverians by their battle record, and having expunged any lingering Jacobite tendencies through service and sacrifice, were coming home with a reputation for loyalty and a bright vision of a new future. Scots who had remained at home had developed contacts of a business kind with Scots abroad, which helped to finance and drive a new spirit of commerce and industry. The Commissioners for the Annexed Estates now began to return the forfeited estates to the families of their former owners. As Scottish self-confidence re-emerged, sentiment towards the north was improving, and this was largely because of the proven loyalty of the highland regiments. Prior to American independence, and during the notorious clearances, the flow of highland emigrants to the New World went largely unlamented in London, and may even have been of some advantage to England. While ridding the Highlands of an unreliable and rebellious element, it provided colonists to settle and develop England's overseas possessions. With American independence, England was not well pleased by the ingratitude of the colonists in opposing and shaking off their ties to the erstwhile motherland. In these changed circumstances, a continued outflow of highlanders could only damage British interests by the inevitable loss it caused of industrious men of proven fighting quality to the newly independent American Republic. It became important to make all-out efforts to ensure that they now remained at home, and this meant that the economic development of the

Highlands took on new political urgency. The general wisdom held, rightly, that economic development should be based on the natural resources of the country, which, in the case of the Highlands, were agriculture and fisheries.

A number of private individuals and companies took up the new economic challenge. From 1784 onwards, John Knox and others wrote influential treatises extolling the Scottish fisheries and the great potential of the natural resources of the Highlands. These tracts, pamphlets and books were influential in bringing the state of the fisheries and the plight of the poor to public notice and sympathy. As a result, Parliament appointed in 1785 a 'Commons Committee to Enquire into the State of the British Fisheries'.[50] A few years previously (1778), the Highland Society of London had been formed and while its main functions were literary and artistic, it had among its aims the improvement of the northern part of the kingdom. It immediately added the Chairman and members of the new Commons Committee to its list of honorary members, so that the respective interests of the Commons Committee and the Highland Society in the improvement of Scottish fisheries and industry overlapped.

John Knox, in a lecture to the Highland Society soon after the Commons Committee had been set up, proposed that about forty fishing villages, complete with all necessary services and fitments, should be built with public money. Since the Treasury was unlikely to provide the necessary resources, and many local landholders were unable to do so, it was incumbent on gentlemen like the members of the Society, he said, to raise the money by subscription. The Commons Committee supported this proposal in its report to the House in May, and by 1786 a bill setting up the 'British Society for extending the Fisheries and improving the Sea Coasts of this Kingdom' – a joint stock company known usually as The British Fisheries Society – was signed into law and the company was duly formed. The Secretary of the Highland Society took on the role of Secretary of the Fisheries Society.

The British Fisheries Society set about creating the infrastructure of villages needed to encourage the Scottish fisheries, the construction of Tobermory, Ullapool and Pulteneytown (Wick) being the main results of its efforts. Meanwhile, Knox made further expeditions into the northern counties and pointed to the complete absence of roads serving many of the fishing stations and the disastrous state of those other roads, hardly more than tracks, then in existence. 'Through a considerable part of the year, the inhabitants of each respective glen or valley may be considered as prisoners strongly guarded by impassable mountains on one side, by swamps and furious torrents on the other,' he wrote,[51] and the Fisheries Society was soon persuaded to widen its remit to include building necessary roads. The Society's efforts inevitably involved extensive surveys and civil engineering works, and for these it needed to engage the best civil engineer it could find. In 1790 William Pulteney asked his good friend Thomas Telford – then Surveyor of Public Works in Shropshire

– to survey the Society's proposed new settlements in the Highlands. Telford, himself a Scot, agreed and this commenced his life-long relationship with the Society, for which he eventually refused, altruistically, to accept any payment.

Notwithstanding the good intentions and the early success of the Fisheries and the Highland Societies' intervention, life in the Highlands had not improved to any significant extent. It was, in fact, getting worse. Destitution, despair and emigration continued, with many supposed causes and just as many supposed cures, so that no one course of action could be confidently agreed and acted upon. But everyone agreed on one thing: the lack of reliable means of communication lay at the root of many of the problems. Without good communications, no new ideas or raw materials could ever get into the region and no agricultural, fishery or manufactured goods could get out. At best, a system of new roads would open up the area permitting the development of the western fisheries and stimulating agriculture-based industry (like woollens, hides, timber, etc.). At worst, road works would give employment to many and relief from the poverty that was eating away the very soul of the Highlands. It was long past time the Government did something.

In July 1801 the Lords of the Treasury requested yet another survey. On this occasion the person appointed to carry it out was none other than Telford, who was probably the best practical person with intimate, personal knowledge and experience of the area and with no personal vested interest in it that might colour his opinions. His instructions, too, were sharp and focused. He was to select the most suitable sites for fishing stations on the west coast; plan safe and convenient communications between the mainland and the islands; and consider the possibility of inland navigation between the east and the west coasts. Telford lost no time in setting out for Scotland, completing and reporting on his first survey by October of 1801. His news was encouraging: the objective of creating road networks and forming a canal through the Great Glen joining the east and west coasts was feasible, and would have a striking effect on the welfare and prosperity of the British Empire. Moreover, in order to leave out nothing that might please, he stated his opinion that 'the fisheries would not fail to be improved and much extended'.[52] The Lords of the Treasury were impressed and encouraged by this, but what, they wanted to know, about the causes of emigration? They directed him to make a second survey and report on that matter.

Telford completed his second survey in 1803, devoting his report to plans for the roads and bridges that he proposed should be constructed, and to the Caledonian Canal he proposed to drive through the Great Glen. As to emigration, his opinion was that the introduction of sheep and the changes that sheep farming precipitated in the management of the estates were causing the high numbers that were emigrating: 'about three thousand persons went away in the course of the last year, and if I am rightly informed three times that number are preparing to leave the country in the present year'.[53] Seriously alarmed now, the Government lost no time in responding and by July 1803 the Royal Assent was given to two Acts of Parliament setting up two separate commissions. One was the Commission for making Roads and building Bridges in the Highlands of Scotland, the other the Commission for the Caledonian Canal. Crucially, Telford was appointed engineer to both commissions and John Rickman was appointed secretary. The latter was secretary to Charles Abbot, Speaker of the House, who was made chairman of both commissions. Abbot had been chief secretary for Ireland for a brief period from May 1801 to February 1802.

By 1803, then, the parlous state of the Highlands, long underscored by the Highland and the Fisheries Societies – one of destitution, distress, apathy and emigration – was widely acknowledged, and an appropriate response, that is, public works involving road building and construction of the canal that would give employment and develop the deficient infrastructure, had been formulated. Most importantly, the practical day-to-day implementation of the policy was placed in the hands of an experienced, talented engineer rather than those of political amateurs alone. Few can have fully realised the magnitude of the tasks that faced the commissions and their engineer, since few really appreciated the difficulties of communication, the absence of trained personnel, the lack of construction skills, the shortage of tools, the inclemency of the weather and the generally isolated and economically depressed state of the Highlands. Nevertheless, by spring 1805 planning, surveying and engineering works were already under way, making Inverness and Fort William bustling centres of construction activity. It is this that may have acted as the magnet that drew Nimmo to Inverness and its Academy. Even Nimmo himself may not have realised the magnitude of the effect that exposure to this activity was about to have on his life. Nor could anyone have realised the extent to which the commissions' proposed solution to Scotland's ills, which Nimmo would experience at first hand, would become the model for addressing Ireland's problems within a few years, or that Telford's role in Scotland would be emulated in Ireland by the young Rector of Inverness Academy. But first, the schoolmaster would need to don the apron of the apprentice and learn the craft of the surveyor/engineer.

NOTES

1. R. N. Smart, *Biographical Register of the University of St Andrews 1747–1897*. (Fife: University of St Andrews Library, 2004).
2. *The Commercial Directory of Ireland, Scotland and the four most northern counties of England for 1820–1 and 1822* (Manchester: J. Pigot and Co., 1820).
3. J. Morrell, 'John Playfair', in H. G. C. Matthew and B. Harrison (eds) *Oxford Dictionary of National Biography* (Oxford: Oxford University Press, 2004).
4. Cupar Town Council, Minutes of Council Meeting, 2 November 1796.
5. University of St Andrews, Matriculation and Graduation Roll, 1739–1888.
6. University of St Andrews, Minutes of United College, 2 November 1796, p. 39.
7. University of Edinburgh, Arts Matriculation Records, 1799.
8. D. Stevenson, *Life of Robert Stevenson* (Edinburgh: A. and C. Black, 1878).

9. R. Goring, *Chambers Scottish Biographical Dictionary* (Edinburgh: Chambers, 1992).
10. R. Mudie (ed.), *The Surveyor, Engineer and Architect for the year 1842* (London: Bell and Wood, 1842).
11. Presbytery of Chanonry, Session Minutes, 2 February 1802. (NAS, MfilP/CH2/66/6), hereafter referred to as Session Minutes.
12. *Caledonian Mercury*, 15 April 1805.
13. Session Minutes, 17 April 1805.
14. Session Minutes, 18 June 1805.
15. Ibid.
16. J. Playfair, in Minutes of the Royal Society of Edinburgh, 3 December 1804 (NLS, Acc 10000/4).
17. Inverness Academy, Minutes Books of the Directors of Inverness Academy, vol. 2, 1798–1820. Meeting of 13 July 1805 (Highland Council Archive, Inverness Library, C1.5.4.0), hereafter referred to as Academy Minutes.
18. Ibid.
19. A. Robertson to A. McGregor, 29 March 1803 (NAS, GD 128/10/5).
20. Anonymous (1801), Report of number of students 1800/1801 at Inverness Academy (NAS, GD 128/34/1).
21. J. Mitchell, *Reminiscences of my Life in the Highlands*, 2 vols (London: privately printed, 1883, 1884).
22. Anonymous, *A Plan for Inverness Academy* (NAS, GD 128/34/1/46).
23. Anonymous, *Relating to the Academy at Inverness*, 1810 (NAS, GD 150/34/20).
24. Academy Minutes, 2 May 1807.
25. Academy Minutes, 26 May 1811.
26. R. Moodie to Provost Gilzean, 12 April 1808 (NAS, GD 128/34/1/46).
27. Academy Minutes, 29 July 1808.
28. A. Nimmo to J. Rickman, 12 October 1806, in A. Nimmo, *Journal along the North, East and South of Inverness shire* (NLS, Adv MS 34.4.20).
29. Mitchell, *Reminiscences of my Life in the Highlands*.
30. Academy Minutes, 8 February 1810.
31. A tawse was a leather strap split into strips at the end, used to chastise young miscreants.
32. Advertisement in the *Inverness Journal*, 28 July 1809.
33. Mitchell, *Reminiscences of my Life in the Highlands*.
34. Fourth Report of the Commissioners for the Bogs of Ireland, appendix 1 (B.P.P. 1813/14 [131], vol.VI, Pt 2, mf 15.33–36).
35. Academy Minutes, 11 June 1811.
36. Academy Minutes, 7 January 1811.
37. Academy Minutes, 28 May 1811.
38. Academy Minutes, 11 June 1811.
39. M. Adam to W. Fraser, October 1812 quoted in R. Preece, 'Some pages from the History of an Inverness School', *Occasional Papers of Inverness Academy No. 2. Matthew Adam* (Inverness: Royal Academy, 2003).
40. T. R. Marchant, and G. D. Sevastopulo, 'Richard Griffith and the Development of Stratigraphy in Ireland', in G. L. Herries-Davies and R. C. Mollan (eds), *Richard Griffith 1784–1878*. (Dublin: Royal Dublin Society, 1980).
41. A. Nimmo, map entitled 'Nimmo's Connemara' (GSL 999).
42. G. L. Herries-Davies, 'Richard Griffith – His Life and Character', in G. L. Herries-Davies and R. C. Mollan (eds), *Richard Griffith 1784–1878* (Dublin: Royal Dublin Society, 1980).
43. M. G. Storrie, 'William Bald FRSE, c. 1789–1857; Surveyor, Cartographer and Civil Engineer', *Transactions of the Institute of British Geographers*, 47 (1968), pp. 205–31.
44. A. Nimmo, Sketch of Fort George (Highland Folk Museum, Kingussie, call no. C.c.17).
45. J. Playfair, in Minutes of the Royal Society of Edinburgh, 3 December 1804 (NLS, Acc 10000/4).
46. J. A. Steers, *The Coastline of Scotland* (Cambridge: Cambridge University Press, 1973).
47. A. Nimmo, 'On the Application of the Science of Geology to the Purposes of Practical Navigation', *Transactions of the Royal Irish Academy*, vol. XIV (1825), pp. 39–50.
48. Stevenson, *Life of Robert Stevenson*.
49. A. R. B. Haldane, *New Ways through the Glens* (Edinburgh: Thomas Nelson and Sons, 1962).
50. J. Dunlop, *The British Fisheries Society 1786–1893* (Edinburgh: John Donald, 1978).
51. J. Knox, *A Tour through the Highlands of Scotland and the Hebrides Isles in 1786* quoted in Haldane, *New Ways through the Glens*.
52. T. Telford to N. Van Sittart, 7 October 1801, quoted in Haldane, *New Ways through the Glens*.
53. T. Telford, Survey and Report, 1803, quoted in Haldane, *New Ways through the Glens*, p. 33.

CHAPTER

2

The Inverness Perambulation and the *Edinburgh Encyclopaedia*

The works of the two commissions were major and expensive. The costs of the Caledonian Canal were to be borne entirely by the Government, as it would likely be a boon to the whole country. But it was relatively easy to determine which estates would be enhanced and which counties opened up by the proposed roads and bridges, so in their case half of the cost was to be levied on the counties and the beneficial proprietors and half on the central government. County Assessments Acts were introduced so that the money could be raised 'to meet the liberality of Government without imposing on individuals an unequal burden'.[1] This, naturally, required that the boundaries of the different counties be established with accuracy. Ainslie's map of Scotland, the best at the time, was not sufficiently accurate in this matter and the individual estate maps, of which a plethora existed, were too estate-specific to answer the general need. Also, as Knox had found, extensive moor and mountainous tracts were still virtually *terra incognita* to all but the few people living there. The commissioners therefore decided that Aaron Arrowsmith, a London-based cartographer, should be contracted to engrave a new map that would indicate the boundaries in a clear and accurate way. A reliable person, competent in surveying, was needed for this, a task of no mean difficulty. The county of Cromarty, for example, was then a conglomeration of small, geographically separate enclaves rather than one contiguous area. In many places the boundaries were in mountainous country, or crossed desolate moors. In another case, the uninhabited island of St Mungo's in Loch Leven contained a graveyard used by parishes of different shires on both shores of the loch. Pasturage there of £25 per annum was paid alternately to the proprietors of Glencoe and Callart on the opposing shores. Therefore, the island had to be considered properly as belonging in alternate years to the shires of Inverness and Argyll.[2]

Whether Nimmo deliberately sought out employment with the roads and bridges commission, or the commissioners sought him out, we do not know for certain, but the parties were in contact within months of Nimmo's arrival in Inverness. According to Arrowsmith, 'Telford found on enquiry that Mr Alexander Nimmo, Rector of Inverness Academy, was a very competent person to be employed by the Commissioners on this service,' which may indicate that Telford sought him out for the task.[3] There seems no reason why Telford would have known of Nimmo prior to this, and it may have been Arrowsmith himself who drew Telford's attention to the rector. Arrowsmith acknowledges that Nimmo had sent him a useful sketch of the Black Isle (the peninsula of Cromarty on which Fortrose is located) and had also communicated to him 'an account of the Monocleagh [Monadhliath] Ridge of mountains, which is not marked as continuous high ground in the military survey'.[4] The survey work for these may have been done separately from, and maybe before, the eventual boundary survey. The commission certainly did not pay Nimmo for them, as they did for his other work. However he came to Telford's attention, the latter must have been convinced of his ability, because we read in the third report of the commissioners: 'We ... appointed Mr Nimmo, Rector of Inverness Academy, to investigate and perambulate the boundary of the shires of Inverness, Ross, Sutherland, Caithness and Cromarty.'[5] Had they decided to contract Ainslie rather than Arrowsmith for the new map, it is possible that William Bald would have been the one to do the survey. He was already surveying in Argyll on Ainslie's behalf and was familiar with the area. But since the contract went to Arrowsmith, Ainslie's man was not really available, fortunately for Nimmo. Maybe Bald had, in fact, been approached to do the task and perhaps it was he who suggested that Nimmo would be appropriate for the job? It was, after all, intended to be only a short, 'once-off' survey, and so Bald may well have directed the task the way of his fellow-Kirkcaldian.

In the spring of 1806, less than eight months into his rectorship, Nimmo was approached to undertake the boundary survey. Anticipating that he would be successfully appointed, he wrote to the Convenor of the County of Inverness in April, 'expressing my sentiments concerning the method which might be adopted in communicating *to the person appointed to survey* that information which he might require' (italics added here).[6] Since there were large areas of commonage along the borders, he suggested that the various adjacent counties be invited to appoint some commissioners to agree on the proper line of division between them. In the event, the Inverness County administration decided on 20 April that a circular letter should be sent to the border proprietors inviting them to 'take such methods as might be necessary to give the surveyor such answers as were necessary'. Replies were received from some, but most did not bother to answer. In these circumstances, Nimmo regarded it as fortunate that the boundaries of many properties that coincided with the county boundaries, were marked with landmarks 'perfectly well known to the farmers and shepherds on the spot', so that their knowledge could be relied on, at least regarding those lands close to where they lived.

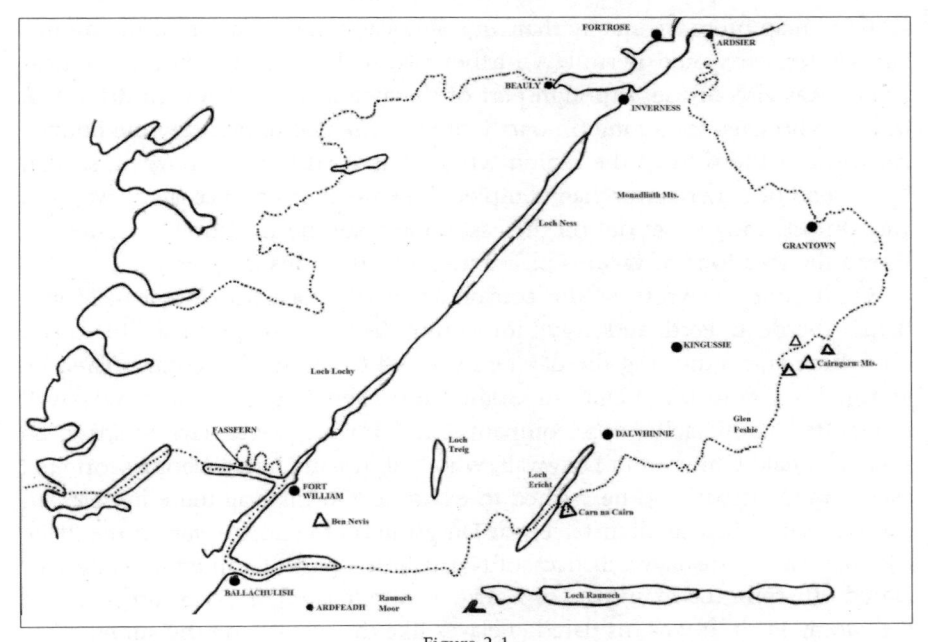

Figure 2.1
Boundary of Inverness-shire after Nimmo's survey in 1807. (Based on Arrowsmith's map of Scotland drawn for the Commissioners for Highland Roads and Bridges.)

On 20 May he received Arrowsmith's skeleton map and instructions, together with surveying instruments necessary for his purpose. Already he had almost completed a concise historical account of the successive divisions of the northern shires and their origins, drawn from esoteric sources, historical tracts and old Acts of Parliament that the commissioners subsequently published as an appendix to their third report in 1807.[7] He was now ready to commence his 'perambulation'. First however, he came under pressure from Sir George Steuart Mackenzie to show him the skeleton map, at least insofar as it concerned the boundary with the shire of Ross. Nimmo rode out to Sir George's seat at Coul in compliance, and he also showed the skeleton map to Fraser of Lovat residing at Beaufort Castle; both men were influential persons in the shire. Mackenzie was a noted explorer and mineralogist as well as a respected land improver on his own estate. Nimmo became friendly with him and remained firmly in his good books, so much so that it was Mackenzie who proposed him for membership of the Geological Society of London in 1809.

The meeting with Frazer was particularly profitable: Frazer had extensive plans of his estates in the Barony of Glen Strathvarer on the border of Ross and Inverness. These detailed plans had been drawn up in 1758 under the direction of the Board of Exchequer during the annexation of the forfeited estates. Nimmo realised that they allowed him to insert the county boundary on the skeleton map more accurately than any physical perambulation of the mountainous terrain would permit. A further plan exhibiting the boundary near Beauly was also of use, although part of the area it covered was in dispute. A farmer who lived for a long time in Strath Glas helped in defining the boundary towards the west of the region where it abutted Kintail in Argyll, so that by the end of May Nimmo had completed the northern part of his survey and was almost ready to set out on the eastern and southern sections. Figure 2.1 shows the locations of various places mentioned in this chapter.

On 1 June he wrote to the convenors of the counties of Nairn, Moray, Banff, Aberdeen, Perth and Argyll informing them that he proposed to set out on 6 June and indicating the day he expected to be in each county. Then he engaged a young man, Duncan Grant from Glen Urquhart, who was well acquainted with Gaelic, to accompany him. Having some days to spare, he visited Donald Cameron in Dingwall, who had in hand some plans of forfeited estates in Cromarty that he wished to examine. On his way there he stopped off in Ferrintosh, a small district near Dingwall on the north-west of the Black Isle that was an 'insulated district' of Nairnshire – part of, but completely separated off from, the main part of Nairn, a shire located on the south shore of the Moray Firth. It was insulated enclaves like this that made the survey both necessary and difficult. Nimmo made a map of the enclave and moved on to Dingwall the next day, where he examined plans of the estate of Strathpeffer, another insulated enclave, this time part of the county of Cromarty but located in the county of Ross. This, too, he mapped before returning to Inverness, where replies were awaiting him from Nairn and Moray. On 5 June a telescope arrived from Aberdeen, the instrument he found of most use in his travels. A sextant and horizon had arrived earlier by sea, but he found them very inconvenient to use. Because of their bulk they were difficult to carry and particularly difficult to unlash from the horse's back when needed. In any event, the skeleton map was already proving to be remarkably accurate so that confirming the existing outline boundaries might not demand any very complicated triangulation. Nimmo came to rely almost exclusively on the telescope and a small pocket compass for most of his needs.

At last, on 6 June, the end of term at the Academy, he set off, with a first stop at nearby Ardersier, where he commenced surveying along with two persons from the county of Nairn. As would become obvious during the survey, the boundary was often very much more complicated in fertile lowlands like this than in the mountainous interior. This was because some farms straddled natural boundaries such as rivers, so that parts were located in one county and parts in another. Sometimes adjacent farms were assigned to different counties in almost a random patchwork manner. The situation around Cawdor and Cantray was particularly intricate.

As he became more immersed in the survey a pattern developed in his mode of progressing. First, he had some idea of the general boundaries from his historical study of the shires. Then he examined various books of modern records: the cess books (a tax record) of the various counties; the Rolls of Assessed Taxes held by Provost Grant of Inverness; and the 'Book of Splittings'. In cases of the partial sale or transfer of property, a committee of Commissioners of Supply was called to split or divide the valuation affixed to the whole estate into parts proportionate to the conceived value of each lot. A record was kept of these proceedings and that was the 'Book of Splittings'. But far and away the best information came from his third source, the proprietors, factors, tenants, shepherds and other persons who met him and who walked with him, or pointed out the boundaries to him. Finally, their information could usually be confirmed by the existence of marker stones along the marches of the various estates.

On 11 June he was back in Inverness, where a letter awaited from the Clerk of Supply of Perthshire, couched in terms rather singular. Not only did the clerk regard the notice he had received of Nimmo's impending visit as too short, he protested sternly at the very idea of such proceedings as were proposed. Nimmo made no reply, luckily as we shall see. He left Inverness again the next day for Grantown, intending to take a route by Strathspey to Perthshire and thence across country to Ballachulish in Argyll, on to Fort William, crossing by ferry to Kintail and returning to Inverness by Strath Glas. The route sounds simple enough. In reality, it represented a journey on horseback of approximately 200 miles through some of the highest, most isolated, inaccessible, trackless, treacherous landscape in the whole of Britain, among a sparse populace whose language he did not speak, with only one young man for company, facing an unknown reception whenever he came across local habitants. His route would pass through the heart of the stupendous Grampian Mountains, cross the awesome Moor of Rannoch, pass down weeping Glen Coe and back into the rugged north-western highlands, north of the Great Glen. Even today those parts of it that can be travelled in a comfortable car along roads that have been surfaced, widened, graded and straightened, is a magnificent journey, but not one to be undertaken lightly in unpredictable weather.

Early on the first leg of the survey, he commented on the rough nature of General Wade's military road from Grantown to Fort George, criticising Wade, who, in his opinion, was 'no engineer, although what he did was no doubt much for the benefit for the north'.[8] Wade (1673–1748) was born in Westmeath, Ireland of a Cromwellian family. Between 1720 and 1740 he built over 240 miles of road and thirty bridges in the Highlands, 'a remarkable feat of engineering' according to his biography.[9] That Nimmo felt prepared to criticise the engineering skills of the man who first opened the Highlands suggests that he was already thinking of himself as an accomplished engineer; that he could have thought that Wade had conferred much benefit on the Highlands is a more contentious opinion.

In the survey of Speyside the precise markings of the boundaries were less certain because of the nature of the farm practices in the area. The counties of Nairn, Elgin, Banff and Aberdeen interdigitated along this eastern boundary of Inverness, some farms actually straddling the boundary. Topography was not the only problem: one particular farm, for example, which was one-third in Inverness and two-thirds in Moray, was 'three aughten parts' in size, that is, three eights of a davoch, which latter was 'a portion of land which required four team of oxen to plough it in a season – usually about 100 acres'. It is easy to appreciate just how complicated, and contentious, allocating the boundaries could be when land was still measured in so arcane a manner. Comparing today's boundary with that of Groome (1890)[10] and with Nimmo's account, illustrates the extent to which the limits and extent of Inverness-shire have altered over the two centuries since Nimmo first surveyed them.

In the mountain country things were different. In those times, tenants kept as many cattle on the lower home farm as they could winter on the fodder of their arable land. In the spring they moved the herds on to the nearby pastures (while they cultivated the arable tracts) and in the summer moved with them to distant and sequestered high glens where they pastured the cattle and lived themselves in small, rude huts called bothies or shielings. In the autumn they returned to the lower ground. In the upland pastures the people respected the positions of the various shielings as effectively marking the boundaries of the different farms to which they belonged. These boundaries were well known to the tenants themselves and to the factors of the different estates without whose help Nimmo could never have determined their precise locations and whom he acknowledged with gratitude. Higher up again, the boundaries usually ran along the mountain summits, being marked by 'the wind and weather shear', that is, the highest ridge from which the water ran off in opposite directions.

The Grampians were so devoid of roads or tracks that Nimmo was unable to perambulate properly the boundary separating Perthshire and Inverness in this region, although he did attempt to mount the course of the Feshie River through the glen of the same name. The sheer isolation and impenetrability of the upper reaches led him to propose his first-ever road, one that would connect the valley of Strathspey with the Spital of Glenshee. This, he proposed, should cross the summit at Lairig Ghru and join the turnpike road from Aberdeen near Braemar. This unlikely and probably impossible line was never developed and today there is still no route that traverses the heart of the Grampian Mountains from east to west at this difficult place. To the west and north of the Grampians the Monadhliath Mountains are another impenetrable waste without gap or pass between the valleys of Strathspey in the east and the Spean in the west.

On 20 June Nimmo left Aviemore for Kingussie, then a town only three years in existence but already possessed of stone-built houses and buildings 'that may even be styled elegant'.[11] It was well adapted, in Nimmo's view, as an inland manufacturing town and 'afforded a pleasing anticipation of what the Highlands of Scotland may become, when once laid open with good roads and the industry of the people directed to useful and valuable purposes'. The only person in the locality who was familiar with the shire boundaries was a Mr McPherson of Ralia, who agreed to accompany him in the hills of Drumochter. They spent the night in McPherson's house and the next day travelled to the boundary near Dalwhinnie. To the east, the boundary marches followed the summits; to the west, they extended to Ben Udlamain near Loch Ericht where there was some dispute as to the exact division between the lands of the Duke of Atholl (Perthshire) and those of the Duke of Gordon (Inverness-shire). The futility of such disputes is neatly captured in Nimmo's mordant comment: 'it is scarcely necessary to remark that being on the ridge of the Grampians, the very *Domum Britanniae*, there is no habitation nearer than five miles'.[12] Equally worthy of remark is the fact that his new guide, McPherson of Ralia, was then an 80-year-old man who proved more than well able to 'perambulate' such elevated, difficult terrain.

Nimmo's reporting of this section of the survey is rather deadpan:

> After dining at Dalwhinnie with Mr McPherson Ralia we proceeded along the side of Loch Eruch [Ericht]. We were flattered for some time with the appearance of a path, but it gradually disappeared and the mountains hanging precipitously over the lake, we had some difficulty in getting forward to a shooting quarter belonging to the Lord Chief Baron on the east bank named Dal'n Lynchart where we remained that evening.[13]

The following day, Sunday, he 'found it necessary to push on by the lake side although we found considerable difficulty in getting our horses forward among the rocks and stones; there was indeed no choice of road between the mountain on the one hand and the lake on the other'.

They followed the boundary up the lake to the waters of Ailler, where it swung up-river to the north-west before turning south again to Rannoch Moor. Having completed their observations and being now only about five miles from Loch Rannoch, they decided to go there. The short journey took an incredible four hours, over terrain comprised mostly of bog and quicksands. Arriving eventually at 10 o'clock, they were fortunate to come upon the house of Sir Robert Menzies' factor, a man also named Menzies, who put them up. Here Nimmo was informed that, contrary to what he had been told by the Clerk of Supply of Perthshire, his survey was regarded as one of far greater consequence than just the recording of the boundary. They thought it involved a definitive adjudication of the disputed county boundary, which was very much desired, and a committee comprising the Duke of Atholl, the Earl of Breadalbane, Colonel Robertson of Strowan and Sir Robert Menzies had been deputed to deal with Nimmo on the matter. These men were, without doubt, the 'big guns' of the district. As it happened, both Atholl and Breadalbane were then absent in Parliament and Sir Robert was at Bath, so the matter was left to rest.

Next day, while the horses that had lost their shoes on the trek were sent down to the head of Loch Tummel, Nimmo climbed a nearby hill with a guide who pointed out the line of the boundary to the head of Blackwater. March cairns ensured that there could be no doubt of the exact line. Facing him, Nimmo saw the great Moor of Rannoch. Nearby, Bridge of Gaur lay about seventy miles from Perth by poor county roads and the only connection this district had to the sea was at Perth or Inverness, both routes necessitating long and costly land carriage. Yet the district was only about twenty-seven miles from the sea at the west coast, across Rannoch Moor's trackless and pathless waste. Nimmo proposed that a new road should be laid across the moor, creating an east–west highway that would ultimately link Perth with Ballachulish, thus opening up the 'centre of the Kingdom from the Eastern to the Western Sea'. The link from Perth to Tummel Bridge was already in place, as was the western section from Ardfeadh to Ballachulish, (these were old military roads made by Wade), so that only about eighteen miles of new road were needed. Whether a road could be constructed successfully across the moor was an unanswered question; these were no ordinary miles to be traversed and only a person boundlessly optimistic, as Nimmo was, would ever make such a proposal. Already he was exhibiting a propensity to view things on a grand scale, not circumscribed by the precise task, resources or difficulties at hand.

He set out to test the feasibility of such a road the very next day. In order to get to King's House (a government inn near Ardfeadh on the far side of the moor) he was faced with sending the horses by Glen Lyon, Loch Tay and Glendochart, a circuitous route of ninety miles, or taking them as best he might directly across the moor. He adopted the latter course, remarking afterwards: 'had not the weather been uncommonly dry for some time before, I must have certainly lost them. I was several times in danger of doing so.' His route lead from Bridge of Gaur via Dunan to Loch Eigeach, thence to Loch Lydon and the southern flank of A'Chruach (Stub na Cruaice). Three counties – Inverness, Argyll and Perth – met in this spot, and it needed a local shepherd to point out the precise boundaries where 'the hill of Cruach divides a little slip of Argyll from a similar slip of Perth interjected between this hill and Loch Lydon'.[14] The boundaries were no less zealously and meticulously recorded in this wilderness than in the fertile lowlands. Here, at the westernmost limit of Perthshire, the horses unexpectedly sank up to their bellies in the soft ground. The surveying instruments broke free from one horse, and its rider was slightly injured. After that scare, they carried the instruments themselves, and the horses were lead rather than ridden for the rest of the journey.

To the south and west of Loch Lydon the moor was 'covered with an unprofitable moss and exhibiting only a wide and desert waste'. Interspersed with lochs, the area was one of unutterable desolation, although one also of great beauty. Having previously said of Loch Ericht that 'the view of the lake [Ericht] and Ben Aaler [Alder] at this place has much of the sublime', Nimmo went on to remark of Loch Lydon that it was 'as beautiful a sheet of water as I have seen'. Clearly he had not lost his sensibilities despite the rigours of the expedition. But the prospect of the moor was daunting nonetheless:

> The south side of the lakes however is a mere desert being an entire bog perfectly impassable. It is quite destitute of trees. The north side on the other hand is pretty well wooded but they are going quite into decay. The trees in the western part of the Cruach exhibit a ruinous picture – numerous trunks lie half rotted on the ground serving as dams to obstruct the hill torrent and lay the foundation of a further moss.[15]

To the south the summits of Ben an Dothaidh, Ben a'Chreachain and An Meall Buidhe rose like a forbidding boundary wall.

Entering Argyll, the party found a sort of rough track that would lead them eventually to King's House. Although it was stony, uneven and 'winds in a strange manner', it was the only solid ground on this, the summit of Rannoch Moor. The tract all the way south-west to the Black Mount was covered with bogs, morasses and lochs with many roots of trees in every direction but not one standing trunk to be seen. The King's House inn was bleak and solitary and in so sorry a condition that it 'did no honour to His Majesty'. While they were glad enough to reach it in the evening, they moved on smartly the next morning down Glencoe, which Nimmo described in as fine a piece of descriptive travel writing as one could wish from any schoolmaster.

> The hills on the south of Glenco [sic] may be termed the Montferrat of Scotland being full of naked peaks and perpendicular craggy chasms. Down these the various streams, at all seasons but particularly early during thaws, pour immense torrents of coarse gravel and stones, covering the whole valley with this destructive kind of flood. The road in particular has been in many places strangely disfigured by it. It is by no means uncommon for a bridge over one of the streams to be literally buried several feet under the surface. The torrents of stone form perhaps the most striking feature in Glenco. They are indeed well known over every part of the Highlands, but I have seldom seen them so numerous or so vast as in this quarter. It must appear a singular scene to the traveller to see whole mountains, as it were, in motion – or going to wreck. I should imagine it little inferior to the avalanches of the Alps, and as wanting only fire (the most terrific part without doubt) to have all the destructive appearance of a torrent of lava. The falling of these stones is often dangerous to people passing beneath. Several have been wounded in

> Glenco. It is not confined to seasons of rain but occurs at all times in a greater or less degree. The day on which I passed through Glenco was very fine and warm yet I heard the trickling of stones, so to speak, in various places of the mountain beside me.[16]

Even as he described the phenomenon in such dramatic terms, he was quite certain of its ultimate, mundane cause.

> It seems these landslips have increased much since the hills were depastured with sheep. On purpose to improve the pasture, the heather and young wood whose roots bound the loose soil of the hill have been burnt off and eradicated, while the sheep form tracks and roads on the sides of the mountain which, serving to convey the water of a considerable tract to one point, form falls that cut upon the loose soil till at length the bank slips down and sends many tons of this loose, friable stuff into motion. There is an appropriate Gaelic phrase for this. The highlanders call such a slip a *Scridan*. It will always form a great obstacle to the complete preservation of roads conducted through a country of this description.

He maintained this mixture of romanticism and practicality as he approached Ballachulish: 'The lower end of Glenco,' he wrote, 'is beautiful and romantic. It is also pretty well cultivated.' Glencoe is in the district known as Lochaber, so he had now, in fact, traversed the Road to the Isles: '*sure by Tummel and Loch Rannoch and Lochaber I will go.*' His romanticism was short-lived and his practicality soon reasserted itself. The next day he made various observations on the slate quarrying, boat traffic and local cultivation at the mouth of Loch Leven and even managed an excursion down the Appin coast as far as Connel on Loch Etive, making complimentary comments on the state of the roads in that part of Argyll.

On Saturday, 28 June he set off again on survey towards Fort William, where he arrived at Glen Nevis House. Angus Rankin, tenant of a farm on the Leven River and a forester of the Royal Forest of Dalness, met him there and helped him complete his sketch of the river and its accompanying lakes, the accuracy of which Mr Stewart, barrack master at the fort, and Captain John Cameron confirmed.

From the mouth of Loch Leven the border of Inverness ran up through the sea loch named Loch Linnhe to the mouth of the River Lochy, at which point Loch Eil, a continuation of the sea loch, extends to the west. The northern shore of Loch Eil was then 'by some singular arrangement' included in Argyll and the border of the counties was quite complicated in the area. Nimmo set off to see Mr Cameron of Fassfern, a settlement located in this northerly strip of Argyll, the road to which was already well made with a number of bridges. That did not save him from some trying experiences:

the streams were high until I came to the waters of Fassifern [sic]. This had the appearance of a mighty river. The night was dark and no one near me. I rode a spirited little highland pony and I made two ineffectual attempts to ford the stream. At length however, by discovering a place where it spread to a great width on the old road near the shore I got safely over – not however without some alarm and a good, hearty ducking. I can only say that if the Moor of Rannoch demonstrated to me the importance of good roads I was equally convinced of the necessity of bridges at Fassifern.[17]

The journey was worth it. Cameron was factor of the estate of Locheil and he knew the area intimately. Better still, he had a plan of the estate, a copy of the survey that had been drawn up at the time of the forfeiture, which could be relied upon for accuracy. The boundaries of the several farms were distinctly marked and Cameron pointed out which of them were included in the shire of Inverness and which in Argyll. In the upper part of the Glen of Loy there were also several shielings attached to the lower Argyll farms on the loch side, so that it was necessary to consider these as part of the county to which the lower farms were allocated. By now, of course, Nimmo was familiar with situations of this kind and only one farm in Glen Loy, of which it was known that only one-sixth lay in Inverness, created any difficulty. Further on from Fassfern, the marches of the Glen Finnan estate (in Inverness) were pointed out to him, after which he returned to Fassfern. There is no evidence that he ever went as far as the coastal extremity of Inverness-shire at the Sound of Sleat, or that he journeyed down Loch Sheil to Moidart to trace the boundary towards the south-west.

Nimmo's journal ends abruptly at Fort William, and it is possible that the document may be incomplete. On one hand, its title page does state that the survey 'ends at Fort William', suggesting that this is the complete record. On the other hand, while 1 July is the date of the final journal entry, Arrowsmith says that Nimmo did not return to Inverness until 1 August, having continued to survey Kintail and the remaining boundary of Inverness with Rosshire, together with the boundary of the latter with Sutherland and Cromarty.[18] In a letter to Rickman dated 12 October appended to the journal, Nimmo claimed that by the time he had completed his tasks he had exceeded the term of absence allowed him by the Inverness Academy.[19] Since the new term started in mid-July, he may indeed have continued for some weeks more, but not kept up the journal.

Between 2 June and 1 August, therefore, Nimmo had perambulated the greater part of the boundary of Inverness and his journal, however cursory, is a valuable record of his progress. It was not, he tells us, 'at first drawn up with any methodical or even connected arrangement' and 'As the inconveniences to which I was sometimes subjected can be of no consequence to the public

I shall not therefore touch upon them unless when they appear necessary to elucidate my subject.'[20]

The commissioners stated that he performed the task 'with a zeal and intelligence surpassing the expectation of the Commissioners', a sentiment with which Arrowsmith thoroughly agreed:

> He delineated the boundaries with the greatest care and minuteness, which indeed was necessary, especially in the case of Cromarty, a county consisting of about twelve or fourteen patches scattered irregularly in Rosshire, and unnoticed in any former map, augmenting the old shire of Cromarty fifteen-fold. Mr Nimmo also transmitted a journal of his proceedings, which was of great use to me, besides that the whole of his perambulation really and necessarily formed a severe scrutiny of the accuracy of the Military Survey, which was found more correct, except in a few trifling instances, than was expected[21]

The commissioners paid Nimmo the quite substantial sum of £150 for his trouble.[22] Compared with the annual salary of £200 given to Rickman as Secretary of the Commission, and the total of £300 paid to Arrowsmith for his completed map, Nimmo could have had no complaint – it equated to almost a year's salary from his rectorship and it took only two months of his holiday to earn.[23] Indeed, one can see how it might have convinced him of the pecuniary benefit likely to result from a permanent change of career.

The journal is much more than just an account of a perambulation and a description of the technical and other observations carried out. These we might expect from any technical journal. But Nimmo's journal wanders far from the rather limited objective of a boundary survey. In it we can see the progressive expression of his wider interests in much more diverse fields. For instance, he went to some effort to record accurately the meanings of Gaelic place names. One strangely prescient example was his record: 'Mam or Maum – is usually applied to those high and elevated paths which run across the summits that divide parallel glens.'[24] Years later he would build an office (now Keane's public house) at a place called Maam in Connemara. Of greater interest are his thoughts on the social and agricultural economics of the Highlands. At the beginning of the journal his comments related almost exclusively to boundary issues. As he progressed, observations relating to agriculture and social conditions grew in prominence. We have mentioned his views on the new village of Kingussie; neither did he fail to notice the effects of the Highland clearances and the spread of sheep farming. The size of the flocks that could be pastured on the higher slopes of the hill country was limited, he judged, by the quantity of 'wintering' available to the owner in the lower ground. This accounted, in his opinion, for the policy that induced the landowner to remove as many small tenants as he could. The places that the tenants inhabited 'are green, having been long under the plough. They are not liable to "storm" being usually situated in the bottom of the valley. They form, of course, his most valuable winter pasture. Having been already cleared he may lay much of them under the plough and raise artificial green crops. Besides, he removes neighbours who may be troublesome to him but whom he does not expect ever to become useful.'[25] At this stage Nimmo was not entirely against sheep farming: 'I cannot think the sheep farming system must necessarily be unfavourable to population. Nay, in the very instance before us it is certainly the reverse. On the north side of the [Rannoch] Moor are several houses inhabited by shepherds where there formerly were only wretched shealings for the shelter of herds during a few weeks of summer.' By the time he arrived at Glencoe his opinion of the benefits of sheep farming was much less positive, as we have already seen. He was constantly aware of the improvements that could be achieved if only the bogs were tackled, invoking Arthur Young: 'some patches of potato ground scooped out of the moss show us that much might be done in the agricultural way even in this bleak district of the Highlands. Arthur Young, I think, has said that no improvements are so valuable as those made on a bog.' 'But the farmers and proprietors of the Highlands,' he went on dryly, 'have this, in a great measure, yet to learn.'[26] What had started out as a delineation of the boundary had come, within a few short weeks, to encompass a commentary on agronomy verging on rural economics. Here, in the heart of the Scottish Highlands, the seeds of Nimmo's later approach to the problems of the west and south-west of Ireland were well and truly sown.

By 12 October 1806, the survey had been completed and Nimmo sent back to Rickman the skeleton map with the boundaries drawn in, the counties coloured and the proposed insertions and deletions noted, together with extra detailed sketches of those parts of the map – especially the boundary with Argyll and Perth – where particular attention was needed to update the original. He also forwarded his journal. His accompanying letter discussed many of the points still at issue and explained the general approach that he had taken.[27] But it went a lot further. Nimmo sketched the outline of a road from Dingwall to Bonar Bridge that could be constructed easily along an existing drove road, thereby considerably shortening the distance between the two towns. On the skeleton map he inserted the outline for a proposed road 'that is much wished for' leading from Glen Garry by Glen Spean, Loch Treig, Rannoch Moor and on to Killin, where it would join the military road to Stirling. 'This would become the great cattle road,' he said, 'and would be the surest means of opening a communication with the Western Highlands.' Now in full flight, as it were, he went on:

> Another road which seems a branch of this has been proposed to lead from Rannoch to King's House (Ardfeadh) thereby giving the upper parts of Perthshire a ready communication with the sea at Loch Leven.

> Good roads already exist from Rannoch to Perth and from Dunkeld to Montrose. The proposed line would complete and open a direct communication across the Island through the Central Highlands. There are only 18 miles to be formed through an open and (though high) a level valley from Rannoch to King's House.

Here again was his proposal, first made in the journal, to traverse Rannoch Moor from east to west and now, in addition, to cross it from north (Glen Spean) to south (Killin). Having made these daring suggestions, he quickly concluded: 'But I am wandering from my proper subject.'

Nimmo was far too astute to wander unintentionally from anything. Fully conscious that the commission was, after all, specifically charged with building roads and bridges, his 'wandering' was quite calculated. It also proved to be successful: in the fifth Report of the commissioners we read that Mr Nimmo was paid £50 for a survey of 'High Bridge to Killin and Loch Tummel', which Telford called the Rannoch Road.[28] The proposed line went from High Bridge (now Spean Bridge) east to Tulloch, where it crossed the River Spean and ascended to the foot of Loch Treig, then ran south along the east shore of the loch.[29] Thence it crossed Rannoch Moor near the east end of Loch Lydoch (Laidon) by a morassy track to the pass of Gual-Vearan and on to the head of Glen Lyon. From there it went by the pass of Larig na Loon to the head of Glen Lochy, down the north side of which leads to Killin. The line was fifteen miles shorter than the existing military road via King's House and Tyndrum and the estimated cost was £17,283. Telford presented the new line as his own proposal, but justified it using Nimmo's calculations:

> The main advantage will result to the breeders of sheep and black cattle; and has been estimated by Mr Nimmo of Inverness, on the supposition that 80,000 sheep are annually driven in the direction of the new road, and that a saving of 3 or 4 days in droving expenses and the better condition of the animal at market, is equal at least to one shilling each, or £4,000. And that a similar saving, and augmentation of value, will take place on 20,000 black cattle at 8 shillings each, the amount of £8,000.[30]

According to Haldane, Telford had long hoped to make such a road and had contracted Nimmo to do the survey for it.[31] Whether Telford's hopes had pre-dated the proposal for this road made in Nimmo's letter to Rickman in 1806 is not certain. The plan was apparently supported strongly by the commissioners but was never put into effect. With hindsight, Haldane doubted that even Telford's genius would have sufficed to make and maintain a sound road over the boggy miles of the moor. Interestingly, the line of the existing West Highland Railway from Spean Bridge via Tulloch Station to Rannoch Station and some miles further south follows uncannily Nimmo's proposed line for the upper part of the Rannoch Road. The track of the railway, regarded as one of Britain's best-engineered, had to be laid down on a mattress of tree trunks and many thousands of tons of hardcore material. To this day, the now upgraded military road to the west, between Tyndrum and King's House (the modern A82) remains the only road skirting Rannoch Moor; no road runs through it.

Life as a schoolmaster must have seemed staid after the great perambulation. Apart from the survey of the Rannoch Road, probably done in 1809, Nimmo is not recorded as having worked again with either of the two commissions, nor does he appear on any list of Telford's assistants. In 1809 he received some singular good news: he was nominated for membership of the Geological Society of London on 5 May, the same day that Telford was elected to membership.[32] Nimmo was subsequently elected, member number 152, on 2 June 1809.

A new outlet was soon to arise for his many talents. His contemporary at Edinburgh University, David Brewster, had taken up scientific research and tutoring in and around Edinburgh from 1800 onwards. While engaged in these activities, he turned to a career in the Church and was licensed as a minister in 1804. This was not a good career choice for a man who was incapable of speaking in public due to an excessively nervous disposition, and he did not last long as a preacher.[33] Like his mentor Playfair before him, he resigned his ministry, took up a post as a private tutor and in 1805 applied for the Chair of Mathematics at Edinburgh University when Playfair moved to Natural Philosophy. His application proved unsuccessful, as did that for a similar post at St Andrews University in 1807. No teaching post would ever appear to have been suitable for one with his particular nervous disposition, but he was in serious need of a steady income. Departing from the *Edinburgh Magazine*, he turned his efforts to a new editorial project. This was the *Edinburgh Encyclopaedia*, which he undertook to edit with the assistance of Thomas Telford, Peter Nicholson the mathematician and architect (1765–1844), and other gentlemen of eminence, many of whom chose to remain anonymous. The first part appeared in April 1808 and other parts followed, sometimes after long delays, so that the whole series was not completed until 1830. The early parts were a great success. The first was out of print within weeks of publication and was reissued in June along with the first editions of parts two and three.[34] Ten parts had been published by March of the following year, 1809.[35]

Nimmo contributed a number of articles, alone or in cooperation with others, to the *Encyclopaedia*. They are notable for their remarkable familiarity with written sources in the classical and modern continental languages, as well as for their clear understanding and exposition of scientific and civil engineering principles. His first article, initialled A.N., on 'Boscovich's Theory of Natural Philosophy', describes the ideas of this renowned physicist that are now largely supplanted by modern atomic theory.[36] His second, 'Bridge',[37] and third, 'Carpentry',[38] were written conjointly with Telford and Peter Nicholson respectively. In introducing the 'Bridge' article, Brewster noted: 'In this valuable and original article on the theory and practice of bridge-building, which is the only complete treatise on the subject that has yet been published, the Editor is indebted to Thomas Telford FRSE civil engineer and to Alexander Nimmo FRSE.' Telford was already a renowned engineer and his name would have added immeasurably to the value of the article. Nimmo, on the other hand, was then a relative unknown, his FRSE having only just been awarded. Because of this, it is hard to be certain exactly how far the views and opinions expressed by Nimmo in the article were influenced by Telford.

The 'Bridge' article is divided into separate sections, Nimmo's on the 'Theory of Bridges' and Telford's on the 'Practice of Bridge Building'. Nimmo writes on the history of bridges from ancient times up to the then current time. He considers the modern development of bridges – from the eighteenth century forward – to owe nothing to advances in mathematical treatment:

> It is our intention to put out a new way of considering this subject ... It may indeed be deficient, if not in some respects erroneous ... Indeed, though we highly value the sublime geometry, we are inclined to think that the unnecessary parade of calculus in the application of science to the arts, has been one of the chief causes of the dislike, which many able practical men of our country have shown to analytical investigation.

Scientific theory, he insisted, constrained vision without solving the real practical problems of bridge construction. This was a most remarkable thesis, coming from a man who was both a teacher and expert in mathematics and physics. According to Billington, 'all structural artists since Telford have argued this thesis and its corollary that the simplification of analyses liberates the imagination'.[39] The debate as to the true nature of good design, whether an art or a science, is no less relevant or less heated in our own time, as new structures and new materials become more exciting, more daring and visually more attractive.

The thesis was actually propounded by Nimmo, not Telford, but we can ask was it really his own opinion? In all that he did previously and subsequently, Nimmo placed inordinate importance on prior mathematical analysis. He himself was entering civil engineering from a mathematical, scientific background with virtually no prior practical experience. The seeming distrust of science and mathematics in his Theory of Bridges is therefore strikingly anomalous, but is fully in keeping with the views of Telford, who had little use for science, was untrained in mathematics, and is reputed to have had no knowledge of geometry, not to mention calculus.[40] To say this is not to disparage the great engineer; rather, it highlights the brilliance and intuitive genius of his empirical designs. Perhaps Nimmo was acknowledging and celebrating this in his anomalous thesis? Telford was, after all, the acknowledged expert in bridge design. It would hardly have been easy, or diplomatic, for a young man to diverge too sharply from the views of so eminent an expert and joint author, especially one who could be very influential in the life of any would-be engineer. Telford was a major shareholder in the *Encyclopaedia* and may have been important in the selection of Nimmo as author for so many disparate articles.[41] If Nimmo felt compromised by this there is no evidence here of any unease, so maybe he had really absorbed Telford's views as his own. The main content of Nimmo's section goes on to treat proficiently and in an illustrative manner the mathematics of arches and flow patterns around piers, foundations, and so forth. In all, his theoretical contribution ran to thirty pages whereas Telford's contribution on the practical aspects ran to twenty-six.

Nimmo's third contribution to the *Encyclopaedia* was a joint work with Peter Nicholson entitled 'Carpentry'. Here too, Nimmo was the relative unknown and Nicholson, who had just been appointed Surveyor for the County of Cumberland on Telford's recommendation, was the acknowledged expert. Brewster again made a short introduction: 'The Editor has been indebted for the theoretical part of this valuable article to Alexander Nimmo FRSE, civil engineer and for the part on constructive carpentry to Mr. Peter Nicholson, architect, whose attention has been long and successfully directed to this interesting subject.' Nimmo's contribution, over seventeen pages, deals mainly with an analysis of the strength of different kinds of timber and the strains to which it is subjected in use. He relies heavily on the works of continental writers but also on his own treatment of the topic. Nicholson's section was much more practical and applied.

The 'Boscovich' and the 'Bridge' articles were published in 1811, the 'Carpentry' article in 1812. The first two must have been written before Nimmo left for Ireland in June 1811, and the third was probably also completed at the same time. His fourth article, also initialled A.N., was on 'Draining'.[42] In this he reviewed the drainage of different types of soil and how best to achieve it. In discussing Ireland, he mentions that the Bogs Commissioners 'have already published two reports on the subject'. These had been published in March 1811 and this reference proves that Nimmo's article was written after that date, but probably before he joined the commission in June (otherwise it might have included more personal observations). It indicates the relevance of his theoretical knowledge to his new Irish appointment and it certainly proves his acquaintance with the commission's work. He also refers in his paper to two other articles that he may have assisted with, or perhaps written. At the very beginning of 'Draining' he states: 'We have already, under that article ["Agriculture"] detailed some of the modes of affecting this object [draining].' The 'Agriculture' article is initialled 'B'; it is massive and comprehensive, with one chapter given over entirely to drainage, but there is no clear indication of any direct input from Nimmo. It was probably written by Brewster himself, possibly with input from Nimmo.

Later in 'Draining', Nimmo refers to the article on 'Inland Navigation' in a way that indicates he had an input to that contribution, although he did not sign or initial it. Others also have claimed, without adducing any evidence, that he was responsible, in part at least, for 'Inland Navigation', first published in 1821.[43] While there is no signature to the article, it reads like the kind of contribution Nimmo would write. It reviews the historical aspects of inland navigation in Egypt, China, India, Persia, Greece and Turkey before going on to continental Europe, Britain, Ireland and America. It then discusses bars at river mouths; the manner of propelling boats; sails, oars and steam; canals and still-water navigation; canal cuttings and embankments. Internal evidence shows that it was written between 1819 and 1821. Its intimate knowledge of work on the lock at Fort Augustus on the Caledonian Canal, and its reference to the foundations of the lock at Corpach (already fully covered in detail in the 'Bridge' article), suggest that Telford or Nimmo could have written it; reference to Monnikendam as 'the oldest boat gate *we have seen* [italics added]' suggests that it was written by Nimmo, who had visited Holland in 1814. The conclusion of the article, elaborating on the thesis discussed earlier, fits either party, if our analysis of the 'Bridge' article is correct: 'In this country we find that the most valuable inventions and improvements originate not from the depths of science, but from ingenious practical skill, engrafted upon a few general principles, and proceeding from observation and experiments.' It is unbelievable that Nimmo would be so trenchantly opposed to scientific analysis. The closing statement regarding the source of the opinions expressed in the article is ambiguous: 'What regards Egypt, Holland, the Low Countries, Sweden, America, Ireland and Scotland must be admitted as entirely original; also much that regards many of the most important navigations of England. The scientific discussion and practical applications are chiefly original.' Confirmation that he had an input to 'Inland Navigation' was given by Nimmo himself in reply to a question in his evidence to the House of Lords Enquiry into the State of Ireland in 1824.[44] Having said that he had given an account of all the canals of Ireland, he was asked where that account was given, to which he replied: 'The account was given by me to Mr Charles Grant; but an abridged account of it is to be found in the article "Inland Navigation" in the *Edinburgh Encyclopaedia*.' So, like the 'Bridge' article, it appears probable that 'Inland Navigation' was another joint effort by Nimmo, possibly with Telford and/or William Jessop, whom Ruddock believes, without giving evidence, was also involved.[45]

Nimmo may have been scheduled for one other article that does not appear to have been published. In 'Boscovich' he wrote: 'We intended to have concluded this article with a few general observations ... but as we shall have occasion to resume the subject under the head of "Corpuscular Philosophy" these observations may be introduced in that part of our work.' The entry under 'Corpuscular Philosophy' in volume 7, published in 1813, refers the reader to the prospective article on 'Mechanics'. The volume containing 'Mechanics' is dated 1819 and it makes no mention of Nimmo, nor is it written in his style.

Perhaps an article on 'Corpuscular Philosophy' was planned originally, but was dropped or replaced by the time publication got that far. While it is possible that Nimmo's inability to complete this commitment was caused by his departure to Ireland, it could also reflect changes in the understanding of atomic structure that arose in the period since the article was first mooted.

By the time he left Inverness in 1811, therefore, Nimmo had a significant corpus of technical writing to his credit, considerable background knowledge and some good practical experience of land surveying, geology, hydrography and agronomy. More importantly, he had earned the confidence of some of the leading scientific and engineering experts in Scotland and was widely regarded among influential persons as a man of sound judgement and great erudition. Appropriately enough, his transformation from schoolmaster to engineer was accompanied by his admission as a Fellow of the Royal Society of Edinburgh on 1 January 1811, in whose roll he is inscribed as 'engineer'. His nomination was proposed by Sir George Steuart MacKenzie, seconded by Professor Christison and Mr Allen, an Edinburgh mineralogist. In all probability it was Sir George, himself a member, who had nominated him in 1809 to the Geological Society also. Nimmo's visit to his residence at Coul in 1807 was therefore well rewarded.

His break with school-teaching was to be thorough and complete: as a token of this he referred to himself as 'civil engineer' from January 1811 onwards. At that stage of the Industrial Revolution it was not uncommon for a person to describe himself at one time as one thing and at another as a completely different thing, one day a stonemason, another an engineer.[46] Nimmo was therefore not unique in taking upon himself the title, and there is no suggestion that anyone thought anything unusual about it. His uniqueness lay in the route he took into the engineering profession. Until then, that route had been traditionally (and informally) through apprenticeship, the candidate progressing subsequently to pupilage under an established engineer. Nimmo's route was from a strong academic background in formal science and mathematics, with no formal apprenticeship or pupilage in the practical skills of engineering. This would colour all his subsequent activities and may explain his reduced attention to the more mundane, practical aspects of his designs and plans. Today, all civil engineers are academically trained, and the very best courses in civil engineering are now grounded in mathematics, physics and computing. Nimmo's route represented a turning point for the profession. New structures, new materials, new means of motive power, new modes of transport and so on would all increasingly demand a different kind of engineer after about 1800, one that depended less on the individual genius of one person and more on established, repeatable principles of science. Nimmo was one of the very first of a new kind of engineer that would take the Industrial Revolution on to its full maturity.

ALEXANDER NIMMO, MASTER ENGINEER

NOTES

1. A. Arrowsmith, *Memoir relative to the construction of the Map of Scotland published by Aaron Arrowsmith* (NAS, G 09/40). (Arrowsmith's map is archived in NAS, RHP 14008.)
2. A. Nimmo, *Journal along the North, East and South of Inverness Shire, ends at Fort Willliam*, p.45. (NLS, Adv MS 34.4.20), hereafter referred to as Inverness Survey.
3. Arrowsmith, *Memoir relative to the construction of the Map of Scotland*.
4. Ibid.
5. Third Report of the Commissioners for the Highland Roads and Bridges (B.P.P. 1807 [100], vol. III, mf 8.12–13), hereafter referred to as Third Report HRB.
6. Inverness Survey, p. 3.
7. A. Nimmo, *Historical Statement of the Erection and Boundaries of the Shires of Inverness, Ross, Cromarty, Sutherland and Caithness*, Third Report HRB.
8. Inverness Survey, p.16.
9. S. Brumwell, 'George Wade 1673–1748', in H. G. C. Matthew and B. Harrison (eds) *Oxford Dictionary of National Biography* (Oxford: Oxford University Press, 2004).
10. F. Groome, *Gazetteer of Scotland* (Edinburgh: T. C. Jack, 1890).
11. Inverness Survey, p. 28.
12. Ibid., p. 30.
13. Ibid., p. 31.
14. Ibid., p. 40.
15. Ibid., p. 39.
16. Ibid., pp. 42–5.
17. Ibid., p. 49.
18. Arrowsmith, *Memoir relative to the construction of the Map of Scotland*.
19. A. Nimmo to J. Rickman, 12 October 1806, in Inverness Survey, pp. 52–4.
20. Inverness Survey, p. 2.
21. Arrowsmith, *Memoir relative to the construction of the Map of Scotland*.
22. Fourth Report of the Commissioners for the Highland Roads and Bridges, appendix O, p. 63 (B.P.P. 1809 [167], vol. IV, mf 10.28).
23. Arrowsmith, *Memoir relative to the construction of the Map of Scotland*.
24. Inverness Survey, p.8.
25. Ibid., p. 39.
26. Ibid., p. 41.
27. A. Nimmo to J. Rickman, 12 October 1806, in Inverness Survey, pp.52–4.
28. Fifth Report of the Commissioners for the Highland Roads and Bridges, appendix U (B.P.P. 1810/1811 [112], vol. IV, mf 12.24).
29. T. Telford, *Report and estimates relative to a proposed road in Scotland from Kyle-rhea in Inverness-shire to Killin in Perthshire by Rannoch Moor* (London: Luke Hansard and Son, 1810).
30. Ibid.
31. A. R. B. Haldane, *New Ways through the Glens* (Edinburgh: Thomas Nelson and Sons, 1962).
32. Geological Society of London, Meeting of 5 May 1809 (Nimmo nominated, Telford elected); Meeting of 2 June 1809 (Nimmo elected member no. 152) (Geological Society of London, Ordinary Minute Book No. 1).
33. A. D. Morrison-Low, 'Sir David Brewster 1781–1868', in H. G. C. Matthew and B. Harrison (eds), *Oxford Dictionary of National Biography*. (Oxford: Oxford University Press, 2004).
34. *Caledonian Mercury*, 11 June 1808.
35. *Caledonian Mercury*, 16 March 1809.
36. A. Nimmo, 'Boscovitch's Theory of Natural Philosophy', in *The Edinburgh Encyclopaedia*, ed. by D. Brewster (Edinburgh: Blackwood, 1811), hereafter referred to as Edinburgh Encyclopaedia.
37. A. Nimmo, and T. Telford, 'Bridge', in *Edinburgh Encyclopaedia*.
38. A. Nimmo, and P. Nicholson, 'Carpentry', in *Edinburgh Encyclopaedia*.
39. D. P. Billington, *The Tower and the Bridge* (New York: Basic Books, 1983).
40. A. Nimmo, and T. Telford, 'Bridge', in *Edinburgh Encyclopaedia*.
41. R. Paxton, 'Alexander Nimmo', in Matthew and Harrison (eds), *Oxford Dictionary of National Biography*.
42. A. Nimmo, ' Draining', in *Edinburgh Encyclopaedia*.
43. A.W. Skempton, *Biographical Dictionary of Civil Engineers*, vol. 1, 1500–1830 (London: Thomas Telford, 2002).
44. A. Nimmo, Evidence, in Minutes of Evidence taken before the Select Committee of the House of Lords to examine into the Nature and Extent of the Disturbances which have prevailed in those districts of Ireland which are now subject to the Provisions of the Insurrection Act and to report to the House (B.P.P. 1825 [200], vol.VII, mf 27.60–63).
45. T. Ruddock, 'Alexander Nimmo', in Skempton, *Biographical Dictionary of Civil Engineers*.
46. T. S. Ashton, *The Industrial Revolution, 1760–1830* (Oxford: Oxford University Press, 1969).

CHAPTER 3

The Bogs Commission, 1809–1813

Long before the Highland Society and the British Fisheries Society came into being, the Dublin Society was active in Ireland. Founded in 1731, its full title was The Dublin Society for improving Husbandry, Manufactures and the other Useful Arts. Later on, 'Sciences' would be added to its title and in 1820 the epithet 'Royal', which it retains to this day. The Society undertook in Ireland many of the roles that were played by the former two societies in Scotland. Its objectives were to improve the agriculture and manufacturing base of the country by encouraging the application of the most modern scientific principles and practices in all aspects of Irish enterprise. Membership was drawn from the landed and establishment classes and the Society was especially interested in discovering and exploiting the supposed mineral wealth of the country.[1] Such wealth, especially the coal and iron resources, underpinned the Industrial Revolution in Britain, and Irish landowners anticipated that similar resources would underlie their Irish estates. The Society therefore took a particular interest in mineralogy and by 1795 it had appointed a Professor of Chemistry and Mineralogy to give lectures on these topics at the Society's rooms, where a chemical laboratory was also fitted out.

The more mundane aspects of practical agriculture and manufacture were supported with premia and prizes for innovation and development. For instance, as a result of the Society's encouragement, some 25 million trees were planted in Ireland between 1786 and 1806.[2] Nothing was too insignificant for the Society: it offered, for example, a premium of ten shillings for every stock of bees preserved through the winter, over and above ten stocks. Fisheries were encouraged by distributing charts of the coast, aiding the building of quays, translating fishery treatises and fostering the salting and distribution of fish. It highlighted the general lack of piers suitable for the fisheries all around the coast. From 1761, Parliament made regular grants to the Society that helped to fund its numerous premia and prizes and this parliamentary generosity continued long after the Act of Union.

In the year of the Society's foundation, the English Parliament had passed an act encouraging the improvement and cultivation of bogs in England. The act did not apply to Ireland, but Irish interests tended to emulate developments in Britain as far as possible. So, from the very beginning, the improvement of the Irish bogs by drainage and fertilisation became an important topic on the Society's agenda. At one of its first meetings Thomas Prior, the moving spirit in the foundation of the Society, presented a treatise on 'A new method of draining marshy and boggy lands' that was inscribed in full into the minute book and is one of the earliest treatises of the Society.[3] Another of its earliest members was Archbishop Theophilus Bolton of Cashel, an acknowledged improver of land 'by draining bogs which were large and useless and turning them into pasture and tillage', and at least two articles on bog reclamation were published in the Society's transactions by 1800.

When agricultural and statistical surveys of Britain began to be published in the last decade of the eighteenth century, the spirit of emulation was kindled in Ireland. One of the last actions of the Irish Parliament prior to its dissolution was to request the Society to undertake surveys similar to those under way in Britain, of all the counties of Ireland. The Society responded immediately and surveys of twenty-three of the thirty-two counties would be completed and published between 1801 and 1832. Each was designed to assess the economic needs and agricultural status of the county, but they covered many other topics as well, including the social conditions of the inhabitants.

By the early nineteenth century Britain was reaching a stage where it could no longer produce sufficient grain to feed its expanding population and war on the Continent made it impossible to obtain supplies from there.[4] Legislation granting bounties on the export of corn to England, and imposing duties on its importation from abroad, had considerably encouraged tillage in place of pasturage on Irish midland farms, which became the great grain producers and exporters of the time, so much so that they were once known as 'the granary of England',[4] Appreciating that the high price of corn made tillage land very valuable, and that Ireland had over one million acres of unproductive bog land which, once drained and improved, might yield untold wealth to its owners, John Foster MP introduced into Parliament in March 1809 a measure to cultivate the Irish bogs, bringing forward a bill 'to appoint Commissioners to enquire and examine into the Nature and Extent of the several Bogs in Ireland; and the Practicability of Draining and Cultivating them, and the best means of effecting the same'.[5] It received parliamentary approval and was passed into law the following May, only weeks after it had first been raised.[6]

There is a popular misconception that the Act was intended to alleviate famine or shortage of food among the Irish. Its preamble puts paid to that idea:

> Whereas there are large tracts of undrained bog in Ireland, the drainage whereof is necessary for their being brought into a state of tillage: And

whereas the adding of their contents to the lands already under cultivation would not only increase the agriculture of Ireland, and contribute much to its resources for the sustenance of the British Empire and its profitable export of corn, but is highly expedient towards promoting a secure supply of flax and hemp within the United Kingdom for the use of the Navy and support of the linen manufacture, independent of foreign nations and of the interruptions arising from the influence of political events upon foreign trade ... Be it therefore enacted etc.

The purpose of the act had nothing whatever to do with any perceived plight of the native Irish (neither distress, nor poverty are mentioned) but was really an attempt to secure the supply of grain and hemp for the sustenance of Britain, the equipping of the Navy and the well-being of the Empire. The act laid out unambiguously the duties of the commissioners:

> ascertaining the extent of all such bogs in Ireland as shall by repute or estimation exceed the extent of 500 acres; and for enquiring and examining into the practicability of draining each such bog and into the best mode and probable expense of effecting such drainage; the depth of bog soil, the nature of the strata immediately underneath; the nature and distance of the manure best fitted for their improvement, and the expense of making the necessary roads or canals for conveying such manure into and through the same from the nearest or most convenient public high road or canals, and for the carrying out the future produce of the bog land when cultivated to the nearest or most public roads or canals; together with the opinion of the said Commissioners of such measures as they shall deem necessary or expedient for carrying into speedy effect the drainage, cultivation, and improvement of all such bogs, and the future increase of timber in Ireland, by providing for the plantation and preservation of trees in such parts thereof as shall be best fitted for the purpose.[6]

Since an estimated 15 per cent of the landmass of Ireland (a gross underestimate) was to be investigated, the magnitude of the task imposed on the commissioners can be imagined. The number to be appointed and the date of their commencement were left to the Viceroy, a safe enough provision since 'all and every person or persons who shall act as a Commissioner or Commissioners under this Act, shall so act without Salary, Recompense or Reward whatever'. The commission was granted the small sum of £5,000 for its operation, which was originally to last only until 1811. Given that there were then estimated to be over one million acres of bog (ignoring the numerous tracts of mountain bog in the west and southwest and all bogs under 500 acres), the grant amounted to less than one penny per acre.

The commissioners were empowered to summon and examine under oath all persons concerning any aspect of their brief and to employ such engineers, surveyors and others as were necessary to execute their obligations under the act. They and their employees were

> authorised and empowered from time to time to enter into and upon all and every or any such bogs, and to survey the same and to ascertain the boundaries thereof, and the nature and strata, soil and composition thereof, and also to bore, dig, cut, trench, get, remove, take and carry away any earth, stone, clay, soil, rubbish, trees, roots of trees, gravel or sand, or other matters and things as may be deemed requisite, necessary or proper for the ascertaining of the nature, strata, soil and composition of such bogs; and also to enter upon any lands or grounds which [they] shall conceive to be expedient for the purpose of carrying a drain for the purpose intended by this Act

These were draconian powers to confer on any unelected commissioners and would never have been acceptable unless balanced by distinct advantage to the landowners. The balance of advantage indeed lay decidedly with the owners, who would have their bog land surveyed by competent engineers at no cost and would have plans drawn up for its development, improvement and cultivation, along with proposals for access roads, all of which might significantly enhance rental values.

Maps were to be prepared of all the bogs showing the areas encompassed and the engineers' plans for drainage and cultivation. Every attention was to be given to the improvements necessary to increase the production of grain, flax and timber as efficaciously as possible. However, any actual development or improvement would be the responsibility of the landowner alone, since the Government had absolutely no intention of funding anything other than the surveying, planning and estimation work. The commissioners were therefore further charged with ascertaining 'the names of the parties who were proprietors of, or interested in or entitled to, or who claim any right or interest in any such bogs or any part thereof and to what extent' with a view, no doubt, to identifying those likely beneficiaries who might later be asked to contribute towards the survey expenses.

The Warrant of Appointment was issued on 15 September 1809 and the commission held its first meeting on the nineteenth of that month.[7] The first commissioners were a somewhat motley group of eminent men of differing affairs: General Charles Vallancey, soldier, surveyor, engineer, Fellow of the Royal Society and vice-president of the Dublin Society, then 85 years old and nearing the end of a long life enriched by eccentric views and enlivened by controversial opinions (The Dictionary of National Biography said unkindly of him that he was 'ignorant of Irish; published worthless tracts on Irish philology and history'[8]), was to be Chairman; Richard Griffith Snr, landowner, late of the East India Company, man of Ascendancy business and affairs, member of the Royal Irish Academy and the Dublin Society, member of the late

Irish Parliament, director of the Grand Canal Company and father of Richard, later Sir Richard, Griffith the famous engineer; J. S. Rochfort; William Gore; and Henry Hamilton, all members of the Dublin Society. John Leslie Foster, barrister and MP, secretary and later Vice-President of the Dublin Society, replaced Rochfort on the commission very soon. Vallancey retired early and died in 1812; Foster replaced him as Chairman and Hans Blackwood was appointed a new ordinary commissioner. Since he never signed the third and fourth (final) reports, it seems that Griffith, too, may have retired before the commission completed its work.

Just as the two Scottish commissions were intimately linked to the Highland and the Fisheries Societies, it was only natural that the Bogs Commission would be intimately associated with the Dublin Society. No other body was as well acquainted with the state of Irish agriculture, society and infrastructure as this Society. The Bogs Commission, set up with no premises of its own and with no provisions for any, met in the rooms of the Society in Hawkins Street, Dublin; the assistant-secretary of the Society took on the duties of Secretary to the Commission; the superintendent of the Society's repository became the commission's assistant-clerk and the Society's prospective mining engineer became one of its first engineers. All analyses of the peat sampled from the bogs were performed in the Society's chemical laboratory.[9]

Working from a copy of Vallancey's map of Ireland,[10] the commissioners determined that the greater part of the bogs (85 per cent) formed one connected whole, comprising about one-fourth of the total landmass of the country. They lay in a broad belt formed by drawing two lines across the country, one from Wicklow Head to Galway and the other from Howth Head to Sligo. The River Shannon bisected this belt into a western region containing about 600,000 acres (70 per cent), and an eastern region containing 260,000 acres (30 per cent). Outside this belt the remaining bogs were estimated to comprise about 140,000 acres, making a total of one million acres for the whole country. In the eastern section of the great belt lay the Bog of Allen, which, they emphasised, did not constitute one great morass but a series of independent bogs, perfectly distinct from one another and separated by ridges of cultivated ground. They proceeded to divide the eastern sector of the bog belt into six districts: the north-eastern, south-eastern, north-western, south-western, Westmeath and Longford. A seventh district was recognised in Tipperary, centred between Roscrea and Cashel.

One year after the act it seemed to Parliament that nothing was being done, so a committee was appointed to enquire into the apparent lack of progress. Its scrutiny evoked an immediate response from the commissioners.[11] They explained what they had already done, giving as their reason for concentrating on the Bog of Allen its proximity to the capital and the routes through it of the Grand and the Royal Canals. In fact their approach may have been influenced also by the exigencies of the time. On 23 September 1809 they had appointed Richard Griffith – son of Richard Griffith the commissioner – as their first engineer. He was already engaged on a survey of the Leinster coalfields on behalf of the Dublin Society and since he had frequent business in the capital, the north-eastern district of the Bog of Allen suited him. Richard Lovell Edgeworth, engineer, inventor and polymath, had been appointed a week later: extending the survey into Longford suited him as his seat was at Edgeworthstown in that county, and so he was given responsibility for that district. On 9 October Thomas Townshend, on 15 November James Alexander Jones, and on 6 December John Longfield, had been appointed engineers to the north-western, Westmeath and south-western districts respectively, followed on 3 January 1810 by David Aher. The latter, engineer to the Grand Canal Company, would only work in the area of Tipperary 'on account of his other engagements', which probably explains the early extension of the survey to that county. Richard Brassington had been appointed in November 1809 to the south-eastern district, but he left the post and was replaced by Thomas Colbourne in September 1810.[12] All these men were established surveyors or civil engineers with reputable practices that also needed their attention.

The commissioners told the parliamentary committee that they hoped to extend their surveys into the remaining districts as soon as possible. These were the districts lying between the River Shannon and the Western Ocean, as well as considerable tracts in other parts of Ireland. These had yet to receive any attention, and no engineers had been assigned to them, partly because of their remoteness from Dublin and their generally impoverished reputation. As well, there was a shortage of suitably qualified Irish-based applicants willing to undertake the commission's work: 'We have employed every engineer we could find in Ireland who would enter our engagement and of whose qualifications we approved … We have endeavoured to procure others from England and Scotland, but without success upon the present rate of our allowances…'[13] Obviously the salary was not attractive, something we should remember when we come later to discuss Nimmo's accounts. Worse still, the initial grant of £5,000 was almost expended by March 1811 and costs were running at £280 per week. Unless a further grant was forthcoming the survey could not hope to continue even with the work already in hand. The parliamentary committee agreed about the shortage of funds and the matter was duly attended to by a new Act, which also extended the remit of the commission for a further two years.[14] The shortage of engineers was another matter entirely. The commissioners were obliged to continue looking for staff wherever they could if they were ever to tackle the remoter western and south-western bogs then 'in a state of desolation and abandonment'.

One British surveyor immediately came to attention as a possible suitable addition to the survey team. William Bald had been commissioned by the Mayo Grand Jury in 1809 to prepare a map of that county and he had come from Scotland to take up the task. The commission now recruited him, allocating the Mayo district to his care on 22 May 1810. Bald was not really an engineer at all, but a trained surveyor and cartographer. At this stage the commission could hardly be too choosy, being already into the second year of its four-year life with nothing yet started in the west. Bald, who was already *in situ* in Mayo, would have to combine his work for the Grand Jury with his Bogs Commission work. Others of the engineers were able to combine their commission duties with their ongoing businesses, and Bald would be no different to them in that regard. All of them were engaged on a daily basis and they were not expected to attend to the commission's work every day. It was sufficient that they swear an oath regarding the exact number of days they actually worked for, and charged to, the commission. They were paid two guineas for every day worked and, in *lieu* of all expenses, one guinea extra for each day actually spent on survey. Surveyors working under them received one guinea per day, with one guinea for expenses. Staff-men, chain-men and labourers were to receive whatever sums the engineers could agree with them, not exceeding three shillings a day – a quite handsome pay for labourers, when road menders could be had for eight pence a day.

The last of the engineers to be appointed was Alexander Nimmo, who reportedly entered the commission's employment on 5 January 1811 and was allocated a dispersed area in the south-west that included most of County Kerry along with part of counties Cork and Limerick.[15] He cannot have taken up his post straight away: as we have already seen, he continued in office at the Inverness Academy until June 1811. On the other hand, he had been nominated to Fellowship of the Royal Society of Edinburgh on 3 December 1810, and was inscribed in the society's roll as a civil engineer, not as a rector, in January 1811.[16] So maybe in spirit, if not in body, he had already turned by January to fresh fields and pastures new. Since this was the first time that he was titled 'engineer', it may have resulted from the post offered by the Bogs Commissioners, or from a commitment he made to them late in 1810. While there is no evidence that the Directors of the Academy were directly aware of his impending resignation until June of 1811, some suspicion may have been in the air during their meeting of 28 May when they instructed him to pay certain long outstanding debts. On the day of his official resignation, 11 June 1811, Nimmo was already in Dublin *en route* to Kerry, and the County Treasurer for Kerry (at Dublin Castle) was asked on that very day to let him see the Kerry county map.[17]

Many years later, John Foster said that it was Telford who had recommended Nimmo to the Bogs Commissioners, as he apparently had done earlier to the Commission for Highland Roads and Bridges, but no conclusive evidence has been unearthed yet to confirm this.[18] (Foster had not been a commissioner when the commission first started, so his evidence is not conclusive.) Telford certainly knew Nimmo and his capabilities at first hand, but no evidence has been found that he was in direct contact with the Bogs Commissioners in

1809 or 1810. John Rennie, on the other hand, was definitely undertaking harbour works at Dalkey and Dublin Port at the time the commission started, and he was certainly consulted (for which he was paid) before the commission's work started in earnest.[19] A recommendation from either Telford or Rennie would undoubtedly have validated a claim that Nimmo was a suitable 'engineer' rather than a surveyor (or still less, a mere schoolmaster!), for which his experience at that time was more appropriate. However, having already taken on Bald as an engineer the commission could hardly quibble about Nimmo's much stronger, albeit mainly academic, qualifications for an engineering post. In any event, no other suitable engineers in Britain or Ireland appear to have offered themselves for consideration. The commissioners were desperate, so once again Nimmo benefited fortuitously from the exigencies of the time. If it were Telford who recommended him, then that great engineer had really given Nimmo not one, but two, starts in engineering – first with the Highland Roads and Bridges Commission and now with the Bogs Commission.

The circumstances that brought Nimmo and Bald to the attention of the commission after its earlier failure to recruit anyone from Scotland could well have had a personal explanation. As indicated previously, it is very likely that Griffith, Bald and Nimmo knew each other in Scotland only three years earlier. What would be more natural than that Griffith, already engaged by the commissioners, would have contacted Bald after the latter came to Mayo, and invited him to join the bogs survey? Later on Griffith, or possibly Bald, might have told Nimmo of the commission's difficulty in recruiting engineers and the latter, seeing his chance to change careers permanently, may have jumped at the opportunity presented, with the support of Telford. Whatever the actual case, Griffith, Bald and Nimmo were the three engineers who gave the greatest number of days to the service of the Bogs Commission (1,054, 598 and 720 days respectively up to the end of 1813), a service so full that it calls into question the manner of their working or of their accounting. Herries-Davies is suspicious of Griffith's claim to have worked 1,054 days from a total of 1,333 weekdays (79 per cent) between his appointment in September 1809 and 31 December 1813.[20] What, then, must we make of Nimmo's claim to have worked 720 of a possible 780 weekdays (92 per cent) between early June 1811 and 31 December 1813? All his life he appears to have been extraordinarily busy and this confirms it – if the figures can be believed. When we remember that Bald and Griffith had engagements for other clients during their tenure at the bogs survey, and their proportionate days worked are not very much less than Nimmo's, then we may be inclined to give Nimmo the benefit of the doubt. His output by way of maps and reports certainly confirms that he expended a considerable amount of valuable and productive labour in the task, as indeed the others did. He appears to have been what we would today call a 'workaholic' and his early death at 49 years of age should be a warning to his imitators. On the other hand, he was rather cavalier in his accounting duties and slow to pay up throughout his entire professional career, as the outstanding bills at the Academy already presaged. The commissioners for their part may not have been too scrupulous about the accounts because of the acknowledged poor remuneration they offered.

The engineers had to appoint their own surveyors and other assistants for whose qualifications they were held responsible. No fewer than forty-five surveyors were engaged during the four years of the commission, seventeen of them joining during the first six months. Many would go on to make reputations as engineers or as surveyors, contractors or architects in future years. Each engineer was responsible for the district allocated to him, and worked to a set of instructions drawn up by the commissioners with the advice of Rennie.[21] In addition to surveying each bog, the engineer was expected to ascertain the proper line of direction of one or more drains needed to release its waters into the rivers most convenient to it. The main drains were to be designed for use eventually as canals or navigable waterways, the instructions specifying the dimensions necessary for that purpose, namely, a width of fourteen feet at the bottom and a depth of five feet. Where appropriate, locks were to be installed, seventy feet long and over seven feet wide. Only the largest bogs could ever sustain canals of such dimensions, and the low rate of flow through them would soon result in massive deposition of silt. Main drains were to be laid out to permit collateral drains to communicate with them in such a way that the greatest possible area of bog was reclaimed.

Most importantly, a map was to be prepared of the whole area delineating the existing bog boundaries; the nature of the soil; the situation of springs, rivers and lakes; the courses of intersecting rivers, streams, roads and canals; the drains proposed; and the lines of new roads that were required to bring in manure and take out agricultural produce. Levels were to be ascertained; the general topography of the area marked out; and 'all remarkable objects which are likely to be permanent, such as raths, castles, towers, cairns, hill tops, market houses etc.' were to be laid down. The maps were to be accompanied by suitable written reports containing generally whatever occurred to the author regarding the drainage of their districts, especially estimates of the probable expense of improvement works. They were to recommend the mode of agriculture best suited to the district, the best locations for corn mills and the probable value of the land when reclaimed. In order to connect the maps of the various districts, levels were to be ascertained accurately so that they could eventually be joined and ultimately be referred to 'the level of the platform on the capital of the column erected in the memory of Lord Nelson in Dublin'. Finally, the names of all the parties with interests or rights in the bogs were to be established as far as that was possible. In its mapping role, therefore, the commission was endeavouring to lay the groundwork for a national topographical map of the bogs that would be of immense use for developmental

TABLE 1.
DATES OF COMPLETION AND OF PUBLICATION OF THE VARIOUS REPORTS OF THE
ENGINEERS OF THE BOGS COMMISSION

DISTRICT	ENGINEER	REPORT DATED	REPORT PUBLISHED	REPORT OF COMMISSION
Bog of Allen (east)	Griffith	June 1810	June 1810	First, Appendix 4
R. Brosna	Longfield	Oct. 1810	Mar. 1811	Second, Appendix 6
R. Inny and L. Ree	Edgeworth	Oct. 1810	Mar. 1811	Second, Appendix 8
Bog of Allen, (west)	Griffith	Feb. 1811	Mar. 1811	Second, Appendix 4
R. Shannon	Townshend	Feb. 1811	Mar. 1811	Second, Appendix 7
Meath, King's Co., River Boyne.	Jones	Feb. 1811	Mar. 1811	Second, Appendix 5
Kildare and King's Co. and R. Barrow.	Brassington	Mar. 1811	Apr. 1814	Third, Appendix 5
West Clare	Colbourne	July 1811	Apr. 1814	Third, Appendix 4
Roscrea/Killenaule	Aher	July 1811	Apr. 1814	Third, Appendix 2
Mayo, (Castlebar)	Bald	Dec. 1811	Apr. 1814	Third, Appendix 7
Iveragh, Kerry	Nimmo	Dec. 1811	Apr. 1814	Fourth, Appendix 5
L. Gara, Sligo/Roscommon	Longfield	May 1812	Apr. 1814	Third, Appendix 1
Mayo (North-west)	Griffith	May 1812	Apr. 1814	Fourth, Appendix 11
Queen's County	Aher	June 1812	Apr. 1814	Third, Appendix 3
R. Suck, Galway and Roscommon	Griffith	June 1812	Apr. 1814	Fourth, Appendix 8
Kerry, Dunkerron and Kenmare	Nimmo	1812	Apr. 1814	Fourth, Appendix 6
Kerry, R. Laune and River Maine	Nimmo	1812	Apr. 1814	Fourth, Appendix 6

cont

Kerry, Upper River Maine	Nimmo	1812	Apr. 1814	Fourth, Appendix 6
Kerry, River Cashen	Nimmo	1812	Apr. 1814	Fourth, Appendix 6
Kerry and Cork, Slieve Luachar	Nimmo	1812	Apr. 1814	Fourth, Appendix 6
Mountains of Dublin and Wicklow	Griffith	Jan. 1813	Apr. 1814	Fourth, Appendix 10
Mayo, (East)	Bald	Jan. 1813	Apr. 1814	Third, Appendix 9
Mayo, (South-west)	Bald	No date	Apr. 1814	Third, Appendix 8
L. Neagh and River Bann	Townshend	Mar. 1813	Apr. 1814	Third, Appendix 10
R. Suck, Mount Talbot area	Griffith	Apr. 1813	Apr. 1814	Fourth, Appendix 9
East of L. Corrib	Jones	May 1813	Apr. 1814	Third, Appendix 6
Leitrim and Longford, East of R. Shannon	Edgeworth	May 1813	Apr. 1814	Fourth, Appendix 7
Connemara	Nimmo	1813/4	Apr. 1814	Fourth, Appendix 12

as well as cadastral purposes, an ambition that presaged the eventual institution of the Ordnance Survey of Ireland.

In drawing up the instructions, the commissioners gave both direction and freedom of action to the engineers in meeting the requirements of the act; achieving all that was envisaged would be no small task. Nevertheless, Griffith had produced his report and map of the north-eastern district within nine months and the surveys of all the bogs were successfully completed by the close of the commission's short life at the end of 1813. A list of the dates of completion and the dates of publication of all twenty-eight reports is shown in table 1. Nimmo produced seven reports, Griffith six and Bald three; the others produced one or two each. Each engineer took his own distinctive approach, reflecting his own personal skills and the precise nature of the bogs assigned to him. The whole suite of maps and reports constitutes an invaluable account of the various localities as they stood before the spread of the Industrial Revolution in

Ireland and the ravages of the nineteenth-century famines. Because the proposed triangulation of the individual regions was never carried out (due to the shortage of time), the individual maps could not be tied to one another and no integrated national map resulted from the exercise. Having no permanent premises of its own, the commission donated all the original maps and reports to the Dublin Society, and they are now archived in the National Library of Ireland. A full evaluation of the commission's work and achievements still awaits.

By completion, it was estimated that the bogs totalled more than 2.8 million English acres (1.75 million Irish acres), so the survey achieved an examination of a significant portion of the landmass of Ireland. In the two following decades the population grew rapidly and the period was punctuated by frequent, sharp famines. Such conditions should have made the improvement of the bogs necessary, cheap and worthwhile. But very little large-scale reclamation or drainage were ever attempted, although small, localised reclamation projects based on the engineers' recommendations were undertaken, as they always had been, by a handful of enlightened proprietors. The reasons for the failure were clear enough. Firstly, it was never the intention of the Government to take on the expense of actual improvement works, and the proprietors were either reluctant or unable to make the capital expenditure necessary to implement the recommendations. Secondly, the end of the French wars and the peace of 1814 occurred within months of the commission's final report, and with the peace treaty of 1815, imports to Britain of cereals from the Continent and timber from the Baltic were soon restored. The result was a dramatic rise in cereal supplies in Britain and a concomitant fall in prices. The rapidly increasing use of coal or coke for smelting and other purposes reduced the demand for charcoal and hence for fast-growing, coppiced timber. The economic circumstances that had favoured the utilisation of marginal land in Britain to produce these necessities during the war years no longer applied; *a fortiori*, it was no longer necessary to press into service the bogs of Ireland, which constituted truly marginal land, and they were therefore allowed to remain in neglect.

The reason why the bogs had been allowed to fall into, and continue in, neglect from the very beginning was obvious to the commissioners, and in stating their opinion on this matter they were, unwittingly, putting the significance of their own endeavours into a much broader historical context. The neglect appeared to them to have three related causes: uncertainty as to the limits of individual farms due to the absence of clear mearings; uncertainty as to the rights of others on the bogs due to the tradition of extensive pasturage; and shortness of tenure that did not encourage occupiers to improve rented land.[22] Under the circumstances, they did not feel they could recommend Government action on improvements unless it was known who owned the individual bogs, what exactly their boundaries were and what their value was.

The commissioners' approach, in effect, presaged not just the foundation of the Ordnance Survey, but also of Griffith's Valuation and the Geological Survey, all of which would be set up within the following fifteen years. Unsurprisingly, Griffith, Nimmo and Bald, the main engineers of the Bogs Commission, would be pivotal in the emergence of these essential institutions.

The good intentions that underlay the Bogs Commission Act were patently obvious; it was certainly no emergency measure to relieve poverty, or to alleviate an imminent distress. It was much greater than that, and the approach of the commissioners to their task was much more akin to that of the Scottish commissions: it involved nothing less than the design of an integrated development plan, based on a major natural resource – the bogs – and the optimisation of the infrastructure towards exploiting that resource. This required awareness of the locations, both inside and outside each local district and away from the bogs themselves, of the various sources of imports and likely destinations for exports. In short, the plan for each district envisaged its integration into the wider regional landscape, even far from the particular bog in question. In a sense, the outcome of the Bogs Commission could have been the first real fruit of the Union; it certainly was the first attempt to address a major infrastructural deficit in Ireland by the application of the engineering skills of the Industrial Revolution, *sensu lato*. If it failed, it was not necessarily the technology that was at fault, but the exigencies and social mores of the time.

As it happens, two of the districts assigned to Nimmo – the Iveragh peninsula in Kerry and the district west of Lough Corrib in Galway – were almost self-contained regions in the modern socio-economic sense. Both were discrete areas cut off from the wider hinterland by mountains in one case and lakes in the other. Nimmo quickly appreciated this fact and his proposals for these two areas would turn out to be true 'regional development plans' like the kind we know today. Prior to this, individual proprietors managed their own extensive holdings, including whatever infrastructure they possessed, with little regard to adjacent properties or to the benefit of integration with the needs and resources of the wider region. Estate management involved essentially internal piecemeal development, if it involved development at all. As we shall see, Nimmo in particular espoused a broad, holistic approach to the problems of these districts and initiated the process of their integration into the economic and social fabric of the rest of Ireland, a process still not complete today.

NOTES

1. H. F. Berry, *A History of the Royal Dublin Society* (London: Longman, Green and Co., 1915).
2. E. Neeson, 'Woodland in History and culture', in J.W. Foster and H.C.G Chesney (eds), *Nature in Ireland: A Scientific and Cultural History* (Montreal: McGill-Queen's University Press, 1997).
3. H. F. Berry, *History of the Royal Dublin Society*.

4. D. A. Chart, *An Economic History of Ireland* (Dublin: Talbot Press, 1920).
5. *A Bill to appoint Commissioners to enquire and examine into the nature and extent of the several Bogs in Ireland; and the Practicability of Draining and Cultivating them, and the best means of effecting the same* (B.P.P.1809 [212], vol. I, mf 10.6).
6. 49 Geo III C. 102. *An Act to appoint Commissioners to enquire and examine, until the first day of August one thousand eight hundred and eleven, into the nature and extent of the several bogs in Ireland and the practicability of draining and cultivating them, and the best means of effecting the same*, 15 June 1809.
7. First Report of the Commissioners appointed to enquire into the nature and extent of the several bogs in Ireland; and the practicability of draining and cultivating them (B.P.P. 1810 [365], vol. X, mf 11.60), hereafter referred to as First Bogs Report.
8. S. Lee, (ed.), *Dictionary of National Biography* (London: Smith, Elder and Co., 1906).
9. First Bogs Report.
10. No copy of this map has been found.
11. Report from the Committee Respecting the Draining of Bogs in Ireland (B.P.P. 1810 [148], vol. IV, mf 11.23).
12. First Bogs Report, appendix 2.
13. Second Bogs Report of the Commissioners appointed to enquire into the nature and extent of the several bogs in Ireland; and the practicability of draining and cultivating them (B.P.P. 1810/11 [96], vol. VI, mf 12.32–5t), hereafter referred to as Second Bogs report.
14. 51 Geo III C. 112. *An Act to continue, until the first day of January 1813 an Act for appointing Commissioners to enquire and examine into the nature and extent of the several bogs in Ireland and the practicability of draining and cultivating them, and the best means of effecting the same*, 2 July 1811.
15. Second Bogs Report, appendix 1.
16. Minutes of the Royal Society of Edinburgh, Meeting of 3 December 1804 (NLS, Acc 10000/4).
17. Personal communication, J. H. Andrews to Arnold Horner, in A. Horner, *Iveragh Co. Kerry in 1811. Alexander Nimmo's map for the Bogs Commissioners* (Dublin: Glen Maps, 2002).
18. J. Foster, Evidence, p. 30, in Select Committee on the Survey and Valuation of Ireland. Report, Minutes of Evidence and Appendices (B.P.P. 1824 [445], vol. VIII, mf 26.47–51, p. 75).
19. Second Bogs Report, appendix 2.
20. G. L. Herries-Davies, 'Richard Griffith – His Life and Character', in G. L. Herries-Davies and R. C. Mollan (eds), *Richard Griffith 1784–1878* (Dublin: Royal Dublin Society, 1980).
21. First Bogs Report, appendix 1.
22. Fourth Report of the Commissioners appointed to enquire into the nature and extent of the several bogs in Ireland; and the practicability of draining and cultivating them (B.P.P. 1813/14 [131], vol. VI, pt.2, mf 15.33–6), hereafter referred to as Fourth Bogs Report.

CHAPTER 4

A Kerry Apprenticeship

Nimmo was allocated the whole of County Kerry by the Bogs Commissioners. He saw instantly that the county could be divided topographically into three distinct areas that shared little geologically and whose bogs were in a great measure unconnected with one another. These comprised first, the low-lying northern plain towards the River Shannon, which drains by the rivers Gale and Cashen, that seems to have been added to his brief almost as an afterthought.[1] The second, his Middle District, was from Tralee eastwards to the boundary with Cork and Limerick and southwards to Killarney. This was largely upland country, mainly draining southwards to Dingle Bay via the rivers Maine and Laune. It expanded greatly at its eastern end and contained extensive limestone beds. South of these two parts lay the third area – the mountainy heartland of the county, comprising mainly the Iveragh peninsula and contiguous areas around Killarney and Kenmare and into the Derrynasaggart Mountains.

Coming to Iveragh cannot have been an entirely easy or comfortable experience for him. He had never been outside Scotland previously and this, his first foray abroad, was to an area so remote that even the Bogs Commissioners made the mistake of locating the barony in County Roscommon![2] But bogs did not bother Nimmo, nor turf trouble him. His previous encounter with Rannoch Moor had prepared him for the mountainous, boggy, impoverished landscape he was about to enter and where his *Encyclopaedia* writings on drainage would be put to the test in real earnest.

Iveragh is completely cut off from easy intercourse with other parts of Ireland, being surrounded on the east and south by high mountains and on the north and west by the wild Atlantic. The only roads in Nimmo's time consisted of bridle paths; there were no wheeled carriages in the whole barony so he had to travel everywhere on horseback. The best maps he had to guide him could not be relied on for accurate detail; large tracts of Kerry were truly *terra incognita* to all except those who lived there. The commissioners referred to such places as being 'in a state of desolation and abandonment'. Nimmo was a stranger in a wild country, far from his engineering colleagues, starting a new career and still only 28 years of age. Nevertheless, by December 1811 he had completed a detailed survey of Iveragh and produced a map that de Courcy considered 'quite magnificent'[3] and which Horner has described as 'probably the most beautiful and cartographically sophisticated of all the maps produced for the Bogs Commissioners'.[4] It was accompanied by an account of the district and its bogs, a theory of bog formation, proposals for drainage, reclamation, cultivation with crop rotation, new roads, a canal and even the suggestion of draining Lough Currane. No ordinary person starting from scratch could possibly have surveyed all this so comprehensively in so short a time – only seven months. A team of surveyors working under him carried out most of the requisite probing, measuring and digging. Their method was to take levels and make borings with long rods down to the hard sub-surface of the bogs. This was repeated along parallel traverses spaced about a quarter of a mile apart.[5] Chain-men and labourers assisted the surveyors, with Nimmo himself supervising overall.

The first part of his report describes the area and its bedrock and soils. The general agricultural aspect was, he wrote, 'rude, cultivation having made as yet but small progress. The valleys are entirely occupied with bog; round the upper margins of these and along the margins of the streams are narrow stripes of cultivated ground.' The population lived in small settlements dotted around these margins. In the whole barony, containing about two hundred square miles, he found that the plough was completely unknown. All cultivation was by spade, of a type he had not encountered before. Ireland was, he believed, in the same state with regard to agricultural implements as Scotland had been 'before the introduction of English husbandry'. There was no industry of any note, excepting some surface quarrying for slate at Caher (now Cahirsiveen), Beginish and Valentia. The slate was 'for the most part shivery and small, much of it however is of equal quality to the Easdale and Ballachulish ... it is blue, purple and green ... splits readily and bears piercing, is slightly foliated or wavy, harder and more silicious than the Bangor slate, and very durable.'

His account of the remaining geology was not quite as accessible as his account of the slate and it shows a pronounced schoolmasterish tone:

> Indurated schistus of this kind has been named Grauwacke by the German Mineralogists but they have not been able to assign any limit to the shades of difference in its composition and structure; accordingly this term, which is at best little better than barbarous, is now become quite vague and indefinite, as almost any rock of unusual aggregation may be so called. So that we have Grauwacke of all varieties of structure, from the most perfect slate to a complete porphyry. I have therefore rejected this unmeaning term; and when I use the word *indurated* I

> beg to be understood, that it is independent of any theory for the explanation of induration; though I do think a well founded theory of that kind would be of incalculable benefit to the engineer and miner, but merely as indicative of the fact that the rock is there harder than elsewhere; I do not say more compact, for it is often very shivery. Now the structure and hardness of the rock, are almost the only properties that interest the engineer.[6]

Quite what the commissioners, not to mention the ordinary proprietors, made of that kind of comment is hard to imagine.

Nimmo divided Iveragh into southern and northern sectors. The southern sector comprised the valley of the River Eeny from its source to the sea at Ballinskelligs Bay. A low ridge of hills along its east–west spine subdivided the valley into the course of the River Eeny north of the ridge and that of the Cummara River south of it. The Cummara drains into Lough Currane and from there to the sea at Waterville. The Lough is only four or five feet above sea level, and in Nimmo's opinion it was once an inlet of the sea. A mountain stream enters its exit channel at the Lough mouth and he believed that this may have carried down such quantities of gravel and boulders over time that the exit became gradually occluded and the inlet converted into a freshwater lake. This is not an unreasonable explanation, in part, for a lake at this place, and it echoes Nimmo's first research paper on the coastal movement of boulders, read to the Royal Society of Edinbugh.[7] He considered altering the existing exit to lower the level of the Lough in order to improve the farmlands upstream, but realised that the greater improvement to be gained by directly draining the vast extent of bog elsewhere in the district made this idea premature.

The northern sector of Iveragh contained three major bogs. The first was that between Teermoyle Mountain and Cahirsiveen, essentially the valley of the Valentia River and its tributaries the Fahrta and the Cahrun [Carhan] rivers. The second was the low land south of Valentia Island drained by the River Dirreen. The area around Ballinskelligs, drained by the Imlagh River, comprised the third distinct bog in the sector. North of the Valentia River, under the shelter of Knocknadobar, he observed the largest population and the greatest extent of cultivation already under way. Here, bog land was dressed with shell sand and seaweed, producing rough pasture and some small tracts of tillage land. Further east, the bogs of Glenbeigh and Glen Caragh appeared to him, from their greater slopes, higher elevation and proximity to the mountains to be less amenable to reclamation.

He conceived that the bogs of Iveragh required irrigation rather than simple drainage. Because the existing cultivable tracts and meadows lay at the upper edges of the bogs and along the banks of streams, he concluded 'that mere draining these bogs, or at least that deep draining, is not enough for their reclamation'. But he was careful not to be misunderstood, insisting that 'the benefit of irrigation is necessarily connected with an effectual drainage'. As he saw it, water needed to be led through the bog in a managed way, calculating that one-tenth of any bog could be irrigated with the prevailing water sources for a total cost of one guinea an acre. The commissioners thought his calculations were 'ingenious' – they were, in fact, based on empirical observations he made on the flow of water in the rivers Maine and Laune and on the observed slopes of the bogs. The plan he proposed was simple: a catch drain was to be cut across the top of the bog and shallow drains, about a foot wide and ten inches deep, were to be cut from this down the slope at a distance of one perch apart. If necessary, smaller drains were to be dug at an oblique angle across the bog, joining one down-drain to another. Small weirs and cuts could be used to direct water to any part at pleasure. A tail drain was then needed to gather the waters from the down-drains and deliver it, together with excess water from the upper catch drain, to a stream lower down. Such control and management of the flow over and through the bog were the keys to avoiding winter engorgement on the one hand and summer drought on the other, as well as being essential to the proper distribution over its surface of those beneficial substances supposedly carried by the water from higher parts. He then introduced a new, fantastic, idea regarding the supposed benefit of controlled inundation:

> I think there is another point of view, in which we may conceive the very purest water to be useful to vegetation, provided it can be supplied in sufficient quantity, and discharged at pleasure. The water, if so considerable as to reach the leaves of the grasses, may act by its detergent quality, in washing off their various fecalencies [feculences], and promote the perspiration, and consequent health, of the plant. In passing through boggy soil, it will tend to dilute and carry off the astringent principles, which otherwise check the decomposition of the vegetable matter.[8]

The merits of deep drainage versus shallow irrigation would exercise the commissioners on many further occasions, but Nimmo's fantastical 'fecalencies' were quietly forgotten. Plants do, in a manner, perspire, although that process is now called *transpiration* rather than *perspiration*, because the terms signify two quite different functions. Plants do not excrete 'fecalencies' by transpiration, and bogs are not full of 'astringent principles' that prevent the decomposition of vegetable material.

Once the irrigation was controlled, then top dressing with lime and manure, worked in with a suitable system of crop rotation, would lead to the eventual cultivation of the bog. Deep drains were not a solution for Iveragh: 'whatever my opinion might be as to the value of deep drains in certain situations, I cannot accede to their being necessary in this district where, in a few yards, I have seen a difference of level of from ten to twenty feet in the surface of deep bog'. As to liming, 'there is not an atom yet found of limestone in Iveragh', he remarked, and shell sand would have to fill the deficiency.

Because the fall of the bogs was readily apparent, and therefore the general direction of their drainage perfectly obvious, Nimmo decided it would be an unwarrantable misspending of time to continue the levelling survey beyond a few places. The resultant paucity of levels is obvious on his map, when compared to the maps of his brother engineers in other districts. In consequence, the routine work of the surveyors and their assistants made less demand on his time and attention, so he was able to turn to other matters and look beyond the confines of the individual bogs to the overall topography and general agricultural economy of the whole region.

To come here first day, he had to travel the only road into the barony, the bridle path from Killorglin to Cahirsiveen. This wound its way over the hill of Drung at an elevation of 800 feet and while it was serviceable for horses, it was not adapted to wheeled carriages, of which there was not one in the whole barony in any case. For all its inadequacy this was the main route in and out of Iveragh and it carried so great a trade in butter that it was often called 'the butter road'. Salted butter, packed in small barrels slung on a horse's back, was carried the seventy-five miles from Iveragh to the city of Cork, then the centre for butter sales in the south. Farmers travelled together for the journey, which took them over Drung hill to Killorglin, then through Killarney on to Cork. The round trip took about seven days all told. Trying not to return 'in ballast' as it were, they brought back whatever necessities they could purchase in Cork. Many authorities wondered at the long and arduous road journey that the Iveragh farmers undertook, especially since the butter could have been carried more easily by boat from Dingle, Valentia or Kenmare. But that apparent solution ignored the extreme poverty of the farmers, who could neither pay for shipping nor risk the potential loss of their butter while it would be out of their sight.

The significance of the butter trade was not lost on Nimmo and it brought him face to face with a paradox at the heart of the commission's remit. The commission's stated purpose was to survey the bogs with a view to increasing the cultivation of corn and flax and the growth of trees. The very existence of an established butter trade suggested a different approach that might be more in keeping with whatever potential the Iveragh region had for improvement:

> In a grazing country as this is, and is so particularly the case with the barony of Iveragh, the extension of the meadows is perhaps even of more importance than the increase of the corn culture ... Destitute of a market, of roads and wheel-carriages, this barony seems to have little temptation to increase its corn culture. The nearest grist mill that

deserves the name, I had almost said the only one yet in Kerry, is at Killarney, thirty miles off. There are indeed two mills in Iveragh; they may be worth 10 shillings each.[9]

He clearly favoured the conversion of the bogs to pasture for cattle, thereby sustaining the existing dairy economy rather than exclusively pursuing the more uncertain aim of increasing tillage and cereal production, for which the district was not naturally suited. His method of irrigation was meant to favour the increase of water meadows such as were commonly found in England and Scotland. 'The greatest difficulty our grazier has to contend with,' he stated, 'is the supply of winter and especially of spring food, for those cattle he can graze in summer on the upland pastures. It is precisely on this account that irrigated meadows are so much valued in Wilts, Gloucester and other parts of England.' That, however, was not what the commission was charged with, and Nimmo was very meticulous to ensure that he met exactly the requirements laid down by them in their instructions, his personal views notwithstanding. He therefore provided information and estimates, as instructed, on all aspects of reclamation and conversion of the bogs to tillage. His suggested crop rotations for Iveragh would, nevertheless, provide such fodder as would be necessary and suitable for the winter needs of cattle, with some production of oats and sufficient potatoes to meet human needs. He approached each bog from its own particular aspect so that his detailed plan for each one was largely unique.

In a country lacking ploughs his suggested improvements had to be achievable by spadework alone, something he readily took into account. Potatoes were then the principal crop raised, but some degree of crop rotation was not uncommon. The following rotation was, he said, 'almost the constant routine throughout Kerry': plots of about one to two acres were subdivided by trenches and surface-treated with shell sand and whatever other manure the farmer could obtain. Cultivation was then started with potatoes grown in lazy-beds for the first two years. For this, the seed potatoes were laid on the surface of the ridge and material dug from the trenches alongside was turned over onto them. The young plants were earthed-up later with more bog material dug up from the trenches. The second potato crop was usually better than the first, and after the second year oats or wheat were grown without manuring for one or more years. Then the land was allowed to go to grass as coarse pasture and the drains and trenches were allowed to choke gradually. In a few more seasons the original improvements were effaced and the land was once again barely distinguishable from the unimproved bog.

Why was the improved land, reclaimed at so great a cost in labour, allowed to revert like this? Nimmo had no hesitation in identifying the Barren Land Act and the burthen of tithes as the guilty culprits. Legislation exempted reclaimed bog land from tithes for a period of seven years and in that way encouraged initial reclamation. By the end of seven years, following the rotation outlined above, it was time for the land to be broken up from pasture and laid down again in potatoes with further manure. As the exemption would have ceased by then, tithes of one to two pounds per acre would become payable on the tilled land. This, naturally, was a pronounced disincentive to further cultivation. The farmer was therefore led to take his next crop off a new piece of uncultivated land, if he had it, since that required the same expense of labour and manure, but was exempt from tithes. In a few years more, the second cultivated tract would be in the same predicament as the first. This obstacle to the continuous improvement of the bog would be at once removed, Nimmo said, if the allotment of part of the waste land were substituted as an equivalent forever for the tithes of the remainder. The problem with tithes was just as pernicious in England, where it was partly solved by money compositions, and where the tenants were more able and more willing to pay. In Ireland, especially in Munster, the tithe system was particularly onerous and obnoxious: pasture land was exempt from it but it was levied even on the smallest potato plots that were the sole means of subsistence for the poor; it went to support a Church not embraced by the majority of those paying it; and the very poorest had no way of avoiding it. In Scotland, Nimmo said, the revenues of the Church were raised in a different way, and he seems to have had no sympathy at all for the Irish tithe holders.

His frank comments, which were partisan in such a bitter controversy, cannot have been popular with those who benefited from the tithes. They were hardly the sentiments either that the commissioners would wish to hear from an objective engineer. They were, however, very much in the train of thought of the Knight of Kerry, Sir Maurice Fitzgerald, a trenchant opponent of tithes and a supporter of Catholic emancipation all his life. Nimmo probably became friendly with him during this time, and they would act in consort in many future undertakings.

Nimmo's comments on the tithe were his first venture into the political arena and he would make increasingly more such forays as time went on. When, for example, the administration prepared a bill for the improvement of the bogs in 1820, the chief secretary sent a draft to him for comment. He replied immediately, outlining the obstacles that, in his opinion, needed to be addressed in the legislation, among which was the tithe issue: 'the tythe [sic] in particular would require much more detail and I doubt not but it has often attracted your attention. It appears however to me that less issue will arise from it in this case than perhaps any other with which it is connected – the tythe interest in the bogs of Ireland being evidently so very small.'[10] The bill was not enacted that session and in the November following Nimmo sent 'some strictures on the bogs bill as drawn up for last session' to the chief secretary, offering his assistance in drawing up the heads of a new bill, should he be required.[11]

The extent to which the bogs could ultimately be rehabilitated was clearly dependent, in his opinion, on the amount of capital available. The existing occupiers could not afford to do much, whereas individuals or companies possessed of extensive capital might undertake greater works, such as canal building. He himself was dead set against canals in Iveragh: 'The construction of a navigable canal,' he wrote, 'involves with it such a host of extra expenses … that I have no hesitation in considering it a useless and premature expenditure of capital in such a country.' Despite this trenchant view, he continued (discretion being the better part of valour): 'nevertheless I have ventured, in compliance with the fifth article of Instructions, to suggest the propriety of converting one irrigating drain into a navigable communication'. However, he compromised by designing it for use only with small boats rather than with larger vessels, as envisaged by the commissioners. His canal plan, which formed a separate appendix to his report, therefore proposed a navigable waterway eight feet wide at the bottom and fifteen at the top, extending from the River Eeny at Keaning farm (near Foildrenagh Bridge today) almost to the southern arm of Valentia harbour.[12] The likely expense of a complete lockage down to the tideway seemed too great, so he proposed to terminate the canal half a mile from the sea. It would be 22,664 yards long (about thirteen miles) and it would cost over £3,464 in total. A branch from the summit level to Ballinskelligs, if such were required, would add another £5,196. As for its usage, the main purpose he envisaged was the transport of shell sand from the sea coast to the valley of the Eeny River. It would be expected to make considerable savings on the local transport of sand and other materials, which at the time were normally carried in sacks or in panniers by horses, each load being about two hundredweight, at a cost of one shilling and a penny for a man and horse per day. Where that cost was too great, as it often was, transport was on the backs of men and women.

Although he prepared his usual cost–benefit analysis of the navigation, Nimmo's heart does not seem to have been in this project and his head certainly was not. Because the canal was not to be taken to the tideway it would require the shell sand from Beginish to be carried half a mile up from the landing place on the shore to a wharf at the canal head. Ferried along in small boats pulled by men or horses, it would need to be unloaded again and taken by horse to the farms at the inland end. There was no other cargo, critically no return or 'back' cargo, and the canal's secondary function of drainage was not essential. Had it been built it would have been the greatest white elephant ever constructed in Kerry – essentially a thirteen-mile canal to nowhere, from nowhere, with no traffic demand. The local tenants in the Eeny valley preferred to get their shell sand from the shore at the nearby river mouth, which, they maintained, fostered 'red and white clover', rather than from Beginish, which gave only 'daisy', so even the intended beneficiaries did not want the canal. Other than satisfying the commissioners that he had followed their instructions, Nimmo had no good social, agricultural or commercial reason to pursue the idea. In obedience, however, he laid down a line for a canal that is clearly indicated on his map. It was to commence above Mastergeehy, run along the north bank of the River Eeny, then turn north via Kennagh to cross the summit bog and terminate close to Valentia harbour near where Kilkeaveragh Bridge is today.

It was never built. Its only advantage was that it would have cost less than a railway through the district, and Nimmo was even more opposed to that idea in Iveragh than he was to a canal. The slightest railway that would be of any use would cost £500 per mile, he said, and in a country 'where even wheeled carriages of the rudest kind have hardly penetrated, it is not to be expected that the people could be easily familiarised with the benefits and management of a railway'. So he made no further proposals in this particular matter. However, this was his first mention of railways in Ireland, a topic he would return to and in which he would become deeply involved in later years. At that early stage in railway history, a railway was not the kind that we know today. Nimmo would have had in mind a timber rail track, most likely temporary, along which a horse or a man would pull a wheeled wagon. He showed a familiarity with railways that was very modern for its time and presaged his later interest in rail transport: 'I may observe, by the way, that in making railroads, iron is much to be preferred to wood: it is, on the whole, nearly as cheap, more durable, and, as now made, by locking together, is much less exposed to depredation.'

The other trade then ongoing in Iveragh was the sale of turf. Boats already plied between Valentia and Dingle and there was no reason why the harbour of Valentia might not be drawn further into maritime service. Nimmo surveyed and sounded it, confirming its suitability for much larger vessels and establishing the general accuracy of Mackenzie's earlier survey.[13] So suitable was it that years later Maurice Fitzgerald, helped by Nimmo, would propose its use as a transatlantic steamer base. Lesser places, too – like the mouth of the River Inny – could be turned, he believed, into small trading ports.

As to roads, he proposed to construct an extensive network through the bogs, with a new line of main road from Cahirsiveen to Killorglin around the base of Drung Hill that would be more suited to wheeled carriages than the old line over the summit (Plate 1, figure 1). It would give much improved communication between the barony and the country around Killorglin and Killarney. Many of his proposed roads would traverse the bogs and were entirely new; their lines were determined by the general lie of the bogs and the location of the farms around their periphery. Generally speaking, they would create access to the bog itself for the supply of manures and the eventual removal of the anticipated crops. Other proposed roads, that would improve the general circulation of the wider area and improve access to the sea and to the hinterland, were realignments and upgrades of existing bridle paths and farm tracks. These were not new in the sense that their general routes were already dictated by broad topographic features such as rivers, mountains, anchorages and the location of the main centres of commerce and social activity. When examining Nimmo's maps and his depiction of pre-existing 'roads', it should be remembered that these were mere tracks, barely passable on foot in the best of conditions and absolutely impassable quagmires in winter. The roads he proposed in Iveragh and elsewhere in Kerry are listed in table 2. In total, he proposed more than twenty-eight miles of road in Iveragh at an estimated cost of £3,765.

The new line of carriage road that he laid down around the base of Drung Hill was eventually built and it is now part of the northern section of the road known as the Ring of Kerry (N70). One of the bridges on it (K2-N70-024-00) is named Nimmo's Bridge in the six-inch Ordnance Survey map of the area (surveyed in 1842), suggesting that it was most probably designed by him.[14] In recent years the road here has been widened and Nimmo's Bridge is no longer easily noticed. Its downstream face, completely hidden from view and difficult to access, shows a bold, confident, stilted single round-arch structure splendidly constructed on a very difficult site spanning a narrow deep gorge (figure 4.1). The less said about the modern upstream face, a throw-together mix of field stones and concrete, the better. Further west on this part of the N70, O'Connell's Bridge, now overlooked by a later railway viaduct, crosses the Gleensk River. This road bridge has been rebuilt since its first construction, but the original had a plaque that read: 'O'Connell Bridge A.D. 1820. Messrs Meredith and McSweeney, contractors'.[15] Ben Meredith worked with Nimmo for many years and we shall meet him again in later chapters. The remaining bridges on the upland part of the N70 between Gleensk and Cahirsiveen are generally small, artisanal structures, probably designed by Nimmo and built by local contractors.

By any measure, Nimmo's report and map are a *tour de force*, presenting an incomparably comprehensive description of the barony as it was at that time. For instance, it gives the soundings in Valentia harbour, the heights of the mountain summits on land and the depths of the bogs in different places – three distinct vertical metrics on one map. Visually, the map has a certain dark, brooding appearance that comes mainly from the depiction of the extensive mountains. His second map – that of the southern part of the barony of Dunkerron on the southern shore of the peninsula – shows a much lighter touch, not entirely explained by its smaller size and less mountainous coverage.[16] His other bog maps of the Middle and Northern districts of Kerry are even less dramatic than that of Dunkerron, approaching more the flat style of the maps drawn by the other Bogs Commission engineers. Unquestionably, his Iveragh map and report stand apart visually and narratively as a topographical, social, agricultural and developmental commentary on an entire region until then largely unknown.

His second district was that part of the Barony of Dunkerron that lay on the southern side of the Iveragh peninsula; taken together, Iveragh and Dunkerron comprise the whole peninsula, which really can be approached as one integrated whole. The main routes out of the peninsula, even today, are by the northern – Valentia to Killorglin – and the southern – Waterville to Kenmare – coastal roads. These constitute what we now call the Ring of Kerry, one of the most dramatic and scenic routes in the country.

Nimmo started the Dunkerron survey in 1811, even before he had completed the Iveragh report. Once again, he commenced with the geological features, which were not greatly dissimilar to those of Iveragh. The bogs could be divided into three separate basins, the Bally Bog centred on Sneem, the Blackwater Bog further east and Finihy Bog, a small region north of Kenmare.

Bally Bog comprised about 5,500 Irish acres. It had almost no subsoil and was underlain by bare rock that rose in ridges to the peninsula's mountainous spine; it would be necessary to cross-cut the rock in places if the individual bogs within it were to be properly drained. Altogether this was a most unpromising region for cultivation in his view, which was not to say that no improvement was possible. The success of the Bland family of Derryquin, whose improvements 'form a pleasing contrast to the dreary scene around him', did not fail to impress. Stands of larch, Scots fir and birch marked the success of their efforts over generations in what must have been a particularly arduous task, cultivating an intractable area that lacked a single road. There were, of course, some bridle paths and cattle tracks, and the family made efforts to convert these into useable roads wherever possible.

Blackwater Bog to the east was far more promising. The subsoil was earthy gravel with less rock, and the streams ran freely down the declivity so that drainage was natural and did not require cross-cutting the rock. As well as that, limestone was available from Cappanacush, only four or five miles away. Finihy Bog near Kenmare was a very level, wet bog about twenty feet deep, surrounded on all sides except the north-east by high ground. The Finihy River, its natural drain, was dull and crooked and considerably obstructed by brushwood; it would require straightening and scouring, or indeed an entirely new channel might need to be cut alongside it.

Apart from some rather general recommendations for drainage and improvement, necessary to meet the commission's remit, Nimmo dwelt little enough on the bogs of Dunkerron. Indeed, his overall treatment of the area was cursory, not to say skimpy. For example, while his Iveragh report was thirty-one pages long, that on Dunkerron was only five. He took relatively few measures of bog depths, confining his observations to general values given in a table accompanying the map.[16] The map itself rivals the Iveragh map in attractiveness, with a far less deeply brooding character than the latter. Hydrographic soundings for Kenmare River and Sneem harbour were not his own; he was satisfied to use values taken from Mackenzie's surveys. A modern reproduction of the Dunkerron map was published in 2003.[17]

It was unlike Nimmo not to be sanguine about an area, especially when there was good evidence – like the achievements of the Bland family – to indicate potential for development. But if his Iveragh survey exercised his talent as a rural economist, his Dunkerron venture gave rein to other aspects of his skills. In his report on the two hundred square miles of Iveragh, he had proposed just over twenty-eight miles of new or upgraded roads; in the much less extensive Dunkerron he proposed over thirty-seven miles (table 2). What this barony particularly needed, he stated, was the improvement of the existing, and the formation of new roads. Francis Christopher Bland, with whom he would become a firm friend, was already making a line of road from Derryquin to the shore at Cove for the collection and distribution of shell sand. Nimmo helped by laying out a new line for him, avoiding some difficult ground and skirting the steep slopes near Staigue Fort. Their cooperation may well have led him to concentrate on roads rather than on the bogs themselves. On the topic of Staigue Fort, the two men shared mutual interest. Nimmo speculated that it was a very ancient solar observatory and predicted that its doorway would prove to be aligned exactly due south to catch the meridional sun. So it proved when Bland completed his own survey of the monument. Bland reported his observations, together with Nimmo's comments, in a paper read to the Royal Irish Academy in 1821 and subsequently published in 1825.[18]

In an eastward direction the road from Derryquin towards Kenmare was little more than a track as far as Blackwater Bridge, so Nimmo laid down a new line that ran closer to the shore and was nearly level all the way. Work actually commenced on this during 1812 and he was able to mark it on the map as 'proposed road laid out by A.N. and now executing', thus stamping engineering authenticity on it and on himself. (Plate 1, figure 2). Taken together with the road from Kenmare to Blackwater and the new line of road from Sneem to Cove, it constituted the whole line from Kenmare almost to Caherdaniel. All that was now needed was a new road from Caherdaniel to Waterville (completed in the 1830s) and the Ring of Kerry circuit from Kenmare to Killorglin via Sneem, Waterville and Cahirsiveen would be complete. Much later, Nimmo laid out the road from Kenmare to Killarney, effectively closing the Ring. There is little doubt that a complete road circuit of the Iveragh peninsula would inevitably have been constructed even if Nimmo had never set foot in Kerry. Nevertheless, the Ring of Kerry route stands as a tribute to his early endeavour, but this is nowadays largely unacknowledged in Kerry, where Nimmo is virtually forgotten.[19]

Nimmo laid down other lines of road through the Dunkerron bogs, the principal one of which was an inland road from Sneem to the Blackwater valley and on to a location about two miles south of Moll's Gap on the present-day

TABLE 2.

The location, length and cost of the various roads proposed by Nimmo in county Kerry in 1813. Place names as given in Nimmo's report.

REGION	ROAD	LENGTH (yards)	COST
Iveragh (Fahtra bog)	Gortnagree to Gurrane	4,180	
	Liss to Tirenehily	1,650	
	Coarse to Gurrane	1,815	590-15-0
Iveragh (Carhun bog)	Bridge of Fahrta to Coom of Cambryn	5,500	
	Behagh to Kepoghs	2,750	637-15-0
Iveragh (Summit level bog)	Kennagh to Imlagh Bridge	2,750	127-10-0
	From Imlagh up the bog	2,750	212-10-0
Iveragh (Imlagh bog)	Bridge of Kinnard to Coom	2,200	170-0-0
	Dungeagan to Cooles.	3,025	233-15-0
Iveragh (Eeny River)	Kinnah to Achnakettin. Crossroads to Coomleagh	12,100	945-0-0
	9 bridges	4,950	382-10-0
Iveragh (Aghter)	Achter to Eeny mouth 2 bridges	4,400	346-13-0
Iveragh (Valentia)	Valentia Island, road from Coarhubeg	1,650	120-0-0
Dunkerron (Bally)	Blackwater Bridge to Deriquin	7,975	1087-10-0
	Deriquin to Glanlough	12,210	555-0-0
	Glanlough to Cove	4,540	618-0-0
	Sneem to Lettermuniel	6,810	464-5-0
	Tahalla to Letterfinish	3,025	206-5-0
Dunkerron (Blackwater)	Blackwater Bridge to above Dromin	2,080	189-0-0
	Old road to L. Bryn	5,973	135-15-0
	L. Bryn thro' Bealagh Beami	8,063	733-0-0
	Branch to Lettermuniel 2 bridges	4,543	309-15-0
	East branch to head of valley 1 bridge.	5,500	375-0-0
Dunkerron (Finihy)	Road to Dunkerron	4,125	187-10-0
	New road to Bawn Nedeen 1 bridge	2,145	146-5-0
Middle District (Killorglin)	New line from Killorglin towards Iveragh	3,850	262-10-0
	Killorglin towards Annagarry	1,194	162-15-0
	Annagarry to Glangulough	2,717	185-5-0
	Glangulough to Tullig and the shore	3,531	440-15-0
	Glangulough to the shore	2,750	187-10-0
	Road by Loughnabragh-derrig	3,795	258-15-0
	Dooch to Lough Cara	4,290	292-10-0
Middle District (Awinigarry)	Awinigarry to Glen Cuthane and Glen Carra 2 small bridges	5,500	375-0-0

cont

Location	Description	Yards	Cost
Slieve Luachar (Annagh)	From Scrahanfadda to mail-coach road at Knockanimirish 7 culverts	7,260	660-0-0
Slieve Luachar (Deanagh and Spa)	From Spa across the bog	3,575	243-15-0
Slieve Luachar (Mynass)	New road over the bog	2,475	168-15-0
Slieve Luachar (Quarries bog)	Gowlane to mail-coach line	3,520	240-0-0
Slieve Luachar (Beheena and Rathbeg)	Awnascartan to Droum Stream Droum stream to six mile bridge 2 bridges	6,105 1,860	416-5-0 126-15-0
Slieve Luachar (Curreal and Headfort bogs)	Stream of Droum to the Flesk 1 bridge	3,575	243-15-0
Slieve Luachar (Cahirbarna)	From Awnascartaan river to the mail-coach road 2 bridges	5,192	354-0-0
Slieve Luachar (Awinaghloor)	Road across the bog 1 bridge	2,541	173-5-0
Slieve Luachar (Bohurbuoy)	Road to lime quarries at Tanre 2 small bridges	5,995	408-15-0
Slieve Luachar (Rathcoole bog)	Necessary lines of road exclusive of farm roads	–	4038-6-0
Northern District (Cashen) (Ballyenessy)	From bridge at Gale to Ennismore Branch to the Feale at Dysert Two roads across Moinveanlough	6,325 2,200 6,050	862-10-0 300-0-0 825-0-0
Northern District (Ballylongford)	From Gunsboro' to Murhure A branch from Tullamore A branch to Tarbert Road Road from Kilgarvan to Tullyhanagh	8,250 1,320 5,500 2,750	2445-0-0
Northern District (Glinn)	Road across bog	2,200	300-0-0

Total 221,029 yards = almost 127 miles

FIGURE 4.1
Downstream face of Nimmo's Bridge on the N70, the Ring of Kerry road.

Kenmare to Killarney road. From here he intended to continue it to Kenmare through the Finihy Bog. The modern R568 road from Sneem to Gearha Bridge and beyond runs along part of Nimmo's line. From it he proposed bye-roads from Letterfinish south to Tahilla and from a junction above Dromin (old Dromore) south to the mouth of the Blackwater, both of which were later built. From Dromin he proposed to perfect the track north to Lough Brin; this road was subsequently built. He proposed a spur off this road, up through 'Bealagh Beami' mountain pass, where no previous road existed. This is now known as the road through the Bealagh Beama Pass, forming a communication between the Blackwater valley in Dunkerron and Glencar in Iveragh, one of the most impressive and neglected road passes in the whole peninsula. This road, too, was eventually built, although its lower reach, between the pass and the R568, is not exactly on Nimmo's line, which was more to the east. From Finihy Bog he proposed a road south to Reen on the coast and another, completely new, to Kenmare, a town he sometimes referred to as Bawn Nedeen. The latter road was to cross the Finihy River near modern Sahaleen Bridge and run south to Kenmare on a line east of the modern N71 road.

In 1812 he commenced his surveys of the Middle District, a large tract stretching from Mangerton and the Derrynasaggart Mountains in the south to the Glanaruddery and Stack Mountains in the north, encompassing all the land from the sea to North Cork.[20] The most elevated part is known as Sliabh Luachra, largely an upland moor at an elevation of eight to nine hundred feet. Otherwise the district comprises low-level alluvial plains from which some deep valleys, such as that of the rivers Flesk and Brown Flesk, penetrate into the uplands. In describing the geology of the area and comparing it with other districts, which he did with his customary skill, Nimmo exhibited a good knowledge of the geology of places extending as far north as Mayo and east to Kilkenny. His eye, as usual, was keen for exploitable resources such as slate, limestone, culm and ore-bearing rock. Strangely enough, he made no mention of the ore then being mined, with his active participation, at Ross Island near Killarney: perhaps he did not wish that this be known at the time to the Bogs Commissioners.

There were many individual bogs to be surveyed in this extensive district, for each of which he proposed a distinctive scheme of improvement and cultivation. They included the bogs around and just west of Killorglin, the most significant of which was near Lough Yganavan, between Castlemaine harbour and the northern slopes of Macgillycuddy's Reeks. Noting the extensive sand dunes between the neck of Cromane peninsula and the mouth of the River Caragh, and surmising that sand blown from these had already helped to convert a nearby bog into arable land, he conceived a plan to improve a much wider area. This involved making an inland navigation towards and through Lough Yganavan, whereby 'sand from Dooagh Point, shell-sand from the North Bull, sea sleech and sand from near Douglas and lime from the Killorglin side' could be transported around the area. This navigation would cost almost £8,000, he estimated; unsurprisingly, nothing was ever done about it. Today there is a golf course (Dooks Golf Club) at the dunes.

Exactly the opposite approach was his preferred strategy for the bogs at the mouth of the River Maine. There the river had little fall, spring tides flowing to a point above the bridge at Castlemaine and a great deal of ground regularly flooding every winter. To secure the bog from winter inundation required an embankment. This would provide an opportunity to adapt the river for navigation, which in turn would facilitate the distribution of sea manure by barge into the bog. Useful as that would be, Nimmo realised that the bog would subside and contract as a result of the drainage he recommended. Since the Maine was a particularly turbid river, lowering the height of the bog would permit warping, a process he much favoured, to be carried out. In this, water is permitted to flood the land and in so doing, it deposits its suspended particles, adding alluvium to the flooded surface. The excess water is then released back to the river. Dams and sluices were essential for warping, which depends on controlled management of the water, whether sea- or fresh-water. Warping would work well below Castlemaine, he believed, because the fall of the tide would facilitate the return of the warp water to the river. Above Castlemaine Bridge, it would be necessary to use water wheels to raise the water from the low-lying land and deposit it back into the Maine, which flowed at a higher level. It would also be essential to straighten the river, improving its outfall and facilitating navigation by barges up to the head of tide and beyond. The navigation to Castlemaine was crooked and inconvenient and a sandbar at the harbour mouth did not help either. Therefore, he sounded Castlemaine Bay and found a safe channel five fathoms deep, with a minimum of ten feet over the bar. All that it needed was suitable buoys and markers, provided at a trifling expense, in order to make the river entrance perfectly accessible. As to the crooked river, he would make a major cut across the narrow neck near Milltown, cutting out a reach of one mile and lowering the outfall by eighteen inches. Upstream of this, the river was to be embanked to four or five feet in height. Small craft could then carry limestone and shell sand to most of the bogs in the catchment. On the upper part of the catchment he favoured warping with flood water taken from the Brown Flesk River. The excess waters would subsequently drain back down to the Brown Flesk, which was swift-running, unlike the Maine, into which it debouched. Above the confluence of these two rivers near Currans he thought that 'perhaps it would be very proper to cut off some of its water and send them [sic] down to Tralee, both to slake its own floods, and to increase the water power in the sea.' It seems he meant by the latter purpose that the extra flow would help scour the estuary in the channel from Tralee to Blennerville quay. Siltation prevented vessels of burthen from coming alongside there and they had to be serviced by lighters. Nimmo's proposal came long before the construction of the Tralee ship canal, which was completed in 1848.[21]

Moving south from the valley of the Maine, he examined numerous small bogs on the way to Killarney. The main enterprise he conceived in this area was much more ambitious and would have had significance well beyond the immediate district, had it ever been implemented. Sea-going vessels drawing ten feet already entered the River Laune as far as Killorglin, and flat barges could continue some miles further up. He was convinced of the value of extending the navigation all the way to Killarney. Acting, he wrote, 'in obedience to the instructions of the Board I have examined how this could be connected with the improvement of the bogs'. It seems strange that he should so carefully invoke the commissioners' instructions for the proposal he was about to make, but its real significance would only become clear later. His proposed Laune navigation would begin at the tideway at Dunmanaheen, where one double-lock of twelve-foot lift would raise the level to that of the river as far as its confluence with the River Gweestan. A second lock there of sixteen-foot lift would carry the line to Ballymalis Castle, from whence there was a long reach of the river perfectly navigable without any alteration. A single lock at Carhue Beg and two at Beaufort bridge, would complete the navigation to Lough Leane. The whole would cost about thirteen thousand pounds.

Another, much more imperfect method might be proposed, he continued, for navigating the river, by using flushes from the lake. Presumably he meant to manage the flow in the river by regular, controlled releases of water through sluices at Lough Leane. It would be interesting had he indicated exactly how this was to be achieved, and how the boats would manage the upstream journey against the flow; he probably envisaged them being pulled by horses or men, as was usual in canals at the time.

Most of the remaining bogs in the Middle District were in the valleys formed where the River Flesk and its tributaries exited the high ground of the Derrynasaggart Mountains and the Slieve Luachra uplands. All that was needed for these were various cuts and drains that made access to the streams more direct. Limestone was available from the quarries at Masurour (near Gneevgullia), and he considered how best this might be delivered to the bogs at the lower level. He wrote:

> I have been at some pains to enquire whether the limestone from some of the quarries, and particularly that of Masurour, could be transported into these various bogs by means of inland navigation. This, if ever attempted, must be by a series of level reaches connected by inclined planes, since lockage over such elevations is altogether out of the question. As there does not appear any through trade for such a set of canals, the transportation of the limestone (entirely a descending trade)

may be effected without any communication between the levels, by having convenient discharging places, where it can be lowered from one reach to another, or it may be removed to and from the canal by wagons and railways.[22]

Thoughts of inclined planes must have been in his mind when he extended his survey through Rathmore Bog into North Cork around Millstreet, opening out his perspective to the good lands of Cork, Limerick and on to Tipperary. Here his thoughts took on a grand aspect:

> The northern parts of the County Cork, and a great part of Limerick, are nearly destitute of fuel; culm for lime-burning they get from Kanturk, and turf may be brought from Kerry. For this purpose a level canal with railways should proceed across the upper Blackwater through the culm district. Beyond Kanturk it enters the great limestone field and may be continued indefinitely by Buttevant, Doneraile etc. or through the hollow of Ballyhaura and by Charleville into the county of Limerick. A canal in that direction has often been in contemplation and about three miles of one, upon a great scale, from Mallow towards Kanturk, were formerly executed; its extension into Kerry would be of great benefit for the mutual accommodation of the districts with fuel and limestone. By means of three inclined planes, near Stagmount, Brewsterfield and Lisavigeen, it may descend to the lake of Killarney, and from thence by lockage, to the sea in Dingle Bay.[23]

This grand scheme, joining the Golden Vale of Munster to the sea in Dingle Bay, far exceeded anything that the commission might ever have anticipated or even contemplated. Nimmo had no instruction to extend his considerations outside of Kerry and the contiguous part of North Cork. This scheme was far too grand in scale, but to their credit the commissioners did not condemn it overtly, even if by ignoring it they effectively consigned the idea to oblivion. However, it certainly explained and tied in nicely with his proposal for a navigation of the River Laune from Killarney to the sea. Without directly referring to any grand scheme, Nimmo presented a profile of the Middle District illustrating the elevation all the way from Millstreet in Cork to Dingle Bay. The line is remarkably low-lying, nowhere exceeding four hundred and fifty feet, a level he made to appear completely insignificant by showing in the background the elevations of some of Kerry's highest peaks. He had calculated these high mountain elevations from geometric or barometric observations, which he regarded as sufficiently accurate for the purpose at hand. (They were, in fact, remarkably accurate and his calculation of the height of Carrauntouhil, for example, is within four feet of the value given in the current OS maps.) Even if a real will to proceed with this scheme should ever have arisen, it is doubtful that it was feasible with the technology of the time. Nimmo never failed to see and promote the 'big picture' like this, and if one were to criticise him it would be that his vision too often exceeded his brief and also the resources of his clients *at the time*. Suffice it to say that his route for the canal and inclined railway became the line of the Mallow to Killarney locomotive railway constructed later in the century. Nimmo finished the Middle District by surveying 'the romantic scenery of Glen Flesk' and the bogs to the south and west of it.

Cashen and the northern area of County Kerry are bounded on the south and east by uplands, on the north by the River Shannon and by the sea on the west. Two basins, one north and east of Listowel draining by the Feale and the Gale rivers, the other south and west of the town, draining partly by the River Brick, make up the region. Despite its maritime location, the district was almost cut off from water-borne trade, lacking usable shipping wharves and safe harbours, although those of Tarbert and Ballylongford were capable of being made into two of the most useful ports on the Shannon. With the exception of Knockanore and Kerry Head, the land is generally very low-lying, a unique aspect in a county as mountainous as Kerry. This lends a certain dreariness to the landscape, especially in places where the soil is boggy. Nimmo must have felt this after his traverses of the dramatic southern and south-western mountains, and he therefore introduced his account of the Cashen bogs factually but smacking of a modicum of *ennui*: 'The bog and marsh on the Cashen ... This extensive morass contains 5146 Irish = 8,336 English acres.'[24] Later in his report, he would say of the area: 'the fine limestone vale of the Cashen is particularly backward though provided abundantly with manure which can be multiplied at pleasure. The inhabitants seem perfectly ignorant that the chief sustenance of the southern part of the county is derived from a soil, which either is, or has been, a bog of as forbidding an aspect as any in their neighbourhood.'

He regarded the lower part of the Cashen as ideally suited to warping, and efforts in that direction by a Mr Mason of Aghermore had been successful previously. Indeed, of all the places he had visited in Kerry, the Cashen district had experienced the most attempts at early drainage and bog improvement. With the passage of time all the artificial drains had been allowed to choke up so that the early improvements had been entirely negated. Nimmo seemed to draw little lesson from this fact and his recipe was really more of the same, that is, improved drainage, warping and embankment, with constant maintenance of the drains. In addition, he recommended that the bogs be laid out in regular fields by means of ditches. He proposed to make a cut through the spit that lay across the mouth of the Cashen river, thereby shortening it by about four hundred yards and increasing the scour. The net result of the increased outfall would be a lowering of the bog, and wherever necessary the river could be embanked to control the warp.

There were another 200,000 English acres in the uplands around the rim of the Northern District and extending into the Dingle peninsula that he did not survey in detail. In his own words, 'In these tracts I have contented myself with a general reconnoisance [sic] or eye survey.' As for roads, he proposed only twenty-two miles in the whole Northern district. He gave some consideration to a navigation canal from Barra (Barrow) Harbour to Listowel, going as far as to mark out the line of level between the two places. The line meandered over the area so that to construct a canal would, by his estimate, have cost about £30,000. Needless to say, he took that proposal no further.

Nimmo had arrived in Kerry in June 1811 and had completed his entire survey of the county eighteen months later, at the end of 1812. He had climbed mountains, dredged harbours, gone down mines and criss-crossed bogs with incredible tenacity and dedication. In that short period he had surveyed eighty-one separate bogs comprising 150,000 English acres, proposed more than one hundred and twenty miles of roads, sounded two harbours, proposed canals and inclined railways to intersect the county, recommended river straightening, embankment, drainage and warping and commented on relevant social, agricultural, economic and political issues. In effect, he had drawn up an outline county development plan where none was asked for or even contemplated. In a way, he had done far too much. Yet he was about to do the same in Connemara, undaunted that others did not share the breadth of his vision, the liveliness of his enthusiasm or his boundless optimism.

NOTES

1. A. Nimmo, 'The Report of Mr Alexander Nimmo on Bogs in the Barony of Iveragh in the county of Roscommon', in Fourth Bogs Report, appendix 5, p.11.
2. Ibid.
3. S. de Courcy, 'Alexander Nimmo: Engineer Extraordinary', *Connemara. Journal of the Clifden and Connemara Heritage Group*, 2 (1995), pp. 47–56.
4. A. Horner, *Iveragh Co. Kerry in 1811. Alexander Nimmo's Map for the Bogs Commissioners* (Dublin: Glen Maps, 2002), p. 4.
5. T. Ruddock, 'Alexander Nimmo', in A.W. Skempton (ed.), *Biographical Dictionary of Civil Engineers*, vol. 1, 1500–1880 (London: Thomas Telford, 2002).
6. Fourth Bogs Report, appendix 5, p. 28.
7. Minutes of the Royal Society of Edinburgh, 3 December 1804 (NLS, Acc 10000/4).
8. Fourth Bogs Report, appendix 5, p. 32.
9. Ibid.
10. A. Nimmo to C. Grant, 22 March 1820 (NAI, CSORP 1820 N).
11. A. Nimmo to C. Grant, 16 November 1820 (NAI, CSORP 1820 N 16).
12. A. Nimmo, 'A Navigable Canal of derivation from the Eeny River towards the Harbour of Valentia', in Fourth Bogs Report, appendix 5, pp. 52–5.
13. M. Mackenzie, *A Maritime Survey of Ireland and the west of Great Britain undertaken by order of the Rt. Hon the Lords Commissioners of Admiralty by Murdoch Mackenzie Snr, F.R.S* (NLI, Maps, 16 C 17).
14. Ordnance Survey, first edition, Kerry sheet 79.
15. R. McMorran, 'Alexander Nimmo in Kerry', *Kerry Magazine* (1995), pp. 37–9.
16. Fourth Bogs Report, appendix 6, pp. 59–65.

17. A. Horner, *Kenmare River Co. Kerry in 1812. Alexander Nimmo's Map for the Bogs Commissioners* (Dublin: Glen Maps, 2003).
18. F. C. Bland, 'Description of a remarkable Building, on the north side of Kenmare river, commonly called Staigue Fort', *Transactions of the Royal Irish Academy*, vol. XIV (1825), pp. 17– 30.
19. R. McMorran, 'Alexander Nimmo in Kerry', *Kerry Magazine* (1995), pp. 37–9.
20. Fourth Bogs Report, appendix 6, pp. 65–90.
21. R. Delaney, *Ireland's Inland Waterways. Celebrating 300 Years* (Belfast: Appletree Press, 2004).
22. Fourth Bogs Report, appendix 6, pp. 81.
23. Ibid., p. 82.
24. Ibid., pp. 90–102.

CHAPTER 5

First Foray into Connemara

In 1811 Nimmo was largely a book-learned surveyor with little enough practical experience. By the time he came to Connemara in 1812 he was more experienced, better practised and was already making a name for himself through his involvement with road works, mining and harbour surveying. From his Inverness survey in 1807 onwards, he had shown a keen eye for landscapes and their potentialities. This reflected his profound geological understanding of place and the way in which the underlying bedrock determined the surface topography, the soil and ultimately the way of life of the people. 'The various soils, also, and means of improvement, will be best understood by being previously acquainted with the various rocks which form their base and with their particular position', he wrote.[1] It also reflected his ability to look on an area of land and set it in a far wider geographic and economic framework than even the landowner could envisage. In other words, he could see and evaluate detail within a locality and make appropriate connections from it to areas and resources outside. We have seen, for example, how he protracted his vision for Kerry into the arable lands of the Golden Vale of Munster, just as he had earlier protracted his vision for Rannoch Moor to the eastern and the western seas of Scotland. By the time he entered Connemara these skills had been finely honed and his general outlook was determinedly sanguine. Recognising straight away that Larkin's map, with which he had been provided, was neither accurate nor adequate, he remedied the deficiency by a quick but exact survey of the area between Galway and the western ocean, starting even before his full instructions from the commissioners were received in November 1812.[2] More significantly, he decided to undertake a major geological survey of this most diverse and complex district.

Prior to Nimmo, various mineralogists and amateur geologists had visited Connemara on surveys that were scant at best and usually of limited value. One was Roderick O'Flaherty, known now for his *Chorographical Description of West or h-Iar Connaught*, who made some of the earliest (1684) observations regarding various rock features in the area.[3] Another was Richard Kirwan (1733–1812), who lived near Galway and had an international reputation as a chemist and a geologist specialising in mineralogy.[4] He had a laboratory at his home where he analysed minerals, some without doubt originating in nearby Connemara. Unfortunately, his unquestioned intellectual skills were accompanied by abounding eccentricities and a religious fundamentalism that branded him largely a figure of fun, and he made little lasting contribution to Irish geology.[5] These two were hardly propitious antecedents for Nimmo. More auspicious was the work of the Frenchman, Mons. Subrine, hired by Richard Martin to evaluate the lead ore deposits near Oughterard in 1802. His was the first professional study of Connemara rocks that we are aware of.[6] The botanist Joseph Woods, who penetrated as far as the Twelve Bens in 1809, also made some valuable mineralogical observations *en route*.[7] Apart from these, the field of the geology of Connemara was still virtually untilled when Nimmo came to it. There were no existing geological maps, even of the crudest kind, and no evidence that the general rock structure of the area was at all well known, even in scientific circles.

In order to appreciate the quality and extent of Nimmo's survey and what he achieved geologically, we should consider our modern knowledge of the geology of the area. Professor Mohr's simplified map (Plate 3) gives a brief summary of this, and the main features to note for our purpose are as follows:

- The predominant structural layout consists of four east-to-west bands of rock, respectively granite, gneiss and amphobolite, quartzite with schist, and shale with sandstone, succeeding one another in a south to north series.
- The eastern limestone plain that covers central Ireland extends across the Corrib in a belt from Galway to Oughterard (between the main N59 road and the lake) but does not cross the Corrib River in Galway city.
- The southern end of Gorumna and Lettermullan Islands, a small area called the South Connemara Group, are comprised of rock quite distinct from those of the surrounding basins.
- The granites of Iar-Connacht and the basins are separated from the quartzite, schist and marble mountainous band further north by a distinct, low-lying and lake-filled band extending from Slyne Head to an area south of Oughterard.
- The broad zone between Lough Mask and the sea at Killary is separated from the mountains to the south by a narrow, variable zone of sandstone and shale.
- Tully Mountain is a hill of quartzite entirely enveloped by the surrounding schists and marble. (This detail is not shown in the map here.)
- Omey Island and Aughris point are composed of granite, unlike their immediate surrounds.[8]

Nimmo's broad description of this terrain and its underlying structure is as follows:

> The country from Galway to Hinehead [Slyne Head] is a sheet of granite, or rather syenite, with few mountains of any remarkable elevation. To the north of that tract, a hollow valley runs through the whole extent of Connemara, distinctly marked by a chain of narrow lakes from Lough Corrib to Mannin Bay; its greatest elevation is only 164 feet above the sea; ... The country strictly mountainous is from Lough Corrib to Aghris Point, where the summits are from 1,200 to 2,500 feet, they are composed of quartz; around their bases ... are gneiss and mica slate, with bands of hornblende and primitive limestone. Along the north side of Lough Corrib and to Ballinakill, the mica slate and hornblende rises [sic] but the limestone disappears. From Lough Mask to the Killary, a transition country of greenstone and grawacke slate, covered by the old red sandstone or glomerate, which also forms the Hill of Mulrea in Morisk. The upper beds of this and of the greenstone are frequently porphyric.[9]

Allowing for the archaic terminology and the absence of stratigraphy, this is as good a general description of the area as any modern geological writer has achieved. There are, naturally enough, some features that the expert would disagree with: for example, the rock between Lough Mask and Killary is not old red sandstone, so-called. But overall, few experts would quibble that Nimmo's account, almost two hundred years on, still illustrates accurately the fundamental geological structure of the region. His, obviously, was no scant or superficial survey.

Neither was it one compiled from previous surveys by others unknown. He went on to describe the region in great detail, describing the nature of the rocks, their colour, stratification, bedding, orientation and other features. Quite clearly he climbed the various summits, recording outcrops of marble and primitive limestone even from high elevations, and remarking that the view from Coolnacarton summit was the best in the whole mountain district. The fine detail of his observations can be judged from his description of the hill of Glan outside Oughterard:

> the western end, like the hills on the border of Joyce's Country, is composed of quartz; the northeast side is mica slate; the middle is penetrated in a winding manner by beds of mica slate, containing hornblende and granular limestone, covered by thick beds of pyritous greenstone. On the south and east are granite and syenite, which runs [sic] under the sandstone conglomerate towards Oughterard, and this again passes under the fletz limestone, which subsequently passing Lough Corrib, occupies the greater part of the provinces of Connaught and Leinster.[10]

On the isthmus between Loughs Corrib and Mask he noted that limestone was confined to the eastern end (east of Cong) and to the south shore of Lough Mask. While it continued almost to Westport, he noted that it did not reach that town but veered to the north-east towards Castlebar. On the western coast, his observations were equally detailed and precise:

> The insulated hill of Renville [Tully Mountain] is quartz; on the northern side at the shore, the mica slate passes into a very peculiar kind of porphyry. On the south side we have a little limestone ... Omey Island and Aghris Point are low fields of granite; some veins of granite are found traversing the mica slate; but the particular situation of that rock, and of its foreign beds, will be described hereafter.[11]

The detail in which he described the 'primary limestone' running through the central valley can be gauged from comments such as the following:

> the position of this limestone is usually vertical, with conformable beds of hornblende slate, very variable in its thickness, from the fraction of an inch to 126 feet and frequently contorted and interrupted; it is in general white or grey, sometimes striped green; grains of lead, copper and iron pyrites sometimes occur in it. It is granular and micaceous, and sometimes there are thin plates of silica running through it; the quantity of carbonate is various, according to the proportion of foreign matter, but is sometimes as high as ninety-six percent and it is in general easily calcined.[12]

He noted, too, that the limestone sheet did not cross the Corrib in Galway town but stopped short of the harbour on its eastern side.[12]

When we compare Nimmo's detailed observations with modern descriptions of the region, there can be no doubt that he completed an accurate, thorough and detailed personal survey sufficient to enable a useful geological map of Connemara to be constructed. He got the major bedrock features correct, and his reference to small but important details such as the Cong isthmus, the Omey granite, the Tully mountain quartzite, the Oughterard granite and the absence of limestone immediately east of Galway harbour prove beyond any doubt that he physically surveyed the region thoroughly and with precision.

But now we encounter a mystery: why was his published report universally ignored by geologists and geological commentators of the time, an omission that continues right up to this day? His name seems to have been expunged from the Irish geological consciousness, although various quotes by him regarding bogs, the benefit of road building and the plight of the poor are still encountered. The three short modern biographies of him, by Ruddock,[13] Paxton[14] and de Courcy,[15] never mention his geological achievements. More seriously, the important historical works of Herries-Davies on Irish geology either ignore him completely or damn him with faint praise, without ever adverting to his Connemara work.[16] 'Nimmo is known to have been interested in geology,' was Herries-Davies' comment on him, before going on to make some further short remarks on his contribution to sub-marine geology. By no means was Nimmo's report the last word on the geology of Connemara, but it was an excellent first word that deserves greater recognition than it has received until now. Marchant and Sevastopulo are among the exceptions who have acknowledged the significance of his role and his influence on later workers.[17] Herries-Davies, however, does point us very perspicaciously in a direction that may explain the mystery.

Nimmo's Connemara report appeared as appendix 12 in the fourth report of the Bogs Commissioners; alongside it, Griffith's report on the mountainous regions of north-west Mayo appeared as appendix 11, and his report on Dublin and Wicklow as appendix 10. All three reports were published together in 1814, although the three were written at different times, Griffith's two in 1812 and Nimmo's in 1813 (see table 1). Incredibly, Griffith never made any public reference to Nimmo's geological work at any time ever. It was as if it had never been reported. The story from here can best be taken up in Herries-Davies's own words:

> On 2 July 1811 the Commissioners directed Griffith to go to northwestern County Mayo to make a reconnaissance survey of the region's bogs, drainage, population, soil and communications. There was also the following direction: *Mr Griffith is to make such mineralogical or geological observations as may incidentally arise from his inspection.* In addition, Griffith was told to include in his report the most accurate map that he could achieve of northwestern Mayo, and this map was completed in 1812 ... It carries just one geological line, bearing the inscription 'Line of the junction of the Primary and Secondary countries'. The line is a somewhat crude plotting of the boundary along which the Carboniferous strata in the east overlap onto the Dalradian rocks to the west, the boundary running from the north Mayo coast near Balderig to the shores of Lough Conn ... Griffith admitted that he had not been able to plot this line with any precision ... Neither his map of Mayo nor that of Wicklow can be hailed as a geological masterpiece, but they are both of interest as Griffith's earliest surviving attempts to represent aspects of Ireland's solid geology in cartographic form.[18]

They are indeed of interest: appearing alongside Nimmo's contemporary, comprehensive written account of the geology of Connemara, they make Griffith – Mining Professor of the Dublin Society, surveyor of the Leinster coalfields – appear a rank geological amateur. Griffith's written accounts accompanying the maps were informative as far as they went. Quite how far that was, can be gauged from Griffith's own words in the Mayo report: 'I had neither the time nor opportunity of making any very particular observations ... my observations being chiefly made in going out or returning from the business of my survey.'[19]

It stretches credulity to the limit to suggest that Griffith was unaware after 1814 of Nimmo's detailed Connemara report and its implication: it must have surprised, if not shocked him that this relatively unknown Scotsman, having spent just over two years in Ireland, could upstage him so comprehensively at a time when he (Griffith) was putting himself forward as the obvious, best-qualified person to undertake the geological mapping of Ireland then being mooted. Griffith was therefore most unlikely to acknowledge Nimmo's work, then or later; others working under or with Griffith, or dependent on him for patronage, would take their cue from him and ignore Nimmo.

Now the plot thickens. It might be contended, as justification for ignoring him, that there was no independent confirmation of the accuracy of Nimmo's observations. Such a contention would be quite wrong. From 18 July to 2 September 1813, George Bellas Greenough (1778–1855), founder member and first President of the Geological Society of London, surveyed the west of Ireland. He had been constructing his own Geological Map of England and Wales since 1808 and was an undoubted geological authority. He was a good friend of Griffith's to boot. Nimmo describes what happened:

> I have spent a good deal of last summer [1813] in that district [Connemara], and had the satisfaction of finding my ideas of its geological structure, as well as that of Ireland in general, confirmed, during a tour with a gentleman deservedly eminent in that department of science.[20]

In a footnote Nimmo identifies the gentleman in question as none other than Greenough.

Greenough's Irish tour was organised by his friend Robert Hutton of Dublin, who wrote to him in May 1813:

> I believe the beginning of July will be the best time for our Western expedition. Nimmo is still here [in Dublin]. I saw him yesterday. He is going to Galway in a few days and intends then to go to London. He thinks it is not improbable that he may be in Connemara or its neighbourhood in July. At all events, his assistants will be there and he says one or two of them have been over the district with him, and are very intelligent.[21]

Meeting Nimmo was to be an important feature of their trip:

> [Y]ou should see him [Nimmo] before you present your map of Ireland. I should think he could render considerable assistance in the southern and southwestern parts. He appears well acquainted with Cork, Kerry, Clare, Tipperary and Galway. He has been in Connemara and speaks of it as very interesting. I hope before he leaves Dublin to get some information respecting it which may be useful to us.[22]

When Greenough and party arrived in Oughterard on 26 July, Nimmo joined them and accompanied them through Galway, Mayo, Sligo and Fermanagh.[23]

Theirs was a geological, not a tourist, expedition and throughout its duration (they were in Connemara until 5 August) they had opportunity to record and discuss the geological features of the districts through which they passed. They collected rock samples that they packed into casks for Nimmo to forward to London in due course. When the casks were late to arrive, Hutton wrote enquiring about them. It transpires that the Excise Officer in Galway had impounded them, believing them to contain 'potsheen'! They were soon released (sooner, no doubt, than if they had contained the drink) and forwarded to Greenough.[24] Hutton had been inclined early on to dismiss Nimmo with faint praise: 'He appears to be a very clever, intelligent man but he does not profess to be a good mineralogist.'[25] No doubt his opinion had changed before Nimmo left the party on 18 August, two days before Griffith joined it.

In the face of this corroboration of his geological ability it might be thought, however unlikely, that Nimmo had produced no map of his findings, meaning that he himself may not have regarded them as really worthy of serious consideration or publication. But he had produced a map. In his letter of 11 May 1813, Hutton reported: '[Nimmo] showed me a large map of Connemara and offers to have a reduced copy taken for us.' Here we unearth a wonderfully interesting cartographic item, neglected to date. In the library of the Geological Society of London there is a small (approx. 23 x 35 cm), colour-washed geological sketch map of Connemara, reproduced here for the first time, with the permission of the society.[26] (Plate 2) The accession card gives the following information: 'Geological Map of Connemara, Ireland. No date. Notes: captioned by Greenough "Nimmo", "Connemara"'. Is it not reasonable to speculate that this could be the reduced copy mentioned by Hutton, given by Nimmo to Greenough when they met in Connemara? How else could Nimmo have described and discussed adequately with him the structural complexity of the whole region? How else would such a map have come into Greenough's possession with Nimmo's name inscribed on it by Greenough?

As to Nimmo's additional claim that Greenough had confirmed his (Nimmo's) views of the geological structure of Ireland in general, another map discussed by Herries-Davies takes on added importance.[27] This is a copy of Taylor's 'A new map of Ireland' (1793), somewhat crudely colour-washed to represent eight geological categories ('granite', 'slate clay', etc.).[28] On the reverse, in an old hand, is the inscription: 'This map belonged to Mr Nimmo the Engineer who surveyed the Kingdom by order of the Government.' Herries-Davies does not doubt that it really had belonged to Nimmo, and he sees three possibilities regarding the geological colouring: either Nimmo did it from his own notes, observations and study of pre-existing maps; or Nimmo copied it from maps used by Griffith, in his Dublin Society lectures that started in spring 1814; or finally, it may be an early draft given to Nimmo by Griffith in which case the colouring is Griffith's rather than Nimmo's. Herries-Davies favours the third explanation and never entertains the possibility that Nimmo had coloured it entirely from his own notes and observations. Knowing of Greenough's and Nimmo's meeting, is it unwarranted to suggest that Nimmo may have coloured his copy of Taylor's map in order to elaborate for Greenough his own ideas on the general geology of Ireland, and this is that map? If so, then the honour of producing the first, however tentative and crude, Geological Map of Ireland must fall to Nimmo, not Griffith. There is now, surely, sufficient ambiguity to shake Griffith's pre-eminence with respect to Ireland's first geological map?

If Greenough had Nimmo's Connemara sketch map in his possession in 1813, and if he had examined a geologically coloured map of the country done by Nimmo, it raises some interesting questions regarding the early development of the first Geological Map of Ireland. Griffith variously stated that he started this in 1809, 1811 and 1812.[29] Herries-Davies favours 1811 as the start of the project, which, he thinks, may have started out as a cooperative venture between Griffith and Greenough. We know from Hutton's letter of 3 April 1813 that Greenough was then engaged in drawing up a geological map of Ireland on his own, if not assisted by Griffith. Since that proposed map of Ireland came to nothing, maybe Greenough had come to realise that Nimmo was well ahead in the endeavour and the former therefore gave up on the Irish project to concentrate instead on his map of England and Wales. Greenough probably discussed Nimmo's ideas and observations with his friend Griffith before returning home from his Irish tour, and this information may have stimulated Griffith to renew his own geological mapping endeavours. The January following (1814), Griffith wrote to Greenough saying that he hoped to give 'some idea of the general geology of Ireland' at the lecture course he was about to give at the Dublin Society and, most tellingly, he asked Greenough to pencil in all the geological observations he had made during his Irish tour of the previous summer.[30] Did he (still) not know of Nimmo's findings, or could he not have asked Nimmo directly? Or was a query through Greenough a way to gain the information indirectly, thereby avoiding the need to acknowledge the primacy of Nimmo? Whatever the case, Greenough proved resistant to giving away any information.

In the end, the first version of Griffith's Geological Map of Ireland would not be published until a quarter of a century later, in 1838, followed by the real masterpiece version in 1839.[31] These and subsequent revised editions relied heavily on the work of many surveyors and correspondents. Whatever his other characteristics, Griffith was not generous in acknowledging the help he received from others, even to the extent of presenting their observations as his own without any reference to their efforts and contributions. He liked to give the impression that he did it all himself, and the main contributing surveyors and assistants, men like Ganly and Kelly, would later complain of his niggardliness in this matter, thereby taking a little of the gloss from Griffith's achievement.[32] As for Nimmo, Griffith did credit him with providing certain sub-marine information to the great map, but he never mentioned Nimmo's Connemara study. Ironically, when Greenough's geological map of England appeared in 1820 under his sole authorship,[33] the publishers added a codicil which stated: 'On the basis of the original map of Wm. Smith, 1815.'[34] Griffith's disregard of Nimmo and other colleagues was not, it would appear, an entirely unique or isolated occurrence in the world of geological publication.

Returning briefly to Nimmo's geological sketch map of Connemara, it shows one feature at least that went unnoticed for a further fifty years: Nimmo recorded that the southern part of Gorumna Island and all of Lettermullan differed from the surrounding area, being comprised of 'gneiss' whereas the surrounds were syenite. He was, in fact, distinguishing the South Connemara Group, a structural feature that did not appear again in geological literature until the 1860s and which is absent in Griffith's *magnum opus*. This alone disproves any suggestion that Nimmo's map could have been Griffith's.

Geology took up nine of the twenty-three pages in Nimmo's report to the commissioners. Returning now to its wider aspects, the instructions given to him were similar to those given to Griffith regarding north-west Mayo and Wicklow. The commissioners treated all these mountainous areas, 'which, in their present state of desolation and abandonment, did not appear to us to justify the expense of a particular survey', differently to all the other bog districts. All they required was a general examination and general reports to be compiled accompanied by such maps as the engineers could furnish. Apart from their significance in any other respect, the remaining fourteen pages of Nimmo's published report give an informed and valuable insight into Connemara life and its conditions at the start of the nineteenth century, unparalleled in its objectivity by any other government report.

To many, the whole area stretching from Lough Corrib to the Atlantic Ocean south of Killary harbour is regarded as Connemara. In fact there are three separate regions in that general geographic locality. Connemara proper (the Barony of Ballinahinch), lies west of a line drawn from Screeb to Maam Cross and thence along the Maam Turk mountain range to Leenane in Killary harbour. North and east of the line from Maam to Leenane the area is called Joyce's Country and encompasses the Half-Barony of Ross. The remaining district, bounded by Lough Corrib on the east, the Oughterard to Maam Cross road (N59) on the north, the Maam Cross to Screeb road (R336) on the west and Galway Bay on the south, is known as Iar-Connacht, being mainly the Barony of Moycullin. The area known today as Ceanntar na nOileán, the labyrinth of islands, small inlets and peninsulas lying to the west of the Screeb to Casla Bridge road (R336), is often included in Connemara proper, although it rightly belongs in Iar-Connacht. Nimmo used the term Connemara to denote the whole district from Lough Corrib to the ocean.

The only place where there was significant cultivation at the time was the strip of land beside the Corrib river and lake, from Galway city to Oughterard. This is part of the limestone plain that extends from central Ireland, and was therefore arable to a good degree in those places where the overlying rock had been removed. There was also some cultivation along the coastal strip and on the north-facing slopes of some of the hills. Nimmo estimated that just 7per cent of the total land area was arable, 34 per cent was bog, 57 per cent mountain and upland pasture and 2 per cent bare rock. Is it any wonder that he should make the following remark: 'The district appears, not undeservedly, to be considered as one of the most uncultivated parts of Ireland. On a general view, indeed, it seems one continued tract of bog and mountain, the quality [sic] of arable land not amounting to one-tenth, perhaps not one-twentieth of the whole'?[35] 'Where cultivation has made the greatest progress on the south shore of Lough Corrib,' he went on, 'the arable or dry land is interspersed with extensive tracts of naked limestone rock of a most desolate aspect; and it appears to be only after incredible labour, that a few patches of soil have been won from the general waste.'

Nimmo was nothing if not irrepressibly and resolutely confident. His opinion was: 'Though the general improvement and cultivation of Connemara would seem an undertaking of the most arduous description, it is not without facilities, which might, upon a candid consideration, make it appear a subject more worthy of attention than many other of the waste lands of the Kingdom.'[36] There were not then, nor are there now, many who would espouse this positive sentiment. Nimmo went on to list the advantages of the district. The climate was mild; the mountains gave some shelter; limestone, shell sand and seaweed were available in abundance, with no part of the whole district being more than four miles from an easy supply of one or more of these manures; and fuel supplies were inexhaustible. Although Connemara was mountainous, it was not like Wicklow: at least three-fourths of Connemara proper was lower than one hundred feet above sea level, he pointed out, and most of Iar-Connacht under three hundred feet. Joyce's Country, on the other hand, was an elevated tract with flat-topped hills of heights of up to 2,500 feet intersected by deep and narrow valleys. There were upwards of twenty safe and capacious harbours fit for vessels of any burthen; about twenty-five navigable lakes in the interior of a mile or more in length, and hundreds of smaller lakes; there were four hundred miles of sea shore and at least fifty miles of lake shore. Lough Corrib was navigable as far as Galway town and the navigation could easily be extended to the sea. The multiplicity of inland navigations meant that transport on water need never be a problem anywhere in the district. Many of the lakes were separated by narrow rocky ridges and where they communicated, usually had an excellent fall for the powering by water of machinery to pound limestone or do other useful work.

Seen, therefore, through Nimmo's optimistic eyes, Connemara was not the field of desolation so often described and recorded with fastidious relish by those who passed through it with eyes wide open and minds tightly shut. Passing through it successfully was, of course, the real problem: while coastwise navigation was common, and boats also plied on Lough Corrib as far as

Maam, there were no carriage roads through the interior. Even twenty years later, when Maria Edgeworth travelled through its heart, Connemara was as impenetrable to wheeled carriages as any desert.

About 30,000 persons lived there, one-half in Connemara proper. More than 90 per cent were living on the coast; less than three hundred families lived inland, mainly along the routes of the bridle roads. About 10,000 lived in Iar-Connacht, either on the sea coast or on the northern slopes of the hills near the limestone country. The uplands of Joyce's Country were uninhabited. Potato growing, fishing, kelp-making and turf-cutting were the main occupations of the men; the women kept whatever animals the family possessed, helped with the potato patch and looked after the family.

Cultivation along the coastal strip was well advanced, with some barley and wheat growing in suitable patches. Nimmo suggested that the use of seaweed as manure was the reason for this success. There was no limestone in the area, but considerable quantities were brought from Aran and County Clare as ballast in returning turf boats and thrown out on the shore. While such stones were used for building, they were not applied much in agriculture, although they could be burnt on the shore to make lime for less than one shilling a barrel. The cultivated strip extended all around the coasts of Kilkieran and Bertraghboy Bays and into Mannin Bay. The vast banks of shell and coral sand in these bays were largely untouched; on the occasions when coral was used, it was raised from low down on the shore. A boat was beached at low tide and the coral dredged into it; as the tide rose, the boat was floated up the shore and beached at the high-tide mark, where it was unloaded. However, the sheer abundance of red seaweed that was cast ashore, or that could be cut in deeper water, made all other soil improvers practically redundant. Two or three boatloads of about six tons each were usually applied over an acre of potato ground and a simple crop rotation was followed. After manuring, a first crop of potatoes was grown, followed by oats or barley the next year. The land was then left as natural meadow for four or five years, following which the rotation started again with seaweed as before. After a second cycle, the surface was pared and burned, processes that Nimmo felt were destructive to a soil that was really just a layer of thin red bog over bare granite. His recommendation was to spread the sea-ouze or sludge from rocky creeks on the land. This sludge was 'partly decayed marine vegetables, and partly mud or bog-stuff which has been transported to the sea'. He even suggested it might be worthwhile to float off bog into some of the creeks, where the sea would convert it into manure.

Apart from their potato patches, people on the coast were not particularly interested in extending cultivation. Kelp-making and fishing were more congenial to them, despite the hard work involved. Nimmo estimated that about four thousand tons of kelp were then manufactured each year in Connemara. Cutting the black seaweed for burning commenced in May and continued until Michelmas. About twenty days of work were needed to cut and land enough weed to make one ton of kelp. A good kelper could make three tons, but the average was two; three men in a household could make seven to eight tons in a season. From this, Nimmo calculated that no fewer than two thousand men were engaged in kelp-making. Since the coast-dwelling population – men, women and children – was estimated at about 25,000 it would seem that most able-bodied male adults contributed to the work. The reason was obvious: in 1808 kelp sold for thirteen pounds a ton in Galway, with five shillings for its freight to the town. That was no small contribution to a family's income, especially important since the occupiers of land close to the shore were subjected to higher rents because of their favourable location near sea-manure.[37]

Nimmo expressed some surprise that the proprietors of the land (not the tenants) did not ensure that the seaweed was used for manure instead of making kelp. To make 4,000 tons required, on his calculations, about 50,000 tons of weed, which, in the raw state, was sufficient to fertilise 4,000 acres of land. Unprocessed seaweed sold in Galway as manure made half a guinea per ton and was regularly shipped through Lough Corrib to farms in the interior. Land improved with seaweed would yield two to three pounds per acre in rent. By 1812 the price of kelp at Galway had declined to four pounds per ton or less and Nimmo felt that as long as this low price prevailed, the case for using the raw weed as manure was strong, and the potential benefit to the land and to the landowner was measurable. But what was the likely benefit to the tenant? Why should he improve the soil only to see the rent raised in consequence? Nimmo's agronomic theory might have been impeccable, but his sense of social justice was still imperfect. He was fully aware that the tenants' rents included a seaweed supplement, but he did not seem to appreciate that his proposal would deny them the cash benefit of the kelp. The actual cutting and burning of weed to make kelp, carried out on the foreshore, added nothing permanent to the rental value of the land, and in making kelp the tenant was not cutting a stick with which the landowner would undoubtedly beat him. Nimmo failed to notice that his strictures on the operation of the tithes in Kerry were just as relevant when applied to other forms of pernicious exploitation by some landowners in the west. Indeed he recounted, not without a sense of approval, the following incident:

> Mr O'Flaherty reclaimed a large tract of bog at Renville etc. to the extent perhaps of 1,000 acres; he removed the cottagers from their old stations, and settled them on the bog: this they reclaimed with potatoes and seaweed, treating it afterwards with the sand of the shore, which contains no calcareous matter; the effect has been very great. A 'collop' [in Ireland, a small measure of grazing land] there is worth about one guinea; but these tenants pay all duties, services etc.[38]

Another explanation for the low level of agriculture practised in Connemara presented itself to him. The population had always been concentrated on the coast, a conclusion he deduced from the fact that the ancient churches and chapels were all on the shore. Also, since very few persons on the coast could be considered as farmers alone, he surmised: 'Farming and fishing, it is well known, do not assort well together; and however active the natives appear in the latter occupation, they are little inclined to exertion in the former.' This, of course, simply expresses the same rationale as outlined already: catching and selling fish yielded a cash return that did not increase the rental value of the land. Because the cash often went towards helping to pay the rent, the landlord could hardly complain. Nimmo reported that the number of large boats in the district was considerable − 'scarce a farm but has one or two of them, besides smaller' − a claim that is difficult to believe. The size of boat he meant was about nine tons burthen and cost about forty pounds, exclusive of tackle.[39] It is most unlikely that this much capital was tied up in tenants' boats at the time, although his point − that they put more capital into fishing and navigation than into agriculture − was probably true: the only implement of husbandry they possessed was the spade. The large boats, hookers especially, were useful, among other things, for the sun-fish (basking shark) hunt that took place in April and May and for which Connemara was then a centre of activity. Sun-fish oil was a valuable commodity not only for local uses but also as another item that generated cash.

Further inland, along the valley that is now the route of the Galway to Clifden (N59) road, the tenants' activities took a slightly different course. Outcrops of marble gave ready access to material for lime-making. Turf for the kiln was easily available alongside, so burning the marble in temporary kilns was a common activity. The kilns, each big enough to produce fifty barrels of lime, were often located in the bog itself and were re-used repeatedly. The fire was first kindled at 7 in the morning and was in full heat by 10 the next day. To feed and attend it, Nimmo estimated that it took three men to quarry the stone, six to break it, one to cut the turf, two to 'rear' it and four to draw it to the kiln and burn it. Three horse-loads of stone produced two barrels of lime. At a wage of one shilling each, he estimated the total cost of the operation was seventeen to twenty-one shillings, about five pence per barrel when fifty barrels were made. The lime was carried out on horseback at two pence per barrel per mile. Needless to say, the area was more grassy than the southern coast because of the limestone, but the tenants generally were just as reluctant to engage in improvements. Turf and lime took the place of kelp and fish as 'cash crops' for the inland residents. In Joyce's Country, further north again, cultivation was practised even less, due to the elevated nature of the land and the sparse population. By the shores of Killary harbour some tillage was pursued and because there was no limestone in the area, sea sand and coral were brought in by boat from Ballinakill Bay.

The presence of a market at Westport had acted as a spur to agriculture north of Killary, and Nimmo was confident that new towns and markets would also arise in Connemara once decent roads were constructed. In 1812 John D'Arcy, a landowner who was in the course of building a new castle for himself at Clifton, conceived the idea, possibly encouraged by Nimmo, of starting a new town and seaport nearby.[40] Nimmo also recommended that a new seaport village be established at the mouth of the Ballinahinch River on the Martin estate at Roundstone, and another at Cleggan on the Blake property, so maybe D'Arcy was trying to pre-empt such rival developments in order to ensure the pre-eminence of his own new seat at Clifton. As late as 1822 'Newtown Clifton' – now called Clifden – still consisted of only one slated house and a small number of cabins, but Nimmo had already marked the location on his geological map.

His initial approach to Connemara was therefore one of great discernment and certitude and must have been music to the ears of those improving landlords who were attempting to make something of their estates. He would become friends with many of them – notably the D'Arcys of Clifden, the Blakes of Renvyle and the Martins of Ballinahinch – and was accused later of favouring landlords against the poor, a judgement that seems altogether too harsh.[41] As an engineer, he could hardly afford to alienate the landed class on whom he would depend for his employment, either directly or through engagement by local (e.g. the Grand Juries) and national (e.g. the Fisheries Commissioners) bodies on which only the landed and moneyed classes were represented. To his credit, Nimmo did not criticise the poor as indolent and stupid like many others did, nor did he patronise them. When he mentioned them at all he wrote well of their ability as workers, especially given the want of capital that was essential if any progress was ever to be made with land that was so derelict. As in Kerry, the spade (and its bog-cutting equivalent the *sleán*) were the only agricultural implements in use in Connemara; transport was mostly on the backs of men and women; houses were mere mud or turf cabins; formal education was almost unknown and the wet, windy climate was not always conducive to work on the sea or in the bogs. We will see later that he was very supportive of the poor when giving his evidence to various select committees of Parliament, which seem the right places to make such observations effectively. But in 1813 that was all still in the future.

On further, more detailed inspection of the individual bogs the potential of the whole district took on a more sombre hue. Those on the limestone strip beside the Corrib, and around Cong and the south-east shore of Lough Mask, were straightforward enough and could be improved by drainage, embankment and all the other remedies he had proposed in Kerry. The 'Moor of Sillermore' – his name for central Iar-Connacht – comprised about 50,000 Irish acres much roughened by naked blocks of granite and was mostly inaccessible and practically uninhabited. Its main artery, the coast road from Galway to Casla Bay, was not passable for carriages, and the only other roads were mere bridle paths. Only the hills above the limestone field were tolerably cultivated, and apart from improvements to them he had nothing to propose for this great moor. The basins surrounding the bays of Kilkieran, Casla, Camus, Casheen and so on, effectively Ceanntar na nOileán, contained 14,000 acres of bog and about a hundred lakes. It was very low-lying and extensively penetrated by the sea in shallow inlets. From the eastern side of Casla Bay to the western side of Kilkieran Bay was only eight miles as the crow flies, but there were one hundred miles of seashore, excluding the shores of uninhabited islands. An acre of potato ground here cost three guineas to rent and a cow sold for less than two pounds.

Proceeding westwards through south Connemara, things did not get much better. Excepting the coastal fringe, the land was uncultivated and virtually uncultivable. At best it made rough pasture for animals permitted to roam at will. The Bertraghboy basin had 9,500 acres of bog and ninety-six lakes. Orrismore, despite some significant agriculture around Bunowen, Mannin Bay and Clifden Bay, was even worse, as described by Nimmo in his own words:

> From Mannin Bay to the river at Ballinahinch is an extensive flat moor of a singular appearance; from one of its principal districts I have named it the Moor of Orrismore. The tract is about seven miles by four and, generally speaking, is a plain not much elevated above the sea; it is however intersected by many low ridges of mica slate, and in hollows between them are a multitude of lakes. I have [made] an exact survey of the moor, and find the number of lakes to be about 143, of different sizes, many of them having numerous and intricate arms. When viewed from an elevation, this appears to be a complete labyrinth, in which it is difficult to perceive the direction of drainage.[42]

North of Clifden, in the Ballinakill area, things were better. Some drainage had been undertakes and the abundance of coral sand in the locality had contributed to the success of farming – helped also, no doubt, by the coerced displacement of tenants on to the bogs to force their reclamation. But south and east of Ballinakill, towards the central vale of Connemara, lay the Twelve Bens and their surrounding lowland. The mountains were unusable; the low ground around the Inagh valley was rough, but some low ridges and knolls had tolerable soil affording some pasture, and the Martins and some other resident principals were planting trees around their properties. Nimmo was convinced that trees were the best possible items for cultivation in Connemara, an opinion that found favour in earnest only in modern times, but has once again fallen from use in the last few years. Of bogs, there were over 17,000 acres, and 115 lakes of differing dimensions along with many small pools dotted the landscape. As regards the southernmost part of the central valley of Connemara, Nimmo said it all with finality: 'The southern part is a desert.' It was, and remains, a granite platform sloping gently to the sea covered in bog and lakes, drained by the Invermore and Owengowla rivers.

The existing roads were mere bridle paths and we can form a good idea of their number and locations from Nimmo's geological sketch map (Plate 2). The coast road from Galway passed through an Tulaig (Inverin today) and continued to Casla, where it met the bridle path coming from Bunnagippaun near Oughterard. The north road from Galway to Oughterard and on to Clifton (Clifden) formed the main communication spine through the interior, for what it was worth. A path from Screeb joined it at what is now Maam Cross and proceeded from there to the ferry at the foot of Leckavrea. Another path entered from Rosmuc at Leahy's Bridge and proceeded along the western shoulder of the Maamturk Mountain to Finnies Glen and the coast at Killary harbour. From there it went to the Head of Killary and on to Westport. At Garomin, a path off the main central road veered south to Cashel, first giving off a spur down the peninsula to Ardmore near Carna. From Ballinahinch, a path passed south over Orrisbeg Hill, above the site of Roundstone village today, and around the Orrismore peninsula through Bunowen and on to Ardbear. The main road continued from Ballinahinch via Lettershea to Clifton and on to Streamstown, Cleggan, Ballinakill and Renville. From Renville a path ran along the shore of Killary to meet the road from Leahy's Bridge, and another ran nearer to Kylemore meeting the Leahy's Bridge road at Finnies Glen. In the eastern end of the district, a path from the ferry at Drumsnave (opposite the base of Leckavrea) passed through the site of today's Cornamona on its way to Cong and onwards to the large towns of Headford and Ballinrobe. A spur from this circled Benlevy to the village of Bothaun. Excepting parts of the main central road and parts of the coast road, all of these routes were passable to foot traffic only. Many of the villages we know today, such as Spiddle and Moycullin, are not shown on the map and probably did not exist at the time. Places that are mentioned on the map such as Clifton (D'Arcy), Streamstown (D'Arcy), Tullia (Blake), Ballinahinch (Martin) and Renveel (Blake) are shown probably because of the importance of the named families who owned them and had their residences there.

The detailed picture was, therefore, enough to discourage anyone who hoped to devise a plan for the development of Connemara. Undaunted, Nimmo decided that opening up the whole district was warranted, feasible and likely to be profitable. His plan involved new and improved roads, new seaports and improved navigation from the Corrib to the sea (figure 5.1). Looking outside the immediate district, he noted the importance of Westport as a market outlet for the Clifden/Ballinakill area, with Headford and Ballinrobe playing a similar role for the area around Cong. Galway town was the obvious focus for Iar-Connacht and it could also service the Ballinahinch and other coastal districts by sea. He decided to drive a new line of main road through the heart of Iar-Connacht from Galway to Clifden, with branches to

FIGURE 5.1
The roads proposed in Connemara by Nimmo in 1813. (Sketch drawn from information in table 3).

various localities. It would exit Galway by Rahoon, through Boleybeg and onwards, crossing the Spiddle River (Owenboliska River) about four miles from its mouth. From there it would go to Fermeel (Fermoyle today) and on to the bridge of Screeb. It would run to Cashel, continue to Toombeola, where it would cross the Ballinahinch River, and proceed to Derrygunla or Ballyconneely. Two new roads to Westport were planned: one from Clifden through Kylemore and the head of Killary harbour and another in the east, from the ferry at Drumsnave through Cromglan and Maamtrasna up the west side of Lough Mask to join the Mayo road from Headford. It would have a branch over the mountain to join the Westport road from Killary. From the new central Connemara road, cross roads would proceed south and north, generally along existing bridle paths, but along completely new lines where there were no pre-existing, improvable routes. The bridle road along the coast from Galway to Screeb was to be improved and made passable for carts.

Gone were the short distances in which Nimmo measured the proposed Kerry roads. Here he proposed no less than 343 miles of new or improved roads costing an estimated £42,000. This was an astonishing sum, one most unlikely ever to be made available to improve Connemara. Table 3 gives their details, indicating which were completely new, which were improvements on existing tracks, which roads were never built and the relevant present-day representation of those that were; his scheme in 1813 was an outline one, so that the precise lines cannot be exactly delineated, but the general outline, based on his written account, is accurate as far as it goes. Even this outline can be seen as the very first regional development plan for Connemara, one that set out a road, harbour and village network that he would later pursue and that is still evident today.

In order to open the region by water, Nimmo proposed to construct a number of new seaports. One such was essential for the improvement of Ceanntar na nOileán, he decided. For this he chose Bealadangan, across from the south point of Rosmuc, as the most suitable place with sufficient depth of water and being a centre of communication by land and water for the whole district: 'The formation of a quay there, and beaconing and clearing parts of the channel, are necessary steps in the agricultural improvement; a good boat-creek on the spot may be enlarged into a tolerable wharf.'[43] The clearing and beaconing of Bealadangan would take place long after his death.

He proposed another harbour village for Ballinakill Bay:

> The establishment of a regular village on Ballinakill Bay seems advisable. It is an excellent harbour, a common station for the herring fishery and the neighbouring lands are well calculated for improvement. The best site seems to be at Creggaun, on the south side of the middle peninsula; this spot is accessible by vessels of burthen and not ill situated for communications by land.

A village might also be built on the Killary, he stated, either at Bunowen or at Bealnacul (now Leenane). For Connemara proper, he proposed a village at the mouth of the Ballinahinch river,

> as a market, and centre of improvement, for this district; having a safe roadstead, and being accessible by large vessels, it may hereafter become a place of some importance. Behind it are the lakes of Ballinahinch and Derryclare, which admit of being easily converted into one piece of navigable still water six miles in extent; and at present, though this may not be readily (if ever) connected with the tide-way, yet a good road of two miles is only necessary to give access to it.

He himself would develop a village that became Roundstone during his second sojourn in Connemara.

Nimmo's final word on the whole district was a short comment on Lough Corrib. Draining it was out of the question, he believed, but lowering the level by a few feet would greatly improve the surrounding lands, relieving them from flooding. Since he realised that the mill interests of Galway would object to this, he proposed to clear and deepen the channel and remove some eel-weirs and shoals from the river to give them some compensating benefit. The

FIRST FORAY INTO CONNEMARA

TABLE 3.
The location, route and pre-existing state of the roads proposed by Nimmo in Connemara in 1813, together with their modern equivalents.

SILLERMORE DISTRICT

No.	Name	Route via	New or Existing	Modern Road No.
1	Galway – Screeb	Rahoon, Boaleybeg, Spiddal Bridge and Fermeel	New	No road
2	Woodstock – Barna	Boaleybeg	New	R
3	Drimcong – Bealadangan	Spiddle Bridge	New	No road
4	Spiddle – Bunagippaun	Spiddle Bridge	New	No road

KILKIERAN DISTRICT

No.	Name	Route via	New or Existing	Modern Road No.
5	Bealadangan – islands	Annaghvean and Lettermore	New	R374
6	Bealadangan – Killeen		Part new	R343
7	Oughterard – Screeb – Cashla Bridge	Glentrasna	New	R336
8	Screeb – Cashel and Kilkieran	Crossing Invermore River	New	R340 to Kilkieran
9	Rosmuck – L. Shindela	Boheeshul	Part new	No road

BERTRAGHBOY DISTRICT

No.	Name	Route via	New or Existing	Modern Road No.
10	Cashel – Toombeola	Salmon weirs	New	R342
11	Awingoula Bridge – Mweenish and Ardmore		New	R340
12	Lettercamus – Rossrua		Part new	
13	Orrisbeg road		RE-align to lower ground	R341

ORRISMORE DISTRICT

No.	Name	Route via	New or Existing	Modern Road No.
14	Salmon weir – Bally – Conneely and Bunowen	Derrycunlagh	New	
15	Ballyconneely – Orrisbeg		New	R341
16	Munga – Ballinaboy		Part new	No road
17	Gowlane – Head of Roundstone Bay		New	

BALLINAKILL DISTRICT

No.	Name	Route via	New or Existing	Modern Road No.
18	Clifden – Head of Killary	Kylemore	New	N59

cont

CENTRAL VALLEY DISTRICT

19	Invermore – Killary	Boheesul and Finnies Glen	Part new	No road
20	L. Bofin – Clifden	L. Ourid, L. Garomin	Realignment	N59
21	Oughterard – L. Bofin	Glengowla	Realignment	N59
22	Glentrasna – Maam	Maam Cross	New	R336.
23	Maam – Lecavrea		New	R336
24	Bunscanive – Derry	L. Elan	New	

JOYCE'S COUNTRY DISTRICT

25	Lecavrea – Leenaun	Maam	New	R 336
26	Drumsna – Mayo Road	Cromglan, Glantrasna	New	
27	Road over Mamean		Improve Existing track	No road

only serious difficulty in the river navigation was a shoal at Newcastle that could be easily deepened for a small sum. The river had very little fall to Wood Quay in Galway town, and boats carrying fourteen tons, drawing about four feet, plied it; he wished to carry this navigation into the sea at the town. That would require a canal with only two locks, large enough to admit vessels of burthen, at a cost of £6,000. Boats from the sea entering the river with seaweed for the interior would bring turf down as a back cargo. In addition, two weirs constructed in the river at the upper level of the locks would secure a much better supply and greater fall of water to the various mills. That proposal, of course, would really meet the approval of the mill owners. He did not indicate it at the time, but his proposed route for the navigation canal was from Wood Quay along beside what is now Eglinton Street and through the west side of Eyre Square, to enter the sea at the east side of the present docks.

Little could Nimmo, or anyone else for that matter, have known that one day he would be charged with public works that would enable him to put flesh on the bare bones of the plan that he had conceived for Connemara. That charge, with the opportunities that accompanied it, was still almost a decade in the future (see chapter 10), but for the present his work for the Bogs Commission was at an end. While Griffith's maps of Mayo and Wicklow were published, Nimmo's large-scale map of Connemara was not, and we are left only with his written account and his geological sketch map. The commissioners can hardly have failed to appreciate the importance and significance of his thorough description of a district that they regarded poorly, but one that nevertheless supported 30,000 souls. It may not be entirely base or baseless to suggest that Griffith could have influenced them not to publish Nimmo's map as a means of lowering the latter's competitive profile. Griffith's father was a commissioner, who, even if he had retired his commissionership by then (he did not sign the fourth report of 1814), might still have had influence with his erstwhile colleagues. The parsimony of the commissioners in this matter deprived the country of what would have been a unique cartographic record, if Nimmo's Iveragh map is anything to go by. There is convincing evidence that Nimmo did actually produce a large-scale map of Connemara in manuscript: the Government was billed for its production, and paid up, after his death in 1832.[44] Numerous commentators have bemoaned the absence of this map; to be positive, if not realistic, it may still exist but be languishing unrecognised in some archive, as his manuscript geological sketch map was until now, and like it, may one day come to light again.

NOTES

1. A. Nimmo, 'The Report of Mr Nimmo on the Bogs in that Part of the County of Galway to the West of Lough Corrib', in Fourth Bogs Report, appendix 12.
2. Fourth Bogs Report, appendix 12, p. 187.
3. R. O'Flaherty, *A Chorographical Description of West or H-Iar Connaught*, ed. J. Hardiman (Dublin: Irish Archaeological Society, 1846).
4. R. Kirwan, *Elements of Mineralogy* (London: McKinley, 1810).
5. G. L. Herries-Davies, *Sheets of Many Colours* (Dublin: Royal Dublin Society, 1983), p. 17.
6. P. Mohr, *Wind, Rain and Rocks*. (Galway: the author, n.d. [c. 2001].
7. J. Woods, unpublished diary, 1809 (Cambridge University Library, Add MS 4342).
8. P. Mohr *Wind, Rain and Rocks* pp. x–xii and plate 1.
9. Fourth Bogs Report, appendix 12, pp. 191 – 5.
10. Ibid., p. 193.
11. Ibid., p. 192.
12. Ibid., p. 195.
13. T. Ruddock, 'Alexander Nimmo', in Skempton (ed.), *Biographical Dictionary of Civil Engineers*.
14. R. Paxton, 'Alexander Nimmo', in Matthew and Harrison (eds), *Oxford Dictionary of National Biography*.
15. J. W. De Courcy, 'Alexander Nimmo: Engineer Extraordinary'. *Journal of the Clifden and Connemara Heritage Group*, 2 (1995), pp.47–56.
16. Herries-Davies and Mollan (eds), *Richard Griffith 1784–1878*; Herries-Davies, *Sheets of Many Colours*; G. L. Herries-Davies, 'The History of Irish Geology', in C. H. Holland (ed.), *The Geology of Ireland* (Edinburgh: Dunedin Academic Press, 1981).
17. T. R. Marchant, P. and G. D. Sevastopulo, 'Richard Griffith and the Development of Stratigraphy in Ireland', in Herries-Davies and Mollan (eds), *Richard Griffith 1784–1878*.
18. Herries-Davies, *Sheets of Many Colours*, p. 20.
19. R. Griffith, in Fourth Bogs Report, appendix 11, p. 177.
20. Fourth Bogs Report, appendix 12, p. 187.
21. R. Hutton to G. B. Greenough, 11 May 1813. (Cambridge University Library, Greenough Papers, CM 02006), hereafter referred to as CM 02006.
22. R. Hutton to G. B. Greenough, 3 April 1813 (CM 02006).
23. G.B. Greenough, Diary for 1813 (University College London, Greenough Papers, 8/18).
24. R. Hutton, to G. B. Greenough, December 1813 (CM 02006).
25. R. Hutton to G. B. Greenough, 3 April 1813 (CM 02006).
26. Geological Map of Connemara, Ireland. Notes: captioned by Greenough 'Nimmo', 'Connemara'. No date. (Geological Society of London, LDGSL 999).
27. Herries-Davies, *Sheets of Many Colours*, p.53.
28. J. Taylor, 'A New Map of Ireland'. (NLI, Maps, 16 B 3[15]).
29. Herries-Davies, *Sheets of Many Colours*, p. 38.
30. Ibid., p. 39.

31. R. Griffith, 'A General Map of Ireland to accompany the Report of the Railway Commissioners showing the Principal Physical Features and Geological Structure of the Country', 1839.
32. Herries-Davies, *Sheets of Many Colours*, p. 69.
33. G. B. Greenhough, *Geological Map of England and Wales, on six sheets with an accompanying memoir* (London: Geological Society of London, 1820).
34. Linnean Society of London, 'The picture quiz: George Bellas Greenough', in *The Linnean*, 22, (2006), p. 30.
35. Fourth Bogs Report, appendix 12, p. 187.
36. Ibid., p.188.
37. R. Griffith, *A Practical Guide to the Valuation of Rents in Ireland*, ed. J. V. Fitzgerald (Dublin: E. Ponsonby, 1881).
38. Fourth Bogs Report, appendix 12, p. 191.
39. Ibid., p. 190.
40. K. Villiers-Tuthill, *History of Clifden 1810–1860* (Clifden: the author, 1992).
41. T. P. O'Neill, 'Minor Famines and Relief in County Galway 1815–1825,' in G. Moran (ed.), *Galway History and Society* (Dublin: Geography Publications, 1996), ch. 14..
42. Fourth Bogs Report, appendix 12, p. 202.
43. Ibid., p. 201.
44. Accounts of A. Nimmo for the years 1831 and 1832 (Nat. Archs. UK, Works of A. Nimmo, AO 17/492).

CHAPTER

6

Major Harbour Engineering: Dunmore and Cork Harbours

Nimmo had received a total of £1,720 for 720 days work with the Bogs Commission, a handsome reward indeed, compared with what he might have earned had he remained a schoolmaster. In 1814 it was time for him to take a well-earned holiday; the scope and depth of his reports were unprecedented, testament to over two years of unremitting work and travel. In other ways matters were much less satisfactory. For all his plans for Kerry and Connemara, it was most unlikely that anything much would ever be done to implement them, at least in the immediate future. There simply was not sufficient private capital around for such works and the Government was not about to make any available. They would remain just plans, and as such, Nimmo's practical engineering credentials would remain untested for a little longer. Even if the roads he already had in hand in Kerry and Galway were completed, they would not amount to much. Located in remote districts, they were narrow and rude, not at all like the fine, graded, turnpike and mail coach roads then being constructed between the main cities of England and in Ireland. The Irish Mail Coach Roads Act of 1805[1] had charged the Postmaster General to employ competent surveyors in laying down the lines for these roads. One such was Sir Charles Coote, a landowner and an established surveyor who had carried out the surveys of counties Cavan, Monahan, Leitrim and King's County for the Dublin Society. Among the lines he laid out were the mail coach road from Dublin to Derry and others through County Louth. He seems to have become friendly with Nimmo and helped to secure for him his first commission with the Post Office. Nimmo joined the surveyors working in Mayo and Roscommon and by early 1815 he had surveyed, levelled and mapped a number of lines of new road – Ballinrave to Carrick-on-Shannon, Battlebridge to Drumsna and Boyle to Ballina via Tubbercurry. These were parts of the proposed great mail coach road from the River Shannon to Ballina, Nimmo's manuscript maps of which are now in the National Archives of Ireland.[2]

DUNMORE HARBOUR

Unless Nimmo could do some 'real' engineering, with some tangible structural outcome, he would never fully justify the title 'civil engineer' that he so assiduously added after his name. He needed a big commission to sustain that title and once again he was to prove lucky. His association with Coote secured for him the opportunity to apply for a serious engineering challenge, one with the potential to establish his credentials among his English and Scottish colleagues. All the great civil engineers had one or more monumental works that proclaimed their skill and eminence in this young profession. Therefore, when the Post Office sought applications for the design and construction of a new packet station in the estuary of the River Suir in Waterford, Nimmo saw immediately that this could be his chance to join the greats with his own monumental harbour. Up until then, the mail boats arriving from Milford in Wales had to sail upriver to dock in Waterford city. That was not always easy for sailboats when the winds were contrary and they would often need to lie at anchor for days until conditions favoured the 'zigzag' run up the estuary against the tide, the river and sometimes the wind. A packet station lower down would permit the mails to be landed without delay and be taken by road to Waterford, a far more reliable and faster mode of transport than beating up the estuary under sail.

The prospect of gaining the commission certainly warranted postponing all thought of a holiday until he made sure his application was submitted. Therefore he immediately surveyed a suitable site in Waterford Estuary and prepared, together with Coote, a plan for a harbour, pier and lighthouse at Portcullin Cove near Dunmore East, to be the new packet station for the Irish mails. The Postmaster General forwarded the plan, under the authorship of Sir Charles Coote and Alexander Nimmo, dated March 1814, to the chief secretary, Robert Peel, who transmitted it to the Government in London.[3] The Government sent the plan and estimates to Rennie for his comments and expert opinion, but published the map and estimates straight away in May.[4] Coote's name was not mentioned in these, and he dropped out of the project almost immediately; he is never again mentioned in connection with Dunmore. Perhaps he was only named in it in order to ensure that the submission would be given consideration in the first instance. According to Nimmo, it was a simple mistake that the actual plan was never published at the time;[5] it was published much later, in 1846, under both men's names but erroneously dated 4 March 1825.[6] Nimmo later expressed his personal gratitude to Peel for 'preferring me before persons of greater name and experience to conduct the work',[7] from which we can conclude that others would have been happy with the commission had they got it.

The proposed harbour plan is shown in figure 6.1(A). The small cove of Portcullin was to be enclosed by two moles surmounted by piers. The southern was to extend in a north-north-east direction for 250 feet before turning through 60° and continuing almost north-north-west for 100 feet. A lighthouse would dominate the sea end of this pier. The northern pier would extend 200 feet from the end of the island Eilan na Glioch in a south-east direction before turning south for its final thirty-five feet. The harbour opening between the piers was to be 145 feet wide, facing north east. The actual masonry piers were to be forty feet wide at low water and about twenty-seven feet at high water level, giving a pronounced batter to the seaward faces. The inner quay walls were also to be masonry, and a six-foot parapet topped with squared coping was planned along the seaward face of both piers. An approach road from the main Waterford carriageway would lead down to the south pier along a stretch of rock that would be partly cut away. Other small rocks, exposed at low tide, would be cut out from the new inner harbour. A new bridge was to cross the chasm from the mainland to Eilan na Glioch, continuing as a road extending along the north pier. Provision was made for a railway that would be used in the construction phase as well as later when the harbour was in operation. Altogether more than 630,000 cubic feet of rubble stone and 98,000 cubic feet of masonry would be required and the total cost was estimated at £19,384-16-0.[8]

With his plan submitted and under consideration, Nimmo was ready to take his long-awaited holiday, using it to widen his experience of practical engineering, especially regarding piers and harbour works. Immediately after the declaration of peace in June 1814 he took the opportunity to visit the Continent on something of a busman's holiday, visiting France, Holland and Westphalia, where he 'became acquainted and exchanged information with various foreign engineers of merit ... in the hope of acquiring professional information'.[9] The Dutch, he said, were far advanced in works to withstand the sea, and the pier at Den Helder, which was in sixty feet of water, particularly impressed him. He remained on the Continent until late September: Hutton reported that he was in Amsterdam that month, 'but what he is doing I know not'.[10] He returned home through Scotland: according to Ruddock, he visited Kirkcaldy at this time, during which he was witness to the baptism of his niece Mary.[11]

His standing as a man of science was at a peak in his native Scotland and had he so wished he might well have settled comfortably into the scientific circle of Edinburgh with relative ease. His 'Boscovich', 'Theory of Bridges' and 'Carpentry' articles were already in print or in course of publication; his maps and reports for the Bogs Commissioners had been published in April and would have enhanced his reputation enormously. His account of the geology of Connemara would have ranked alongside the best contemporary essays on the regional geology of the United Kingdom, and would have been very appropriate for presentation, had he been so inclined, at meetings of the Royal Society of Edinburgh and the Geological Society of London. Had he prepared

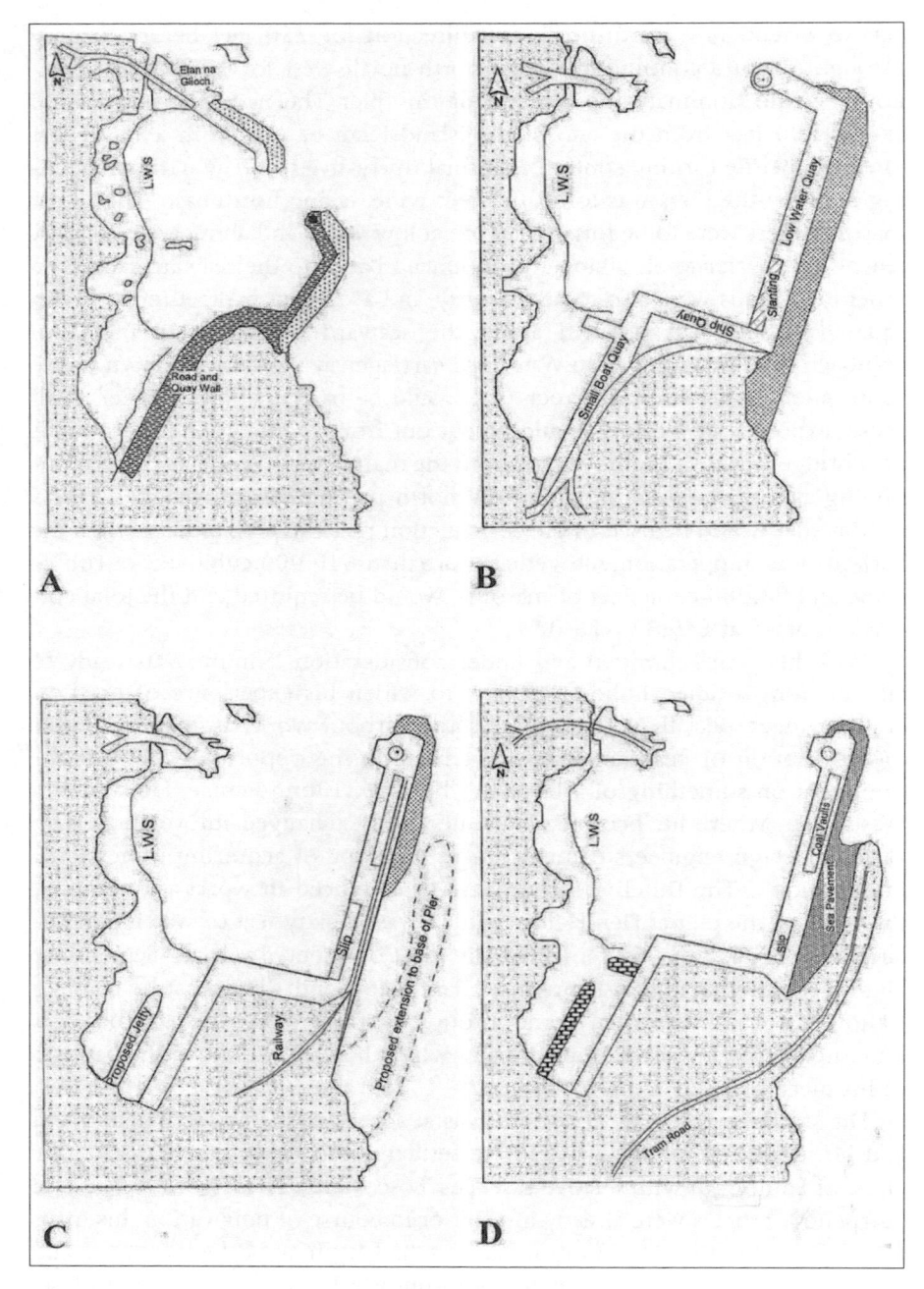

FIGURE 6.1
Sketches of the early evolution of Dunmore Harbour based on extant maps and plans.
A. The original plan of Nimmo and Coote, 1815 (based on ref. 4).
B. Nimmo's plan of 1818, (Based on ref. 15).
C. Plan as implemented by 1832, with proposal for new protective mole (Based on ref. 50).
D. Layout in 1835. (Based on ref. 51).

his geological map of Connemara for engraving and publication on a reliable base map, he might have established himself as one of the foremost field geologists of his day. But it seems that, having taken up the life of a practising engineer in 1811, he was not about to reverse that decision. Three years in the wettest, most remote and mountainous parts of Ireland had not cooled his ardour for the applied work of the surveyor and engineer. Now that he was also within sight of a commission that could establish his civil engineering credentials, he was not in any mood to turn away from Ireland. That year therefore, would prove to be a major turning point in his life. From then on, he was drawn increasingly into professional activities that he had little experience of but succeeded in mastering quickly. These included the drafting and promotion of parliamentary bills, the preparation of Grand Jury presentments and the provision of expert evidence to select committees of Parliament – all this on top of his main engineering activities. He would learn, too, that his plans and proposals would not necessarily receive the assent or approval of his clients or his brother engineers in all cases. Second opinions, alternative proposals, structural amendments and downright rejections were as much part of engineering consultancy in those early days as they have been ever since. We must remember, too, that at the time much of what needed to be done in Ireland was entirely new, or involved difficult, remote locations, or needed skills and tools not readily available. There were added difficulties also, created by the ethos of jobbing, nepotism, political partisanship and corruption that pervaded all public life. He was to be thrown, albeit of his own choice it must be stressed, into this milieu the instant his sojourn abroad ended.

On arrival back to Dublin in November, a letter awaited him from Peel. Rennie had reported and it appears he had some serious comments on the Dunmore plan. He wanted a more northerly orientation to the entrance and the whole harbour to be increased in size. Nimmo replied to Peel that he had originally shared Rennie's opinion prior to actually surveying the site, but had taken the course he did for reasons already given in the original plan. When he was returning from Scotland he had been delayed three days at Portpatrick, precisely because the mouth of that harbour '[was] not kept to seaward as I propose at Dunmore'.[12] However, he accepted that, as planned, the harbour was small, but it could be enlarged by extending the south mole. Wisely, he agreed with Rennie's advice that some amendments were in order

A work of the magnitude and expense of Dunmore required the consent of Parliament, and Nimmo helped in drawing up the heads of the necessary bill that was introduced into the House in 1815. Meanwhile, he continued his work on the mail coach roads in Mayo and Roscommon and prepared the maps already mentioned. Then he met Peel, who impressed on him the importance of the Dunmore proposal.[13] Nimmo saw the advantage of starting work without delay, if necessary leaving consideration of any extensions of the works until later. Peel appears to have urged him to start there and then, even though the Dunmore Harbour Bill had not yet received parliamentary approval. Ireland was in the grip of one of the famines that were to recur regularly throughout the first half of the century and, although the prevailing political mood was against intervention in the social conditions of the peasantry no matter how destitute or distressed, Peel was not entirely without compassion or economic sense. Public works that gave a wage enabling the poor to buy food was one temporary expedient that seemed appropriate under the circumstances. Works at Dunmore would be a valuable public enterprise in that district and would help to alleviate the worst effects of the distress in the south-east.

Nimmo commenced Dunmore harbour in September 1815, but changed the original plan because of what he had learned in France and Holland the previous year, and in the light of Rennie's comments and his own experience in Cork that summer. In Cork, he had noted that southerly gales raised the tide level by several feet and since winds from that quarter were the very ones that would affect Dunmore the most, he deemed it necessary to give a proportionately greater height and base to the moles. He informed the Postmaster General of this alteration. Because the equinoctial gales were soon to be upon him, forcing temporary cessation of the work, he went ahead without awaiting approval and with some economic justification: the prevailing low price of stone offered an opportunity for considerable savings that he could use to increase the capacity and security of the harbour. His presumption that the alterations would be approved was disastrous: they contributed to the bill's failure in Parliament, compromising and embarrassing Peel – not something that one would lightly wish to do to any influential patron, least of all Peel. Nimmo wrote to him from Galway on 10 March 1816 in explanation of his behaviour: 'As I may appear to have compromised you by making a second application to Parliament necessary may I state circumstances in extenuation of my conduct.'[14] He explained that in his original plan he 'did not venture to deviate from Jessop at Troon, Telford at Aberdeen and Rennie at Plymouth ... I am inclined to think that these engineers would have given a design similar to my own'. He went on to explain why he made the alterations: 'I confess this [increasing the batter of the piers] was without order but I saw it was necessary to adopt the Dutch principle of an inclined plane to the sea. ... And I should imagine no engineer ... could act otherwise.' He now submitted a new plan 'for rendering this harbour more complete, thinking it a pity so much money be expended to shelter Packets only, when other important national purposes may be easily obtained by means of the late improvements in this part of hydraulic architecture'. He finished on a confident note: 'I am ready at any time to have my works inspected by professional men' (something that would happen much sooner than he expected). Already his mind was on 'important national purposes', something that would stay with him all his professional life.

The new plan was quite a change from the original, if Nimmo's map of

1818 is a true representation of his ideas in 1816.[15] Gone were all of the works on Eilan na Glioch, so there would be no north pier. The south pier was extended outwards to about 470 feet from the shore. The final, angled section (containing the lighthouse) was also increased to 150 feet. The net effect was to increase greatly the capacity of the harbour and the width of its opening. The head of the south pier was now outside the point of Eilan na Glioch so that the harbour mouth was oriented more to the north, as Rennie had recommended. Both plans are shown in figure 6.1 (A and B) and the degree of alteration is readily apparent, the original structure looking quite feeble compared with the new. Clearly the 1814 plan had been indicative only, although it was produced following an actual survey, and the 1818 plan had obviously benefited from further on-site study and the comments of Rennie. When we remember that the original proposal was the first-ever harbour designed by Nimmo, who had no known practical experience of such work until then, it is hardly surprising that extensive revision had proved necessary. Indeed, Peel appears to have favoured Nimmo for Dunmore while he was still an unknown quantity in structural engineering terms. Peel favoured 'men of science' and he may have seen in Nimmo one of the new generation of those who applied science to practical problems. For his part, Nimmo was glad to be so favoured and was not beyond a little flattery when he wrote to Peel: 'I am more emboldened ... knowing that I address not only an enlightened statesman but a man of science.'[16] Peel must have been placated by the explanation and the new plan and perhaps the flattery: the work at Dunmore continued and the project remained alive with the Government. As a work for the relief of distress through employment of the poor, it probably had its political uses also.

Concerned lest his earlier actions might still be compromising the project, Nimmo wrote again to Peel on 1 October 1816, enquiring 'whether the Government formed any resolution regarding the completion of Dunmore Harbour'.[17] This must have precipitated a positive response because when Nimmo wrote again on 1 April 1817, he could state: 'I have recently sent heads of Bill to Secretary Lees regarding Dunmore. The bell for Dunmore is nearly ready and about fifty feet of our deep water wharf has been laid in eight and ten feet of water during the spring.'[18] He had requested the use of the Howth diving bell the previous autumn, but it was not in working order.[19] He therefore had to construct his own bell for Dunmore, which John Aird called 'Nimmo and Price's Bell'.[20] This had a new air supply system, a task Nimmo had worked on while he was in Britain in 1818. His assistant in this was Henry H. Price (1794–1839), of Neath Abbey Iron Works near Swansea. According to Chrimes, Price worked on inland navigation on the River Suir and the River Lee, and 'It seems likely he was working for Alexander Nimmo who later gave him some of his books.'[21] Meanwhile, Aird was keeping Rennie informed of progress at Dunmore and he sent a newspaper cutting about Nimmo and the diving bell:

Waterford, Dec. 2nd [1817]

Yesterday the first experiment was made of a diving bell at Dunmore, under the inspection of the superintending engineer Mr Nimmo and the acting engineer Mr McGill. These gentlemen, and several others, went down in the bell and it completely answered all their expectations. There are two bells of equal internal dimensions, sufficient to supply a man with air for 24 hours. Weights: – one of 5½ tons, one of 4½ tons. Inside 6½ x 4½ x 5½ high; capable of six persons comfortably. Calculated to admit of the entry/departure of the diver independent of the people outside, and are otherwise improved from the common bells used in hydraulic architecture. There is not the slightest danger in going down in these bells, for their construction is upon principles of absolute safety. Their operations will commence immediately. – Chronicle.[22]

Now that the Dunmore bill was under parliamentary consideration again, the Irish authorities took a more active interest in the progress of the works. The Post Master General drew up regulations specifying that Nimmo should visit the works every two months and make regular reports.[23] He should be paid three guineas a day when engaged, and one shilling travelling expenses for every mile. By March 1818 the Customs authorities wanted to know when Dunmore would be ready for packets and Nimmo reported that the works would be sufficiently advanced for their accommodation by 1 May.[24] However, everything was not quite ready: 'The berths ultimately intended will not indeed be ready by that time, but there are already good moorings laid in the Harbour, 200 feet of quay having 8 to 10 feet at low water, finished with rings etc. – and the safety of the place has been evinced by vessels riding there during the late gales while some of the Packets hove at Passage.'[25] By now, Peel was anxious to progress the matter through the House and Lees urged Nimmo to forward the draft of the bill that he was working on.[26] Peel instructed Lees that he did not wish the new estimate to exceed the old;[27] Lees was also concerned at mounting costs, so he wrote to Nimmo at Dunmore instructing him to wait on Peel without delay when he (Nimmo) was next in London.[28] Nimmo did as requested and stayed on in London for almost three weeks until the bill had passed, 'being required to give evidence thereon'.[29]

The bill duly passed on 3 June 1818, just before Peel left office as chief secretary.[30] Among its provisions, it appointed a commission to oversee the completion of the harbour and, separately, a secretary to carry on its day-to-day administration; Lees filled this latter post. The commissioners maintained Nimmo as superintending engineer, he and the commissioners remaining in office until replaced by the new Commissioners for Public Works in 1831. A levy was placed on all boats entering Dunmore; those arriving and not subsequently reporting at the custom house in Waterford were specifically exempted from the provisions of a local Act that levied dues on boats entering Waterford city docks.[31]

The ink on the Dunmore Act was hardly dry before petitions were made to change the new harbour. In December the President of the Chamber of Commerce of Waterford submitted a memorial to the Post Office requesting that the angle on the south pier be eliminated, the pier to be 'thrown out in a straight forward way'. Lees forwarded the memorial to Nimmo for his comments.[32] If Nimmo's opinion should not be favourable to the proposal, he was to state what alterations he would consider best to meet the wishes of the merchants of Waterford, and give an estimate of costs, so that the Post Master General could decide upon the merits of the memorial. Charles Grant, the new chief secretary, also contacted Nimmo seeking the current estimate for completion of the existing works and the estimated extra sum that would be needed for the proposal in the memorial.[33] Nimmo replied in January 1819 (rather dismissively it seems: his estimate was written on the back of the letter he had received from Grant and now forwarded to Lees), indicating that £37,000 had already been spent and the estimate for completion that year was £8,000, making a grand total of £45,000. If the south pier were to be extended as proposed, the total cost to completion would increase to £72,000, with £12,000 of that needed in 1819. By March, Grant had decided to ask Parliament for the larger sum of £12,000 for the works that year.[34] In consequence, the south pier was extended outwards to a final length of 550 feet, and the angled section was abandoned. The lighthouse was relocated to the end of the extended straight pier (figure 6.1 (C)).

Lees was growing increasingly worried at Nimmo's expenditure. He wrote to him in February 1819 regarding the high expenses that Nimmo had claimed for the trip to London to meet Peel.[35] Lees had understood that Nimmo was going to London anyway on his own private business and he (Lees) had only asked him to call on Peel while he was there. Nimmo's expenses for 1818 had formed a very great part of the grant for Dunmore for that year, Lees complained. Nimmo was trenchant in his own defence: he always charged one shilling per mile, he said, and he had charged the extra time in London – where he had been requested by Sir George Hill to stay for the passage of the bill – to the Dunmore account. That charge should properly be considered as part of the expense of obtaining the act. He went on: 'Regarding Dunmore, I should feel entitled to 2½% if paid by percentage and by examining my accounts you will see that they by no means amount to that proportion. The works themselves required a large amount of my time. When in London I made it as useful as possible to Dunmore by constructing the new air pump and lamps for the temporary lights for that work.'[36] Lees was not entirely convinced: 'Your amount cannot properly speaking be discharged out of the Dunmore account and unless you procure an order from Sir George Hill, the Post Master General would not be authorised in advancing any pay-out of the fund.'[37]

During the summer of 1820 the rough mole of the south pier was completed almost to its full extent, and the subsequent winter storms bought it to the slope and firmness that meant the masonry pavement could be applied. By then the authorities were seriously concerned about the escalation of costs. William Gregory wrote peremptorily to Nimmo from Dublin Castle: 'I am commanded by the Lord Lieutenant to desire that you will, without delay, transmit to me in duplicate, a detailed report on Dunmore Harbour, its origin, history, estimates, money actually expended, total money yet to be expended and probable advantages, in order that it may be laid before the House of Commons.'[38] Aware of this letter, Lees thought it prudent to cover the position of the Post Office in the matter by seeking independent advice from an acknowledged expert. He therefore requested Telford to inspect and report on the harbour before committing any more money to it. Telford visited Dunmore in February 1821 and reported that the works had been carried out to a very considerable extent into water thirty feet deep at low water springs, and 'they have been performed in a manner which reflects equal credit on those who planned, and those who conducted and more immediately superintended the sundry operations'.[39] For Nimmo he had special praise: 'The performances, by means of the improved diving machine, contrived by the able engineer, Mr Nimmo, will be an excellent model for similar works where a great depth of water is required, or where rocks are to be removed under water.' He did not stint either in praise of Mr McGill, the resident engineer: '[who] has brought the diving bell operations in great depths, to a degree of simplicity, safety and celerity, which is scarcely credible; the workmen actually preferring them to the works above high water mark.' Even the financial aspects of the project received a clean bill of health: 'All that relates to payments and pecuniary concerns have been equally well conducted by Mr Robert Grubb, under the direction of the General Post Office.' No more extensions were necessary and he recommended that the works in hand be completed with all practicable dispatch. He finished with a flourish:

> I cannot close this Report without saying, that when the works have been completed as here recommended, a very valuable station will have been acquired in a very important point of the coast; and this, whether we regard the port of Waterford, and the fertile districts to the north of and west of it, the occasional protection to the revenue cruisers and general coastal traders, or to the connection with the south western parts of England. The accommodation will also be sufficiently extensive for all that can be at present contemplated; and if increased prosperity requires more, the present termination offers no obstruction to extending the works to meet the greatest possible demands.[39]

Whatever about the Post Master General, Nimmo could hardly have hoped for a more positive recommendation from the leading engineer of the day. Ten days after Telford's report, Nimmo reported confidently that the pier could be completed in its proposed dimensions within the estimated sum of £42,500. Since two grants of £12,000 each had already been voted by Parliament, a balance of £18,500 was outstanding, £10,000 of which would be needed in the 1821 season.[40] The works were substantially completed by 1822 and the packet service from Milford came into full operation.

A contemporary account of the works illustrates just how massive they must have been:

> this pier, or quay, is 600 feet in length: the depth of low water at the entrance is twenty-five feet, and at the innermost part eighteen feet. The greatest part of this noble quay under low water has been built by means of the diving bell, of which useful machines there are two here, on very improved principles. Under the superintendence of skilful engineers, the workmen (untaught peasants) soon learned to move vast rocks with admirable dexterity: few of these were less than five or six tons weight, with some exceeding ten tons. These immense mountain masses, torn from the solid rock, were transported with apparent ease, on inclined planes and iron railways, to the place where they were squared with immense exactness; they were then disposed in their places, accurately fitted and joined together without the clumsy iron bolts and bands, which are at the same time laborious and expensive.[41]

From the reports it seems that the diving bells at Dunmore were used in deeper water than was common elsewhere, which is what excited Telford's high praise for them. One was used later in Limerick for the foundations of the Wellesley Bridge, when William Rowan Hamilton went down in it, remarking on the painful effects on him of the compression. When a new packet station was built at Pembroke in Wales in 1828, two diving bells were used that were almost certainly Nimmo's. They were fully described in words and pictures by the supervising engineer at Pembroke and they meet the dimensions of Nimmo's bells perfectly.[42] Since Nimmo had contributed to the design of the Pembroke packet station, the conclusion that the bells used there were his is even more likely.

Nimmo remained involved with Dunmore for another decade, submitting annual reports on the works and their maintenance. In November 1828 the Harbour Master, Mr Anthony, recommended that the slanting slip inside the harbour be levelled to the height of the rest of the quay to create an additional berth for packets. Nimmo was against this.[43] In his opinion, alteration would be injurious, as the slanting pier lessened the rebound of the sea in what was essentially a small rock- and wall-bound harbour. Nevertheless, the harbour was deteriorating under the pounding of the seas, as a letter to Lees (by then Sir Edward Lees) written by the Harbour Master in 1831 describes:

> My dear Sir Edward,
>
> I am in total ignorance of what steps are adopted regarding Dunmore. The unfortunate few poor people retained to prevent the destruction of the works, have not been paid one penny and are in a state of destitution and I know not with whom to communicate except yourself, as I believe you are the only official person looked to by Government, being the appointed secretary to Dunmore Harbour.
>
> Shall I entreat your attention to put me in the way of knowing something about it, and I am sure you will see the absolute necessity of urging the immediate attention of the higher powers in authority when you know that I much fear the destruction of the lighthouse and end of the pier on next storm if urgent and quick measures be not resorted to for its protection. There is not any time to be lost. Do be good enough to stir in this matter. Yours etc., Robert Grubb. 27 Sept 1831.[44]

Three weeks later, on 15 October 1831, the new act setting up Commissioners for Public Works was enacted and responsibility for Dunmore passed immediately to them.[45] Henry Paine wrote a letter on their behalf addressed 'to Mr Nimmo or his principal assistants' on 8 November, informing them that the Viceroy had referred Dunmore harbour to the commissioners and requesting that Nimmo send all maps and plans to the commissioners' office.[46] The letter was addressed in that way because news of Nimmo's precarious health (he was suffering his final illness) was widespread. Nevertheless, Nimmo dictated a reply in which he pointed out that he had dispatched all that they now requested to Sir Edward in 1827 and they should look there for everything. He directed that the plan of the harbour hanging in his office in Dublin (he was dictating in Liverpool) be left at Paine's office, but he wanted it returned in due course. He finished: 'I have not had a final settlement of my account for professional services on this work on which a considerable balance is due to me.'[47] He would be dead within two months of this missive. As for payment, he had received £210 in salary in 1820,[48] and after his death his executors claimed £516 for his services at Dunmore from 1821.[49] Dunmore may have added to his engineering reputation, but it appears to have added little enough to his pocket.

Dunmore's importance as a harbour for packets and trade was under threat almost before it was finished. The introduction of steamers on the packet service changed its importance entirely. The first steam packets were paddle-steamers of relatively low power, which continued to terminate at Dunmore, especially when conditions were adverse. However, as they became more powerful, it was inevitable that they would find the upstream passage easy in all conditions and the packet station eventually transferred to Waterford with its extensive quays, its concentration of merchants and its mail coach connections to points west and north.

FIGURE 6.2
The pointed arch bridge to Eilan na Glioch as shown in a photograph from the Lawrence collection. (Detail. Reproduced with permission of the National Library of Ireland.)

Nimmo was barely dead when the Office of Public Works commissioned George Halpin, his erstwhile apprentice, to examine the harbour in 1832. He proposed to widen the sea pavement and the base of the pier to improve its protection from heavy seas – exactly the kind of extra protection that Nimmo had endeavoured to incorporate after his Dutch visit. Halpin outlined his proposals on a copy of Nimmo's map of the harbour as it then was.[50] A railway ran from the quarries west of the village down to and along the sea side of the packet quay, between the new coal stores and the parapet (figure 6.1(C)). Older residents of the area still remember today the cutting through which the line of this track ran, calling it 'the metal road', even though the track has long ago disappeared. A sketch of the harbour in the third Report of the Board of Public Works shows the works at that time (figure 6.1 (D)).[51] By then the sea pavement at the base of the jetty had been widened and backed by a new rubble mole. The railway had been diverted to run along this new mole, supplying the stone necessary for its construction. On the north side of the harbour a road from near the Commissioners' Buildings led on to Eilan na Glioch, crossing the deep chasm between it and the mainland by a bridge, but there was no pier or jetty on the island. However, the presence there of a connecting bridge at this very early date suggests that Nimmo had erected it. The very first (1814) plan had envisaged this bridge at an estimated cost of £500: 'a bridge will also be necessary into Eilan na Glioch'.[52] A photograph in the Lawrence collection (figure 6.2) shows it as a fine,

FIGURE 6.3
View along Steamer Quay to Nimmo's lighthouse at Dunmore.

pointed-arch structure, reminiscent of Shaughnessy's Bridge in Connemara or Poulaphuca Bridge in Wicklow, both attributed here to Nimmo (chapter 11). It was destroyed in the alterations that were made to the area in the 1960s, when Eilan na Glioch was extensively built over and converted into a quay and dock apron, not entirely dissimilar to Nimmo's original plan. The rough mole that currently extends at an angle from the head of the lighthouse pier (narrowing the sea mouth) was made some time before 1900 and follows Nimmo's original plan for an extension towards the north-west from that point.

Dunmore harbour changed from a packet station to a fishing port as the nineteenth century progressed, and after a series of modifications, rose to be the major fishing station of the south-east, a role for which it was never intended originally. Today, it is difficult to relate the existing chaotic port structures to Nimmo's original plans. Large amounts of rock have been excavated in the inner harbour and the slip and small dock of the earlier plans have been built over. However, the visual prospect along the main quay (the original Packet Quay) towards the fluted lighthouse rising majestically at its head, has a certain boulevard-like quality that is decidedly unusual in fishery harbours, recalling the grandeur of its original purpose (figure 6.3).

CORK RIVER AND HARBOUR

Initially, the port of Waterford had reason to be concerned about losing some of its general trade, as it would all of the southern mails, to the new packet station. As early as 1817, Waterford levied dues on boats arriving at Dunmore, much to Nimmo's annoyance.[53] Pre-emptive steps were obviously being taken to ensure, even before its completion, that the new harbour would cause no loss of revenue to the old city port. These fears were partly realised when the Dunmore Act specifically exempted boats entering at Dunmore from Waterford port dues. The air of concern may have spread as far as the port of Cork, which had extensive trade with British ports, being the chief outport for the whole of counties Cork and Kerry. Despite this, Cork had no floating docks, that is, ones where boats could lie afloat at low tide, such as those proposed for Dunmore. Larger or narrower boats entering Cork harbour came only as far as Cove (Cobh) or Passage, where their cargoes were transferred to lighters for onward transport to the city. Small boats, drawing less than eleven feet, could go up at neap tide as far as King's Quay near Blackrock, about two miles below the city, but very few could work up to the city itself except at high spring tides. This was due to the shallowness of the channel in the riverbed. The riverbanks were unformed and consisted of broad marshes that were inundated at high water. At the city, the quays were mostly imperfect masses of rubble stone on banks of mud and rubbish. Once there, boats had to lie aground at low water. The threat from Dunmore was therefore real enough to concern the authorities in the city, who were already trying to develop the port and to improve the navigation in the river. In May 1813 a local act had constituted the Mayor, the Sheriff and some prominent businessmen as harbour commissioners, granting them one-third of the dues of the Cork Butter Market to fund their activities.[54] A further local act was passed in July 1814, which added to the commission's revenue, this time by levying all boats using the port, excepting those carrying mails or coal.[55]

The commissioners met for the first time in September 1814 and on 1 October they appointed a sub-committee to investigate the state of the river and navigation as a prelude to 'deepening, widening and improving the Harbour and River of Cork', as directed by the acts.[56] The sub-committee examined the river down to Blackrock and reported two weeks later, proposing that the existing navigation wall (still called 'the new wall' at the time) should be extended and repaired at the gap in it near Lapps Island. The navigation was, it concluded, 'a disgrace to the city and ruinous of the harbour', because of the accumulation of ballast and rubbish, and the quays were in a dilapidated state for lack of piling.[57]

At their meeting on 5 April 1815, Gerard Callaghan reported that he had 'availed himself of the presence of the celebrated engineer Mr Nimmo to obtain a survey of the River of Cork and suggestions for its improvement'.

The commissioners resolved without hesitation 'that this Board do highly approve of the same and will defray the expense thereof, Mr Nimmo having stated that his demand would be three guineas a day for himself and one guinea per day for his surveyors, provided the same shall not exceed one hundred guineas; but it is requested that Mr Callaghan will please to enquire more particularly of Mr Nimmo what his terms may be'.[58]

On 10 June Nimmo replied and the commissioners immediately formed another sub-committee to take the necessary steps to progress the matter. Starting without delay, Nimmo commenced a hydrographic survey of the navigation from the city as far as Passage and drew a chart, at a scale of twenty inches to one Irish mile, on which the banks, shoals, channels and depths, together with the locations of properties on both shores, were delineated. He personally supervised the trigonometrical observations and the soundings, so that he could guarantee the accuracy of the work. Unfortunately, no copy of this unique and remarkable chart has come to light to date, but the report that accompanied it, dated 11 September 1815, has survived.[59] Figure 6.4 (A), based loosely on Beaufort's map of Cork in 1801 and on Nimmo's written information, gives a general idea of the state of the river and harbour below Lapp's Island as he found them.[60]

Cork city is, in Nimmo's words, embosomed by the River Lee, the north and south branches of which meet at the downstream tip of Lapp's Island. The main (north) branch of the river was then navigable up to St Patrick's Bridge. Below that, the river was shallow but deepened to four feet along Custom-House Quay. On both banks, the channel was bordered by a slab or bank of mud of varying dimensions and consistency. That at Merchants' Quay was forty yards wide in places – did he mean forty feet? – and none of the quays or docks lay directly in the channel. It would be necessary to excavate and remove 9,000 cubic yards of mud and rubbish to give three feet depth of water over low tide alongside these quays. Underpinning and piling Merchant's Quay would cost £319, and £384 would be needed for St Patrick's Quay opposite.

The south branch of the river was accessible by boats as far as Parliament Bridge, but the quays were still largely unformed. Below Morrison's Island, for example, there was no quay adjoining what is now the end of the South Mall; to build a quay there would cost £375. On the opposite bank below George's Quay, the 450 yards of riverfront had no quays at all; to make and perfect quays there would require over £2,000. Clearing the slab and rubbish to give three feet of water in the south channel would take a further £1,740, or £2,740 to give four feet up to Morrison's Island.

Around 1765, a wall, called the navigation, or new, wall had been erected in the waterway to divert the united streams towards the north bank. It extended from opposite Lapp's Island to about a mile downstream, creating a navigation channel between it and the north bank of the river for boats to pass up to the quays. Its construction had altered the original flow of the river, diverting the south branch away from its natural course, thereby causing the old, wide south channel below the city to become occluded and useless for navigation. The navigation wall had been intentionally breached opposite the custom-house, so that some water from the south branch of the river could continue to enter the old south channel and flow through the adjacent marshes; the rising tide also flowed up the old south channel, so that its slob land was entirely under water at each high tide, as indicated in figure 6.4 (A). Thomas Soutell Roberts's watercolour, *View of Cork City* (Plate 4[61]) shows the wall dividing the approach to the city, illustrating the remarkably large expanse of water to the right (south channel) and the much narrower navigation channel to the left (north channel).

Nimmo was able to propose an instant solution to some of the port's problems, based on the existing wall and the old south channel. If an embankment were built to join the sea end of the wall to the south shore, but built low enough that it would be over-topped at high tide, then the sea pouring into the enclosure at high tide would fall on the ebb into the river through the breach near the custom-house. That would provide a powerful scour, occurring at the ebb twice a day, keeping the river channel clear below the breach. Alternatively, by completely closing off the sea end of the south channel with a much higher embankment, the tide could ebb and flow into the south channel only through the breach. The flood of tide coming up the navigation channel would be vastly increased (the enclosure forming a 250-acre basin that would be filled each tidal cycle, requiring more water than otherwise to be drawn up the navigation channel), thus materially assisting the navigation. On the ebb, the draining of the enclosure through the breach would have a highly beneficial scouring effect on the north channel. Neither proposal could ever have been anything but a temporary and partial expedient; the south basin would eventually fill with silt and the benefit would disappear. But they certainly warranted some consideration, if the commissioners wanted a quick solution. If the old south channel could not be put to good use like this, Nimmo said it would be just as well to reclaim it completely.

To extend the wall as far as the King's Quay near Blackrock would cost over £22,000, a sum Nimmo did not think was warranted at the time, although it would result in the enclosure and reclamation of over 300 acres. On the other hand, running a dyke from the existing end of the wall over to Chiplee on the south shore would cost only £6,000, enclosing about 182 acres. Later on, the enclosure could be extended to Clifton, gaining a further eighty acres, but he would not recommend making the final extension to the King's Quay (enclosing the final forty-four acres) until he saw the effect of the reduction of the old south channel backwater on water depth at King's Quay (figure 6.4 (B)).

Below Lapp's Island the united streams flowed mainly between the navigation wall and the north bank. When the water retreating from the old south channel joined it at the sea end of the wall, the depth increased to about six feet and remained so until the King's Quay. Below that, the set of the ebb across the channel and the prevailing westerly winds caused many vessels to go aground on the north-east shoal between Tivoli and Dunkettle. He therefore proposed to build a new embankment, this one along the north shore, commencing at Woodhill and continuing as far as Tivoli. Material dredged from the channel and shoal could be used to strengthen this embankment, which would be faced with rubble stone, battered to one-half its height. At first the foot of this wall would need to be secured by small piles and wattling, or by large baskets filled with stones. He estimated the cost at almost £4,000, calculated at four pounds ten shillings per running yard.[62] The likely total cost of all the works from Lapp's Island to Blackrock was estimated at almost £20,000.

Below King's Quay, the water maintained a depth of eight to eleven feet until it reached the baths at Blackrock. After that, the current swept around Blackrock Point and the channel widened into Lough Mahon, an expanse of water 2,000 yards wide and 5,000 yards long, causing the river to slow and lose its power to maintain a definite and distinct channel. This was the greatest problem besetting the navigation to Cork, since it caused the lough bed to be thrown into numerous shallows and shoals. By boring all over it, Nimmo determined that the shoals were composed, not of rocky outcrops, but of deposits of gravel and mud, their locations reflecting the varying influences of the river and tides. A narrow channel on the west side of the lough offered the best way of securing a passage through the shoals without huge expenditure. To open this up it would be necessary to excavate a shoal 300 yards long, using the excavated material to build an embankment on the lough's western shore that might eventually be suitable as a towpath. Later, the embankment could be extended by degrees to the Douglas River, and thereafter over to the south shore near Rochestown, letting off the waters of the Douglas creek through a sluice and enclosing another large area of land (figure 6.4 (B)). All that, including the dredging of the channel, would cost £30,000, and £18,000 more if it were decided to extend the dredged channel further down the harbour.[63] The estuary would then permit ships of twenty-foot draft to come up to the King's Quay.

Nimmo was wont to take the broad view, and he did so in Cork's case also. Not content with what he had already recommended, he went on with a 'Proposal for a System of Wet Docks at the City with a Canal for Large Vessels to the Deep Water', which he considered would be a radical cure for all the problems of the existing river navigation.[64] He envisaged two floating docks that could be constructed in series as funds permitted. The first, upper dock was to be formed at the back of the navigation wall opposite the new customhouse. It would be 200 yards long and 110 yards wide, covering four and a half acres. This would provide 600 yards of quay fifty feet wide and would hold forty-five boats of over 200 tons in double-tier alongside. The quay would be surrounded by an enclosure of warehouses fifty feet wide, the backs of which would form a high wall so that the whole dock could be secured from outside interference. He envisaged a light iron bridge, of three arches, to communicate with the custom-house on Lapp's Island; the arch nearest the dock would open by turning on a pivot, so that boat access upstream to the south branch of the river would be assured as needed. At other times, when the bridge arch was closed, the dock and the custom-house would be part of a single enclosure and all goods moving from the warehouses to the city could pass through it.

East of this dock he proposed a second dock. This was to be much larger – 1,120 by 350 feet – enclosing nine acres and capable of holding up to ninety sail. He pre-empted any objections on the grounds of scale by pointing out that the area of the two docks would be no greater than that of the existing quays. If funds were short, the smaller upper dock could be made first. Later, the second dock and its associated works could be made by degrees without interfering with the first, or obstructing the navigation. The two could then be joined, with each also having independent locks into the river. From the second dock, he proposed that a canal, 110 feet wide at the surface, fifteen feet deep and 2,300 yards long, should pass down alongside the river, entering it near the King's Quay through a lock forty feet wide and 150 feet in the chamber. The southern quay of the lower dock could, if required, be extended along the side of the canal, thereby increasing the area of floating dock made available.

Fully implemented, his scheme would permit the passage in and out of one hundred sail during two hours before and after high water without significantly lowering the surface level in the docks or canal. Boats drawing fifteen feet could pass all the way to the docks, answering every purpose that could be looked for. Surrounding the docks, Nimmo laid out a regular plan of streets and buildings. One main street, 100 feet wide, was to skirt the docks and connect with a new road then being formed across the marsh. From it, a branch would run between the two docks to the sea wall with an iron swivel bridge over the communication between them. Various cross streets, forty feet wide, were envisaged, with back lanes thirty feet wide.

Unfortunately, no sketch of Nimmo's plan has been unearthed to date and we can only conjecture from his written description what the full layout was really like; one indicative interpretation is shown in figure 6.4 (B). The estimated cost of the floating docks and canal was £69,000. He was sanguine that 'liberal aid will readily be granted towards carrying so beneficial a scheme into execution'. It was not inordinately expensive compared with the estimate for the works at Dunmore, and, in addition, it would have constituted a significant source of employment for the poor of Cork, who were suffering famine and hardship in 1815 and 1817.

Nimmo appeared before the commissioners with his map and plans on 11 September. They approved highly of his map and, having listened to his report, ordered that 200 copies be printed at their expense to be distributed as a future

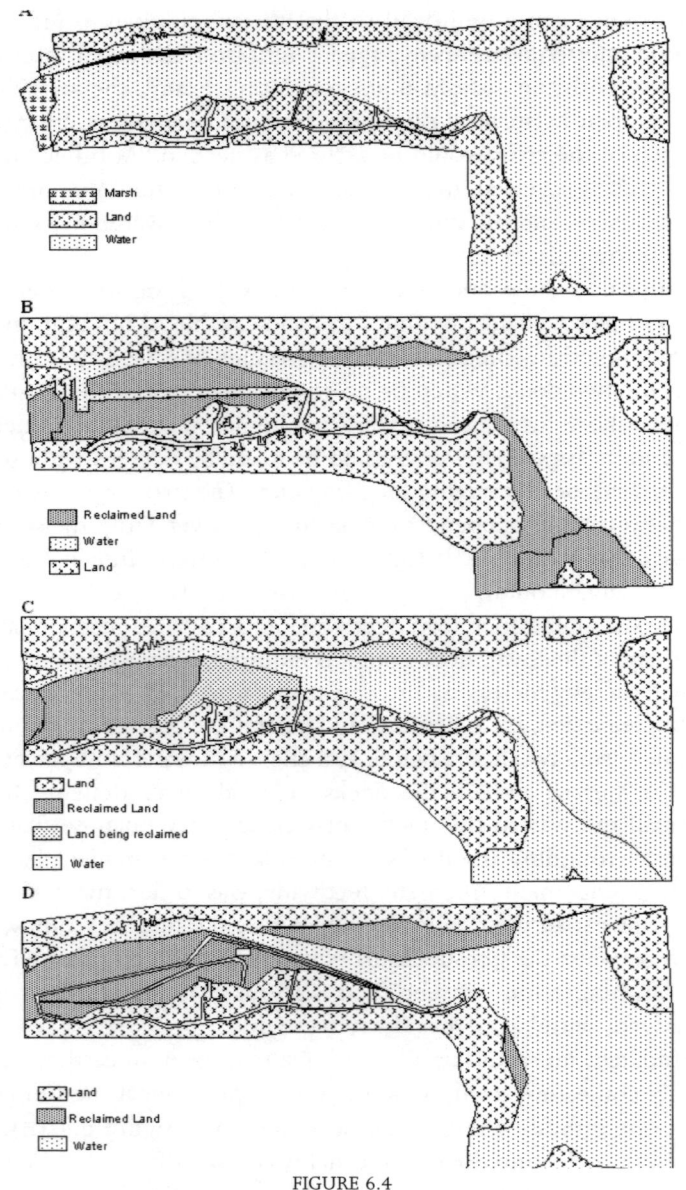

FIGURE 6.4
Sketch of the River Lee and Inner Cork Harbour.
A. As it was in 1814. Note the Navigation Wall extending downriver from Lapp's Island, close to the north shore and the wide area of water to its south. (Based on Beaufort's map of Cork, 1801, ref.60, and Nimmo's description in ref. 59).
B. Nimmo's complete plan. Note the reclamation of the area south of the Navigation Wall and the new canal through the reclaimed land to the proposed floating docks near Lapp's Quay. Note also his proposal to reclaim the slob at Tivoli and from Blackrock to Rochestown. (Based on Nimmo's report and ship canal proposal, 1815, ref. 59).
C. The progress of reclamation in 1843. The section to Chiplee was already dry; the next section, to Clifton was still under reclamation. (Based on Wolfe and Church's survey of 1843, ref. 80.)
D. The area today.

board should think most advisable.⁶⁵ When Robert Porter, who had worked with Nimmo, offered to make copies of the map on a reduced scale the commissioners approved, paying him £20 for the copies, one for the House of Commons and one for the clerk who was to draw up any application that might be made to Parliament for funds for the works. Nimmo presented his bill for £201-7-6, which the treasurer was directed to pay.⁶⁶ Meeting again on 11 November, the commissioners resolved that it would be highly beneficial to extend the 'new wall' all the way to King's Quay and to have an embankment wall built from Tivoli Quay along the north shore down to the northeast shoal. Yet another sub-committee was formed to put this resolution into effect.⁶⁷ By December, the mayor, Mr Newson, and Sir Nicholas Colthurst, MP for Cork, were agreed on applying for a grant for the improvements, and on a deputation to bring the proposal in person to Peel and Vesey Fitzgerald, the chancellor, in Dublin.⁶⁸ The deputation was a joint venture between themselves and the Council of the City of Cork, and in due course a memorial agreed by both bodies was drawn up and approved.⁶⁹

The memorial sought ten thousand pounds for the extension of the navigation wall only, omitting all the other improvements recommended by Nimmo. That was unfortunate: when the deputation met him in February 1816, Peel was able to say that the proposal was too confined in its effect, too local, to warrant placing before the Imperial Parliament, but that he would act on the advice of the Lord-Lieutenant. When the deputation met the Viceroy he claimed that while he 'would be glad to do what he could for Cork', having spoken earlier with Peel he supported Peel's view. So did Vesey Fitzgerald, who said the proposal was not expedient.⁷⁰ This rejection was a bitter disappointment, made especially galling when it came to attention that Waterford city had been given a grant for its port improvements. Waterford's successful application was not, in their view and that of their MP, superior to that of Cork, which had failed. Peel wrote explaining that the Waterford case was recommended by the Board of Inland Navigation whereas Cork had had no such support.

Things were made no better three months later. On 18 June 1816 the harbour commissioners met, 'for purpose of taking into consideration the best mode of appropriating part of their funds in such useful work (consistent with the intentions of the acts) as may give employment to a proportion of the labouring poor in this time of distress'.⁷¹ The failed memorial had stressed that their plans would give just such employment as they were now humanely proposing. It seems they believed they had been harshly treated by the authorities and they resolved to commence, from their own resources, some of the more urgent aspects of the work, such as the repair of the gap in the wall, the extension to King's Quay and the reconstruction of Merchants' Quay. The latter had recently collapsed into the river and they accepted the tender of William Hooper to make the necessary repairs.⁷² They were hopeful also that the Council of the City of Cork would contribute to the repair of the navigation wall, and maybe even its extension.

After much correspondence, the Council of Cork agreed in 1817 to repair the gap and also to fund the extension of the wall, on condition that the commissioners relinquished all claim to property in the wall.[73] The commissioners agreed, and resolved to send another request to the Lord-Lieutenant seeking an advance of ten thousand pounds, this time under the Loan Act of 1817, mortgaged on the security of their income under the Weigh House Act (the short name for 53 Geo III cap. lxx). This time, the loan was for the extension of the navigation wall and for various other improvements, including the uniform rebuilding of all of the city quays. The mayor wrote the application on 22 May, emphasising that the works would provide employment to numerous and distressed poor.[74] In September the Loan Commissioners enquired as to the number of poor persons who would be employed if the loan were granted? Four hundred, for as long as the loan lasted, was the reply from Cork.[75] The application was rejected by the Loan Commissioners in December. Undaunted, the Cork petitioners wrote back asking for five thousand pounds.[76] How could that sum not be granted? The new response from the Loan Commissioners was not quite what was expected in Cork:

> plans and estimates for the proposed works should be prepared to be laid before the Loan Commissioners' engineer who should report thereon; the engineer should come from Dublin to inspect places where intended works were to be carried out; the [Port] Board was to pay him three guineas a day; if his report were positive the Loan Commissioners would advance five thousand pounds as requested.

Amazingly, since the offer represented success at the third attempt, the Port Commissioners turned it down: 'the Board having taken this letter into consideration', they reported, 'however anxious they are to give increased employment to the poor, yet under all the circumstances stated, they think it better at present to confine themselves to the application of their own funds, so far as they will extend, towards the objects for which they are empowered'.[77]

Was this just pique, or the normal Cork reluctance to submit to supervision from Dublin, especially in a matter entirely local? Or could it be that, knowing Nimmo was an engineer with the Loan Commissioners, they did not wish to have him further involved? They had, after all, given the contract to design and construct uniform quays in the city to Mr Thomas Deane, a local architect, and the contract to clear the river to Mr Porter (Nimmo's assistant on the harbour survey), while at the same time they ignored Nimmo's more grandiose scheme for the floating docks and canal.[78] It might not, therefore, have suited them for Nimmo to be the one to adjudicate their rather limited ambitions for the navigation wall. The Loan Commissioners' other engineer, John Killaly, would have been no better for Cork: he was also the engineer of the Board of Inland Navigation that had supported Waterford's earlier successful application, probably because of Waterford's centrality to the Suir and Barrow navigations, which they controlled. So for Cork it really was Hobson's choice between these two engineers!

Had Nimmo not put forward the more elaborate part of his scheme in the first place, he might have had a greater influence on the Port Commissioners' plans, and his name might be better known and respected in the southern capital than it is today. On the other hand, had the commissioners been prescient enough to implement them, Cork could have rivalled Liverpool as a transatlantic steamer terminus, with who knows what benefits. Nimmo's recommendations for the quays, the river branches and the navigation (excluding everything else) would have cost about £40,000, no small sum in 1815, but not entirely impossible either. His recommendations for the north and south branches of the river within the city were carried into effect later by a Mr Foley (working to plans of Mr Deane) and Mr Porter, whether in accordance with Nimmo's ideas or simply because they were the obvious answer to the city's needs. Extensive as they may appear, Nimmo's main recommendations for the river were, in reality, just aspects of embanking, reclamation and dredging that were commonly implemented in many coastal locations. According to Marmion, in 1826 the Port Commissioners purchased a dredger of twelve horsepower that had previously been employed in Waterford harbour and with it removed 183,000 tons of mud and gravel from the river.[79]

The embankment to Chiplee, enclosing an area of 307 acres, was completed by 1836 and the intake was subsequently developed into a public park. Later on, after a period as a race track, the park would become the site of Ford's, Dunlop's and other industrial plants. The area from the Chiplee embankment down to Clifton embankment was still slob land in 1843 but was completely reclaimed later (Figure 6.4 (C)). Today it is the location of the Agricultural Society showgrounds and the Páirc Uí Caoimh stadium. In Cork sporting parlance, 'down the park' is often used to refer to this stadium, or to its immediate predecessor on the site, the Athletic Grounds; the expression is, in fact, much older and derives from the original public park further east. The expression has now, understandably of course, been taken over for the modern stadium. The Atlantic Pond alongside it is a remnant of the old south channel (as is the small drainage ditch that runs along Monahan's Road to the Pond) and it drains through a (modern) sluice at its eastern end and eventually to the sea through the water treatment plant below it. The modern sluice from the Pond is located near the site of the rude causeway shown in the Wolfe and Church map,[80] and the Pond itself occupies the area marked 'gravel and mud' on that map. These enclosures are exactly the ones that Nimmo had proposed almost thirty years earlier. The final intake, from Clifton to King's Quay, came later again and a roadway for carriages, now known as the Marina, was eventually constructed along the general line of the old navigation wall. A later, final, extension carried the line to the small cove at

Blackrock. The King's Quay has been built over, its precise site, and remnants of the quay itself, being located in the gardens of some houses at the locality. The revenue house that stood at the Revenue Wharf still stands occupied. The small lane that ran from Blackrock Road to the King's Quay, formerly called King's Quay Lane, or simply Quay (pronounced 'kay') Lane, remains, but is now gentrified by the name Church Avenue.

Nimmo's recommendations therefore did not go entirely unheeded, but they were financially of little reward to him. Much of the reclamation and embankment that he proposed for the north shore of the river down to and well beyond Tivoli came much later, some only within the lifetime of many of today's readers. Many, too, will remember the sub-division of the slab below Tivoli into reclamation compartments by wattling fences before the container port and industrial centre were constructed there, on what Nimmo called 'the north-east shoal'. The western shore of Lough Mahon has also been embanked around Mahon Point and partly up the Douglas estuary, as he proposed. However, the embankment was never carried over to the south shore of the lough, and therefore the Douglas estuary has not (yet) been reclaimed. Figure 6.4 (D) shows the area approximately as it is today. All the reclaimed land is now fully dry and occupied, but a small drainage ditch feeds water to the Atlantic Pond beside Páirc Uí Caoimh. These are the aquatic remnants of the old south channel. Much of the north-east shoal at Tivoli has been reclaimed and is in use as an industrial park.

Nearer the city, it is interesting that a network of small streets was laid out much later in the present Albert Road area, just south of Nimmo's proposed upper dock, and a bridge (de Valera Bridge) was erected in recent times beside the Port of Cork offices (the old custom-house) close to the site of his proposed bridge. However, the more innovative scheme for floating docks and a ship canal remained just plans. A new Harbour Commissioners Act was passed in 1820 and the new commission then formed remained in charge until very recent times. Before it left office in 1820, the old board had paid £900 for the repair of the gap in the navigation wall and agreed 'that the boring irons and machines that were purchased for Mr Nimmo's use when surveying the river shall be deposited with Mr Deane for the use of the Board in carrying out the works [at Merchants' Quay]'.[81] The board would eventually pay Deane £1,895 for rebuilding 400 feet of Merchants' Quay. This sum, together with 300 guineas paid to Porter for clearing the river and the cost of repairing the gap brought the total to over £3,100, which makes Nimmo's consultancy charge of £201 seem reasonable.

NOTES

1. 45 Geo III C. 43. *An Act to amend the Laws for improving and keeping in repair the Post Roads in Ireland, and for rendering the conveyance of Letters by His Majesty's Post Office more secure and expeditious,* 17 May 1805.

MAJOR HARBOUR ENGINEERING: DUNMORE AND CORK HARBOURS

2. A. Nimmo, maps of sections of the Boyle to Ballina road, signed by Nimmo and dated 1815, archived in the National Archives of Ireland (NAI, OPW 5HC/6/0108 [Boyle to Tubbercurry]; OPW 5HC/6/0425 [Tubbercurry to Ballina]; and OPW 5HC/6/0105 [Boyle to Ballina by Foxford]).
3. A. Nimmo to R. Peel, 13 November 1814 (BL, Add 40240 fo. 172).
4. A. Nimmo, Estimate of the Expense of Forming a Packet Harbour at Portcullin Cove, near Dunmore, May 25 1814. (B.P.P. 1813–1814 [200], vol. VII, mf 15.42). Hereafter referred to as Estimate 1814.
5. A. Nimmo to R. Peel, 13 November 1814 (BL, Add 40240 fo. 172).
6. C. Coote and A. Nimmo, Messrs Coote and Nimmo's report on a harbour at Dunmore, 1815. Royal Commission on Tidal Harbours. Second Report and Minutes of Evidence, p.100 (B.P.P. 1846 [692], vol. XVIII, mf 50.169–77). (Coote and Nimmo's report is erroneously dated 1825 in the text of this), hereafter referred to as Tidal Commission Report.
7. A. Nimmo to R. Peel, 10 March 1816 (BL, Add 40253 fo. 187).
8. Estimate 1814.
9. A. Nimmo to R. Peel, 10 March 1816 (BL, Add 40253 fo. 187).
10. R. Hutton to G. B. Greenough, 15 September 1814 (CM 02006).
11. T. Ruddock, 'Alexander Nimmo', in Skempton, *Biographical Dictionary of Civil Engineers*.
12. A. Nimmo to R. Peel, 13 November 1814 (BL, Add 40240 fo. 172).
13. A. Nimmo to R. Peel, 10 March 1816 (BL, Add 40253 fo. 187).
14. Estimate 1814.
15. A. Nimmo, Plan of the Works of the Harbour of Dunmore (BL, maps, 13748 [6]).
16. A. Nimmo to R. Peel, 10 March 1816 (BL, Add 40253 fo. 187).
17. A. Nimmo to R. Peel, 1 October 1816 (BL, Add 40258 fo. 301).
18. A. Nimmo to R. Peel, 1 April 1817 (BL, Add 40264 fo. 103).
19. A. Nimmo to E. Lees, 27 August 1816 (NAI, OPW8/HOW/2384).
20. J. Aird to J. Rennie, 9 December 1817 (NLS, MS 19791 fos. 122–4).
21. M. Chrimes, 'Henry H. Price', in Skempton, *Biographical Dictionary of Civil Engineers*.
22. Extract dated 2 December from Chronicle enclosed with letter of Aird to Rennie, 9 December 1817 (NLS, MS 19791 fos. 122–4).
23. Post Master General, Instructions to A. Nimmo regarding Dunmore, 28 January 1818 (NAI, OPW8/134/1).
24. A. Maclean to E. Lees, 3 March 1818 (NAI, OPW8/134/1 [2]).
25. A. Nimmo to E. Lees, 10 March 1818 (NAI, OPW8/134/4).
26. E. Lees to A. Nimmo, 3 April 1818 (NAI, OPW8/134/1 [5]).
27. R. Peel to E. Lees, 8 April 1818 (NAI, OPW8/134/1 [6]).
28. E. Lees to A. Nimmo, 13 April 1818. (NAI, OPW8/134/1 [7]).
29. A. Nimmo to E. Lees, March 1819. (NAI, OPW8/134/11).
30. 58 Geo III C 72. *An Act for improving and completing Dunmore Harbour*, 3 June 1818.
31. 56 Geo. III cap. lxiv. *An Act for improving the Port and Harbour of Waterford and for other purposes relating thereto*, 20 June 1816.
32. E. Lees to A. Nimmo, 28 December 1818. (NAI, OPW8/134/1 [10]).
33. C. Grant to A. Nimmo, [Dec.] 1818. (NAI, OPW8/134/1 [13]).
34. Irish Office to A. Nimmo, 22 March 1819. (NAI, OPW8/134/1/ [12]).
35. E. Lees to A. Nimmo, 28 February 1819. (NAI, OPW8/134/1 [1]).
36. A. Nimmo to E. Lees, February/March 1819. (NAI, OPW8/134/1 [11]).
37. E. Lees to A. Nimmo, 27 February 1819. (NAI, OPW8/134/1 [11]). There is some confusion in the dates of the correspondence between Lees and Nimmo at this time. This letter predates that in reference 35, but the content post-dates the latter.
38. W. Gregory to A. Nimmo, 2 May 1820. (NAI, OPW8/134/1 [17]).
39. T. Telford, Report upon the Works of Dunmore Harbour, 10 February 1821 (B.P.P. 1821 [492, vol. XI, mf 23.64).
40. A. Nimmo, Report upon the Works of Dunmore Harbour, 20 February, 1821 (B.P.P. 1821 [492], vol. XI, mf 23.64).
41. R. H. Ryland, *The History, Topography and Antiquities of the County and City of Waterford* (London: John Murray, 1824).
42. W. J. Savage, 'Description of the Landing Wharf at Hobbs' Point, Milford Haven for the accommodation of H.M. Post Office Steam Packet Establishment at that Station, and of the diving bells and machinery used in the erection. Written May 1836', in *Papers on subjects connected with the duties of the Corps of Royal Engineers*, 2nd edition, vol. 1, article no. XVI, pp. 102–115. (London: John Weale, 1844) (BL, P.P. 4050.i).
43. A. Nimmo to Sir E. Lees, 3 December 1828 (NAI, OPW8/134/2 [28]).
44. R. Grubb to Sir E. Lees, 27 September 1831 (NAI, OPW8/134/2 [33]).
45. 1 & 2 Wm IV C 33. *An Act for the Extension and Promotion of Public Works in Ireland*, 15 October 1831.
46. H. R. Paine to A. Nimmo, 8 November 1831 (NAI, OPW8/134/2 [34]).

47. A. Nimmo to H. Paine, 10 November 1831 (NAI, OPW8/134/2 [34bis]).
48. Eleventh Report of the Commissioners for Auditing the Public Accounts of Ireland (B.P.P. 1823 [199], vol. X, mf 25.83–5).
49. Select Committee on the Post Office Communication with Ireland, Report and Minutes of Evidence, p. 352 (B.P.P. 1831/1832 [716], vol. XVII, mf 35.136–140).
50. G. Halpin, 'Survey of Dunmore Pier and Harbour for the Commissioners of Public Works 1832, showing proposed extension of base of the Pier and Sea Pavement' (Nat. Archs. UK, MPD 1/30).
51. Commissioners for Public Works. Third Report, plan no. 2 (B.P.P. 1835 [76], vol. XXXVI, mf 38.296–9).
52. Tidal Commission Report, p.100.
53. A. Nimmo to R. Peel, 1 April 1817 (BL, Add 40264 fo. 103).
54. 53 Geo III cap. lxx. *An Act for reviving, amending and making perpetual an Act ... in the Parliament of Ireland [40 Geo III cap. c] for the better regulation of the Butter Trade of the City of Cork and the liberties thereof, and for other purposes*, 21 May 1813.
55. 54 Geo III cap. cxcvi. *An Act to raise a fund for defraying the charge of commercial Improvements within the City and Port of Cork in Ireland*, 14 July 1814.
56. Commissioners for the Port of Cork, Proceedings, meeting of 1 October 1814. (Port of Cork Authority archive), hereafter referred to as Proceedings, Port Commissioners.
57. Proceedings, Port Commissioners, 15 October 1814.
58. Ibid., 5 April 1815.
59. A. Nimmo, *Report on the means of improving the River and Harbour of Cork*. (Cork: Dix and Co, 1815, reprinted 1973 in limited edition by Fercor Press, Cork), hereafter referred to as Nimmo's Report.
60. W. Beaufort, Map of Cork, 1801 (Cork City Library, map no. 23).
61. Reproduced with permission of the Crawford Municipal Gallery, Cork.
62. Nimmo's Report, p. 8.
63. Ibid., p. 18.
64. A. Nimmo, 'Proposal for a System of Wet Docks at the City with a Canal for Large Vessels to the Deep Water', in Nimmo's Report, pp. 20–30.
65. Proceedings, Port Commissioners, 11 September 1815.
66. Ibid., 13 September 1815.
67. Ibid., 11 November 1815.
68. Ibid., 29 December 1815.
69. Ibid., 2 January 1816.
70. Ibid., 2 March 1816.
71. Ibid., 18 June 1816.
72. Ibid., 21 June 1816.
73. Ibid., 19 May 1817.
74. E. Allen to Lord-Lieutenant, 22 May 1817, Proceedings, Port Commissioners, 30 May 1817.
75. Proceedings, Port Commissioners, 13 September 1817.
76. Ibid., 13 December 1817.
77. Ibid., 19 December 1817.
78. Ibid., 2 June 1817
79. A. Marmion, *The Ancient and Modern History of the Maritime Ports of Ireland*. (London: V. H. Banks, 1855).
80. 'The Port of Cork', surveyed by Commander James Wolfe and Lieutenant W. H. Church, RN, 1843 (London: Hydrographic Office of the Admiralty, 1846) (BL, maps, SEC.1. [1773]).
81. Proceedings, Port Commissioners, 19 May 1819.

CHAPTER 7

Grand Juries, Nimmo and the Emergence of the Ordnance Survey of Ireland

If Nimmo erred in addressing Peel as 'an enlightened statesman and a man of science' it was in respect of the first, rather than the second, attribute.[1] Peel had come to Ireland as chief secretary in 1812, aged 24 years. He was a distinguished Oxford graduate with a double-first in Literature and Mathematics, the first such in these combined subjects, which had made his examination performance legendary.[2] He now found himself in a system where patronage was rife in civil affairs, often accompanied by blackmail, and was far more important than knowledge, education or expertise in successfully obtaining appointment or preferment. He found the 'jobbing aristocracy with a "vortex of local patronage" and loyalist associations' distasteful in the extreme. He therefore declared that the first requirement for any post, not being a mere sinecure, was to be the competence of the individual appointed. In his opinion, the improper use of public money was the worst offence of those entrusted with government, and the scope for corruption needed to be eliminated by employing only demonstrably competent and skilled persons.

Nimmo, himself no mean expert in languages and mathematics, could not have failed to be impressed with Peel's mathematical credentials and his regard for the world of science. Politicians of that kind were rare indeed, and it is little wonder that the two men appear to have had mutual respect for one another. Nimmo was exactly the kind of person Peel envisaged for public office in Ireland: he was not from an Ascendancy family (with all the expectation of preferment that such a background would have implied); he had no landed or vested interests to influence his opinions; he held whatever positions he had on the strength of his engineering expertise; he was a man of scientific repute from a good academic background in St Andrews and Edinburgh; he was conscientious and hard-working; and few enough of the engineers in Ireland were willing to work in the remoter districts outside of the Pale, as the Bogs Commissioners had earlier found. In short, Nimmo was a new 'technocrat', and Peel was committed to the greater application of science and technology in all technical aspects of public life.

In Ireland at the time there was no Ordnance Survey, no Geological Survey, no Valuation Office, no Fishery Commission and no Office of Public Works (OPW) (there was an office of public works in Dublin, but that simply maintained the government buildings and Dublin Castle). All of these bodies were to come into being in the period between 1819 and 1831, but their distant origin can be traced to the Bogs Commission. Whatever the stated reason for its institution, that commission represented the first serious attempt at the 'technocratisation' of modern Ireland. In fact it attempted, maybe unwittingly, to introduce the goals and methods of the Industrial Revolution to a country that seemed to be inextricably bound to an older agricultural order. True, the Grand and the Royal canals had been built years earlier but, apart from these, industrial advances had by then made little inroad into Irish life, especially since the Act of Union. The Bogs Commission introduced skilled engineers to address some of the country's poorest agricultural regions, *as if their problems were purely technical*. It quickly became apparent to Nimmo, and later to the commissioners themselves, that the problems were deeper, and were more socio-economic and political than merely technical. Perhaps the main success of the Bogs Commission, therefore, was not its outcome but its process: in introducing professional civil engineers such as Griffith, Bald and Nimmo into Irish public works, it added a technocratic element, hitherto largely absent, to the administration of Irish public affairs. It marked the end of the era when the authorities relied on gentleman amateurs and retired military officers for technical expertise in purely civil matters. The three engineers would contribute immensely to the development of the country in ensuing years and would become driving forces in the emergence of the public bodies mentioned. Once the Bogs Commission had exposed the underlying problems with clarity and focus, a multitude of interrelated select committees and a variety of legislative enactments would eventuate in an evolving ferment of political action that would finally crystallise in the new public institutions.

Three major legislative strands can be identified in this evolving process; they are summarised in table 4. The strands merge and separate at different times; they are, in fact, so inextricably linked that no single aspect, say the emergence of the Office of Public Works, can be fully understood without reference to the others. The recurrence of famine and distress was to influence the evolution of all of them, as later chapters will show, and all were dependent to a greater or lesser degree on the Grand Jury system.

Briefly, one strand comprised the actions that attempted to reform the Grand Jury system. It aimed to make the levying of local taxes more equitable, and the process of Grand Jury presentments – the method by which public works were approved and funded – more accountable. Starting with the reports of the Select Committee on Grand Jury Presentments in 1815 and 1816, it proceeded through the various acts and select committees noted in the table, culminating in the Select Committee on the Survey and Valuation of Ireland, which lead finally to the initiation of the

Ordnance Survey of Ireland and the Boundary Survey in 1824 and 1825 respectively. This strand will be the principal concern of this chapter.

The second strand addressed the question of money advances from the Consolidated Fund to support public works, especially as a means of employing the poor in times of distress. The expense of such works was a charge on local taxation. In the period between their approval by the Grand Jury and the eventual completion of the works, there was a need for loans to fund the activity. The original act that made such loans available out of the Consolidated Fund was the Loan Act of 1817.[3] Money advanced had ultimately to be repaid with interest by the county and was further secured by charges on tolls, profits and so forth. The Loan Act was amended on numerous occasions, largely because of the inability of Irish capitalists to find the matching funds necessary as security for the loans granted. Not every work could be expected to yield the security demanded, and in periods of very serious distress, the Government was forced to offer free grants, and to ease the restrictions on loans for those works that particularly gave employment to the poor. The main act for this aspect of the second strand was 3 Geo IV C. 34 and Nimmo's public works described in later chapters were funded mainly under this Act.[4] This strand matured ultimately in the setting-up of the Office of Public Works in 1831 and will be the substance of the next chapter.

The final strand was that concerning the Irish fisheries. Fisheries were regulated by various early acts of the Irish parliament, predating the Act of Union. Seeming to lack immediate importance after the Union, the old fishery laws were extended on a rolling basis until the Select Committee on the State of Disease and the Labouring Poor brought them to prominence in 1819.[5] A new fishery act was passed that year, which, together with its later amendments, attempted not only to improve regulation of the fisheries but also to address deficiencies in the coastal infrastructure, for instance the shortage of piers and harbours. Fisheries were perceived and later promoted as an important means of employing the poor, and in that way they impinged on, and influenced, the other strands. As a consequence, after eleven years of separate administration, responsibility for fisheries would be transferred, logically enough at the time, to the new OPW, after a brief spell in the care of the Board for Inland Navigation.

The activities that held these three strands together were the various select committees set up by Parliament between 1815 and 1831. Nimmo came to hold office under legislation enacted in all three strands, and he was centrally involved in the evidence that informed the various committees, as well as other committees on the State of Ireland in 1824 and on Emigration in 1829. Of the engineers mentioned, he was the one most clearly and most closely associated with all three strands of development and it is no exaggeration to state that he played a seminal role in the genesis of all the public institutions that were to emerge under them.

Apart from the absence of public institutions in 1814, there was a shortage of reliable, trained surveyors capable of delineating and mapping boundaries, whether of townlands, parishes, baronies or counties. This in turn meant that many of the activities attaching to the land, such as enclosure, drainage, road making, reclamation and improvement could not be addressed properly for lack of precise delineation of the true boundaries of properties. In theory, the size and value of each holding determined the amount of the county rates – called the county cess – and the tithe to be paid. Since holdings could not be measured accurately, or the measurements were based on outdated surveys, the levies were widely held to have become unjust and unfair over time. Theoretically, each county had its map delineating the various properties, but these were very out-dated yet jealously guarded and were of little real use and rarely consulted.[6] Initially set up to administer justice on a local scale, Grand Juries were progressively burdened with the duty of levying the county cess, as well as approving proposals for the construction and maintenance of roads and public works paid out of the cess funds collected. This dual role – the supply-side raising of local taxes and the demand-side supervision of their expenditure on public works – was particularly susceptible to abuse by an unelected, privileged administration. County Grand Juries were appointed by the sheriff of each county; he was appointed by the Viceroy, most often on the recommendation of the local Member of Parliament. Since the jurors were members of important property-owning families, and inevitably of those most sympathetic to the interests of the sheriff and his patron, great potential existed for partisan action. Individual proprietors could bring proposals to the Grand Jury at the Assizes for the construction, repair, diversion or alteration of a road or bridge, either towards or away from their own property as desired, and they were rarely refused. There was little or no consideration of the necessity, or public utility, of the proposal not to mention any examination of the actual plans or of the skills of the overseer, as long as he was sufficiently well connected. Once a successful application was approved, a presentment was granted by the Grand Jury that was then taken to the county treasurer in order to draw down the agreed funds. The charge was then levied on the county cess.

By 1813 pressure was mounting for a thorough examination of the role and procedures of the Grand Juries in raising local taxes. William Vesey Fitzgerald, MP for Ennis and Chancellor of the Irish Exchequer, eventually endorsed calls for such an investigation. Copies of the Grand Jury presentments for 1814 were laid before Parliament on 5 April 1815 and a select committee was appointed to examine them and report back to the House. The Committee duly reported in June. As well as examining the papers laid before it, it had taken evidence from nineteen witnesses, the majority being Irish Members of Parliament, including Vesey Fitzgerald. The inadequacies and inequities of the Grand Jury system were expounded in detail. The committee noted that the burden of work on the criminal side at Assizes was so great that it was not possible for the jury to give all the business at hand, especially on the civil side, the attention it deserved. Therefore, when applications were made for presentments, the habit had grown up of granting or refusing them without *viva voce* examination, relying solely on sworn 'Grounding Affidavits' that were often not sworn or were otherwise improperly drawn up. As to affidavits accounting for public money already granted at former Assizes, these were passed without any examination whatever, good, bad or indifferent.

In order to improve matters, the committee proposed that some means should be adopted whereby competent persons would examine all applications for public money, and the subsequent accounts of expenditure, prior to the Assizes. The members were also of the opinion that some method should be devised to render assessments more equal, 'the defect appearing to them to arise from ... the levy being made in reference to old surveys (which were taken on the measure of land which was deemed profitable at the time of such surveys) which, of course, cannot comprehend the great improvements that have taken place in Ireland since the period at which those surveys took place'.[7] In its report, the committee alluded to the demand-side and the supply-side inadequacies and implicitly acknowledged the need for competent surveyors and engineers to help deal with them. This corresponded closely, on the demand-side at least, with the views of Vesey Fitzgerald who had stated in his evidence:

> The great defects of the system appear to me to be an expenditure for works not necessary, and too great a one for those even which are admitted to be necessary ... I should have wished to enable the Grand Jury to appoint ... a Surveyor General of the roads and public works of each county ... It would be necessary, I think, that he should always be a competent civil engineer and it would be desirable that he should have been previously employed in such works.[8]

His namesake, Maurice Fitzgerald, expressed a somewhat similar opinion: 'I should think it extremely desirable in any new system that the execution of the works should be made through the hands of professional engineers.'[9]

These and other opinions of the committee mirrored the sentiments of Peel regarding the proper management of public money by competent appointees. But the committee did not favour dismantling the old Grand Jury system in its entirety: 'Upon the whole,' they concluded, 'Your Committee are [sic] of opinion that the final control of the presenting and accounting for public money should remain with the Grand Juries; but that the laws upon the subject of Grand Jury presentments require revision and amendment.' However, in remarkably modern-sounding phraseology, the committee thought 'that such a comprehensive measure as is required could not be reasonably expected to be matured for the sanction of Parliament, with the hope of its being enacted into law, during the present session'.[10] All the same, the

TABLE 4.

The three strands of legislation leading to the emergence of the Commission for Irish Fisheries, the Ordnance Survey of Ireland and the Office of Public Works between 1815 and 1831. These strands were linked by a series of Select Committees. In strand 3, funds from the Consolidated Fund were provided as loans for public works and sums set aside for the Lord Lieutenant (LL) were part of the total amounts granted.

DATE	COMMISSIONERS FOR IRISH FISHERIES	ORDNANCE SURVEY AND VALUATION OF IRELAND	OFFICE OF PUBLIC WORKS
1815		Report of Select Committee on Grand Jury Presentments	
1816		Report of Select Committee on Grand Jury Presentments	
1817	57 Geo III C 69 7 July. (Old Fisheries Acts continued)	57 Geo III C. 107 11 July (Presentments to Grand Juries Act)	57 Geo III C. 34 (Loan Act) 16 June 57 Geo III C. 124. 11 July (Amends Loan Act)
1817 (Dec.)		Interviews for County Surveyors	
1818	58 Geo III C 94 10 June (Old Fisheries Acts continued)	58 Geo III C. 2 18 February (Suspends C. 107 above) 58 Geo III C 67 3 June (Repeals and replaces C. 107 above)	58 Geo III C. 88. 5 June (Amends and extends the two previous acts)
1819		Second Report of the Select Committee to enquire into the State of Ireland, as to Disease... and the Condition of the Labouring Poor.	
1819	59 Geo III C. 109 12 July (Fisheries Act)	59 Geo III C 84 7 July (Roads by Juries) Bill re levies by Grand Juries (Not enacted)	
1820	1 Geo IV C. 82 24 July (Amends C. 109)		1 Geo IV C. 81. 24 July Moiety Act. (Amends above three acts)
1822		Select Committee on Grand Jury Presentments recommends whole island survey	
1822			3 Geo IV C. 34 24 May (Employment of the Poor act) 3 Geo IV C. 84 26 July (Temporary advances) 3 Geo IV C 112. 5 Aug. (Amends above Acts Grants £250K for loans for public works)

cont

1823		4 Geo IV C. 42. 27 June. (Grants £100K from commercial Funds for public works)
1823 (June and July)	Select Committee on Employment of the Poor in Ireland	
1824 (Feb.)	£5,000 voted for Irish Topographical Survey	
1824 (May)	Report of the Select Committee on the Survey and Valuation of Ireland.	
17/6/24	5 Geo IV C. 64 17 June (Amends main fishery act) (Money for piers) 5 Geo IV C. 65. 17 June. (Salt duties cease)	5 Geo IV C. 77 17 June
21/6/24	Ordnance Survey of Ireland set up All other surveys stopped	
1825 (June and July)	James Donnell appointed Harbour engineer. 20 June	6 Geo IV C. 35. 10 June (Grants £300K for loans for public works) 6 Geo IV C. 101. 5 July. (Roads to Inland Navigation).
1825 27 August	Boundary Survey set up	
1826	7 Geo IV C. 47 26 May (Certain bounties continued)	7 Geo IV C. 30. 5 May (Interest rate raised from 4% to 5%)
1827		7&8 Geo IV C. 12. 2 April (Extends 1 Geo IV C. 81) (Reduces sum given in 6 Geo IV C. 35) (Sets aside £100K for LL) 7&8 Geo IV C. 47. 23 June (Clarifies earlier Act C. 12)
1829	Select committee on the State of the Poor in Ireland and means of Improving their condition.	
1829	10 Geo IV C. 33. 4 June (New piers permitted)	
1830	1 Wm IV C. 54 16 July Funds for piers Fisheries transferred to Board of Inland Navigation	
1831	1&2 Wm IV C. 33. An Act for the extension and Promotion of Public Works in Ireland 15 October (Office of Public Works set up) Fisheries transferred to Office of Public Works Inland Navigation transferred to Office of Public Works Government engineers replaced by engineers of OPW	

germ of an idea had been well and truly planted, and the acknowledgement that the employment of engineers was essential in order to add competence and accountability to local taxation and to the execution of public works, would crystallise as time went on.

The following year (1816) the committee revisited the matter and recommended that a bill be drawn up consolidating and amending certain of the Irish road acts in so far as they concerned the functioning of the Grand Juries.[11] Among their many recommendations to address the deficiencies that they had previously identified, the committee included ones relating to surveyors:

- That there should be a surveyor of County Works attached to each county, and that no person should be appointed to that office, who shall not have passed an examination before a Board of Civil Engineers in Dublin, and have obtained a certificate of his qualification to discharge the duties of his office; the appointment of which Board of Civil Engineers should be vested in the Lord lieutenant;
- That the County Surveyor should be required to attend to all matters relating to presentments submitted to Quarter Sessions or Assizes, and be present on these occasions,
- That the Surveyor of the county should examine and audit the accounts of public money expended.

The question of the inequity in assessments arising from ancient survey data was left to another time.

The bill passed in July 1817 as *An Act for investigating Presentments to Grand Juries in Ireland for Public Works and Roads and for Accounting for Money raised*.[12] Among its provisions, surveyors were to be sworn and appointed in every county. No surveyor was to be appointed who had not received a certificate of qualification from a Board of Civil Engineers in Dublin appointed by the Lord-Lieutenant. Some surveyors might be appointed to two counties, sure recognition that there might not be sufficient suitable candidates for thirty-two counties. No presentment for any building over £20, or any new road or footpath, was to be granted unless first examined by a county surveyor who was also required to report on the state of all public works in his county and to classify each road. These requirements were essentially the recommendations made by the 1816 Committee.

Peel moved rapidly to implement the Act by setting up the examining board. Major Alex Taylor of the Post Office, Thomas Telford and Alexander Nimmo were his chosen examiners. Nimmo accepted the nomination in a letter to Peel dated 17 December.[13] The following day Telford, who had just arrived in Dublin, also wrote to Peel.[14] He had no problem with Alex Taylor, he said,

> But I beg leave to suggest, without the least reflection on Mr Nimmo's talents or integrity, that a more proper selection might be made. Firstly, he is yet but a young engineer and is a candidate for a surveyorship. Secondly, I think it is desirable that there should be one Irishman in the Commission and thirdly, as a part of the surveyor's duty consists of attention to public buildings, it is desirable there should be one commissioner an experienced architect.

He went on to propose that Francis Johnston, already employed by the Irish authorities, be appointed in place of Nimmo. All this advice sounds eminently reasonable. Nimmo was now 34 years of age – not old, but hardly young, unless Telford meant that he was young in the profession and had yet to produce a 'monumental' work of his own. Nor is there any evidence that Nimmo was ever a candidate for a surveyorship. It does not seem to have dawned on Telford that, if he were really so concerned that an Irishman should sit on the board, he himself could have declined to serve, so that one could be appointed in his place. We do not know if Peel, who was not for easy turning, was swayed by Telford's arguments. (Peel himself was then only 29 and could hardly agree that 34 was too young!) But even by nominating him in the first place, Peel had once again shown his high regard for Nimmo. In the event, Lord Talbot the Viceroy took Telford's advice, and Johnston replaced Nimmo on the interview board, which met on 23 December and reconvened for the period 10 to 19 January 1818.[15]

Robert Hutton, writing from Dublin to George Bellas Greenough on 13 January keeping him abreast of events in Ireland, told him:

> Nimmo is in town. I was present last night at his admission as a member of the Royal Irish Academy. Bald was in town lately and so was [sic] many engineers in Ireland for, by a late Act of Parliament an engineer is to be appointed in each of the counties and no new road or any public works is [sic] to be executed without being first submitted to him. The salary varies according to the size of the county, from £200 to £400 p.a. An examination of 200 candidates took place lately in Dublin by Telford and Johnson [sic]. Nimmo, Griffith, Bald, William Edgeworth, Larkin were amongst the number. The examiners made a report of the qualifications of the candidates to government which makes the appointment. None of them are yet known. Everyone is astonished to find there are so many persons in Ireland who could regard themselves qualified to be candidates for such a situation.[16]

It is unlikely that all the eminent exponents of the engineering and surveying arts named by Hutton – Nimmo, Bald and Griffith among them – were candidates for the posts, as he seems to imply. Four hundred pounds would have been paltry recompense for men of their experience when we consider that the Bogs Commission had great difficulty in attracting skilled and suitable persons with much higher salaries seven years earlier.

As it turned out, an insufficient number of candidates was found suitable for appointment, hardly a surprising outcome. Every half-competent contractor and semi-trained surveyor must have applied for what would have been seen by them as plum jobs, full of potential for discreet patronage and finely tuned favouritism. But Telford and the other examiners were no push-overs, and when the interview process ('this painful and tedious investigation', they called it) failed to yield the hoped-for crop of competent applicants, they refused to grant even a single certificate.[17] Because of this, the Act was suspended by the Act 58 Geo III C. 2, passed on 18 February 1818.[18] It was eventually repealed entirely and replaced by the Act 58 Geo III C. 67, passed in June 1818, which made no mention of county surveyors.[19] In this latest Act, presentments were to be submitted for examination by the magistrates meeting *en bloc*, before being forwarded to the Grand Jury at the Assizes. Vesey Fitzgerald seems to have had more confidence in the impartiality of the magistrates than anyone in Ireland, or anywhere else, did.

Early in 1817 Nimmo was in London while the first bill on Irish Grand Juries, as advanced by the select committee, was being drafted. He wrote to Vesey Fitzgerald offering his help with the bill:

> Having in the course of his studies and when on the continent paid great attention to the mode of conducting public works in France, the Low Countries etc. which bears much resemblance to that proposed for Ireland, [he] would be happy to communicate any information in his power on the subject. Mr Nimmo believes he is, perhaps, the only professional man in town acquainted with the Grand Jury business of Ireland.[20]

He enclosed with this letter two accounts, one entitled 'Arrangements for the *Ponts et Chaussees* in France' as it was in 1760[21] and the other an account of the *Corps des Ingeneurs des Ponts et Chaussees* as brought up to date and greatly improved under the government of Bonaparte.[22] The engineers of the *Corps* had responsibility for the construction of roads and bridges in France and for the education, through its own *Ecole*, of civil engineers. Nimmo had gathered information and collected documents about the *Corps* while on the Continent and clearly favoured the creation of a similar establishment in Ireland, which would train young surveyors and engineers while supervising the whole road network of the country. What he was endeavouring to promote was nothing less than an Irish version of the French establishment – essentially an Office of Public Works for the whole country, only more elaborate – long before that idea occupied the mind of the authorities. It is small wonder that Peel had nominated him as an examiner on the board for surveyors later that same year, a choice suggesting that Peel and Nimmo may have been far ahead of Telford and Lord Talbot in their thinking. Nimmo, Bald, Griffith and Killaly already had teams of young men training and working under them, which would almost certainly have lead to the emergence of a native Irish School of Surveying and Engineering, had circumstances at the time been more propitious, or if Nimmo's advice had been followed.[23]

Peel left office in the summer of 1818, one of the bad, but not exceptional, famine years. Distress was widespread and fever rampant over large areas of Ireland. Charles Grant, the new chief secretary, was appalled by what he saw on arrival – the poor 'obliged to feed on esculent plants, such as mustard-seed, nettles, potato-tops and potato-stalks, a diet that brought on a debility of body and increased the disease more than anything else could have done'.[24] Despite his natural humane sentiments, he was reluctant at first to breach the orthodoxy of leaving economic problems to market forces. While he agreed that the provision of employment in public works would save the great mass of the labouring poor from absolute beggary, he was nevertheless wary of using public works as a general remedy for the problem. Clever political tactics by some Irish Members of Parliament, notably Sir John Newport, MP for Waterford, eventually resulted in a Parliamentary investigation into the state of affairs in Ireland. A new Select Committee was set up with the cumbersome title

> To inquire into the state of Ireland as to disease and how far the measures, remedial and preventive, adopted by the Legislature or otherwise during the last year, have been effective for its removal or mitigation; and also into the condition of the Labouring Poor of that part of the United Kingdom, with a view to facilitate the application of the funds of private individuals and associations, for their employment in useful and productive labour; and to report their observations, together with their opinion on these subjects from time to time to the House.

This committee was to be inordinately influential in the formalisation of the Ordnance Survey of Ireland and Irish public works.

The committee made its first report on the disease aspects of its brief on 17 May, 1819, but this need not detain us here.[25] Nine days earlier, on 8 May it had commenced hearings on the labouring poor aspect, taking evidence from witnesses under the chairmanship of Sir John Newport. It was an exercise that smacked of a real set-up. Only three witnesses were interviewed: Nimmo, Telford and Robert Fraser, the last a member of the British Fisheries Society and experienced commentator on Scottish fisheries. The questioning was gentle, partisan and leading, the questioners already well familiar with the opinions that their questions were aimed to elicit. Sir John, for example, would already have been well acquainted with Nimmo's activities in his own constituency of Waterford and therefore knew what to expect. Neither did Nimmo disappoint. He had recently been in north Kerry and Limerick, he said, where he had seen several hundred labourers hire themselves out, in desperation, at four pence a day, when the going rate was eight pence, a rate he confirmed when checking the accounts on behalf of the Loan Commissioners. The poor people of the village of Hospital, County Limerick were so exasperated that they 'fell upon them and nearly killed several of them, and some afterwards died'.[26] The only new work in north Kerry, from whence many of the migrants had come, was the Tralee mail coach road, but that would only employ local, not migrant, workers for two years. In Connaught there were no public works under way at all. At Dunmore, migrant workers came to his works, travelling up to one hundred miles, but the limited number of workmen required there, and the claims of the distressed local population, prevented them from obtaining much relief. Hundreds would gather on the shore in the evening, fishing for mackerel, their only source of food.

The activity Nimmo thought most likely to give significant employment was drainage of the bogs, something he was familiar with from his Bogs Commission days and from his studies of drainage elsewhere. He delivered and read two papers that he had written – one on the Drainage of Holland[27] and one on the Drainage of the Fens in England[28] – but he was not sanguine that funds would be made available for the drainage of Irish bogs. There simply was not sufficient private capital to fund the initial steps that would leverage money from the Loan Commissioners, exactly the problem predicted earlier by a commentator in *The Times*.[29] Sensing no future in turf, the committee turned its attention to fisheries. 'Have you found in these districts (the mountainous parts of Ireland) many valuable harbours?' he was asked. The best natural harbours were there, Nimmo replied, but they were little used except for the fisheries. Want of wharves or landing places, and of roads through the mountainous and boggy tracts on their shores, meant the harbours were useless for trade and commerce. He and his brother engineers had recommended that the provision of roads was the best means of improvement in districts like these, where the number of fishermen was very large.

These were precisely the answers that the committee wanted to hear. They led it logically to enquire whether the provision of roads in Scotland had opened up the fisheries there, and to receive the expected answer 'that was the chief object of most of those roads'. The man who had built them – Telford – was waiting in the wings to make his contribution, as the committee well knew! Knowing that Fraser, the real fisheries expert, was also scheduled to give evidence, the committee tried to elicit from Nimmo confirmation of the poor state of the Irish fisheries. While he listed off the districts where the fisheries could be improved by the provision of roads and piers, Nimmo was remarkably frank in admitting that he was not accurately acquainted with the state of the Irish fisheries: 'Indeed I am very little acquainted with the fishery of Ireland. I think I can give but very little information upon that subject.'[30] He had, in fact, given more than enough. He had paved the way for the evidence of the others, who would bring to the hearing their first-hand knowledge of what had been achieved in Scotland and would indicate how the Scottish solution could be applied to Ireland.

Robert Fraser's evidence regarding the fisheries of the west coast was mainly inaccurate hearsay. With regulatory and administrative matters he was on surer ground, well aware of the negative effects of the existing salt duties and the likely beneficial effect of granting a bounty on fish landed by Irish fishermen. Peel had shown him a draft of the proposed Irish Fisheries Bill that was being put through the House at the time; Fraser's observations would contribute prominent features to the act, and its amending acts, when they eventually passed. His personal experience of the Irish fisheries was confined to the south and east coasts but he was aware that a great abundance of fish of all kinds was to be found in all the bays of Ireland. 'Ireland,' he said, 'is better calculated for carrying on the fisheries in an extensive manner than any part of the United Kingdom, or perhaps than any other country in Europe … Above all, Dunmore in the county of Waterford is admirably situated for carrying on the fisheries of every description.'[31] This was prophetic (possibly collusive with Nimmo) and was the first time that Dunmore was advanced as a potential fishery port rather than a packet station, even before it received its first packet boat. Clearly he was aware of the parliamentary constituency that the chairman represented.

Telford was examined next. He explained that he had constructed more than 1,000 miles of road and 1,500 bridges in Scotland, together with a very considerable number of harbours and piers, under the Commission for Highland Roads and Bridges. These works had made the fishing stations much more accessible, and boats and fishermen from the east and south coasts could now service the fisheries on the west coast. More generally, the work on the roads and the Caledonian Canal had employed about 3,200 men on average each year with hugely beneficial consequences:

> at first these men could scarcely work at all; they were totally unacquainted with labour; they could not use the tools, but they have since become excellent labourers; of that number … we consider that one fourth left us annually, taught to work; so that these works may be considered in the light of a working academy, from which 800 have annually departed improved workmen. These men have either returned to their native districts, having had the experience of using the most perfect sort of tools and utensils … or they have been usefully disseminated throughout the other parts of the country. Since these roads were opened and made accessible to wheeled carriages, wheelwrights and cartwrights have been established in the country, and, in different parts, the plough has been introduced, and other improved tools and utensils are now used.[32]

He also considered that the moral habits of the great mass of the working classes [sic] were changed by this employment. The costs of the Scottish developments were shared half and half between the Government and local subscription, and in the course of fifteen years about £200,000 of public money had been spent: 'It has been the means of advancing the country at least a hundred years,' was Telford's opinion. He had no doubt that similar benefits would accrue to Ireland by development of its roads and harbours. These trenchant views, coming from the man who had done it all, gave powerful support to the idea of initiating and funding similar public works in Ireland.

These compliant witnesses, all men of high reputation, made the committee's arguments for it. In presenting its report, the committee was prudent and astute enough to emphasise that its deliberations were 'in a great measure controlled by the unquestionable principle, that legislative interference in the operation of human industry, is as much as possible to be avoided', but the case it was about to make was very much an exceptional one. It identified the drainage and reclamation of bogs and the encouragement of the fisheries as the two activities that stood to increase the employment of the people and raise the prosperity of the country the most, particularly if accompanied by the construction of roads through the mountainous and boggy regions to facilitate the transport of manure, fish and agricultural produce. The fisheries received particular attention and favour. 'In whatever way it can be considered,' the report stated, 'whether as a source of national wealth, as a means of employing an overflowing population, or as a nursery of the best seamen, Your Committee cannot too strongly impress on the House the importance of the Irish fishery.'[33] Shortage of capital – occasioned in no small way by the non-residence of a great proportion of the proprietors, especially that portion that could most contribute its rank, its wealth and its moral influence – was at the root of Irish poverty. The expenditure in England resulting from this absenteeism enhanced, in the committee's view, the claims of Ireland on the generous consideration of Parliament: 'On every ground of policy as well as justice, your Committee earnestly recommends the application of the precedent of Scotland to the Highland districts of Ireland.' The precedent in question was the payment by the Government of half the cost of all the public works to be undertaken. As to the other half of the cost, the Grand Juries had the power to tax the land that benefited, in an amount equal to any public grant.

Chief secretary Grant was won over by the committee's report and recommended to Parliament that £100,000 of the money that remained unused from the allocation of £250,000 specified in the Loan Act, be made available immediately for public works.[34] He went on to emphasise that investment in Ireland's natural resources, especially those in the west, would more than repay the Government, as it had done in Scotland. Just as the Commission for the Highland Roads and Bridges had been an implicit model for the Irish Bogs Commission, its success was now being paraded as an explicit example of the likely success a similar approach to Ireland's situation would bring. There can be no doubt that Scottish medicine was to be the remedy for Irish ills, although in smaller doses: we will see that the Fisheries Commissioners endeavoured to pay only one-quarter of the expense of public works in Ireland as against one-half in Scotland's case.

Uniquely in the present case, the Scottish connection was much more interesting than is immediately obvious. All four main characters – Grant, Nimmo, Telford and Fraser – were Scotsmen, and not only that; all four were connected through the Scottish Commission. Fraser was an influential voice in the British Fisheries Society that had canvassed for the establishment of the Commission.[35] Telford was its engineer and Nimmo had worked on the Inverness boundary for it. The nexus was completed by Grant, through an interesting and probably influential coincidence: five years younger than Nimmo, Grant was M.P. for the burghs of Inverness and Fortrose from 1811 to 1818. In the latter year he succeeded his father, Charles Grant (senior) as M.P. for Inverness-shire. Grant senior had represented the shire from 1802 until his death in 1818, right through the period of the Commission's greatest activity and including Nimmo's time as Rector of Inverness Academy and his boundary survey. Both Grants, father and son, would have known Nimmo in his schoolmaster days in Fortrose and in Inverness, and Nimmo could hardly have undertaken the survey of Inverness-shire without the knowledge and cooperation of its influential Member of Parliament. (Grant senior had been born in Glen Urquhart; Nimmo's young assistant on his boundary survey was a Grant from Glen Urquhart. Was this also, one wonders, simply a mere coincidence?) How could chief secretary Grant fail to be swayed by the opinions of his three eminent and well-connected fellow Scotsmen, who had contributed so successfully in a practical way to his own native shire of Inverness?

The report of the committee was followed rapidly by two important pieces of legislation. One was the long-awaited act governing the Irish fisheries, under which Nimmo would play a significant part and which will be considered in chapter 9.[36] It was preceded one week earlier by an act governing Grand Jury presentments as applied to roads and other public works.[37] This was a very comprehensive act, but it deliberately did not address the question of inequity in the assessment process and how it might be made more fair. That matter needed further ongoing consideration.

Grand Jury levies would almost double from £407,000 in 1803 to £750,000 in 1823, so the continuing inequity was no minor matter that could safely be left on the long finger indefinitely. Efforts continued to have this aspect addressed in a separate bill that came under consideration at the time. In January 1819, Richard Griffith had drawn up 'Directions', setting out the duties of county surveyors, should such be appointed by the hoped-for act, which would help them to provide the information needed in order to ensure fairer assessments. It was absolutely necessary, he believed, that the authorities have an accurate map of each barony in the county on a scale of eight inches to an Irish mile:

> These maps should show the boundaries of all parishes and townlands, also all the roads, all bridges and waterpipes ... rivers and streams ... quarries and sandpits ... towns, houses and gardens. The boundaries of all bogs and lakes should be measured and also the boundaries of the different varieties of land ... so as to determine in each townland the extent of the arable and pasture land that is enclosed, of heathy, pasturable mountain land and of rock or boggy mountain land. The distinguishing of these varieties of land ... will, at the time of making the survey, give little additional trouble; but the great benefit will be derived from it, should a new applotment of the land take place so as to equalise the county cess that each landowner should pay per acre, according to the value of his land, and not according to the unequal and very unjust system at present practised in most counties.[38]

Griffith's advice was incorporated into a bill of 12 July 1819, which aimed to provide for a new, more equitable assessment system.[39] The bill did not make it into law; it was followed in 1822 by yet another select committee appointed 'to consider what provisions it may be expedient to establish for regulation of the Grand Jury presentments of Ireland in order to render the levies more equal and for an impartial and correct distribution of the taxation they imposed over the lands to be assessed upon a scale proportional to their value.'[40]

The evidence and recommendations of earlier committees were rehearsed and reprised, the new committee coming to the usual conclusions. These were, that the old modes of assessment required immediate and complete alteration; that the revised valuation system should be uniform all over the island, founded on an accurate survey of the whole acreable content of the country and on a division of the lands into profitable and unprofitable. The committee proposed that a general survey of the whole island should be carried out by proper officers under the direction of the Government and at public charge. By this approach, the pressing need for *local* surveys that facilitated a fair assessment of local taxes (local cadastral surveys) was being subtly enlarged to become a *national, topographical* survey that would require proper triangulation if the various baronies and counties were to be joined cartographically.

Parliament listened, and finally in 1824 it resolved 'That it is expedient for the purpose of apportioning more equally the local burthens of Ireland, to provide for a general Survey and Valuation of that part of the United Kingdom', before going on to vote £5,000 towards the proposed 'Trigonometrical Survey of Ireland'.[41] This, the first explicit grant of government finance for a general trigonometrical survey of the whole island, effectively initiated the Ordnance Survey of Ireland. There is little doubt, therefore, that the origin and rationale of an Irish national topographical survey lay in the need for equity in the Grand Jury system, rather than in any desire to map the country for military or scientific purposes. But the survey would not get under-way formally until, inevitably, yet another select committee sat. This latter, under the chairmanship of Thomas Spring Rice, was appointed in May 1824, 'to consider the best mode of apportioning more equally the local burthens collected in Ireland and to provide for a general Survey and Valuation'.[42] This is commonly referred to as the Survey and Valuation Committee, or the Spring Rice Committee.

The oral hearings of the new committee were attended and addressed by all the important *dramatis personae* of Irish surveying, engineering and cartography, previous and to come – among them the politicians Spring Rice, Vesey Fitzgerald, Maurice Fitzgerald, John Newport, and J. L. Foster (Chairman of the Bogs Commission); the engineers, Richard Griffith, William Bald, Alexander Nimmo, William Edgeworth; the military survey men, Lieutenant-Colonel Keane, Majors Warburton and Colby and Captain Kater. John Wilson Croker, Galway-born secretary of the Admiralty, was also a member. The main issues that exercised the committee were: who should do the trigonometrical survey?; on what lowest level (county, barony, parish, townland or field-by-field)?; at what scale should the map be drawn (in inches to the mile)?; and, how was the valuation element to be pursued? These questions raised important practical issues that would determine whether the survey would be done successfully, in time, and be of value in addressing the inequalities in the local levies. The survey, in other words, was to be cadastral as well as topographic.

Nimmo gave his evidence over four days, prefacing his contribution by stating that it would not be possible for him to take an active share in the survey, even if he were thought qualified, because he was so much occupied in public works.[43] That was an unfortunate opening position to hold, or at least to make public. Having briefly recounted his qualifications and indicated that he had several young men surveying under him, he launched into a schoolmasterish lecture on the history of surveying and cartography. The committee did not seem to mind: a description of the procedures used in the survey of other countries might assist them in deciding what would be best for Ireland. Then the questioning got round to the map of Ireland. There was as yet no correct map of Ireland, he stated, meaning one based on a proper trigonometrical survey. Along with Bald and William Edgeworth, who respectively had completed triangulations of Mayo and Roscommon that were connected to the observatories of Dublin and Elphin, he (Nimmo) had been able to make a tolerable approximation to the position, both in latitude and longitude, of many points on the west coast, where he found the ordinary surveys and charts to be extremely erroneous. As to the survey of the whole island, he had no doubt that the Board of Ordnance was the pre-eminent body to carry it out 'as well for the unity of the [Irish] operation with that in Britain, as from their superior instruments and experience in that particular department'. He envisaged two separate surveys, the large-scale trigonometrical survey that the

Board of Ordnance should perform, about which he had little enough to say, and the valuation, or territorial, survey in which local surveyors would provide information on the fine detail of parishes and townlands and their values. The territorial survey should precede the Ordnance Survey so that the latter would be a check upon the former. 'It is clear', he said, 'the local surveyor would take exceeding great care, for his character, (and, in fact, I would make it a pecuniary loss if he did not), to be correct in his representations, and they should not be carried further than his instruments and his science would permit him to go....I prefer that his work should be independent and that the triangulations of the ordnance should be a check upon his work, not the elements of it.'[44]

In Nimmo's opinion the territorial survey should be pursued at the level of parish, not townland, which he regarded as a very uncertain division of the parish. A surveyor could work townland by townland within a parish to furnish the final parish map and terrier, but he should not be required to connect the various parish maps, which would be the work of superior surveyors of the Ordnance. The reason he favoured this way of proceeding was that the parish was a certain community, the map and survey of which would of itself be a very important and useful document, without waiting for the operation of the Ordnance Survey, which might require many years. He had in his employment, he went on, twelve persons, mainly young men, to all of whom with little instruction he would confidently entrust the trigonometrical measurement of a parish. In fact, he had entrusted the trigonometrical operations for connecting a great many miles of the coast of Ireland to them. Besides these men, he had many more 'of an inferior class'. Local surveyors could work at the most minute levels, indicating individual cabins and even fences on their maps.

As to the actual valuation of properties, he was insistent that it should be given as the money value of the actual holdings on a certain future date, not a calculation from a simple acreable standard or any other standard measure. When asked whether the holding should be valued in terms of the price of corn, a theoretical standard measure of its potential profitability, he was acerbically dismissive: 'I conceive that unless you was [sic] to take your taxes in corn, that your corn valuation would be a more variable thing than the money.'[45] Nimmo's last day of evidence, which related to his coast survey and maritime matters involving his skirmish with the Admiralty, is detailed in chapter 9.

In its report, the committee did not favour Nimmo's approach to the territorial survey. But everyone, Nimmo included, agreed that the trigonometric survey of the whole island should be performed by the Board of Ordnance and the maps, on the scale of six inches to the mile, should represent counties, baronies, parishes and townlands. Therefore, there would be little enough need for purely local surveys once the new Ordnance Survey was operating at this fine level, and the committee was hopeful that the public and the holders of public office would cooperate with the survey in what was a great national work. Indeed, the committee recommended that the new survey should supersede all local topographical surveys, whether under the authority of Grand Juries or otherwise, and therefore every local survey was to be discontinued with immediate effect. This, in fact, was not really a logical outcome of the committee's deliberation: we will see in chapter 9 that it was a decision already made by the authorities, precipitated by Nimmo's coastal survey, long before the Spring Rice Committee ever sat. In making this recommendation, the Committee was simply adopting the Admiralty's position regarding maritime surveys and, from here on, Nimmo's coast survey, already in temporary suspension, would never be restarted. Fortunately for him, it was largely complete in any case, although the charts of the western coast were, at the time, mostly unpublished. We will see that Nimmo might reasonably have anticipated this outcome, at least when he saw that Croker from the Admiralty was on the committee. Because of it, and by claiming at the start that he was too busy to engage in the new survey, he had effectively cut himself off from participating in the only official survey that was permitted to continue into the future, that is, the new Ordnance Survey. His opening position, therefore, betrayed once again his lack of political acumen.

All was now ready; on 22 June 1824 the secretary of the Board of Ordnance set the Irish trigonometric survey in motion and on 16 February 1825 an estimate for the Ordnance Survey of Ireland was included in the Ordnance Estimates for the first time, thereby giving final parliamentary approval to the work.[46] The survey commenced in Belfast under the direction of Major Colby a little later. He had immediate staffing problems, which, even with increased recruitment, made it impossible for him to cover the local boundaries and the valuation aspects of the survey in tandem with the trigonometric aspects. He therefore advised that a separate body should undertake the survey of the local boundaries and the valuation, *in advance of* the trigonometric mapping, exactly as Nimmo had proposed in his evidence. The Government set up the Boundary Survey for this task in 1825, putting it under the direction of the politically astute Griffith. It could well have fallen to Nimmo, and maybe would have, had he not excluded himself from the very beginning. It was, after all, the mode of progressing that both men had proposed, initially *without success, to the committee. Griffith commenced work on 27 August 1825 and soon realised that the task was greater than anyone had anticipated. There were 69,000 townlands in Ireland and only three surveyors in the Boundary Office! Griffith, nevertheless, persevered in his task and, with added staff, had completed the boundary survey for half of the country by 1831, a truly remarkable and admirable achievement.

Further difficulties with the Grand Jury system were to continue, as were further Select Committees. The next one, meeting in 1827, did little more than re-examine earlier documents.[47] It called no witnesses as the parliamentary session was nearing its close, and the committee members thought it too late to re-open so complex an issue. Among the documents it did examine were Nimmo's observations on the western district taken from his report of 1823 (see chapter 10), his evidence to the House of Lords Committee on the State of Ireland that we will deal with later on (in chapter 15), and his description of the system of road making in the western counties.[48] Nimmo's involvement with Grand Jury presentments had now extended over a decade, longer than that of almost everyone else. Through this involvement he had played a seminal role (along with others, certainly) in the emergence of the Ordnance Survey and the General Valuation of Ireland. It was a major step forward in the technocratisation of the national public infrastructure and in the training and employment of surveyors, cartographers and engineers, subjects that he pioneered throughout his professional life.

NOTES

1. A. Nimmo to R. Peel, 10 March 1816 (BL, Add 40253 fo. 187).
2. J. Prest, 'Robert Peel', in Matthew and Harrison (eds), *Oxford Dictionary of National Biography*.
3. 57 Geo III C. 34. *An Act to authorize the issue of Exchequer Bonds and the advancement of money out of the Consolidated Fund to a limited extent for the carrying out of public works and fisheries in the UK and employment of the poor in Great Britain.* (The Loan Act), 16 June 1817.
4. 3 Geo IV C. 34. *An act for the employment of the poor in certain districts of Ireland*, 24 May 1822.
5. Second Report of the Select Committee on the State of Disease and the condition of the Labouring Poor in Ireland (B.P.P. 1819 [409], vol. VIII, mf 20.68–69), hereafter referred to as Second Report on Labouring Poor.
6. J. H. Andrews, *A Paper Landscape* (Dublin: Four Courts Press, 2002), p. 4..
7. Report from the Select Committee on Grand Jury Presentments in Ireland, with minutes of evidence, p.4 (B.P.P. 1814/1815 [283], vol.VI, mf 16. 30–1), hereafter referred to as Select Committee on Grand Jury Presentments.
8. Select Committee on Grand Jury Presentments, appendix 1, p. 61.
9. Select Committee on Grand Jury Presentments, appendix 1, p. 66.
10. Select Committee on Grand Jury Presentments, p. 6.
11. [Second] Report from the Select Committee on Grand Jury Presentments in Ireland (B.P.P. 1816 [374], vol. IX, mf 17.48).
12. 57 Geo III C. 107. *An Act for investigating Presentments to Grand Juries in Ireland for Public Works and Roads and for Accounting for Money raised*, 11 July 1817.
13. A. Nimmo to R. Peel, 17 December 1817 (BL, Add 40272 fo. 188).
14. T. Telford to R. Peel, 18 December 1817 (BL, Add 40272 fo. 203).
15. Report of the Board of Civil Engineers, which sat at No. 21 Mary Street, Dublin from 23 December 1817 to 19 January 1818 (B.P.P.1818 [2], vol. XVI, mf 19.85), hereafter referred to as Report of Board of Engineers.
16. R. Hutton to G. B. Greenough, 13 January 1818 (CM 02006).
17. Report of Board of Engineers.
18. 58 Geo III C. 2. *An Act to suspend, until the end of the present session of Parliament, the operation of an Act made in the last session of Parliament to provide for the more deliberate Investigation of Presentments to be made by Grand Juries for Roads and Public Works in Ireland, and for accounting for money raised by such Presentments*, 18 February 1818.
19. 58 Geo 3 C. 67. *An Act to provide for the more deliberate Investigation of Presentments to be made by Grand Juries for Roads and Public Works in Ireland, and for accounting for money raised by such Presentments*, 3 June 1818.
20. A. Nimmo to W. Vesey Fitzgerald, 1817 (BL, Add 40210 fo. 244).
21. A. Nimmo, 'Arrangement for the Ponts et Chausees in France 1760' (BL, Add 40210 fo. 246).
22. A. Nimmo, 'The establishment of the Corps des Ingeneurs des Ponts et Chausees 1871 [sic]' (BL, Add 40267 fos. 244–9). The title reads 1871, but internal evidence suggests that this should be 1817, the date of the covering note that accompanied it. This item should be read together with references 20 and 21 which, although archived separately, all appear to form one correspondence.

23. Andrews, *Paper Landscape*.
24. Quoted in B. Jenkins, Era of Emancipation (Montreal: McGill-Queen's University Press, 1988).
25. First Report from the Select Committee on the State of Disease and condition of the Labouring Poor in Ireland (B.P.P. 1819 [314], vol. VIII, mf 20.67).
26. Second Report on Labouring Poor, p. 101.
27. A. Nimmo, 'Some account of the Drainages of Holland etc.', in Second Report on Labouring Poor, pp. 106–7.
28. A. Nimmo, 'Short Account of the Drainage of the Fens in England', in Second Report on Labouring Poor, pp. 107–9.
29. *The Times*, 29 April 1817, p. 3.
30. Second Report on Labouring Poor, pp. 103–4.
31. Ibid., p. 116.
32. Ibid., p. 112.
33. Ibid.
34. Jenkins, Era of Emancipation, p. 143.
35. J. Dunlop, *The British Fisheries Society 1786–1893* (Edinburgh: John Donald, 1978).
36. 59 Geo III C. 109. *An Act for the further encouragement and improvement of the Irish Fisheries*, 12 July 1819.
37. 59 Geo III C. 84. *An Act to amend the Laws for making, repairing and improving roads and other public works in Ireland by Grand jury presentments and for a more effectual investigation of such presentments, and for further securing a true, full and faithful account of all monies levied under the same*, 7 July 1819.
38. R. Griffith, 'Notes respecting the Duties of county surveyors, written in the month of January 1819, by Richard Griffith, civil Engineer', in Report from the Select Committee to consider the best mode of apportioning more equally the local burthens collected in Ireland and to provide for a general Survey and Valuation, Minutes of Evidence, appendix F (B.P.P. 1824 [445], vol. VIII, mf 26.49), hereafter referred to as Spring Rice Committee.
39. Bill, 12 July 1819. *A Bill to provide for the more equitable assessment of sums required to be raised off the several counties in Ireland in pursuance of presentments by Grand Juries or otherwise* (B.P.P. 1819 [570], vol. I B, mf 20.15).
40. [First] Report from the Select Committee to consider what provisions it may be expedient to establish for regulation of the Grand Jury presentments of Ireland in order to render the levies [...] more equal and for an impartial and correct distribution of the taxation they imposed over the lands to be assessed upon a scale proportional to their value (B.P.P. 1822 [353], vol. VII, mf 24.46).
41. Estimates, 9 February 1824. An estimate of the sum necessary for commencing a Trigonometrical Survey of Ireland to be carried on and completed under the direction of the Master-General and Board of Ordnance (B.P.P. 1824 [56], vol. XXI, mf 26.135).
42. [Second] Report from the Select Committee to consider the best mode of apportioning more equally the local burthens collected in Ireland and to provide for a general Survey and Valuation (B.P.P. 1824 [360], vol. VIII, mf 26.47).
43. Spring Rice Committee, p.75.
44. Ibid., p. 79.
45. Ibid., p. 81.
46. Ordnance Estimates (B.P.P.1825 [35], vol. XVIII, mf 27.151–2).
47. Report from the Select Committee on Grand Jury Presentments (B.P.P. 1826/1827 [555], vol. III, mf 29.23–4), hereafter referred to as Grand Jury Report 1826/7.
48. A. Nimmo, 'Mr Nimmo's description of the system of road making in the counties of Galway, Mayo, Sligo, Leitrim and Roscommon', in Grand Jury Report 1826/7.

CHAPTER 8

A Labyrinth of Legislation: Famine, Relief Work and the Emergence of the Office of Public Works

Food scarcity, if not outright famine, was a recurrent feature of Irish life in the early nineteenth century. Usually it arose from bad weather causing failure of the potato crop, and 1815, 1817, 1819, 1822 and 1824 were particularly bad years. Potatoes were not necessarily scarce all over the country at the same time, nor were other food items in short supply. However, even when alternative food was abundant, the problem was that the poor could not afford to buy it. Tenant farmers too, when they could not grow potatoes, or when their crops failed, lacked the wherewithal to buy alternatives, a lack often aggravated in their case by mounting rent arrears. The year 1815 was made worse by the ending of war on the Continent and the agricultural recession in England and Ireland that the peace precipitated. Agricultural prices collapsed, so that many landowners switched from tillage to pasturage and tenant farmers were put off the land. Distress was widespread, but whereas England had its Poor Laws to help it cope, Ireland had nothing. Harvests were slightly better in 1816, but famine and disease returned in 1817.

Political orthodoxy did not favour the intervention of the Government in the market, no matter how pressing the need or how distorted the economy. However, even the most heartless Government in Westminster, and the arch-conservatives Peel and William Gregory (the under-secretary) in the Irish administration, could not fail to be moved by the number and plight of the poor in 1817. Since they knew the underlying problem, the strategy they proposed to adopt was to stimulate employment by initiating public works in the worst-affected counties. As we have seen, the cost of public works in Ireland fell on the local landowners through the county cess. Initially, the authorities had no intention, no matter how bad conditions might become, of usurping the role and obligation of the landowners in this matter. However, the Grand Juries were loath to make very many presentments for public works, the jurors (men drawn from the landowning class) being reluctant to cut a stick with which to beat themselves. This was partly for the reason later explained succinctly, (even if it sounds like partisan special pleading) by Archbishop Le Poer Trench of Tuam: 'In the same proportion that the lower orders are reduced, the higher orders are incapacitated to afford them relief.'[1] Peel and Gregory were therefore forced to act, despite their own pronounced reluctance to intervene. In cases of extreme necessity, where people were actually starving and without hope of relief from other quarters, they were prepared (very discreetly, in case their relief effort should become generally known in other places) to ensure that food was secured and was given to poor tenants and not monopolised by their landlords.[2]

By June, the Government at Westminster was fully aware of the plight of the destitute in both England and Ireland and could no longer stand idly by. An Act *To authorize the issue of Exchequer Bonds and the advancement of money out of the Consolidated Fund to a limited extent for the carrying out of public works and fisheries in the UK and employment of the poor in Great Britain in the manner therein mentioned* (57 Geo III C. 34) was passed on 16 June 1817 becoming known by its short title, 'The Loan Act'. It granted £1.5 million to England and £250,000 to Ireland for relief. The money was to be spent 'for carrying on of works of a public nature … upon due security being given for the repayment of the sums so advanced within a time to be limited'. The Lord-Lieutenant would administer the fund in Ireland with the aid of Loan Commissioners. Bodies politic or corporate, companies, proprietors and ordinary persons willing to finance public works, fisheries or collieries, could apply to the commission for a loan from the fund. The commissioners would evaluate each application with due regard 'to the benefit that might arise by employing the labouring classes and the number that would be employed'. The Irish Loan Commissioners, three in number, were empowered to appoint engineers, surveyors and other officers as they thought necessary at whatever salary they, the commissioners, deemed reasonable, to assist them in their assessments.

Already in its preamble, the Loan Act made it clear that the money advanced was a loan, not a grant-in-aid, and would therefore have to be repaid with interest in due course. Even though the conditions of the loans were onerous, the administration anticipated that an early start would be made to relief works. The question of the ultimate repayment was left somewhat vague, but it was intended that repayment would, in the final eventuality, be a charge on the whole county or on the barony.

Nimmo and John Killaly were appointed engineers to the commissioners. Nimmo thus describe their role:

> There is a Memorial sent in the first instance from the various proprietors in the district, stating the difficulties of communication through that county, and applying for aid; in addition to that a presentment is obtained from [the] county for the assistance of the work. This

Memorial is referred by the Government to Mr Killaly and myself, who act as engineers, to report upon the propriety of the plans and estimates of those works, and we then sign a report, stating our opinion of the benefits of the work to the country generally, in a national point of view – that it will tend to improve the country, and of course to encourage the employment of the people; and then, that the estimate and plan is a fair one, unless we find reason to doubt it: very often we have occasion to alter the plan, and frequently the estimate.[3]

As it happened, by the end of 1818, when the immediate famine had receded, very little of the money – less than £70,000 – had been expended, probably because it could only be disbursed by way of repayable loans (rather than free grants) and was restricted to cases where private subscriptions had already been established or could otherwise be guaranteed. For example, money could be advanced to Grand Juries on a mortgage of the county cess, or to the trustees of turnpike roads who, on receipt of an advance, were permitted to increase tolls to fund the repayment. The Times anticipated that the conditions attaching to the loans would create major difficulties: 'our apprehension of the complete success of the measure results from a doubt whether people will be found adventurous enough to borrow money for public purposes upon their own private security.'[4] That any government could have failed to foresee and anticipate this potential problem in Ireland is remarkable. The Loan Act was hopelessly optimistic where this country was concerned: there simply were not sufficient persons of capital willing to risk their personal sureties to undertake public works aided by loans repayable at 6 per cent. Within a month, Parliament had amended it: the amending legislation (57 Geo III C. 124) empowered the Lord-Lieutenant to advance, at his own discretion, part of the £250,000 given by the Loan Act to any private or local enterprise for the employment of the poor that was in receipt of a local contribution or subscription, without any requirement of personal security.[5] The amount to be advanced to any one scheme was limited to one half of the sum subscribed locally. A new set of commissioners called Special Commissioners for Relief in Cases of Extreme Necessity was set up to help administer the act. Once they were appointed the unlimited discretionary power of the Viceroy to make unfettered gifts was restricted to a sum of £40,000.

The works to be supported did not need a Grand Jury presentment, and any person who agreed to undertake an approved project would be taking on himself only half of the total cost, the balance being provided by the commissioners. It anticipated what we might call today 'public–private partnerships'. This was an improvement, but still not enough to make a significant impact and it certainly did nothing to win over the Grand Juries. In June 1818, yet another amending Act was necessary (58 Geo III C. 88).[6] This time, advances of loans could be made in instalments as the work progressed; they could be secured on the credit of anticipated tolls, rates, and so forth; the interest rate was reduced below 6 per cent; and advances could be made in anticipation of presentments that were yet to be granted by a Grand Jury. However, when the time came the Grand Juries, as was their right and duly acknowledged in the acts, would mostly refuse to make the required presentments.

Clearly the legislation was not achieving its intended effect of stimulating employment and very little was expended from the £250,000 originally allocated. The year 1819 had been one of great scarcity, so Parliament passed yet another Act in 1820 to amend the three acts passed since 1817. This latest was the Moiety Act (1 Geo IV C. 81).[7] It did not grant any new funds, but relaxed many of the conditions in the previous acts and extended the grant-in-aid provision. The Lord-Lieutenant was empowered to reduce the interest rate on existing loans below 5 per cent and to issue all further loans at whatever interest rate he considered appropriate, not exceeding five percent. He was empowered to appoint engineers to administer the arrangements, but they were to act 'without salary, fee or reward' in respect of their services under this act. Both Nimmo and Killaly were then appointed to these new engineering positions.

As soon as the engineers approved a proposal, it was lawful for the Grand Jury, and 'they are hereby required' (section 11) to make a presentment for a moiety (usually one-half) of the expense envisaged. A certified copy of the presentment, together with the engineers' certificate of approval and the estimate and plan, were then laid before the commissioners of the Loan or the Moiety Acts, who were required to certify to the Lord-Lieutenant the amount of the presentment. He would then make an advance equal to the whole presentment, or equal to one-half of the estimated cost of the whole proposal, to be paid out of the £250,000 allocated by the 1817 Loan Act. Payment would be made only when it was ascertained to the satisfaction of the engineers that half the presentment sum, or one-half part of the total estimate, had been well and *bona fide* expended in pursuance of the agreed works. The money was paid to the treasurer of the county, who then paid the overseer; the overseers were not liable for repayment of the sums granted, which were outright grants-in-aid. The total to be advanced was not to exceed £100,000 overall.

In terms of distress, 1822 was a truly appalling year, especially in the western and south-western coastal counties, and the Government was obliged to step in again with further legislation. On 24 May 1822, *An Act for the Employment of the Poor in certain districts of Ireland* was enacted.[8] The 'certain districts' were not named in the title of the act, but they were the south-western district of Cork, Kerry and Limerick, the middle district of Clare and contiguous regions, and the western district comprising all of Connaught. A notable and surprising feature of the act was that it did not make any new funds available.

Griffith, Killaly and Nimmo were appointed engineers to the respective districts. The Act permitted the Lord-Lieutenant to advance, from the sums already approved, loans for the full amount of all the presentments for roads and public works that had been passed at the Spring Assizes of that year. The money was to be paid to the engineers and they could use it to pay for materials, labour or any other purposes connected with the presentments. They were accountable for all sums expended by them, and when fully accounted for, the county treasurers would repay the sums to the exchequer in the normal fashion, out of the county cess. In addition, the Lord-Lieutenant could direct each engineer to report on any plan for making a road, which he could then fund at his own discretion (out of the already allocated funds), to the maximum extent of £50,000. Construction of all roads funded in this way was to be supervised by the engineers. On completion, the Lord-Lieutenant would place the accounts before the Grand Jury at the next ensuing Assize for it to make a presentment for the repayment of the whole, or part, of the sum out of the county cess. Once again, the jury was free, if it wished, to refuse to make such presentment. Since the *modus operandi* of the act was one that deliberately and comprehensively bypassed the Grand Jury from the very outset, this freedom to refuse was unfortunate, even if necessary. The Galway Grand Jury, for example, was reluctant to make presentments under the act when requested to do so: such presentments would have the effect of throwing on to the county rates the cost of works that had been approved, completed and paid for by central Government, without the Jury's prior consideration or consent. Few Grand Juries would wish to do this. An example of such reluctance occurred in Mayo, a county whose Grand Jury was normally active in availing of the loans on offer. When asked to make a presentment for the repair of branch roads off the main Erris road the Grand Jury declined, on the grounds that Nimmo had started the branch roads in question with Government funds, and therefore the Jury did not feel it was up to it to finish the task.[9] This reluctance did not affect normal presentments being approved, that is, those routine applications for work on existing county roads paid for from the county cess in the normal way and not requesting loans from the Consolidated Fund.

The Lord-Lieutenant responded to the reluctance of the Grand Juries by ordering the county treasurers to withhold their warrants levying the sums presented (i.e., ordering the treasurers not to debit the county cess with the sums presented). This was meant to encourage the juries to pass more presentments. When the stratagem failed, the Lord-Lieutenant ordered the advance payment of moneys from the Irish exchequer to the county treasurers, so that funds would be readily at hand for any presentments that were, or might be, made. This action, of doubtful legitimacy under the acts, was duly legitimated by a new act (3 Geo IV C. 84), which allowed the sums advanced to be used 'whether any presentment shall have been made for such purposes or not'.[10] The reason for circumventing the Grand Jury in this way was stated unambiguously in the act: 'in some parts of Ireland the Grand Juries, from a sense of the distresses of the country, have declined to make presentments for the repair of useful roads, and other existing public works which stand in need of repair...'

Another potential danger, namely that in the future the Grand Juries would fail to maintain and keep in repair the new roads that were being constructed with direct Government funds, was very real and did indeed come to pass. To avoid the deterioration this failure would cause to the new roads, an act, 6 Geo IV C. 101, was passed in July 1825.[11] Under this, completed new roads that had been funded by Government to at least one-half of their cost were placed in the care of the Directors of Inland Navigation for upkeep and maintenance. The directors would receive money from the Lord-Lieutenant to meet this new responsibility, paid from the sums already approved. However, at the end of the day the Grand Juries were required by the act to pass appropriate presentments *post hoc*, although repayment could be in instalments spread over a period of up to six years, as the Jury thought best.

Returning to 1822, a further act was passed on 5 August, 3 Geo IV C. 112, which approved a new advance of £250,000 from the Consolidated Fund to be used for relief purposes.[12] In addition, anyone who contributed private money for approved works that gave employment was granted interest at four percent per annum on the sums they contributed until the expected rates, tolls or profits of the works would themselves start to yield a dividend of 4 per cent. All of the works undertaken under the Loan Act and these various amendments, including the *Employment of the Poor in certain Districts Act* (3 Geo IV C. 34) and this latest Act, (3Geo IV C. 112), were to be proceeded with immediately. Where such works might require the purchase of lands, the Lord-Lieutenant was empowered to compel sheriffs to summon and empanel a jury to assess the value of the land in question. If any jurors refused to act, the sheriff could empanel '…other honest or indifferent men of the standers by…' to act in default.[13] The legislation seemed to be taking on grotesque dimensions, and in 1823 its reach was extended to funds of a different nature: by a new act, 4 Geo IV C. 42, the sums remaining unspent from various earlier commercial acts (ones that had approved loans for trade and manufactures in Ireland), could be appropriated by the Lord-Lieutenant, subject to a limit of £100,000, for public works carried out in furtherance of the purposes of the Loan Act.[14] For works that had already been agreed under the Moiety Act, the money could be advanced in instalments as the work progressed, provided the engineers certified that the instalments were properly expended on the agreed work. The act allowed an advance to be made of one-eighth of the estimated cost, once one-fourth of the work had been certified, a stipulation that meant the engineers, who were not paid for their services under the Moiety Act, might have to visit the works at least four times. It is little wonder that Nimmo expressed his unhappiness with this arrangement (see below). The provisions of the Moiety Act ran out in 1825, but were extended for one more year by a section of the Act 6 Geo IV C. 101, which transferred the roads and bridges to the Board of Inland Navigation.[15]

In June 1825 a new Act, 6 Geo IV C. 35, advanced a further £300,000

TABLE 5.
Sums granted from the Consolidated Fund in Ireland for Public Works, 1817 to 1827, showing dates granted, the relevant Acts, the total amounts approved for general purposes and the amounts reserved to the Lord Lieutenant for loans and gifts. (The allocation of £100,000 in 1823 under 4 Geo IV C. 42 came from the Commercial Fund; this and the reduction made in 7&8 Geo IV C. 12 are explained in the text).
Data collated from tables in references 20, 27 and 30 and from the text of the Appropriations Acts to 1827.

Date	Act	Sums allocated
June 1817	57 Geo III C. 34, s 5.	£250,000
August 1822	3 Geo IV C. 112, s 1	£250,000
June 1823	4 Geo IV C. 42	£100,000
June 1825	6 Geo IV C. 35	£300,000
July 1825	6 Geo IV C.134, s 10	£45,500
May 1826	7 Geo IV C. 79, s 12	£36,000
April 1827	7&8 Geo IV C. 12, s 47	£100,000
July 1827	7&8 Geo IV C.70, s 22	£20,000
Total allocated		£1,101,500
Less:		
Reduction made in 7&8 Geo IV C. 12	£100,000	
*Sums reserved to Lord Lieutenant,	£325,000	
Total deductions	£425,000	
Net sum available to the Commissioners of the Loan fund		£676,500
Total sum actually expended by them to April 1831[20]		£495,000.
*Sums reserved for the Lord-Lieutenant:		
Date.	Act	
July 1817	57 Geo III C.124, s 1	£35,000
July 1820	1 Geo IV C. 81, ss 11, 12	£100,000
May 1822	57 Geo III C. 124, s 1	£40,000
May 1822	3 Geo IV C. 34, s 4	£50,000
April 1827	7&8 Geo IV C.12 s	£100,000
Total reserved for Lord-Lieutenant (As gifts: £90,000)		£325,000

Notes:
Further sums, totalling £137,800, were granted later for public works in Ireland under the Acts 9 Geo IV C. 95 s 12 (July 1828), 10 Geo IV C. 60 s 17 (July 1829), 11 Geo IV &1 Wm IV C. 63 s 19 (July 1830), 1 Wm IV C. 5 s 14 (December 1830), 1&2 Wm IV C. 33 s 62 (October 1831), 1&2 Wm IV C. 54 s 19 (October 1831). All these were Appropriation Acts except 1&2 Wm IV C. 33, which set up the OPW.

under the usual conditions of repayment, with interest at 4 per cent.[16] The relief granted by the latest acts (4 Geo IV C. 42 and 6 Geo IV C. 35) was soon enough rescinded. For example, in May of the next year, 1826, the interest rate was raised to 5 per cent (by the Act 7 Geo IV C. 30[17]) and in April of 1827 a most confusing act, 7&8 Geo IV C. 12, was passed.[18] Section 1 of this act granted a further £100,000 over two years to the Lord-Lieutenant for public works but section 2 reduced the £300,000 granted in 1825 to £200,000. Half of this latter sum was to be administered by the commissioners, and half was reserved to the discretion of the Lord-Lieutenant. The confusion caused by these sections needed immediate clarification so that, only two months later, an act of clarification (7&8 Geo IV C. 47) was passed: the sum of £100,000 under section 1 of the disputed act was for the discretionary use of the Lord-Lieutenant; the sum of £200,000 in section 2 was divided equally between the commissioners and the Lord-Lieutenant, to be disbursed under the conditions specified in the Act 6 Geo IV C. 35, that is, as loans.[19]

The labyrinthine diversity and enormous complexity of the legislation was becoming a problem for those who administered it, those who accounted for it and those who worked with its support. It also gave the engineers immense power in selecting and approving new works. They could recommend works on their own initiative, or when proposed by others, either for loan support or grant-in-aid. The resources made available to them did not depend on the prior approval of the Grand Juries; these might contribute to some operations, but they were not essential to the powers and authority of the engineers. Prior approval, phased progress and final completion had to be certified by them and they were personally accountable for most of the expenditure. They were truly masters of all they surveyed. Naturally enough, the multiplicity of certification that was required led to severe difficulties in maintaining accurate accounts and records, not to mention the difficulty of site visits and the transport of instalments of cash safely from Dublin to the regions. On the negative side, while the engineers could spend the money as they needed to, they had to wait until the authorities reimbursed them some time afterwards. In many instances, Nimmo spent money quickly, as the law intended and the authorities wished, but had to wait for inordinate lengths of time before the Government finally released funds to reimburse him.

In summary, then, the Government's approach to the relief of distress in Ireland changed progressively from 1817 to 1827. Over the decade a total of about one million pounds was granted for relief, mainly through employment of the poor. Of that amount, the sum of £325,000 was reserved to the Lord-Lieutenant to allocate, at his discretion, to public works for which half the cost was available from elsewhere. Of this reserved sum, he could give up to £90,000 in outright gifts under the terms of the Moiety Act. All of the money administered by the commissioners (about £676,000), and the balance of the Lord-Lieutenant's reserve (£235,000), were to be expended in *repayable loans*. The various sums are shown in table 5. The total allocated, 1.1 million pounds, was not an entirely ungenerous response by Government at a time when non-intervention was the political orthodoxy; if it did not solve the difficulties of Ireland in the distressful decade of 1817 to 1827, that was more a fault of the social and institutional structures through which it was administered than any major inadequacy of the primary financial response. The Grand jury system was too compromised for the task; indigenous capital to leverage the funds was lacking for the most part; proprietors were absent or disinterested in many cases; and the engineers, Nimmo in particular, were left to act as best they could with inadequate institutional support and under severe human and environmental conditions.

Ensuring that all loans were repaid to the Exchequer with interest was the primary consideration initially, and the operation of the Grand Jury presentment system was the vehicle chosen to implement the Government's intentions. In that way, the Government did not have to abandon entirely its cherished policy of non-interference – all that it gave had to be repaid from the county cess so that, with the exception of the Lord-Lieutenant's £90,000, there was no money given as a free gift. As Gregory wrote to Grant (who had seemed to be yielding in the face of the distress he saw on his arrival in Dublin in 1819), 'the Government are not, on all occasions, to be the first resort when any pressure occurs, without the gentry contributing to the relief of their impoverished tenantry'.[21] As far as he was concerned, property had its duties as well as its rights, and the gentlemen of the country were under obligation to make some exertion themselves. Nevertheless, the exigencies of the time forced changes on the Government both in policy and in practice. While all advances were by way of loans with high interest rates at first, the interest was gradually reduced and the initial onerous conditions of surety and guarantee were progressively relaxed. When this failed to spur the Grand Juries to their duty, the legislation by-passed them entirely by giving extensive independent powers to the engineers. Overall, the authorities came to accept that reliance on the presentment system was not adequate for the crises. For example, the records show that over the period from 1817 to 1829 presentments availing of loans under the Loan Act and its amendments were made on only one occasion by the Sligo Grand Jury, two by Galway, three by Mayo, five by Leitrim and five by Roscommon – the most distressed counties during the period. In contrast, there were six made by the Kerry Grand Jury and there were very many applications made by private individuals.[22] In the seven years from 1824 to 1830 inclusive, the named counties levied the following amounts in connection with road works not requesting Consolidated Loan funds: Sligo, unstated number, £4,557; Galway, unstated number, £7,830; Mayo, 80, £4,914; Leitrim, unstated number, £745; Roscommon, unstated number, £2,198; Kerry, 135, £8,460.[23] To encourage private investors the interest rate was even reversed as, almost in desperation, investors were offered short-term interest on their investment in public works to overcome their hesitance in getting started.

By 1831, the total paid out in loans from the Consolidated Fund *for all purposes in Ireland* was £495,000. In fact, little enough of the money actually spent by the commissioners of the Loan Fund seems to have gone to the poorest districts, and the discretionary funds at the disposal of the Lord-Lieutenant formed the main Government support of the people in those districts throughout the period.

Nimmo and Killaly were appointed engineers under the three principal acts in this strand, namely; the Loan Act, the Moiety Act and the Employment of the Poor Act. Works that Nimmo supervised included the road from Bantry to Berehaven, the Cahirciveen to Tralee road, roads in north Kerry, the Killarney to Kenmare road, Poulaphuca Bridge and numerous roads in the western district to be discussed later. Neither he nor Killaly took these duties lightly. For instance, they failed to approve a road from Dublin to Clane, County Meath, because in their opinion it did not come within the purview of the Loan Act. As regards a proposed road from Glenville to Mallow, in County Cork, it did not warrant a grant in their view because 'it does not appear a work of such peculiar difficulty or national importance'.[24] It is certain that Nimmo was involved in many more such projects that may come to light in a future study, but some insight into his involvement can be gleaned from his accounts. He charged £198 for sixty-three days work prior to 24 May 1822,[25] and £431 for 137 days between November 1828 and February 1829.[26] If he worked an average of thirty-six days a year (three days a month) for the Loan Commissioners – surely a conservative estimate – then over the decade, his income under the Loan Act, entirely separate from his income as engineer of the Western District and engineer of the Fishery Commission, cannot have been less than about £1,100. Under the Moiety Act, the records show only twelve gifts of money, totalling over £32,000, made by the Lord-Lieutenant for public works up to 1824.[27] Of these, Nimmo was associated with nine namely Poulaphuca Bridge, Brenogue (Courtown) harbour, the Berehaven to Glengarriff road, the Killarney to Kenmare road, Loughallen road, the road from Celenaveeny colliery to Lough Gill, the road through Corraun to Achill, and the roads from Loughrea to Craughwell and between Dingle and Tralee. Whether he had any association with the remaining three – the roads from Leighlinbridge to Abbeyleix, from Sandyford to Stepaside and the proposed railway from Dromagh collieries along the Blackwater near Mallow, is not yet established. He received no payment for his services under this Act, a situation he did not much agree with, especially when the law allowed payment to contractors by instalments. On 5 June, 1823, for example, he wrote to Goulburn:

> May I draw your attention to 1 Geo IV [C. 81]. By that law engineers are to act 'without salary, fee or reward' in various matters of a professional nature. But they need to visit remote sites before reporting on them and this is expensive. By the proposed increase in the number of instalments this work will be greatly increased. If they could be paid as overseers or surveyors then they could continue to undertake the duties and be paid for them.[28]

Since he was being paid under the Loan Act, by the Fishery Commission, and under the *Employment of the Poor* Act all at the same time, he probably had little enough to complain about. Interestingly, almost all the works carried out by gift of money under the Moiety Act from 1825, totalling over £36,000, were monitored by Killaly and Griffith (Cahir Bridge in County Waterford was an exception[29]), so it is possible that Nimmo may have decided not to continue acting without due recompense.[30]

Under the *Employment of the Poor Act*, each engineer received payment of three guineas a day for the number of days worked. In Nimmo's case that amounted to £8,149 between June 1822 and January 1832, a sum calculated from his manuscript annual accounts as submitted (the precise information on his salary is lacking in published accounts).[31] It included £4,230 for 1,343 days from 1 February 1824 to 20 January 1832, equating to three guineas a day over that period. He charged £3,865 for the period from June 1822 to 1 February 1824, without specifying the actual number of days worked. This latter amount must have included significant expenses, because there were only 591 days in that period which, charged at three guineas per day, every single day including Sundays, would sum to much less than the amount he claimed. A similar anomaly had occurred with his Bogs Commission emolument, so this was not the first time, nor would it be the last, that his accounts failed to balance. Griffith charged £3,893 (including expenses) for 820 weekdays from June 1822 to 31 December 1824, so Griffith also claimed to have worked more than six days a week over the period.[32] A similar anomaly had occurred in Griffith's Bogs Commission accounts also.

The authorities had no doubt that the public works carried out by the engineers had a beneficial effect and were particularly deserving of parliamentary aid, 'as being calculated to give an additional and permanent stimulus to industry, an increased demand for labour, a greater facility of access to markets for all produce, and as a further incidental advantage, making the administration of the law, and the operation of the police more effective'.[33] As the decade wore on, however, and the legislation proliferated, the need for a single authority to control and execute public works, and be accountable to Parliament for the money spent, was becoming glaringly obvious, not least to those charged with scrutinising the accounts. When the Irish estimates were examined by a select committee in 1829 the members expressed the opinion that 'it is expedient … that a fixed, competent and controlling authority be established, acting under the direction of the Government, and responsible to Parliament, upon principles analogous to those adopted in the management of the Highland Roads and the North Wales Roads'.[34] The following year (1830) the Committee on the State of the Labouring Poor in Ireland agreed that public works had indeed been of immense value in public administration and life in general.[35] Like the Accounts Committee, it based its opinion and recommendations mainly on the information given in Nimmo's reports, which it quoted admiringly. It even reprinted in full his entire evidence to the House of Lord's Enquiry into the State of Ireland (1824), more than five folio pages of close print. Although Nimmo had not given evidence in person to either the select committee in 1830 or the committee on the estimates, his influence on them through his reports was considerable and recognisable. The former committee was chaired by Thomas Spring Rice, whose views were at one with those of his friends Nimmo and Maurice Fitzgerald. Fitzgerald now prepared a critique of public works in Ireland, which he transmitted to Wellington, the Prime Minister in June.[36] That document, with its historical and classical references, its sanguine view of Irish natural resources, its rejection of the poorhouse and its embracing of a certain form of support for public works, smacks strongly of his friend Nimmo. It was, in fact, partly a scathing attack on the deficiencies of the Grand Jury system, partly a commendation of the government engineers, partly an evaluation of the Irish national character and behaviour and principally an exhortation against stated, debilitating, aspects of poor law relief (aspects that Nimmo also rejected totally). It went on to propose a new commission for the management of public works.

The select committee reported in May 1830, making an unambiguous recommendation for the establishment of a Board of Works, even indicating the precise functions it should exercise:

> They recommend the revision of the laws under which public works are now carried on; the consolidation, under one fixed and responsible authority, of the functions now performed by the fishery board, the mail coach roads department of the Post Office, the Board of Inland Navigation, and the Government Engineers; the establishment of a system that shall ensure a proper selection of the work to be executed; a scientific formation of plans, surveys, specifications and estimates; a perfect execution of the work; a money payment to the labourers; a due audit and examination of accounts; and the preservation and maintenance of the works when completed.[37]

Nimmo alone of the engineers had carried out all of these functions on his own (together with his 'young men', of course) for more than a decade. By now, the long debate of political economists and politicians regarding the usefulness of public works was turning in favour of government intervention.[38] At long last the extent, the scope and the impact of Nimmo's practical contribution was becoming increasingly recognised and the idea championed by him since 1817, of an Office of Public Works along the lines of the French *Corps des Ponts et Chaussees*, was being mooted in official circles. By July of 1831, Parliament had before it a bill for an Irish Board of Public Works and the resulting *Act for the Extension and Promotion of Public Works in Ireland* passed into law on 15 October 1831, three months before Nimmo died.[39] It set up the new Office of Public Works (OPW) that took on the duties and responsibilities of five pre-existing boards, namely; the Board of Works for Dublin, the Board of Inland Navigation, the board for lending money from the Consolidated Fund, the Commissioners for Kingstown Harbour and the Commissioners for Dunmore Harbour. The Board of Inland Navigation had earlier assumed, pro tem, the powers and duties of the Fishery Commissioners, which were now transferred to the new Office of Public Works.[40]

Among the projects taken over by the new OPW were the Wellesley bridge, Ardglass harbour, the Drogheda harbour works, Courtown harbour, the road from Tarmonbarry to Lung Bridge, Dunmore harbour and the Dublin to Kingstown railway. These were all projects that Nimmo had initiated and progressed significantly, as the new OPW readily acknowledged, and there were very many others that did not warrant separate mention in the reports. Without diminishing the achievements of Griffith, Killaly and Bald, or the efforts of those others who took over Nimmo's public works on his demise, it is fair to say that none of them had made as comprehensive, sustained and dedicated a practical contribution as Nimmo had, in words and deeds, to the emergence of the new Office of Public Works. Above everyone else, it was Nimmo's abundant and overwhelming public service (both its successes and its failures) in fisheries, relief works, employment of the poor and at select committees and parliamentary enquiries that created the conditions for the balance to tip inevitably in favour of an Office of Public Works. When we consider all that he had been expected to do and what he had successfully achieved, the only wonder is that it took so long for the government to arrive at this inevitable conclusion.

There is little doubt that his friends Fitzgerald and Spring Rice had long been acting in consort in their political efforts to promote the formalisation of an Office of Public Works. Why then, we may ask, did they not ensure that Nimmo became an officer, or an engineer, of the new board? Perhaps it was because his health was seriously failing, or that he was spending more time at business affairs in Britain (to the detriment of his Irish commitments), or that he was coming under criticism, particularly directed from Galway, for the failure of some of his enterprises. A cryptic letter of 27 December 1829 from Peter LaTouche to Fitzgerald may have induced Nimmo's friends not to endorse him too publicly:

> I do not know what your plan is as to public works, but let the money be expended under the sole direction and responsibility of the Government – no societies, either Protestant or Catholic; and only from the apprehension of offending [sic] you and my friend Rice, let it be under any engineer but Mr Nimmo.[41]

Whatever LaTouche was implying, this innuendo may have been warranted in some part, but was grossly unfair in the main. After his death, all suspicion attaching to Nimmo's accounts – the only obvious reason for the veiled comment – was dispersed by the Public Accounts Office, leaving his reputation unblemished (see Appendix 4). Although he never served as one of its officers, his role in the emergence of the OPW remains an unacknowledged aspect of Nimmo's considerable legacy to his adopted country. It was not quite the *Corp des Ingeneurs des Ponts et Chaussees* that he had hoped for and worked towards, but it was nevertheless an achievement of great and lasting significance in Irish public life.

NOTES

1. J. D. Sirr, *A Memoir of the Honourable and Most Reverend Power Le Poer Trench, Last Archbishop of Tuam.* (Dublin: William Curry, 1845).
2. Jenkins, Era of Emancipation, p.129.
3. A. Nimmo, Evidence, p. 156, in Minutes of Evidence taken before the Select Committee of the House of Lords to examine into the Nature and Extent of the Disturbances which have prevailed in those districts of Ireland which are now subject to the Provisions of the Insurrection Act and to report to the House (B.P.P. 1825 [200], vol.VII. mf 27.60–3), hereafter referred to as House of Lords enquiry.
4. *The Times*, 29 April 1819, p. 3.
5. 57 Geo III C. 124. *An Act to amend an Act made in the present session of Parliament, for allowing the issue of Exchequer Bills, and the Advance of Money for carrying on Public Works and Fisheries and Employment of the Poor*, 11 July 1817.
6. 58 Geo III C. 88. *An Act to amend two Acts made in the last session of Parliament, for authorizing the issue of exchequer bills and the advance of money for carrying on public works and the employment of the poor; and to extend the powers of the commissioners appointed for carrying the said Act into execution in Ireland*, 5 June 1818.
7. 1 Geo IV C. 81. *An Act to amend several Acts made in the 57th and 58th years of his late majesty for the advance of money for carrying on public works, and for other purposes, so far as the said Acts relate to Ireland (The Moiety Act)*, 24 July 1820.
8. 3 Geo IV C. 34. *An Act for the Employment of the Poor in certain districts of Ireland.* 24 May 1822.
9. D. Bingham to W. Gregory, 18 March 1830 (NAI, CSORP 1830/453 [2]).
10. 3 Geo IV C. 84. *An Act to authorize certain temporary advances of Money, for the relief of the Distress existing in Ireland*, 27 July 1822.
11. 6 Geo IV C.101. *An Act to provide for the repairing, maintaining and keeping in repair certain roads and bridges in Ireland*, 5 July 1825.
12. 3 Geo IV C. 112. *An Act to authorize the further advance of money out of the consolidated fund for the completion of works of a public nature, and for the encouragement of the fisheries in Ireland.* 5 August 1822. Another amending Act, 3 Geo IV C. 86, passed days earlier on 26 July, had not materially affected the Irish legislation.
13. 3 Geo IV C.112, section 5.
14. 4 Geo IV C. 42. *An Act to amend the several Acts for the assistance of Trade and Manufactures, and the support of commercial credit, in Ireland*, 27 June 1823.
15. 6 Geo IV C. 101, section 12.
16. 6 Geo IV C. 35. *An Act to render more effectual the several Acts for authorizing Advances for carrying on Public Works, so far as relates to Ireland*,10 June 1825.
17. 7 Geo IV C. 30. *An Act to amend the several Acts for authorizing advances for carrying on public works and to extend the provisions thereof in certain cases*, section 2. 5 May 1826.
18. 7&8 Geo IV C.12 *An Act to amend an Act of the first year of His Present Majesty for the Advancement of Money for carrying on Public Works in Ireland*, 2 April 1827.
19. 7&8 Geo IV C. 47. *An Act for the further amendment and extension of the powers of the several Acts authorizing advances for carrying on public works*,23 June 1827.
20. Public Works Ireland. An account of all sums of money advanced out of the Consolidated Fund on the recommendation or with the sanction of the Irish Loan Commissioners [to 1831] (B.P.P.1831 [184], vol. XVII, mf 34.128), hereafter referred to as Account to 1831.
21. W. Gregory to C. Grant,1817, quoted in A. Gregory, *Mr Gregory's Letter Box.* (Oxford: Oxford University Press, 1982), p. 137.
22. Account to 1831.
23. Roads in Ireland. Returns of all Sums of Money expended in each and every county in Ireland ... levied by Grand Jury Presentments (B.P.P.1831 [211], vol. XVII, mf 34.128).
24. Report of the Engineers on the proposed road from Glenville towards Mallow, 7 February 1822 (NAI, CSORP 1822/2569).
25. Twenty-first Report of the Commissioners for Auditing the Public Accounts in Ireland, p. 174 (B.P.P. 1833 [102], vol. XVII, mf 36.124).
26. Works of Alexander Nimmo, 1831 and 1832 (Nat. Archs. UK, AO 17 492).
27. Public Works Ireland. An account of all sums of money advanced out of the Consolidated Fund for Public Works in Ireland, by way of Loan, from 16 June 1817 to 5 January 1824 on the recommendation or with the sanction of the Irish Loan Commissioners (B.P.P. 1824 [278], vol. XXI, mf 26.135).
28. A. Nimmo to H. Goulburn, 5 June 1823. (NAI, CSORP 1823/ 7455).
29. Certificate under I Geo IV C. 81, that Cahir Bridge is three-fourths completed (NAI, OP 972 17).

30. Public Works Ireland. Account of all the Sums placed at the disposal of the Lord-Lieutenant of Ireland, for Public Works: Sums granted by way of gift (B.P.P. 1828 [464], vol. XXII, mf 30.150).
31. Works of Alexander Nimmo (Nat. Archs. UK, AO 17 492.)
32. Seventeenth Report of the Commissioners for Auditing the Public Accounts of Ireland, pp. 145–148 (B.P.P.1829 [193], vol. XIII, mf 31.88–9).
33. Report from the Select Committee on the Employment of the Poor in Ireland (B.P.P. 1823 [561], vol. VI, mf 25.49–51).
34. Report from the Select Committee on the Irish Estimates, p.26 (B.P.P. 1829 [342], vol. IV, mf 31.23–4).
35. Report from the Select Committee on the State of the Poor, being a summary of the First, Second and Third Reports, p. 36 (B.P.P. 1830 [667], vol. II, mf 32.47–57). Hereafter referred to as Committee on Poor.
36. An Abstract of some Communications to His Grace the Duke of Wellington in 1829, in Report from the Select Committee on Public works Ireland, appendix 3, pp. 319–329 (B.P.P. 1835 [537], vol. XX, mf 38.143–8).
37. Committee on Poor, p. 40.
38. R. D. Collison Black, *Economic Thought and the Irish Question 1817–1870* (Cambridge: Cambridge University Press, 1960).
39. 1&2 Wm IV C.33. *An Act for the Extension and Promotion of Public Works in Ireland*, 15 October 1831.
40. 1 Wm IV C. 54. *An Act to revive, continue and amend several Acts relating to the Fisheries.* 16 July 1830.
41. P. LaTouche to M. Fitzgerald, 27 December 1828 (PRONI, Fitzgerald papers, MIC/639/reel 7, no. 28, p. 44).

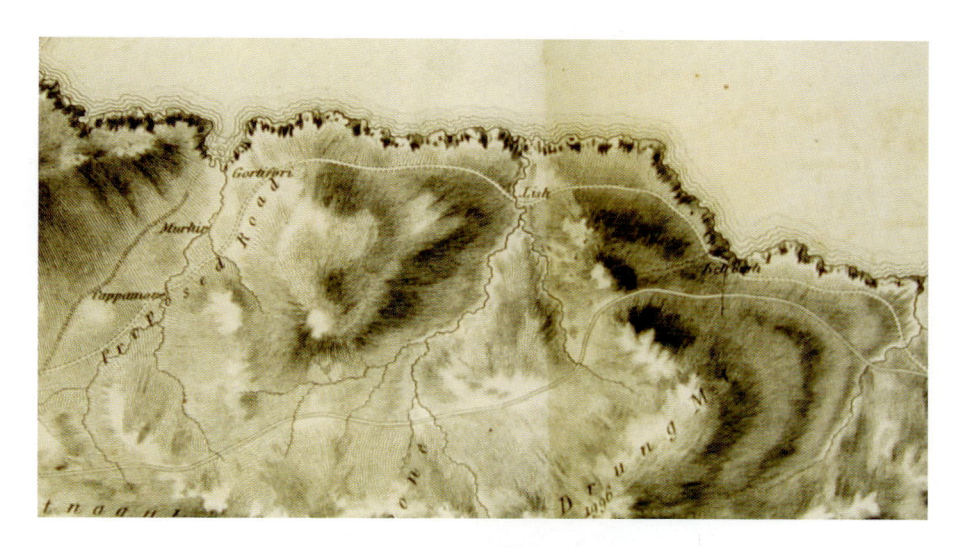

PLATE 1
Figure 1. (Top) Nimmo's proposal for a new line of road around the base of Drung Hill, Co. Kerry, avoiding the existing road over its shoulder. (From fourth report of the Bogs Commissioners).
Figure 2. (Bottom) On his map of Dunkerron, Co. Kerry, Nimmo inserted the line of road he was constructing between Blackwater Bridge and Derryquin. (From fourth report of the Bogs Commissioners).

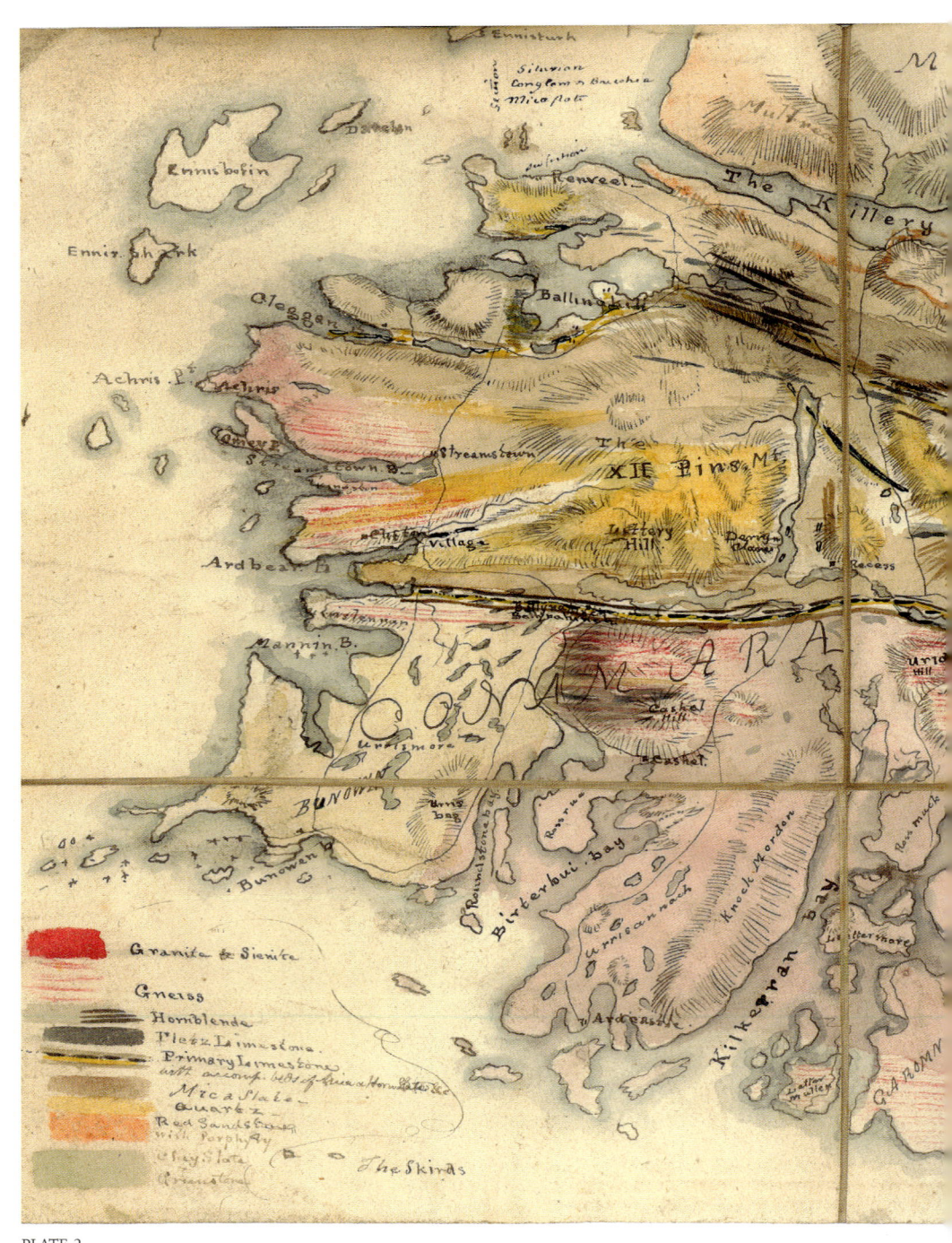

PLATE 2
Geological sketch map of Connemara, Co. Galway inscribed by George Bellas Greenough 'Nimmmo', 'Connemara', and most probably dating from 1813. (Geological Society of London, LDGSL 999. Reproduced with permission of the Society.)

RISK.

Joyce's Country

Lough Mask

Ballinrobe

Cong
Cross

Headfort

Oughterard   Castle

Killeen
Costelo bay
Tullie

ARCONNAUGHT

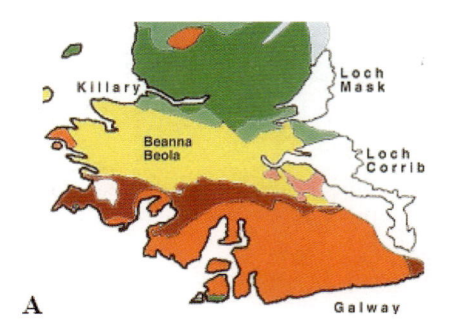

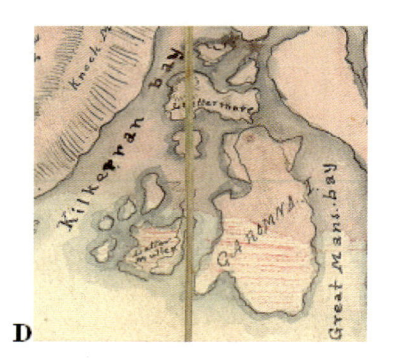

PLATE 3
The geology of Connemara and south Mayo as known today with details of Nimmo's representation of some important features
      A. Simplified geology of Connemara and south Mayo. (From P. Mohr, *Wind, Rain and Rocks*, with permission of the author).
      B. Nimmo's representation of the limestone boundary near Galway town.
      C. Nimmo's representation of the Corrib-Mask isthmus.
      D. Nimmo clearly distinguished the South Connemara Group.

PLATE 4

Thomas Soutell Roberts *A View of Cork*. (Detail of watercolour). The curved line through the centre of the harbour is the Navigation Wall, now the Marina. Note the extent of water south of the wall, an area now fully reclaimed. (Reproduced with permission of the Crawford Municipal Gallery, Cork.)

PLATE 5

Nimmo's map of the sea bed sediments, 1825. (From *Transactions of the Royal Irish Academy* vol. XIV, (1825), Part 1, pp.39–50). (Paper read October 1823).

Granite

Conglomerate
Fathoms
Nymph Bank

WALES

Slate Rocks
Sandstone
Coal
Lias

Granite

50 40

E N G

Red Marl
Lias
Chalk

Granite

60

60

Granite
Scilly I.ˢ

Graywacke or Brown Sand

Calcareous Tract

Red Sand

Grey and White Sand
Granite

Conglomerate or Gravel & Pebbles

Granite

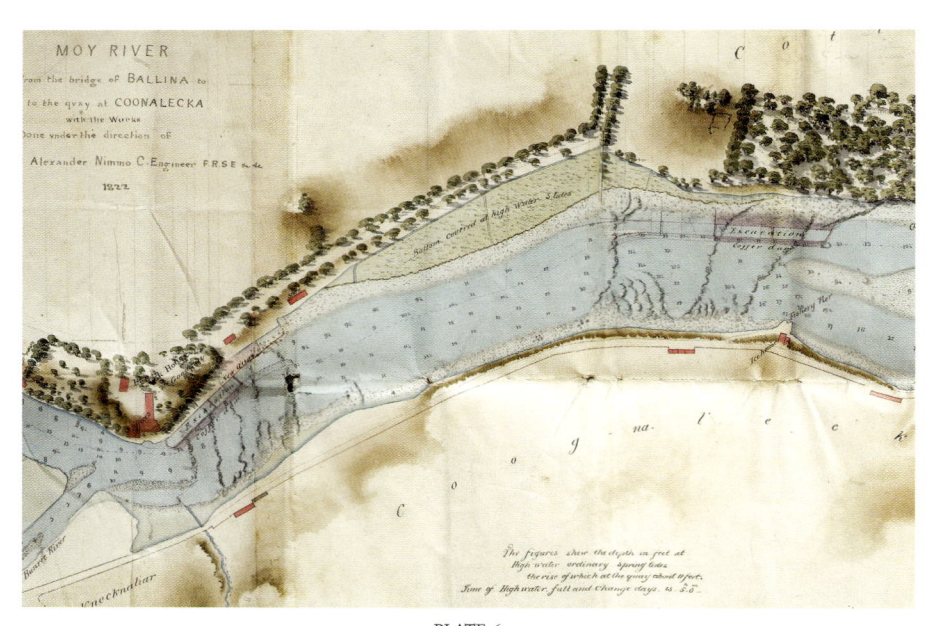

PLATE 6
Nimmo's plan of the River Moy, Ballina Co. Mayo with the works done under his direction, 1822–3. (NAI, CSORP 1823/5144/3, reproduced with permission of the Director, The National Archives of Ireland.)

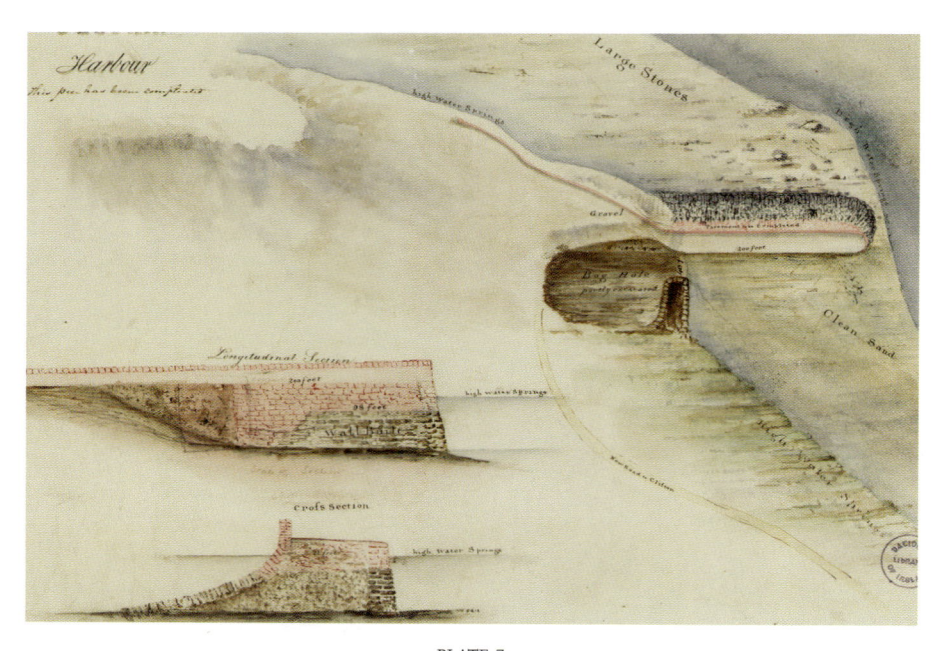

PLATE 7
Detail of Nimmo's plan of Cleggan Harbour, Co. Galway. (NLI, maps, 16 H 5 (5), reproduced with permission of the Director, The National Library of Ireland.)

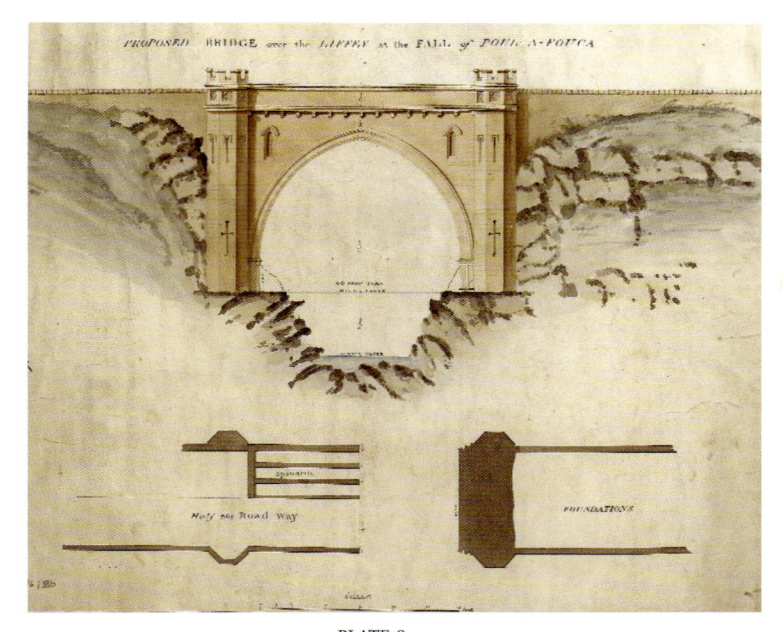

PLATE 8
Drawing of the proposed bridge at Poul-a-fouca (Pollaphuca) Co. Wicklow, attributed here to Nimmo. (NAI, OPW 5HC 386, reproduced with permission of the Director, The National Archives of Ireland.)

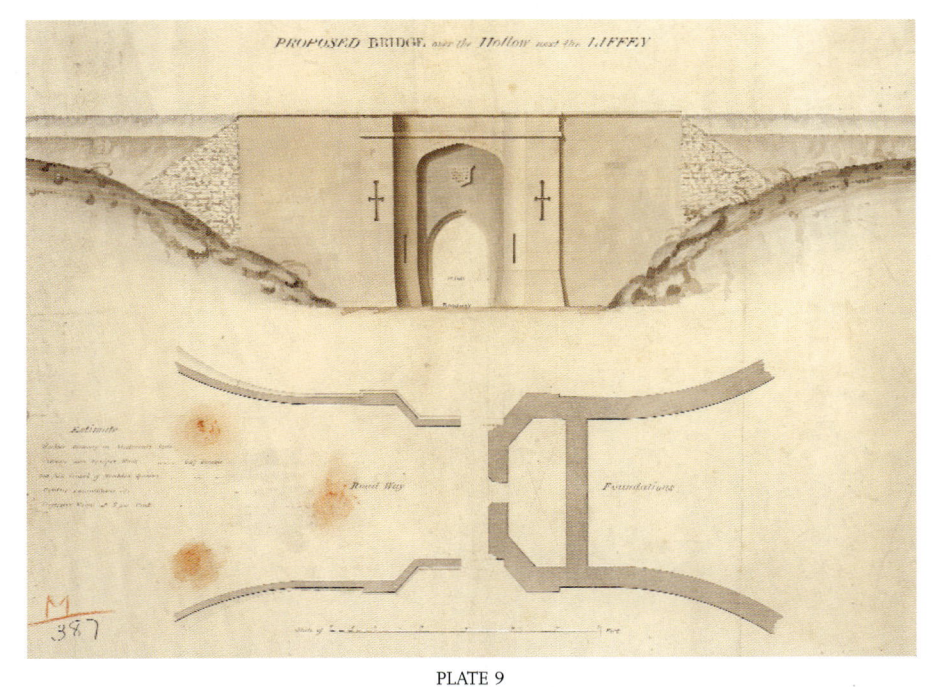

PLATE 9
Drawing of 'Proposed Bridge over the hollow next the Liffey' (Pollaphuca, Co. Wicklow) attributed here to Nimmo. (NAI, OPW 5 HC 387, reproduced with permission of the Director, The National Archives of Ireland.)

PLATE 10
Detail from Nimmo's drawing of the proposed Wellesley Bridge in Limerick city. Note that the original plan was for a seven-arch bridge (not counting the bascule and floating dock section), having an arch at each end that was narrower than the five main arches. (NAI, OPW 5 HC 6 340, reproduced with permission of the Director, The National Archives of Ireland.)

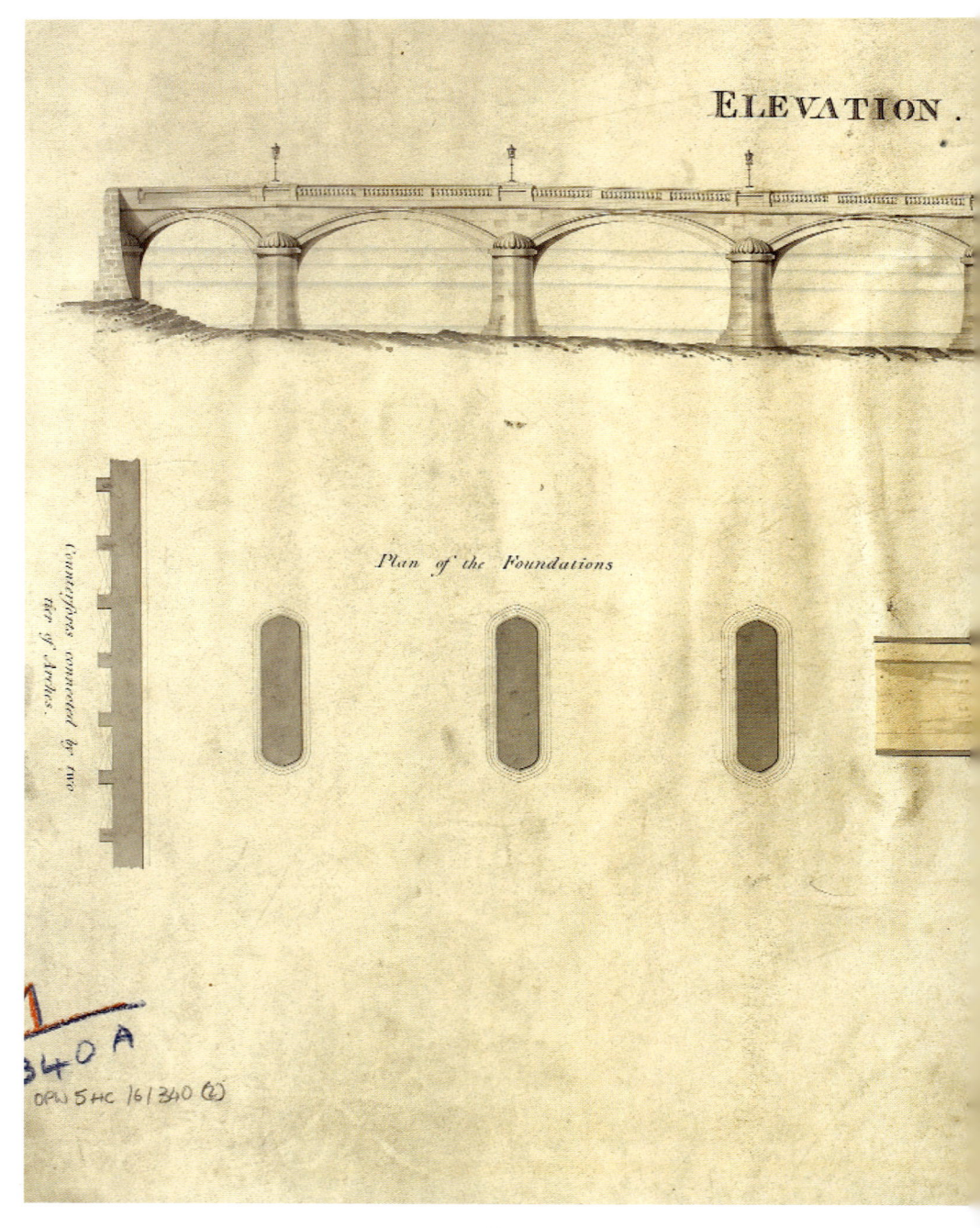

Longitudinal section across t...
the Bascule an...

Plan of the Superstructure

Exterior Quay of Dock.

Dock

dock gate

bascule

Tide

Scale of

PLATE 11

Nimmo's map of the roads in progress in Connemara, Co. Galway in 1828 (NAI, CSORP 1828/1468, reproduced with permission of the Director, The National Archives of Ireland.)

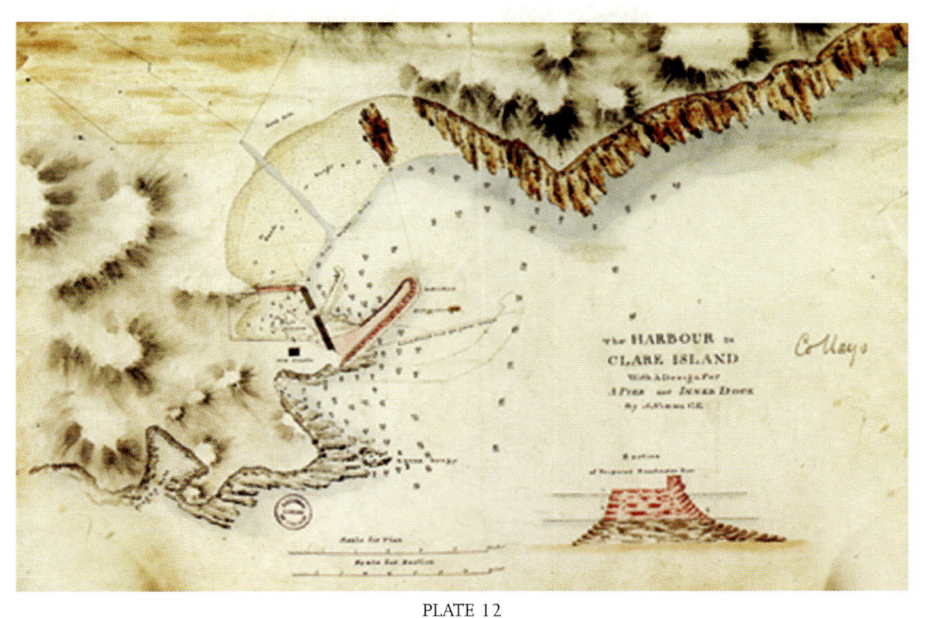

PLATE 12
The Harbour in Clare Island with a design for a Pier and Inner Dock by Alexander Nimmo C. E. (Detail)
(NLI, maps, 16 I 3 (1), reproduced with permission of the Director, The National Library of Ireland.)

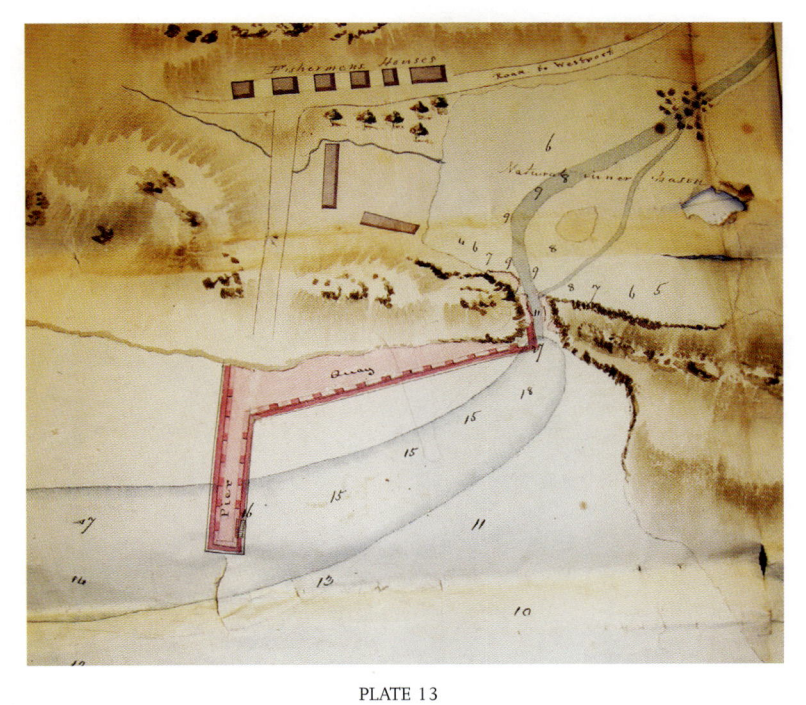

PLATE 13
Nimmo's plan for a quay and pier at Bundorragha, Co. Mayo. (Detail from NAI OPW 8/54/8/8,
reproduced with permission of the Director, The National Archives of Ireland.)

PLATE 14
Detail of Nimmo's plan for Carrigaholt pier, Co. Clare. (NLI, maps, 15 B 9 (2), reproduced with permission of the Director, The National Library of Ireland.)

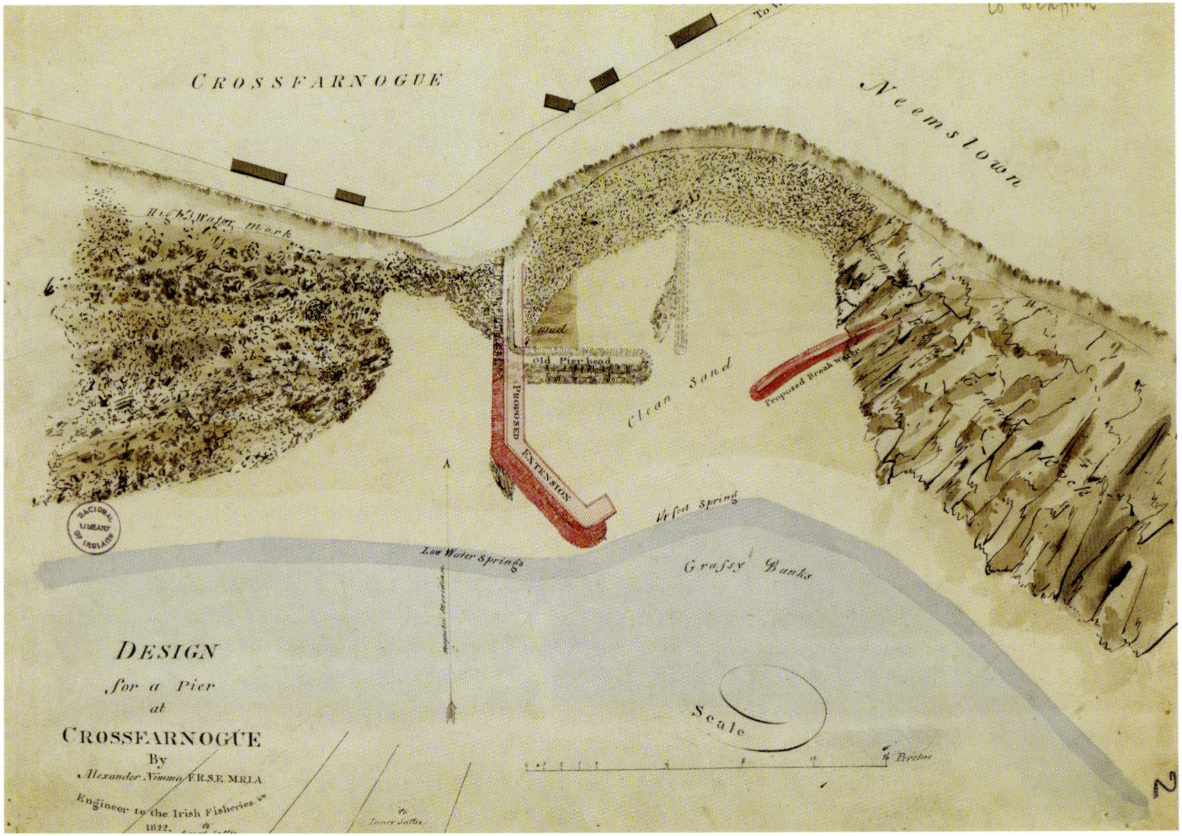

PLATE 15

Nimmo's design for a pier at Crossfarnogue (Kilmore Quay), Co. Wexford in 1822. (Detail from NLI, maps, 16 J 13 (2), reproduced with permission of the Director, The National Library of Ireland.)

PLATE 16
Detail of 'Survey of the Bay and Town of Killough', signed by Nimmo. (PRONI D2581/1, reproduced with permission of the Public Record Office of Northern Ireland.)

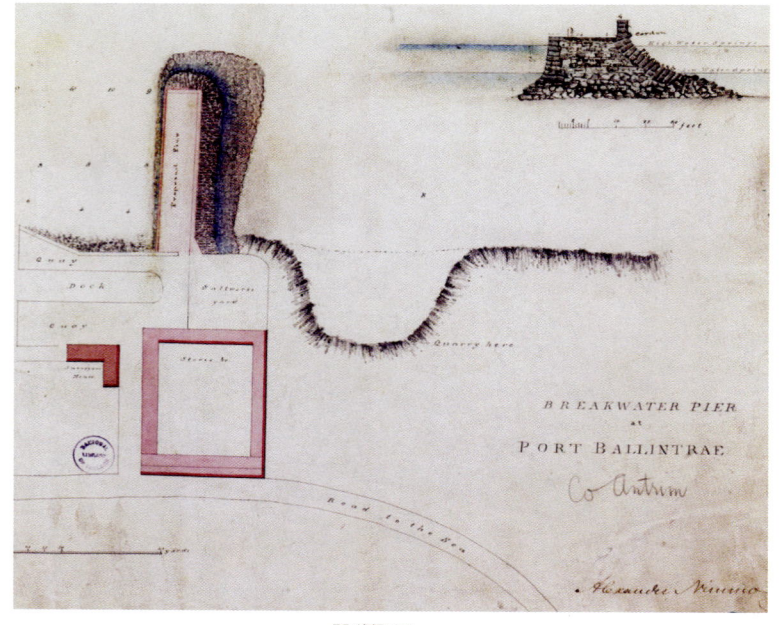

PLATE 17
Nimmo's plan for a breakwater pier at Port Ballintrae, Co. Antrim. This is one of his earliest maritime plans in manuscript. (NLI, maps, 15 B 1 (1), reproduced with permission of the Director, The National Library of Ireland.)

CHAPTER 9

The Fisheries Commission and Nimmo's Coastal Survey

After the Act of Union, the Government did not move very quickly to promote the Irish fisheries. Until the death of Napoleon, the sea coasts of Ireland presented a possible route into the United Kingdom for revolutionary and dissident elements, as events a short number of years previously had proved. Therefore, the less the Irish were encouraged to develop maritime capability the better. In any case, fostering the native Irish fisheries could only damage the fishery interests of Britain, especially the Scottish herring buss fisheries. Herring busses used to visit the coasts of Ireland to buy herrings from the local fishermen, who caught them by seine netting in the shallow bays and inlets. The development of a native Irish herring fleet might therefore cause serious competition for the British fleet. The existing Irish fisheries acts dating from 1734[1] were maintained in place by acts in 1817 and 1818, and it was only when the Select Committee on Disease and the Labouring Poor highlighted their potential in 1819 that some urgency was placed on updating the legislation.[2] By July of that year a new, comprehensive Fisheries Act was passed, setting up the Commission for Irish Fisheries with £5,000 for its operation.[3]

The commissioners met for the first time on 19 October 1819 and proceeded to appoint four Inspectors-General for each of the coasts, with fourteen local inspectors working under them. Regulations and instructions, based on advice received from the Commissioners for the Scottish Herring Fishery, were drawn up for the use of the new officers. The inspectors were required to make regular inspections of their respective districts and to report to the commissioners the number of bays, harbours and creeks, the number and tonnage of fishing boats in each, their fisheries, markets, roads to the coast, headlands, islands and the general state of the population.[4] For a more general survey of the coast, the commissioners requested the Lord-Lieutenant to appoint an engineer to 'make a general survey of the coasts of Ireland; to report the state of the harbours or places of shelter along the coast, the most advantageous sites for fishing stations, and the most useful lines for communication between the principal harbours and interior of the country, through the mountainous districts'.[5] The engineer, it should be noted, was not required to plan, design or construct any harbour, pier or landing place. Nimmo was appointed to the post in August 1820.[6] His friend Maurice Fitzgerald was one of the commissioners and would have favoured him right from the start, although by his own admission to the select committee, Nimmo knew nothing of the Irish fisheries.[7] That, of course, did not stop chief secretary Grant from appointing him, perhaps in reward for his previous help with the select committee on the Poor. Neither did it stop Nimmo from taking up the position with his customary enthusiasm.

On 15 September 1820 he commenced his survey in Sligo, working his way around the north coast in a clockwise direction and completing the circuit as far as Belfast by the end of October.[8] It was, in truth, a rather cursory affair: the weather was bad and the cutter from which he hoped to conduct his observations did not meet him until he arrived in Derry. He therefore carried out the survey from Sligo to Derry by circumambulating the coastline, taking notes of the suitability of the various inlets and harbours, observing the numbers and types of fishing craft and noting the size of the population in different areas and the normal occupations of the people. From local interrogations he determined the approximate locations of the main fishing banks and the availability of markets. On the Sligo and Leitrim coasts he identified very few suitable harbour sites – Raughley and Mullaghmore being the principal ones – but his opinion of the south Donegal coast was very much better. Killybegs inlet and harbour were, in his view, the safest on that part of the coast and the best rendezvous location for the fishing fleet. However, it was then almost abandoned by fishermen, because it was impossible for a vessel to put to sea in a west or south-west wind. Eight miles further on, Tilen (Teelin) had a large number of boats but no pier, and in his opinion one could be made there within the bar. He went on to Rutland Island and described the remnants of the fishing village planned for the place by William Burton Conyngham in 1786 and never completed.[9] Because it was on an island, Nimmo did not favour the location for further development: 'It is inaccessible by land, and therefore could never become a market for the country behind, poor as that is. The island itself is too small to give food to its present inhabitants, and much of it is unproductive.'[10] He did not venture into Sheephaven or Mulroy Bays. There was little that interested him in Lough Swilly except Rathmullan, and his opinion of Buncrana as a possible fishing station was negative: 'it has a bar on which the sea breaks in northwest gales. It would be impossible at any moderate expense to cure these bars by piers.'[11] Greencastle in Lough Foyle held out better prospects: by the construction of two short piers, a tidal harbour of about four acres could be made for about £4,000.

There was little in his report so far that the local inspectors could not have recounted. His sketch plans of the various locations are skimpy beyond words and not a single one gives any hydrographic information except that for Greencastle, which has four soundings. The survey cutter finally met him in Derry, so from there on, he could travel and survey by water except when the exposed nature of the coast prevented it. Portrush was by far his most favoured location for a fishing station on the north coast. The existing port belonged to the merchants of Coleraine and they wanted to improve the shelter at the harbour. Nimmo's suggestion was to form a mole, 240 feet in length, from Rumore Hill southwards with a jetty at its head, and a cross pier to it from the east side of the harbour, forming a mouth ninety feet wide enclosing a much larger basin. The cost would be £6,000 and he made an indicative plan of his proposal.[12] At Port Ballintrae some miles further on, a Mr Spencer had previously built a small harbour to service his salt works and now proposed to build a breakwater to further improve it. Nimmo supported this, which he thought might cost £500 to £600, and wrote to the commissioners saying it was 'judicious' and could be entrusted to the Spencer family.[13] Later on, in 1822, Spencer contacted the commissioners with a new plan but they were slow to respond because Nimmo had shown no evidence that there were any fishing banks in the area.[14] Nimmo continued to support Spencer in 1824, and the proposed breakwater eventually went to contract in 1825.[15] However, the work was soon abandoned when the commission's new harbour inspector, James Donnell, refused to approve the first instalment of the funds.[16] At Ballycastle also, Nimmo thought that the existing harbour could be greatly improved at a cost of about £5,000. Between Ballycastle and Larne he regarded Cushendun as the best site for a pier, which he estimated would cost about £2,000. He ended his first survey in Belfast and submitted his report to the commissioners on 30 January 1821.[17]

From its very commencement the commissioners were enthusiastic for the survey, stating that 'they had every reasonable expectation that the benefits arising from this survey, when completed and published, will not be confined to the national object over which they preside, but will extend to the commercial interests of the country and the internal improvement of vast tracts, which ... still continue in their original state of uncultivated wildness'.[18] Dangerous concepts, like 'national object' and 'commercial interest', were being bruited about and would come back later to haunt them. For the present, however, their mood was upbeat.

Nimmo's observations and plans grew more precise, more detailed, more elaborate and more costly as he moved along the coast, and it is obvious that he was learning rapidly as he went along. In all, he had recommended nearly £20,000 worth of harbour works, four times the gross annual budget of his employers. Something else, too, became obvious: he was not adverse to taking on private commissions for harbour works while he was surveying for the Fisheries Commission. He returned to Sligo in the summer of 1821 ostensibly to enhance his earlier cursory report, but with other things also in mind. The Commissioners for the Port of Sligo had commissioned him to undertake a detailed survey of the bay and to draw up plans for the improvement of the port. He therefore sounded the bay, from the quays out to the bar, and from Coney Island out to the Black Rock. His report, presented on 20 March 1822, recommended extensions to the quays and proposed a cut, one and a half miles long, 180 feet wide and five feet deep at low water, through Cummeen Strand in the inner bay. By embanking part of the inner bay, he proposed to form a reservoir that would help to scour the proposed new channel, keeping it free from sand.[19] He was, as usual, painting the big picture not unlike his proposal for Lough Mahon in Cork, and it had, just like in Cork, a high cost attached. In Sligo's case, it would cost in the region of £15,000, a sum well beyond the resources of the Port Commissioners. Nevertheless, in April the provost of Sligo, with the support of Lord Palmerston and the local traders, brought a Memorial for the harbour improvements to Dublin, seeking financial support.[20] Telford was called in to advise and while he praised Nimmo's 'excellent chart', his opinion was that the site was too sandy to make the Cummeen cut and embankment advisable.[21] He approved Nimmo's other proposals, which were implemented in part later on. The port authorities paid Nimmo £300 for his report and he arranged that the chart of his hydrographic survey of the bay would be published at the expense of the Fisheries Commissioners, thereby saving Sligo the expense.[22]

At the extremity of Sligo Bay, Nimmo had previously identified Raughley Point as the best location for a fishing station. He now proposed that a pier be constructed there in a small, sheltered natural bight. Further excavation at the back of it would make for an admirable inner harbour for small craft. The whole would not only serve the fishery, but also be useful to the trade of Sligo in facilitating access for the pilots and general shelter for boats. It would cost an estimated £1,150. Between Sligo and Donegal, the only other place where fishing boats could come in westerly winds was Mullaghmore, where he now designed a small fishery harbour at an estimated cost of £1,182. It was to comprise a pier and rough breakwater enclosing a small basin on the eastern shore of the peninsula. He had plans for a much bigger pier in deeper water capable of protecting vessels of any size in all conditions, 'but such a work would probably be too expensive an undertaking for any funds that could be expected'.[23] He may have had a sense that Lord Palmerston, on whose land the pier would rest, would take the bait, as it were, and contribute to the bigger proposal, which he did some years later.

Next, he revisited Greencastle, where he proposed two piers to enclose a harbour on a smaller scale and at a lower cost than his original estimate. By now, he was coming round to the view that small piers were what were really needed. 'The benefit of a small pier where none existed before,' he wrote, 'is sufficient to awaken a desire for future improvement; and, at present, it is my opinion that works of that kind should chiefly attract the attention of the Board.'[24] His and the commissioners' conviction regarding small piers was based on a pragmatic assessment of the nature of the Irish fisheries. More has been made of the number, rather than the type, of boats that were engaged in the Irish fisheries before the Great Famine, and reference to the limited number of vessels claiming the various fishing bounties gives no true indication of the total number of boats actually in use by the population. In 1826, for instance, there were 378 decked vessels, 485 half-deckers, 2,334 open sail boats and 7,526 rowboats registered in the fisheries.[25] About 4,500 men worked the decked and half-decked vessels, while over 50,000 worked the open boats. By 1830 there were 1,114 decked and half-decked boats employing about 6,000 men, and 12,000 open boats with almost 59,000 men.[26] Small open boats, rather than any decked craft, were therefore the mainstay of the Irish fisheries. All along the west coast, rowboats and curraghs were by far the predominant boat used in the fisheries and in other maritime activities, something the commissioners well understood. As far as curraghs and rowboats were concerned, what they needed was sheltered, clean shores where they could be drawn up above the tide level, or lie aground safely moored on the slab or shingle. Hence, small piers that sheltered and protected natural coves were the structures most likely to meet the needs of these fisheries. Since most fishing boats would normally lie aground when not in use, the piers did not need to extend below low tide level; this made them easier and cheaper to construct than large floating docks, which require complicated underwater construction methods.

Without question, Nimmo's greatest work in the next season was his survey of the north-east coast from Belfast to Dublin. Having found the survey cutter so valuable in the first season the commissioners, on Nimmo's advice, purchased it outright along with the services of its crew for £1,200.[27] He named it *Dunmore* and, although small and only a sailboat, it ranks as Ireland's first national hydrographic survey vessel.[28] Having the cutter at his sole command added greatly to his ability to progress the survey on the water and led to observations that resulted in some of his finest maritime charts. Four of these – *The Entrance to Strangford River, Killough and Ardglass, Carlingford Lough* and *The Coast of Down from Lee Stone to St John's Point* – were engraved in 1821; three of them (*Carlingford Lough* was excluded) were published by the commissioners in their fourth report for the year 1822.[29] In each, extensive soundings and sailing directions were given, and all four charts were important contributions to the inshore navigation of a region where cross-channel and coast-wise trade was as important as the fisheries. The commissioners acknowledged explicitly the importance of his charts: 'specimens are annexed, illustrating the advantages likely to result from this very important survey; not merely to the Fisheries, but to the general navigation of our Coasts'.[30]

It was no mere chance that they were so useful. Nimmo was aware that no proper charts existed for the area except those of Murdoch Mackenzie, which he knew were in need of revision. The care that he took in drawing up and publishing his own charts is evident in his comments regarding the one he compiled of Belfast and Larne harbours:

> I have made out a draft of the Loughs of Larne and Belfast, from the county maps, aided by Mackenzie's survey. But although the chart so made is certainly superior to this last mentioned [Mackenzie's], and may be sufficient for the purpose of the fishery, yet I find these authorities differ so materially from each other, and also from my own observations, that I cannot recommend it so confidently for publication until an opportunity be obtained of examining the documents above mentioned.[31]

His draft chart of Belfast has survived, and it is indeed far less comprehensive than his other published charts and appears to be of very limited practical utility.[32] The fact that it remained unpublished, taken together with his remarks quoted above, serves to emphasise the caution he exercised when using Mackenzie's observations. In making his own charts, he indicated those soundings that were borrowed from previous surveys, like Mackenzie's, by inserting them at right angles to his own soundings, a precaution that would serve him well later on.

The survey continued into the autumn of 1821, taking in the coasts of Down, Louth and Dublin. Nimmo identified sites where small piers might be useful, but in all cases the plans illustrated were simply indicative sketches. There were, however, some important exceptions, chief of which were Bangor, Ardglass and Killough, which he dealt with in greater detail and which will be discussed later (see gazetteer, appendix I).

By the spring of 1822 the survey had arrived in Dublin Bay, the chart of which ranks with the others mentioned in hydrographic detail and quality.[33] The immense value of the whole undertaking was now obvious. The hydrography in particular was in an eminently satisfactory state and the commissioners had every right to be happy with the achievements of their engineer, who had shown himself to be an accomplished hydrographer. His charts were a major contribution to the safe navigation of the coasts and came to be regarded as important national documents. William Bald considered them to be 'among the finest engraved specimens of our hydrographic surveys yet published'.[34]

Turning briefly to the west coast, Nimmo continued the coastal survey there in 1823, carrying out a 'very careful maritime survey' and laying down the results in a general chart of Connaught and in detailed charts of its harbours.[35] He inserted the hydrography of the Mayo coast on the Grand Jury map of that county that Bald had triangulated accurately but on which only Mackenzie's soundings had up to then been inserted.[36] The whole west coast survey was helped considerably by Nimmo's private commissions: the Commissioners for the Port of Sligo paid for the survey of that bay; merchants of the towns of Ballina and Killala paid for the surveys of the River Moy and adjacent Bay of Killala; and the Marquis of Sligo paid for the survey of Westport harbour. Nimmo later surveyed many of the bays of Connemara in the course of his western district duties, extending his observations more widely to the local creeks and inlets so as to enhance the value of the resulting charts for the use of the fisheries. In this way, he prepared detailed charts of Sligo Bay, Killala Bay, Clew Bay, Bofin Island, Enniskea Island and Greater Galway Bay (from Cleggan to Blackhead in Clare) at little or no cost to the Commissioners for Fisheries. He extended his survey out from the coast to the fishing banks off Erris and Connemara, especially to the Sunfish Bank. He went out almost to the 100-fathom line off Erris: [37]

> It is evident that the bank of soundings, which near Mizen and Brandon Heads, in the south of Ireland, extends only a few miles from the land, gradually expands after passing the Shannon, is of considerable breadth between Connaught and the Western Isles of Scotland, and some say even connects with the Banks of Newfoundland. Some of our soundings off Achill Head extend as far as ninety-seven fathoms, being the deepest yet marked on the western coast of Ireland.[37]

He realised that he was getting out of his depth, literally and otherwise, this far offshore:

> The examination of the banks situated so far in the offing would be more appropriately intrusted [sic] to nautical men than to an engineer and it would, in my opinion, be of the greatest importance to the fisheries and to navigation, if the Admiralty were to direct a sounding survey to be made of that part of the ocean situated outside of the north channel, as far at least as Aitken's Rock, for I am fully convinced our present information of that tract is of a very vague and imperfect nature.[38]

Nevertheless, he continued out beyond the 100-fathom line near the Sunfish Bank, as his report proves:

> This bank is remarkable from the break of the tide on it, with ebb and flood, and is supposed to be a ridge of land extending from the Blaskets to Erris Head, in about seventy fathoms. Half a mile further off we have fifteen fathoms more water, and the increase in depth is also considerable within it; the water outside deepens quickly to 100 fathoms and upwards; and the probability is, that this bank is near the edge of soundings.[39]

Impressive as Nimmo's coastal survey was, it contained no mention of another extensive and novel survey he had in hand, which he described in a paper he read to the Royal Irish Academy in October 1823.[40] When published in 1825, it was accompanied by an engraved, hand-coloured map of the sea area within the 100-fathom line between Ireland and France, including the

English Channel, showing the nature of the sea-bed as determined by the sounding lead. Mariners normally attached a plug of tallow or some other sticky substance to the lead, and when the lead hit the bottom on sounding, some detritus stuck to it. When this was examined, it was possible to deduce the nature of the bottom. Nimmo knew that the very deep sea-bed was comprised of mud and ooze; as one approached nearer the shore, sediments became progressively more sandy, then gravelly, then stony, the heavier sediments being less easily transported away from land to deep water by the tides. Very generally that is true, but not to such a degree that it can be relied on to predict bottom types reliably within the 100-fathom line.[41] We know this now because of the many seabed surveys that have been carried out in modern times, but Nimmo's was the very first marine geological survey and map of this kind.[42] He had taken numerous soundings in his coastal survey and in his other harbour works, extending his observations out to deeper water wherever possible. But even a cursory glance at his map (Plate 5) suggests that it is most unlikely that he personally made all the observations that were necessary to comprehend the whole area covered. He must have compiled data from many other sources, such as existing hydrographic surveys, the observations of fishermen and mariners plying between Britain, the Continent and Ireland, the navigation notes of pilots and other sources.

His observations on the Irish Sea, although not illustrated on the map, were quite detailed and extended northwards to the River Clyde. The banks to the south of Dublin Bay, known as the Irish Ground, were, he reported, composed of sand and rocky ridges extending to the coast of Caernarvonshire, Wales. South of Wicklow, outside the banks, he found that the mid-channel was comprised of shelly sand extending to Tuskar. Off the south coast, the Nymph Bank was a sandy flat plateau. So far, this scenario was probably common enough knowledge to the fishermen who frequented these grounds. But his observations extended far beyond the range at which Irish fishermen would normally work at that time and which they would be familiar with. Generally, his observations encompassed the Sole Bank, the Celtic Sea and the English Channel, with its western and south-western approaches. He identified various accumulations of mud, sand and gravel, outlining their boundaries with a degree of precision and which, if they do not conform precisely to modern knowledge of the area, at least gave contemporary evidence that the sea-bed was not uniform, that shelly material was a large component of the bottom sediment, that there were muddy tracts close to shore and that, broadly, sea-bed material could give some new information of use to the navigator. He concluded that it was possible on rational principles 'to delineate masses of the different soils on the bottom of the sea, as has been done of the land', and that with this information 'a great additional precision will be afforded to this mode of discovering [a] place, when astronomical observations are not to be had and [the] reckoning is doubtful'.

Examination of sea-bed sediments never became an important adjunct to modern navigation, despite Nimmo's conviction that it would. Indeed, rather than geology becoming an aid to navigation, as his paper set out to show, navigation would become an aid to geology, as ships and shipping improved. Almost one hundred years later, when Irish scientists were studying offshore geology by examining rocks and stones dredged up in trawls, they were considered to be remarkably innovative in their approach; the original contribution of their predecessor Nimmo was, once again, unaccountably overlooked.[43] Today we attribute the observed distribution of the sediments to the sorting and winnowing actions of waves and tides acting on post-glacial sediments, rather than reflecting the underlying solid geology of a particular place, as Nimmo mostly thought. Whatever the shortcomings of his efforts when judged against today's greater knowledge, and however imprecise or deficient his appreciation of the complexity of sea-bed deposition processes, Nimmo's was the first real attempt to map the sub-sea deposits of these or any other waters. Real innovation consists not in being correct from the very beginning, but in initiating novelty from which further work may advance towards the truth. For this, Nimmo deserves to be remembered by historians of earth and ocean sciences who, with only a few exceptions, have largely overlooked his pioneering efforts.[44] Griffith, in his great geological map of Ireland, acknowledged Nimmo's contribution to its maritime aspects, showing (to his credit) that he appreciated the pioneering nature of Nimmo's work.[45]

A claim that he was Ireland's first great resident oceanographer would, therefore, rest easily on the evidence of his charts, his practical sailing directions, his harbour soundings, his observations on the nature and diversity of the sea bottom and his soundings to the edge of the continental shelf. But fate would not be kind to him in this matter. The originals of the western charts do not appear to have survived and we know them only from his written reports and especially from the engraved inserts reproduced in his great chart of the coast of Ireland that accompanied his *New Piloting Directions for St George's Channel and the Coast of Ireland*.[46] Copies of that book are not encountered much today and the only copy of the chart recorded in Irish archives is that in the Irish Architectural Archive.[47] The whole chart is very large, approximately 175 cm long and 177 cm wide, dissected into four sections, each approximately 88 cm square, at longitude 8' 30"W and latitude 53' 40"N. The Irish Sea and north-east coast sections extend to the coasts of Scotland, England and the Isle of Man with inshore soundings recorded from Colonsay Island to Preston in Lancashire, in the estuaries of the Rivers Dee and Mersey, and around the Isle of Man. Soundings are extended southwards to 51' 05"N off the south coast of Ireland and out to over ninety-five fathoms off Achill Island. Inserts with soundings of Waterford Harbour, Cork Harbour, Valentia Harbour, Inner Galway Bay, Roundstone Bay, River Moy, Carlingford and Dublin Bay confirm the extent and detail of his surveys on all coasts.

His great chart of Connaught does not appear to have been engraved full size, unless we regard the north-west coast section of the chart of Ireland (section from Inishturk Island to the River Foyle) as fulfilling this role. A small-scale chart of Killala Harbour was published separately by the Fishery Commissioners.[48] The original of his transect of the soundings out to ninety-seven fathoms, and the original manuscript of the soundings at the Sunfish Bank, do not appear to have survived either. The Admiralty survey of the west coast made in the decades after Nimmo's demise never publicly reported depths beyond sixty-five fathoms, so the original manuscript records of Nimmo's achievement are a particular loss. Perhaps it is this absence of the originals of the various charts and documents that explains the diminished awareness among scientists and maritime experts of his contribution to Irish hydrography. It could also be the consequence of an attack on him that occurred late in 1823, which may have dented his reputation and certainly put an early end to his coastal survey.

Nimmo had never intended to encroach on the jurisdiction of the Admiralty, as the quotation above testifies. But he had no diplomatic skills in matters of technical expertise: he was not backward, for example, in pointing out the deficiencies of Mackenzie's charts and in criticising the inadequacies of Captain Bligh's 1801 survey of Dublin Bay, although both were made under the auspices of the Admiralty.[49] The fourth report of the Fisheries Commissioners, which contained Nimmo's survey and charts of the northern and north-eastern coasts, was published in May 1823. His reputation was therefore already at a high point when his lecture on sea-bed sediments, delivered on 27 October of that year, confirmed his primacy in hydrography and oceanography (a word not yet invented at the time). How could one man with one small sailing cutter have done so much, so successfully? Where were the Admiralty and the esteemed Navy, with all their hydrographic and surveying expertise?

Begrudgery being never far from success in Irish life, a letter of complaint was sent to the Admiralty alleging a serious error in his chart of *The Entrance to Strangford Lough*. The Admiralty informed the Lord-Lieutenant, and on 17 January 1824 chief secretary Goulburn contacted the Commissioners for Fisheries. The Admiralty had informed him, he wrote, that the survey of Strangford Lough was very deficient, a rocky shoal being shown one and a quarter miles from its true position. Goulburn's and the Lord-Lieutenant's reaction to the allegation was out of all proportion to the gravity of the alleged error:

> As there is stated to be so mischievous an error in one of the charts issued under your authority the Lord Lieutenant thinks it necessary to [request your explanation] … in as much as, if the statement be correct, His Excellency can have no doubt of the expediency of discontinuing all further survey of the coasts or harbours of Ireland.[50]

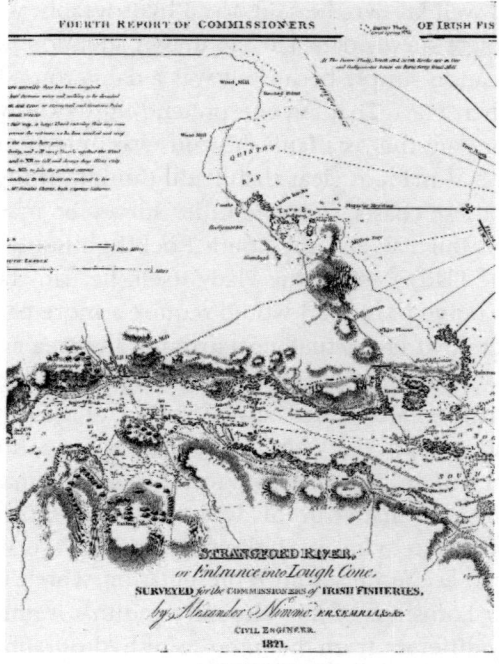

FIGURE 9.1
Detail from Nimmo's published chart of the Entrance into Lough Cone (Strangford Lough). Note that the Butter Plady is shown outside the top margin of the engraving and note the absence of detail in Quinton Bay and Kerney Point. Contrast with figure 9.2. (From plate 1 in ref. 29.)

The draconian nature of this response suggests that someone in authority, maybe in the Admiralty itself, was not altogether happy with Nimmo's maritime achievements, and that the commissioners' high praise for the survey's 'national object' was coming home to roost. The Admiralty was never likely to welcome encroachment of any sort on its perceived 'natural' functions, like hydrographic surveying. Who was this engineer in Ireland who thought he could produce accurate hydrographic charts and carry out sea-bed surveys single-handedly? Who was he to advise the Admiralty what it should, or should not, do outside the North Channel, or anywhere else for that matter? Someone needed to be brought rapidly and thoroughly to heel!

Nimmo replied to the commissioners on 26 January, a letter half of explanation and half of indignation, but full of suspicion.[51] The rocky outcrop at issue, called the Butter Plady, was nowhere close to the entrance to Strangford Lough, he explained, and therefore did not properly form part of his chart entitled *The Entrance to Strangford River* at all; he had inserted it outside the margin of the engraving in a part of the coast that he had not personally surveyed, as indicated by the soundings being laid down horizontally, solely to give an idea of its existence. These details in the original engraving are shown in figure 9.1. Even the landward features illustrating the location were drawn in outline only, a convention well known, he said, to all hydrographers and draftsmen as indicating incomplete survey data. He had drawn attention to the Butter Plady and its general location simply because it was a dangerous shoal that was not shown on any other chart. That anyone pretending to know anything of the matter should represent this as a fault, he said, was 'utterly astonished'. In his written report he had made it clear that 'I still think it of high importance to the navigation of these coasts, that one other survey be made of the numerous rocks between Bur Isle and the South Rock lighthouse.'[52] (That was the general area of the Plady.) As to the Plady itself, he had stated in his report '…that the whole range [of rocks] would require a more particular survey…' and that shipwrecks that had actually occurred in the area even while his survey was ongoing '…imperiously showed the necessity of more minute investigation into the dangers along the coast of Ireland, and our present ignorance of which coast and its harbours, although it employs one third of all the overseas trade of Great Britain, is really disgraceful to us as a maritime nation.'

Having explained the matter in this way, to his own satisfaction at least, he now took the offensive with an indignant stance: 'I cannot help feeling extremely surprised at the uncandid representation which has been made of this matter to their Lordships of the Admiralty and that it must have originated in something very different from the interests of hydrography and navigation.' As his other charts had been examined 'in the same spirit' but without further fault, he took as proof of their accuracy and fidelity, despite their novelty. Having thus engaged the issue he came to the heart of the matter:

> I am told that Captain White in the *Shamrock* has been employed several years, in fact since the Peace [1814], in making surveys on the coast of Ireland, but as no results have yet been made public, I am unable to say how far his operations are calculated to preclude the necessity of any part of the survey we are now carrying on … But on enquiring at the Admiralty for this purpose I was informed by the Secretary that they had in fact nothing yet for the use of the Navy but Mackenzie's charts.

Not content with that trump card, he proceeded to criticise some aspects of Bligh's 1801 chart of Dublin Bay. In brief, while satisfied that it was broadly accurate, he declared that some of Bligh's marks were so imprecise that they could lead boats on to the North Bull, or over rocks. This led him to conclude undiplomatically, recklessly and even dangerously: 'From this instance I am not disposed to place implicit reliance on a survey merely because it has been done by a person employed by the Admiralty.' The whole tone of his letter was now one of the highest dudgeon, and one most unlikely to endear him to the Sea Lords, whatever their view of the original allegation. Relenting a little, he concluded by undertaking to revisit the Plady at the earliest opportunity. He would have done so earlier, he wrote, except that the coastal survey had been diverted to the west coast as a result of which he hoped to show in the spring

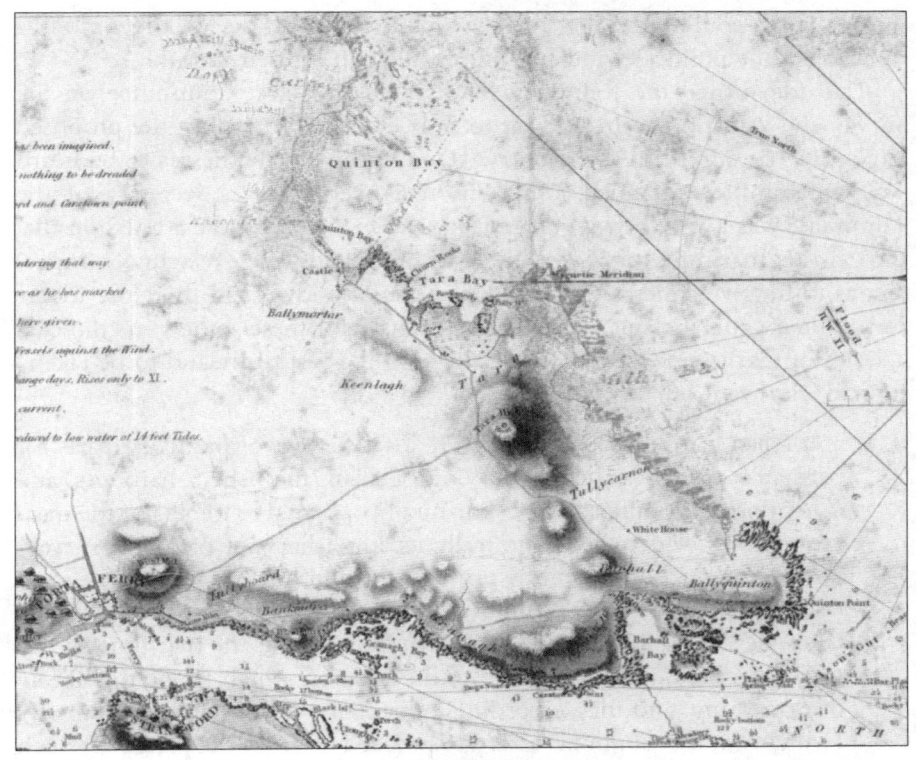

FIGURE 9.2
Alterations (made in red ink) to a copy of the same engraving as figure 9.1. Butter Plady at very top, just out of picture. Note the addition of coastal detail and two compass bearings to locate the Plady. (Detail from ref 54. Reproduced with permission of the National Library of Ireland.)

over 400 rocks and shoals on that coast that were entirely missing on existing charts.

One week later, on 4 February, he wrote again to the commissioners telling them that he had revisited the Strangford area and wished to have some modifications, to an area not previously covered properly, added to the engraved map.[53] This explains the MS modifications inserted in red ink on an engraved copy of the original chart now housed in the National Library of Ireland and shown in figure 9.2.[54] He must have been smarting still from the whole affair because, in a classical 'put up or shut up' challenge to the Admiralty, he finished his missive: 'I gather the Admiralty has recently done a survey. Perhaps on your application the Admiralty would communicate a copy of it, that it may be printed with the others for the use of the fisheries of Ireland.' Since the Admiralty had no such new survey chart, that challenge hardly poured oil on troubled waters. However, if victory would appear to have gone to Nimmo, it was only pyrrhic. No one knew quite as well as the Admiralty when to engage and when to hold fire. While the wind was in Nimmo's sails the Sea Lords could bide their time, which they had to do for a few months only, when the political wind and tide would run in their favour.

The tide turned the following May when the Select Committee on the Survey and Valuation of Ireland sat to hear evidence regarding the proposed trigonometric survey of the country.[55] One of the first witnesses to be heard was a committee member, John Wilson Croker who, as secretary of the Admiralty, was a very heavy gun indeed. He opened with the admission that the existing maritime surveys of the coast of Ireland were exceedingly imperfect as to the principles on which they were laid down and therefore incorrect in their general result. 'Their Lordships have for these some years directed a series of maritime observations relative to the coast of Ireland to be taken,' he said, and went on,

> and I also had heard that there were local surveys of particular parts for particular purposes such as the Bog of Allen, the fishery harbours, and other portions of the country. It naturally occurred to me that if the maritime survey were to be scientifically accomplished, any particular surveys carried on, on a less scientific principle, would become unnecessary and that expense (whatever it might be) might be at once saved ... I stated these views to the Board of Admiralty who, as far as the maritime survey was concerned, fully concurred in the opinion which I took the liberty of expressing, and they accordingly directed, that the maritime surveys (which it was thought necessary to proceed in, from the pressure which we felt on account of the badness of the existing charts) should be concluded in such a manner as would render them hereafter applicable to the Trigonometric Survey of the coast.[56]

As a consequence, he said, he had written in April of 1822 at the direction of the Admiralty to the Board of Ordnance in England proposing cooperation between that body and the Admiralty in linking the ordnance and the maritime surveys. The Board of Ordnance had expressed agreement, he said, and would proceed 'whenever their means admitted' – a convenient enough dismissal. At the time, the Ordnance Survey of England was already thirty-three years in progress and only two-thirds complete, hardly a work of great dispatch, as the select committee rightly observed.[57]

How far 'maritime observations' constitute a proper hydrographic survey of the Irish coast is a moot point, but Croker's gratuitous dismissal of the achievements of the Bogs Commission as a 'local survey of the Bog of Allen' exhibits an *hauteur* of immense proportion, compounded by his disparaging insinuation of 'less scientific principles' in the procedure of all other surveys excepting, of course, those of the Admiralty – all this from a person who, on subsequent examination, showed little real knowledge and no practical experience of hydrography or cartography. When asked whether the hydrographic survey of England was linked to the Ordnance Survey of that country, he had to admit that originally it was not. As regards the Ordnance Survey of Scotland, which had commenced much later than that of England, there was no linkage either, because it 'had been very far proceeded in before the idea which now appears so obvious and simple, of a maritime cooperation, officially presented itself to us'. That, at least, rings true: linkage of the trigonometrical and the hydrographic surveys would greatly reduce the cost and enhance the scope and value of both enterprises, so the proposal was a good one for Ireland, provided both surveys would be progressed with dispatch, and no effort would be spared to accelerate their completion. The progress of the Admiralty survey of the Irish coast up to then was, as Nimmo had indicated, hardly conducive to a sanguine expectation of its early completion. However, in fairness, the Irish Ordnance Survey would get under way within the year, and the hydrographic survey would soon enough be up and running too, partly as a result of these proceedings.

Nimmo was interrogated in depth regarding his coastal survey on 27 May.[58] First he was asked to distinguish which parts of the coast had actually been surveyed by him and which were merely noted. He admitted that the section from Sligo to Derry had only been by perambulation of the coast and that the plans he had given to the commission were in a great measure simply indicative sketches or rough outlines. The only general maritime surveys of the coast then extant, he stated, were those of Mackenzie that were generally used by the Admiralty, and although tolerably correct, they contained a variety of egregious errors that he first discovered by comparison with the county maps. He instanced the issue of the Butter Plady that he had now revisited and laid down properly on the chart so far as that was possible. Others of his charts, for example that of Galway Bay and four or five others for which he had orders for two years, were in preparation for engraving. His charts of the coasts of counties Louth, Dublin and Mayo were based on county maps that were themselves grounded in good trigonometrical surveys. His method was to copy the trigonometrical position of inland features from the county map and then survey the coast (between high water and low water) and the maritime area (below low water) by reference to the inland positions. In the other maritime counties the base maps were of an inferior description and he had found it necessary to make new trigonometrical surveys in those cases. His instruments in these instances were a theodolite measuring to fractions of minutes (seven-inch, made by Nairn; others, smaller, by Jones) and a sextant that took angles to single minutes. In one case, where the survey was very extensive, he used a theodolite which read to one second (made by Troughton, lately owned by General Brown of the Staff Corps and afterwards by Major Taylor, who gave it to Nimmo) and a sextant (ten-inch radius, made by Troughton) reading to ten seconds. Then, 'after the trigonometrical observations which adjusted the whole charts, I have universally measured the whole of the coast with a chain and light theodolite, or sextant, reading to single minutes'. So much for surveys 'on a less scientific principle'! It was a spirited, forceful and trenchant defence of his own professionalism, clearly aimed at countering Croker's dismissive, disparaging and poorly informed opinions.

Nimmo then repeated publicly his claim that a secretary of the Board of Admiralty had informed him that the Admiralty was in possession of nothing other than Mackenzie's charts, but it intended to make surveys of the coast of Ireland when the trigonometrical survey should be complete. When the question was put to him whether it would be advantageous to include the hydrographical detail in the trigonometrical survey, he agreed that it would certainly be of great value to the country. His answer to the separate, but related, question 'In the event of their [hydrographic details] being included, would it supersede the necessity of a more confined and local survey for the same purpose?' was more circumspect: it would depend, he said, on the scale and the object in view; it would supersede the necessity of making general charts of the coast, but not that of harbours. An ordnance map, as he had earlier pointed out, confined itself to the land only, and a nautical chart to deep water, so that the intermediate part, the actual shore, was frequently defective in detail. Therefore, even while agreeing that the linked trigonometrical and hydrographic surveys made sense on a large scale, he could still foresee a role for the continuance of local surveys.

Admiralty secretary Croker, when asked 'Are the existing imperfect surveys of Ireland to which you allude those of Mackenzie and Heather, or whether there are any subsequent?' had replied: 'I know of no others of that general description and none later than those of Heather; but the Admiralty surveys have enabled us to correct those charts in several instances.' He entirely ignored Nimmo's coastal survey and he sailed blithely on, either negligently oblivious to, or deliberately dismissive of, Nimmo's contribution. In the event, the members of the committee, probably unaware of the earlier correspondence and peremptory decision regarding Nimmo's coastal survey, were swayed by the Admiralty's arguments against the continuance of all such surveys and, in consequence, they recommended the immediate discontinuance of all local surveys. Nimmo's coastal survey, in temporary suspension until then, was officially terminated from that moment. Much to their credit the Commissioners for the Irish Fisheries knew exactly who was behind the termination, a fact they felt free to acknowledge four years later: 'The coast survey mentioned in all the former reports of the Commissioners has not been resumed, objections having been raised to the prosecution of local surveys, the value of which would be superseded by the general and minute work now carrying on under the orders of His Royal Highness the Lord High Admiral.'[59] Some years later, the vigilance of the Admiralty would be tested when the Commissioners for Fisheries asked Nimmo to carry out a new survey of Valentia Harbour. Nimmo replied on 2 July 1827, agreeing to do the job for

£180 sterling.⁶⁰ When news of this reached London, Croker proved to be not quite as ignorant of Nimmo as previously. Before July was out, he had written to Gregory who immediately sent a rocket from Dublin Castle to the commissioners: 'Since the Ordnance Survey are doing the Trigonometrical Survey, all local maps and charts are rendered quite unnecessary...', it stated, '...and the Lord Lieutenant requests that the local survey proposed by you be suspended'.⁶¹

Nimmo may have already finished the sounding of Valentia by the time of this latest veto, or maybe it was later rescinded, because on 15 June 1832 the Hydrographic Office of the Admiralty published a chart entitled *The Harbour of Valentia by Alexander Nimmo, civil engineer, 1831*.⁶² This engraved chart comprises three maps: the harbour of Valentia at approx one and a half inches to the nautical mile; an enlarged inset of the north-western entrance; and an enlarged inset of the western entrance. Its publication by the Admiralty, coming six months after his death, indicates some softening of its attitude towards Nimmo. By then the Admiralty, in fairness, had commenced its own hydrographic survey of Ireland and the bitterness of the *contretemps* with Nimmo may have receded entirely once that was under way. From 1830 to 1834 one of the Sea Lords was none other than Maurice Fitzgerald, and he may have played some part in the posthumous rehabilitation of the reputation of his old friend, in whom he had never lost faith: as late as 1834 he was still quoting Nimmo's calculations of the distance between Valentia and Halifax in support of a packet service between the continents.⁶³

There can be little doubt that the success and excellence of Nimmo's surveys and charts of the coast had helped to goad the Admiralty into taking the hydrographic survey of Ireland sufficiently seriously to commence its own full survey in 1830. As a result, Ireland ended up with detailed hydrographic charts of its coastline that were the envy of many maritime nations. Had Nimmo not been so effective, it is doubtful that the coasts of Ireland would have engaged the attention of the Admiralty quite so soon, especially given that Napoleon was dead and the threat of a French invasion had finally died with him.

Hydrography was only one aspect of the coastal survey, which was also intended to identify suitable sites for fishing stations and to recommend appropriate installations and roads associated with them. Nimmo was especially good at this, since he had a particularly keen eye for topography. As we have already seen from his work in Kerry, he made convincing assessments of the lie of the land based only on what he called 'eye surveys'. In addition, he drew all of his plans for piers with a keen awareness of the force and prevailing direction of the winds and tides, and he rarely failed to identify the most judicious sites for the location of piers and breakwaters. By now he had a team of young men working for him, together with others whom he left on site at various places to supervise the installations he had designed. Therefore, he did not necessarily do all the practical work on his own, his *modus operandi*

TABLE 6.

The thirty-three sites chosen for financial support by the Fisheries Commissioners in 1822, indicating the amounts granted by the Commissioners, the Mansion House Committee and the London Tavern Committee. The last column shows the minimum additional sums that were provided to continue the works in ensuing years. To this list the Commissioners later added three more sites – Lambay, Killeany, Duras (Galway), Clifden and Claddagh .

Location	From Mansion House fund	From Castle fund	From London Tavern	From Fisheries Commission	Later sums
County Clare					
Kilbaha	£115	£115		£231	£476
Seafield	£115	£115		£231	£315
Liscannor	£115	£115		£231	£461
County Kerry					
Barra	£57	£57		£116	£230
Dingle	£92	£92		£183	£81
Valentia	£115	£115		£228	£15
County Galway					
Costello	£84	£84		£155	£120
Roundstone	£92	£92		£185	£260
Cleggan	£115	£115		£231	£349
County Mayo					
Old Head	£115	£115		£231	£402
Achill sound	£92	£92		£185	
Ely Erris	£69	£69		£183	
County Galway					
Barna		£79	£76	£131	
Spiddle		£34	£41	£138	£125
Ballinacourty		£69	£76	£131	
St Kitts		£57	£64	£111	£353
New Harbour		£96	£104	£177	£162
County Clare					
New Quay		£46	£53	£104	£219
Carrigaholt		£92	£99	£178	£107
Dunbeg					
County Mayo					
Killala		£42	£42	£84	N A
Buntrahur			£76	£105	
Ennisturk		£92	£99	£191	
Tarmon		£92	£99	£191	
County Kerry					
Brandon Bay		£92	£99	£178	£233
Ballinskelligs		£92	£99	£178	
Cahirsiveen		£92	£99	£178	£15
County Sligo					
Raughly Point		[£190]	[£190]	[£340]	£200
Pollagheeny				[£300]	£198

cont

County Cork
Cullagh		£99	£99	£171	£150
Berehaven		£46	£53	£85	£10
Courtmacsherry		£115	£122	£224	£108
Clonakilty		£115	£122	£224	£176
TOTALS	£1176	£2716	£1712	£5991
PERCENTAGE	10	23	15	52

being that of Telford – he acted as consulting engineer, leaving assistant engineers on site to implement, superintend and direct his plans at the local level. For instance, his assistant William Grant did the surveys of Killala Bay and Inishbofin, and Patrick Knight did that of Enniskea. Some of his assistants were with him from his Bogs Commission days, and he had great confidence in their abilities. Others would join him from time to time for shorter or longer periods. Nor was his work for the Fishery Commission his only professional activity in those years (approximately 1820 to 1825). He continued his involvement, for example, in the road around Drung Hill into Iveragh with Ben Meredith, and his Dunmore harbour activities also continued apace.[64] In modern parlance, his engineering practice had grown and he was now well established as a consultant engineer. But fate, in the form of the return of famine and disease, was about to intrude upon the smooth operation of the Fisheries Commission and to change Nimmo's career once again.

The year 1822 brought another bout of famine and distress, this time principally to the west and south coasts, with people reduced to abject misery in the spring of the year. One year earlier, the Fisheries Commissioners had decided to promote the construction of small piers and breakwaters and had looked to the local proprietors to provide some of the costs involved. Contrary to their expectations, the response to their initiative had been very poor and they now saw in the famine an opportunity to carry their plan into effect, at least in the distressed districts. As well as giving much-needed employment to the starving poor, the construction of fishery piers and harbours would supply the infrastructure they wanted for fisheries. So, when Augustus Fitzgerald of County Clare wrote in May 1822 seeking a pier at Liscannor, the commissioners ordered Nimmo, who was then in Lambay, and the survey cutter which was on the coast of Wicklow, to proceed to the west coast with all dispatch.[65] Nimmo responded instantly and went straight to the west.[66] By 24 May he had prepared a report, plan and estimate for the proposed Liscannor pier.[67] The commissioners next directed him to survey the entire west and south coasts in order to identify suitable sites for immediate pier construction and to furnish estimates of the expenses with the least possible delay.[68] A task of such magnitude was impossible for any one man to complete within a short time. Nimmo himself surveyed the west coast, but as there was not sufficient time for him to visit the south as well, he instructed Alex McGill from Dunmore Harbour to survey the south coast on his behalf.[69] This rapid response soon provided a list of thirty-three sites where the provision of piers could be expected to be of immediate value (table 6).

The commissioners, having taken a decision that half of the cost of each pier must come from a source other than their funds, now contacted the charitable committees in Dublin seeking financial support for their plans, copies of which they forwarded to the Mansion House Committee and to the Commissioners for the Relief of the Distressed Poor in Dublin Castle. These bodies selected, from the original list of thirty-three piers, a short list of twelve that they would support (the first twelve listed in table 6), the gross cost of which was estimated at £5,095. The charities offered one-quarter each of this amount, leaving the Commissioners for Fisheries to make a grant of the remaining half. This arrangement met the commissioners' requirements and they now instructed Nimmo to execute the twelve short-listed piers immediately, according to detailed plans and specifications that he was to furnish. He was instructed to use day labour rather than contract and, in order to be certain that all costs were kept within budget, no pier was to cost more than £500. The commissioners next approached the London Committee for the Relief of Irish Distress (the London Tavern committee), which responded generously and offered them £2,000. The Dublin Castle Relief Commission added another £2,000 to this and the Fisheries Commissioners provided a further £4,000 of their own. This new aggregate sum of £8,000 permitted the remaining twenty-one works to be undertaken as well. Five new sites not listed in the table – Aran Islands, Lambay, Durras, Clifden and Claddagh – were added later, making an overall total of thirty-eight sites to be developed.

As indicated in the table, a grand total of about £11,500 was allocated for the thirty-three listed piers. Seen against the actual contemporary cost of Dunmore (approximately £42,000), or against Nimmo's estimate for Cork (£20,000) or for Sligo (£15,000), the initial sums for the fishery piers were clearly miniscule. Supplementary sums (the absolute minimum amounts of which are shown in the last column of the table) had to be provided by Government within a few years in order to complete the various works. Still, while the amounts look small by today's standards, they must have been a major contribution to the local economy in the distressed districts. About eight pence was the daily rate for a labourer at the time; the average allocation per pier was approximately £350, excluding any other contributions from local sources. Assuming average costs of £150 for stone, timber, transport, supervisor's wages and other contingencies, the balance of £200 would pay for 6,000 man-days of labour, equating to six days' work a week for ten weeks for 100 men, not an insignificant contribution. For the first twelve piers, the Mansion House committee and the Dublin Castle commissioners contributed one-quarter each of the sum necessary, and the Fisheries Commissioners one-half. For the other twenty-one, the London Committee replaced the Mansion House committee in providing one-quarter, the remainder coming from

Dublin Castle and the Fisheries Commissioners. The overall outcome, therefore, was that three-quarters of the total costs came from the public purse, through the Castle Commission and the Fisheries Commission, with the remaining quarter split between the Mansion House Committee (10 per cent of the total) and the London Committee (15 per cent of the total). The full sums offered do not appear to have been fully taken up from either of the two charities, but this may simply reflect the inexact nature of the available data. Despite the urgency, few proprietors proved willing to contribute privately to the costs. Fitzgerald at Liscannor, the Marquis of Sligo at Bundurra and Old Head, and Lord Palmerston at Mullaghmore were among the relatively small number who responded positively and even generously. For their part, the Commissioners for Fisheries, despite their humane sentiments, continued to insist that they would not meet any expenses beyond one-half of the estimated cost. The result was constant financial difficulty with the execution, completion and maintenance of the works at most sites.

This emergency response altered fundamentally Nimmo's role in the commission's service. Where formerly he was engaged only to survey and report on potential fishing stations, he was now expected to take complete charge of the planning and design of piers, site selection, supervision of construction, hire of labour and control of expenditure at thirty-three localities scattered along the coasts of Sligo, Mayo, Galway, Clare, Kerry and Cork. Even today, that would be no small task for any engineering consultancy. Nor did all this include the other piers and harbours, such as Mullaghmore, Killough, Courtown, Drogheda and Dunmore, on which he was engaged but which did not receive support under the emergency response to the famine. As we shall see, it was during this period that he did many of the public works in the western district for which he is rightly renowned. It is little wonder that the more fundamental hydrographic aspects of the coastal survey were put in temporary abeyance before the survey's official termination in 1824. However, he did manage to complete the western leg of the survey in 1823, fortunately as events turned out.

Even Nimmo, with all his energy and dedication and all his young men, could not possibly have hoped to keep close supervision on so much work for the Fisheries Commission. This failure began to have its inevitable effect and he may have had some forewarning that trouble lay ahead on this account. In December 1824 he wrote to Fitzgerald, 'If you hear complaints about me in Dublin, defend me!'[70] The axe fell in June 1825 when the commissioners appointed James Donnell to a newly created post of Harbour Engineer. This action was not a dismissal of Nimmo per se: the decision to terminate all local surveys had simply meant that he could no longer be seen to have a role with the commission if the coast survey was truly closed. But naturally enough it would take a little time to finalise whatever tasks he had already in hand, for example his report on the fishing stations of the west coast.[71] The new title of

'Harbour Engineer' conferred on Donnell signalled that all coastal surveying was over. By then the execution of Nimmo's harbour plans were an integral part of the commission's on-going activities and Donnell took over responsibility for their supervision and completion. Giving the task his undivided attention, and casting a much colder eye than Nimmo on the plight of the destitute, he was effective in ensuring that the works, by and large, were finished to specification and within stringent new budgets.

Before Donnell's appointment, Nimmo had in hand about forty piers and harbours; some were already completed and many were in progress. He had designed almost all of them and the actual construction of most had come under his consultant supervision and the direct supervision of his assistants. In 1829 the commissioners listed all the fishery piers then completed or in the course of construction throughout the whole island of Ireland: the total was 105.[72] Nimmo had contributed to at least fifty-three of these, in most cases in a major way. From 1820 to 1828 he had received £4,550 in salary and expenses; of that amount, almost £4,000 was salary. He had also been involved in at least ten more harbours that had not received support from the commissioners and therefore did not feature in the list, so that he had, in fact, contributed to a minimum of sixty-three of the piers and harbours erected between 1822 and 1829, quite an astonishing contribution from one person. That he is often credited with erecting all the piers in County Galway is therefore little wonder, although grossly incorrect; its corollary, that he is rarely credited with erecting piers anywhere else in Ireland, is also astonishing and just as incorrect: he contributed to nineteen piers and harbours in counties outside of Connaught and Clare, not counting his Dunmore and Cork projects. A gazetteer of Nimmo's most important piers and harbours, giving further details of each and describing their present state, is given in appendix 1. Many are worthy of conservation and some badly in need of it.

In July 1830 the Act 1 Wm IV C. 54[73] granted £13,000 for the completion of the piers and harbours then in progress and transferred the powers and duties of the Fisheries Commissioners, pro tem., to the Board of Inland Navigation. The duties of that board, including the fisheries, were soon transferred to the newly created Office of Public Works in 1831, where they remained until the end of the century. The Irish fisheries never merited a Department of State entirely to themselves, neither under the British administration nor since independence, being administered as a subsidiary part of various other Departments ever since 1831.

NOTES

1. Acts of the Irish Parliament in the eighteenth century that relate to the sea fisheries include: 7 Geo II C. 11; 3 Geo III C. 24 and C. 31; and 5 Geo III C. 7.
2. Second Report from the Select Committee on the State of Disease and condition of the Labouring Poor in Ireland (B.P.P. 1819 [409], vol.VIII, mf 20.68–9), hereafter referred to as Second Report on Poor.

3. 59 Geo III C. 109. *An Act for the further encouragement and improvement of the Irish Fisheries*, 12 July 1819.
4. First Report of the Commissioners for the Irish Fisheries (B.P.P. 1821 [32], vol. XI, mf 23.64).
5. Second Report of the Commissioners for the Irish Fisheries, p. 2 (B.P.P. 1821 [646], vol. XI, mf 23.64).
6. A. Nimmo to W. Gregory, 17 August, 1820 (NAI, CSORP 1820/N/20).
7. Second Report on Poor, pp. 103–4.
8. Fourth Report of the Commissioners for the Irish Fisheries for the year 1822 (B.P.P. 1823 [383], vol. X, mf 25.85–6), hereafter referred to as Fourth Fisheries Report.
9. N. P. Wilkins, 'Some Irish Fishing Villages, Real and Imagined', in B. F. Keegan & R. O'Connor (eds), *Irish Marine Science 1995*. (Galway: National University of Ireland Press, 1966).
10. A. Nimmo, 'Report on the Fishing Stations on the north coast of Ireland' in Fourth Fisheries Report, p. 26
11. Ibid., p. 27.
12. Ibid., p. 28.
13. A. Nimmo to Fisheries Commissioners, 10 December 1820 (NAI, OPW 8/284 [3]).
14. J. Spencer to Fisheries Commissioners, 1 February 1822 (NAI, OPW8/284 [4]).
15. A. Nimmo to Fisheries Commissioners, 19 June 1824 (NAI, OPW8/284 [5]).
16. J. Donnell to Fisheries Commissioners, 6 October 1825 (NAI, OPW8/284 [11] and [12]).
17. A. Nimmo to J. Heron, 30 January 1821 in Fourth Fisheries Report, p. 23.
18. Third Report of the Commissioners of the Irish Fisheries (B.P.P. 1822 [428], vol. XIV, mf 24.109).
19. A. Nimmo, 'Report of Alexander Nimmo, civil engineer on the present State and means of improvement of the Port and Harbour of Sligo, 25 March 1822', in Royal Commission on Tidal Harbours, Second Report and Minutes of Evidence, pp.155–159 (B.P.P. 1846 [692], vol. XVIII, mf 50.169–177), hereafter referred to as Tidal Harbours. An MS extract of this report, signed by Nimmo and dated 20 March 1822, is extant (NAI, CSORP 1822 856 [3]). Copies of the chart of Sligo Bay are archived as follows: Nat. Archs. UK, MPD 1/32/2; BL, maps, 13450 (1).
20. Lord Palmerston to H. Goulburn, 3 May 1822 (NAI, CSORP 1822/ 856 [2]).
21. T. Telford, 'Mr Telford's Report on Sligo Harbour, 22 October 1823', in Tidal Harbours, appendix B 93, pp. 159–60.
22. A. Nimmo, 'Report on the Fishery Stations of the Coast of Connaught', in Sixth Report of the Commissioners for the Irish Fisheries, p.18 (B.P.P.1825 [385], vol. XV, mf 27.124 –5), hereafter referred to as Sixth Fisheries Report.
23. A. Nimmo, 'Further Report on the Fishery Station of Mullaghmore', appendix 2, p. 31, in Fourth Fisheries Report.
24. A. Nimmo to J. Heron, 30 January 1821, in Fourth Fisheries Report, p. 23.
25. Seventh Report of the Commissioners for the Irish Fisheries,appendix 8 (B.P.P. 1826 [395], vol. XI, mf 28.69).
26. Eleventh Report of the Commissioners for the Irish Fisheries, Appendix 8 (B.P.P. 1830 [491], vol. XV, mf 32.118–19).
27. Fourth Fisheries Report, p. 11.
28. Employment of the Poor, Ireland. Reports of the Engineers superintending the Public Works, appendix 11, Report on the western district by Alexander Nimmo, p. 46 (B.P.P. 1823 [249], vol.X, mf 25.86–7).
29. Fourth Fisheries Report, plates 1, 2 and 3.
30. Fourth Fisheries Report, p. 12.
31. A. Nimmo, Evidence in Report of the Select Committee to consider the best mode of apportioning more equally the local burthens collected in Ireland and to provide for a general Survey and Valuation (B.P.P. 1824 [445], vol. VIII. mf 26.47–51) hereafter referred to as Spring Rice Committee.
32. A. Nimmo, Map, Belfast and Larne compiled from various authorities for the use of the Irish fisheries (NLI, maps, 16 B 22 [21]).
33. A. Nimmo, Part of the coast of Dublin, from Howth to Balbriggan, by Alexander Nimmo, C.E., surveyed for the Commissioners of Irish Fisheries, engraved chart (BL, maps, 11795 [1])
34. W. Bald, Evidence, p. 65, in Spring Rice Committee.
35. A. Nimmo, 'Report on the Fishery Stations of the Coast of Connaught', in Sixth Fisheries Report, p.18
36. W. Bald, 'An account of a Trigonometrical survey of Mayo, one of the Maritime Counties of Ireland', *Transactions of the Royal Irish Academy*, Vol. XIV, (1825), pp. 51–61 (Paper read April 1821, published 1825).
37. Sixth Fisheries Report, p. 18.
38. Ibid.
39. Ibid., p. 36.
40. A. Nimmo, 'On the Application of the Science of Geology to the purpose of Practical Navigation', *Transactions of the Royal Irish Academy*, vol. XIV (1825), pp. 39–50. (Paper read October 1823, published 1825).
41. A. J. Lee, and J. W. Ramster (eds), *Atlas of the Seas around the British Isles* (London: Ministry of Agriculture and Food, 1981).

42. P. Wyse-Jackson, 'Alexander Nimmo's *On the application of the science of geology to the purposes of practical navigation* (1825): The First Investigation of Marine Geology and its Bearing on the Geology of Offshore south-west British Isles', *Earth Sciences History*, 15, (1996), pp. 167–71.
43. A. G. J. Cole and T. Crook, 'On Rock Specimens dredged from the floor of the Atlantic off the coast of Ireland and their bearing on submarine geology', in *Memoirs of the Geological Survey of Ireland* (Dublin: Department of Agriculture and Technical Instruction for Ireland, 1910).
44. T. R. Marchant and G. D. Sevastopulo, 'Richard Griffith and the Development of Stratigraphy in Ireland', in Herries-Davies and Mollan (eds), *Richard Griffith 1784–1878*; Wyse-Jackson, 'First Investigation', pp. 167 – 171.
45. R. Griffith, 'A General Map of Ireland to accompany the Report of the Railway Commissioners showing the Principal Physical Features and Geological Structure of the Country', 1839.
46. A. Nimmo, *New Piloting Directions for St George's Channel and the Coast of Ireland. Written to accompany the Chart of St George's Channel and the Coast of Ireland, drawn for the Corporation for Improving the Port of Dublin etc.* (Dublin: A. Thom, 1832).
47. Nimmo's map of the Harbours and Coast of Ireland. (Irish Architectural Archive, cat. 534).
48. Sixth Fisheries Report, plate 2, figure 1.
49. W. Bligh, 'Report of Captain Bligh on the means of improving the Navigation of Dublin Harbour and Bay', pp. 53–64 and plate 2, in Statement of Plans and Reports for the Improvement of Dublin Harbour (B.P.P. 1806 [292–4], vol.XIV, mf 6.93–5).
50. H. Goulburn to the Commissioners for Irish Fisheries, 17 January,1824 (NAI, OPW8/110/2 [29]).
51. A. Nimmo to Commissioners for Irish Fisheries, 26 January 1824 (NAI, OPW 8/110/2 [30]).
52. Nimmo, A. 'On the Fishing Stations on the North East of Ireland', in Fourth Fisheries Report, p. 34.
53. A. Nimmo to Commissioners for the Irish Fisheries, 4 February 1824 (NAI, OPW 8/110/2 [30]).
54. A. Nimmo, 'Strangford River or Entrance into Lough Cone. Surveyed for the Commissioners of Irish fisheries by Alexander Nimmo civil engineer', engraved chart dated 1821, that incorporates Nimmo's MS corrections to the Butter Plady, with new bearings (NLI, maps, 15 B 19 [3]).
55. Known as the Survey and Valuation committee or the Spring Rice Committee (after its chairman).
56. J. W. Croker, Evidence, p. 35, in Spring Rice Committee.
57. Report of the Select Committee to consider the best mode of apportioning more equally the local burthens collected in Ireland and to provide for a general Survey and Valuation. B.P.P. 1824 (360) vol.VIII, mf 26.47.
58. A. Nimmo, Evidence, pp.91ff, in Spring Rice Committee.
59. Ninth Report of the Commissioners for the Irish Fisheries (B.P.P. 1828 [465], vol. XII, mf 30.83).
60. A. Nimmo to H. Townsend, 2 July 1827 (NAI, OPW 8/359 [6]).
61. W. Gregory to Commissioners for Irish Fisheries, 31 July 1827 (NAI, OPW 8/359 [7]).
62. A. Nimmo, engraved chart, *The Harbour of Valentia by Alexander Nimmo civil engineer*, dated 1831. London: Hydrographical Office of the Admiralty (BL, maps, SEC 1. [49]).
63. Committee of Revenue Inquiry referring to Western Harbours of Ireland (B.P.P. 1834 [592], vol. LI, mf 37.362–3).
64. See, for example, the map of the road from Killarney to Cahirsiveen signed Nimmo and Meredith, dated 1821 (NAI, OPW5 HC/6/0307).
65. Sir A. Fitzgerald to Commissioners for Irish Fisheries, 6 May 1822 and reply thereto (NAI, OPW 8/236/1 [1]).
66. A. Nimmo to H. Townsend, 16 May 1822 (NAI, OPW 8/236/1 [3]).
67. A. Nimmo to H. Townsend, 24 May 1822 (NAI, OPW 8/236/1 [4]).
68. Fourth Fisheries Report, p. 8
69. A. Nimmo, 'Report on the fishing Stations of the West Coast of Ireland, from Errishead southwards', p. 36. Eighth Report of the Commissioners for the Irish Fisheries, appendix 10 (B.P.P.1826/7 [487], vol. XI, mf 29.89–90), hereafter referred to as Eighth Fisheries Report.
70. A. Nimmo to M. Fitzgerald, 20 December 1824 (PRONI, Fitzgerald Papers, MIC/639/11/reel 6/78).
71. Nimmo's final report on the Fishing Stations on the West Coast appeared in 1827 as appendix 10 of the Eighth Fisheries Report.
72. Tenth Report of the Commissioners for the Irish Fisheries (B.P.P. 1829 [329], vol. XIII, mf 31.93– 4).
73. 1 Wm IV C. 54. *An Act to revive, continue and amend several Acts relating to the Fisheries,*16 July 1830.

CHAPTER 10

Master of all he Surveyed? Nimmo in Connaught

THE BEST-LAID SCHEMES OF MICE AND MEN ...

The worst of the famine of 1822 was in the west, the south-west and south, and the legislative response was the Act 3 Geo IV C. 34, *An Act for the Employment of the Poor in certain districts in Ireland*. It empowered the Lord-Lieutenant to appoint engineers to initiate and supervise the employment of the poor in public works in certain districts. On 5 June, Richard Griffith was allocated the southern district comprising parts of counties Kerry, Limerick and Cork; John Killaly received the middle district of County Clare and part of east Galway and Tipperary. The western district, a huge area comprising all of Connaught, was placed under Nimmo's superintendence.

When he arrived in Galway on 12 June, the scenes that faced Nimmo were ones of utter desolation and human misery. Making an immediate tour of south Connemara, he confirmed for himself just how bad things were: 'death and distress in several places ... far exceeds anything I have yet witnessed ... the wan, dejected aspect of the poor creatures struck me as more remarkable than any I had seen', he wrote to under-secretary Gregory.[1] On arrival in Clifden, he was temporarily confined by indisposition 'arising, I believe, from the fatigue of travelling along the coast'. He was clearly overcome with horror and dismay at the pitiful scenes he had witnessed, and a less resilient or less determined man might well have abandoned, there and then, his almost impossible brief. Archbishop Trench wrote to Goulburn that 'it would be folly to talk of immediate employment. The people in general are too weak to work and must be fed and strengthened gradually before employment could be available.'[2] Nevertheless, Nimmo set about planning new roads and setting men to work on them. Within weeks, he had hundreds making roads and bridges, building piers and harbours and quarrying stones in every county of Connaught. This instant response put the archbishop more at ease. He wrote to the relief committee in London:

> I have had a long and most interesting interview with their engineer [Nimmo], a very sensible, intelligent man, as he appeared to me, and he

assured me that the bulk of the population of Connemara will be immediately employed, and that many are already so…In respect to employment, the same extensive works are carrying on, as the engineer informs me, in Erris and all the most distressed parts of Mayo.[2]

How could Nimmo have set to work so quickly and what exactly had the Government expected of him? The act allowed the Lord-Lieutenant to advance whatever sums were necessary to fund the presentments granted in the previous Spring Assizes, provided the works were placed under the supervision of the engineer. In addition, the Viceroy could spend £50,000 on public works without presentments, but keeping in mind the need to give immediate employment to the poor. He could take the engineer's advice in allocating this money, and supervision of the actual work was to be put in the engineer's hands. Goulburn explained to the House of Commons on 17 June that: 'Persons had been dispatched at his [Goulburn's] command, into those districts, with full authority to commence such plans as seemed most likely to give relief to the poor.'[3] According to *The Times*, he had asserted also that 'the Engineer in Galway had power to act on his own responsibility' – a sentiment he would soon enough regret.[4] The statement was indeed true, but the devil, as ever, was in the detail. On 16 May, a week before the passing of the act, a relief committee in Dublin Castle comprising W. Disney, T. P. Luscombe, P. LaTouche jnr, and G. Renny, under the chairmanship of under-secretary Gregory, drew up what it called 'Principles to be attended to in the Distribution of the Money granted to alleviate the Scarcity which at present prevails in some Districts in Ireland'.[5] This document identified the chief causes of the distress as arising firstly from the general want of money and secondly, from 'the ignorance and prejudice of many of the lower orders, and the local disturbances' – in short, the famine was the poor's own fault! There might be, the committee grudgingly admitted, a scarcity of potatoes that required a most rigid economy: '[The potato] should, as far as possible be interdicted as food in the distressed districts during the ensuing three weeks and reserved for seed.' Exactly what the poor were supposed to live on during the period of interdiction was left unaddressed.

The 'Principles' themselves were five:

1 To avoid as much as possible gratuitous distribution.
2. As far as practicable to relieve the poor by employment.
3. For this purpose to combine utility in the work with the employment as far as possible; but *still to make the employment of the poor, rather than the accomplishment of the work the object* [italics added here]; to prefer small local undertakings, and those that would not be otherwise carried on, to those on a great scale, or which would certainly be done, though at a more distant period. Works on a great scale have a tendency to invite an accumulation of numbers to a given spot, and probably to disappoint the greater part of them.

4. To give low wages; this is obviously expedient, as well to diffuse the relief among greater numbers, as to prevent interference with the ordinary demand for labourers.
5. To insist on *bona fide* work.

There could be no doubt that the authorities envisaged a short-term strategy and were not intent on any commitment to extended improvements or infrastructural developments. This policy would lead to the censuring of Nimmo later on, but for the moment it was essential that work commence without delay. In an initialled reply written on the reverse of a letter of 19 June from Nimmo to Goulburn, Wellesley, the Lord-Lieutenant, made it clear that Nimmo had 'full powers of acting in the cases which he now refers. He should be admonished of this – W.'[6] Goulburn communicated the thrust of this brevet to Nimmo: 'The anxious desire of his Excellency [is] that you should instantly employ the people in every possible way without any reserve as to the application of any particular funds, it being no longer a legal question, as all funds are to be applied generally to the great objects of the immediate employment and support of the distressed peasantry.'[7]

County Galway had made few enough presentments the previous spring, but the Mayo Grand Jury had made a number. Nimmo soon had these works, and others that had been presented earlier, in hand. It was fortunate for him that William Bald was in Mayo at the time and by recruiting him he could start the works straight away. Bald had prepared a plan in 1821 for a road along the north coast from Killala to Broadhaven, which Nimmo now inspected and to which he recommended a number of improvements. Both men's manuscript annotations are clear on the copy of this map held in the National Archives of Ireland.[8] The Grand Jury had also been constructing a county road, called 'The Right Hon. Robert Peel Road', through the heart of the county from Castlebar to Belmullet, which Nimmo took over.[9] This and its branches would become one of the first, the biggest and the most expensive works that he promoted in that county, costing over £5,699 in the four years from 1822 to 1825 and almost £10,000 overall.[10] Other new roads that he commenced included that from Newport to Achill Island, the Westport to Killary (Bundurra) old road, and the road from Ballinrobe through Claremorris and Ballyhaunis to Tulsk (County Roscommon). In east Mayo, his original intention had been to complete the mail coach road from Ballina to the Royal Canal at Tarmonbarry, the line of which he had laid down as early as 1815 for the Post Office.[11] Presentments already existed for this, but the Lord-Lieutenant countermanded his intentions and he was reduced to undertaking only minor works at the eastern end of that line instead. Nimmo also placed the construction of piers at Tarmon (Blacksod) and Saleen in Bald's hands, and both were reasonably well executed, although at a cost that dismayed him. Bald earned almost £850 with Nimmo before leaving Ireland for France in 1826.[12] Patrick Knight, a

Mayo-born engineer, was also recruited early on and he remained in Nimmo's employment to the end, earning over £1,440 in salary.

The existing Mayo roads had been started as presentment or county roads a long time before and, never having been properly completed, were in impassable condition. Nimmo's priority, once he had initiated work on the main trunks, was to connect them as far as possible with places like Saleen, Tarmon, Achill and Killala, where he was building new piers and harbours for the Fishery Commissioners. Even in those earliest days of his western district duties, he realised the flexibility that his multiple responsibilities carried, and which he would combine creatively as time went on. Each of his separate offices – under the Loan Act, the Moiety Act, the Fishery Commission and the Employment of the Poor Act – gave him very extensive powers. In effect, few public works, no matter how they were funded by Government, could commence in the district without his approval. Some individuals did start works privately with Government loans, but these came under Nimmo's scrutiny also. As he commented in his report for 1822, 'The Acts which permit an issue of money from the consolidated fund for public works, by loan or otherwise, have been found of the greatest importance in those districts ... and considerable exertions have been made by the proprietors to avail themselves of that system.'[13] In his report for 1824 he made it clear how he operated:

> there have also been several new lines laid out under my direction in these counties and carried on by means of loans from the Consolidated Fund. In these I act as an overseer for the Grand Jury without any charge. The detail of superintendence has been intrusted to my assistants, who are paid for this service as the usual overseers; their time, when they are so occupied, being deducted out of their pay from the Government.[14]

He hoped that almost £14,000 of the public works he had initiated could be passed eventually to the Grand Juries of Galway and Mayo for funding under the Loan Act in this way.

It was, however, in County Galway, especially in Connemara, that Nimmo was to have his greatest impact. There, the existing presentments were concerned only with the central Connemara road from Galway to Clifden through Oughterard, a project that was already long in train under the local proprietors. Nimmo decided to keep severely away from it, 'leaving the completion of the central road to the exertions of the country gentlemen', as he put it.[15] Nevertheless, he and his reputation would be dragged into it inevitably (as would many other engineers and 'gentlemen') and, in fact, the road from Galway to Clifden would not be fully completed until one hundred years later. The almost complete absence of presentments in County Galway enabled Nimmo to start with a clean sheet, as it were, and to make a unique, indelible mark on the landscape and seascape of the county. He already had a plan for Connemara dating from his Bogs Commission days, and since Connemara happened to be the poorest, most distressed part of the whole county, it justified his extra attention. Later on, he would make no secret that his Bogs Commission survey and report had informed his approach to the public works that he now determined on:

> Having been employed by the Commissioners for the enquiry into the Bogs of Ireland to survey and report upon this district ... I was well aware, when appointed by his Excellency the Lord Lieutenant to lay out and superintend the public works of the western district, of what operations were most necessary for the improvement of this country, and in conducting them I have been equally anxious to keep in view the benefit of the fisheries as the extension of agriculture.[16]

With the exception of the road he had originally proposed from Galway to Screeb through the centre of Sillermore, which he now abandoned because the central road through Oughterard was being progressed by others, he started work on all the roads that he had envisaged in 1813 (see table 3 and figure 5.1). These comprised principally the coast road (now Bóthar cois Fharraige) through Spiddle, Screeb, Invermore, Cashel, Toombeola, Ballinahinch, Ballinaboy and on to Clifden. From there, he would continue on through Streamstown, with spurs to Cleggan and Renvyle, and then onwards through Kylemore to Leenane. There the road would join a new road he proposed to Cong through the Maam valley, linking Killary harbour and Joyce's Country with Ballinrobe and the new lines of road leading towards the Shannon. He laid down a new line from Leenane up the valley of the River Erriff towards the market town of Westport. At the same time, he sited and designed new piers and harbours at Galway, Spiddle, Casla, Gorumna, Roundstone, Clifden, Cleggan, Derryinver, Leenane and Bundurra, implementing his own priority of ensuring that the new main roads serviced, as far as possible, the new piers and harbours. For their part, the Fisheries Commissioners supported him, earnestly recommending to Government 'the expediency of directing the attention of their engineers, now employed in the execution of public works, to the advantage to be derived from connecting, by short lines of communication, the piers or safety harbours now in progress, with the main roads most contiguous to them.'[17]

Nimmo's attention to piers and harbours was not confined to their use for fishery purposes alone. He saw them as important contributors to agriculture (by facilitating the distribution of sea manure and sand) and to coastal trade and commercial life in the distant regions. For example, when he came to administer the relief funds in Sligo, he recommended that they should be directed to the improvements he had earlier proposed for that harbour. The Sligo authorities agreed, and started to build an embankment and commercial quay below the custom-house. Nimmo did not conceal his disappointment that the work did not go further: 'It is much to be regretted,' he wrote, 'that from certain differences of opinion, the labour of this multitude has not at

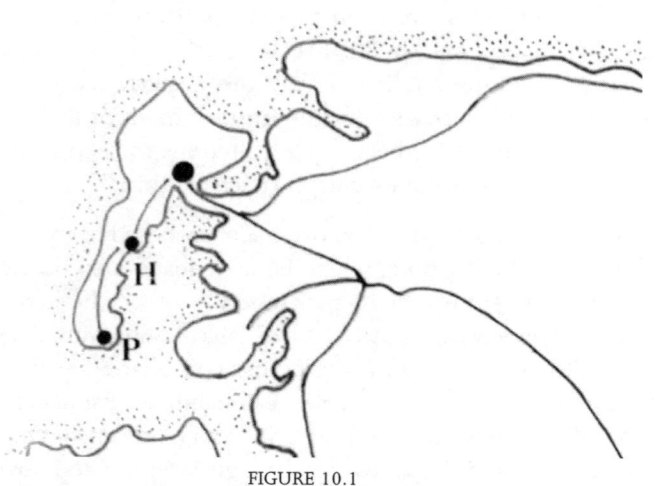

FIGURE 10.1
Nimmo's plan for Connaught involved joining the piers (P) and harbours (H) to the market towns and onwards to the crossings of the River Shannon.

once been directed to the proposed alterations in the river channel below Sligo, as in all probability an improvement of that port might have then been effected, for which so favourable an opportunity may never again occur.'[18] In Ballina, too, he had been commissioned by the merchants to devise improved access from the sea for boats trading with that town. His plan there was to excavate a navigation along the west bank of the River Moy to by-pass some rocky bars in the river (Plate 6). Boats could be hauled up through this navigation either by horses or by men, to a new quay at Belleek and eventually all the way to the town. He spent £1,123 of the relief funds, initiating the works in late 1822, and altogether over £2,860 of relief funds were spent on the Moy navigation between 1822 and 1825.[19] In October 1823 the merchants hosted a public dinner in his honour to acknowledge the benefit he brought to the town.[20] Ultimately, it was all to no real avail: the new quay at Belleek was never much used and the new, excavated navigation in the river was never properly completed due to lack of funds. By 1825 Nimmo's name and fame were being seriously traduced in the local press, after which his works and reputation faded from memory in Mayo.[21] Further details of his Moy works are given in the gazetteer in Appendix 1.

Both Sligo and Ballina were important centres in Nimmo's overall plan for Connaught that was taking very definite shape. The general layout of all the roads and other works he promoted in the province is shown in the map (figure 10.1), which shows how the piers and harbours linked to the local towns and markets, and how these, in turn, joined the great lines of communication towards the Shannon and the Grand and Royal Canals. While all the works contributed, as intended, to the relief of the poor by giving employment, the detailed individual projects he initiated seemed to accord with a higher scheme, of which the Government was not aware at first, and of which it certainly would not approve if it were. Nimmo explained it to the House of Lords in 1824:

> One object I have been attending to in the province of Connaught, is to run main lines through the towns to the principal markets and seaports, and thereby supersede a multitude of crooked cross and branch roads, which form the principal communication at present; we have one main object in view, to communicate between the extremity of the Royal Canal and the seaports of Sligo, Ballina or Killala westwards, and the harbours of Connemara and Galway; another to communicate from the extremity of the Grand Canal on the lines which are now traversed by the mail coaches, to the towns of Castlebar and Westport, Galway and Gort; those are the present lines of great communication.[22]

This clear rationale explains and gives cohesion to his multiple individual works listed in Appendix 4. An account of the individual works is given in full detail by Villiers-Tuthill.[23]

Nimmo's planning of long, purposeful lines of road was his most important innovation in the west. Until then, he noted, lines of communication had been formed 'by a combination of short pieces of road in the most economical way the case admitted, or as it might suit the convenience of the person who obtained the presentment'.[24] This had resulted in many miles of wandering roads with numerous cross roads and junctions. By intentionally designing long lines of communication, and purposely linking them to the piers and harbours of the coast on the one hand, and to the two major canals at the Shannon on the other, Nimmo conceived the first, comprehensive, integrated approach to the development of the infrastructure of the west coast and, indeed, of the whole of Connaught. It would be five more years before the Directors of Inland Navigation, (who by then had charge of the roads) woke up to the wisdom of such a scheme and ventured, with cringing timidity, to declare:

> It may not, perhaps, be altogether irrelevant to observe, that it appears to us most desirable, in the formation of roads (as far as may be practicable) that they should be made available to the lines of inland navigation; and we shall, probably, not be considered as exceeding our duty when we suggest, that in laying out roads for the promotion of intercourse, and the advancement of agriculture in the west of Ireland, attention be given to this object, so that lines of road may lead to the Shannon navigation,

which passes through the heart of Ireland, in a direct course to the city and port of Limerick and communicates with Dublin and Great Britain by means of the Grand and Royal Canals.[25]

Another innovation was his use of money from a diversity of sources to consolidate the different works into a comprehensive whole. This unorthodox merging of different accounts may have been unintended at first and would become his downfall, but it was effective in advancing the integrated scheme he was pursuing. Neither Gregory's 'Principles' nor any of the acts had ever envisaged or intended such a grandiose scheme for which there was no prior Government approval, however meritorious it may now appear in hindsight. Nimmo had endeavoured to mould temporary, emergency relief measures into a regional development plan under the plain gaze of the authorities and contrary to the specific instructions in the 'Principles'. He could do this by interpreting creatively and exercising constructively the multiple roles and duties of his various appointments, but it was mainly the intuitive vision he had conceived for the region on his first foray into Connemara a decade earlier that permeated his plan. When the significance of what he was attempting began to dawn on the authorities, alarm bells started to ring in Dublin and Westminster. On 4 May 1823 Gregory briefed Goulburn who reacted sharply, complaining in the House of Commons that one of the engineers was displaying 'a taste for great national improvements'.[26] Goulburn was strikingly forthright in a letter to Wellesley on 23 May:

> Mr Nimmo certainly misunderstood the views of the Government in his original appointment. He embarked on works not so much calculated to afford employment to relieve immediate distress as to promote ... more what appeared to him great national improvements and the misfortune is that he thereby created an expectation that the Government considers itself bound in all time to come without reference to reasons of distress or of plenty to expend a large annual sum in works which ought in fact to be defrayed by the several counties.[27]

There was no misunderstanding on Nimmo's part. He openly declared that his 1813 plan had informed his actions in Connemara from the start. His knowledge and experience of road surveying for the Post Office in Mayo and Roscommon during 1815 had informed his activities in those parts. Indeed, he much favoured the construction and repair of the mail coach roads especially in the flatter, eastern parts of Connaught, as he wrote:

> To add to the intricate maze of little cross roads in the plain ... seems only an extension of the worst effect of the Grand Jury system, now only to be corrected by the construction or improvement of great leading lines passing directly to the chief markets, through several counties or districts ... Besides, as many such lines have already been designed or surveyed under the authority of the Post Roads Laws, by professional

men, unbiased by local or individual interests, there is the less hesitation in setting them agoing[28]

The chief mail coach road through Mayo was the Carrick-on-Shannon to Ballina road *via* Tubbercurry and Lough Talt. In east County Galway it was that from Athlone to the town of Galway through Ballinasloe, Loughrea and Craughwell. Between 1822 and 1824 the Post Office gave Nimmo £17,500 to develop this road, while his colleague John Killaly undertook the Loughrea to Gort branch.[29] The other main road that Nimmo undertook was the Tuam to Galway road, 'being, after the mail coach lines, the most leading road in the province, and which would have been left very imperfect, if the presented portion only had been repaired'.[30] His eye, as ever, was set on distant targets:

> I have prepared a plan to be laid before the Grand Jury of Galway, at this assizes, for the general improvement of the line from Ballymoe by Dunmore and Tuam to Galway, with a view, in conjunction with the operations in other quarters, to opening a great line of coach communication across the kingdom from Galway to Belfast.[31]

To dismiss these far-sighted enterprises as mere misunderstandings of his brief is a gross misreading of both the vision and the character of Nimmo.

He did not, however, eschew entirely the building of short stretches of road where these gave much-needed employment to large numbers. Examples are the small road from Tuam to Cloontoo and the short road to the sea at Crushua on the south of Galway Bay that facilitated fishermen in that district. Other activities he supported included the employment of tradesmen in Galway to make tools for public works, and projects in most towns of cleaning, levelling and paving streets, filling dung-pits, whitewashing houses and forming footpaths and public walks. In counties Sligo, Roscommon and Leitrim many of his roads were intentionally linked to the coal and iron mining industries.[32] They included the roads around Arigna, the road from Drumshanbo to Lough Allen and the roads leading to the Shannon crossing at Carrick-on-Shannon.

Goulburn's complaint in May 1823 of his taste for 'great national improvements' might have brought him up short, or at least put him on alert, had not the Select Committee on Employment of the Poor come out on his side a few weeks later, on 16 July:

> The continued encouragement of the fisheries, the erection of piers and the formation of harbours, and the opening of roads, are subjects of national interest: the beneficial consequences resulting from both are to be found minutely set forth in the Minutes. The example of the Highland Roads and Bridges is peculiarly applicable to Ireland which will appear in reference to the evidence given on a former occasion by Mr Telford on this subject.[33]

That was trenchant support indeed for what Nimmo was doing. In juxtaposing the Irish public works with those of the Scottish Highland Roads and Bridges and with Telford's evidence in 1819, the committee (chaired by Thomas Spring Rice) was signalling that it well understood what Nimmo was about. It went on to stress that it had taken account of the necessity of avoiding 'artificial encouragements', and that it recognised the danger that such 'public interference' could tend to make the people look to the legislature for relief rather than to their own industry and exertions. This admission, surely, was acknowledgement of the concerns voiced in Goulburn's complaint, and the committee concluded its report invoking again the national interest:

> They cannot however conclude without again expressing their opinion, that the employment of the people of Ireland, and the improvement of their moral condition, is essentially necessary to the peace and tranquility of that Island as well as to the general interests of the United Kingdom.

This support can hardly have pleased the Dublin authorities, who did not much favour expansive and expensive national works, even those of general interest to the United Kingdom. At the time of Goulburn's criticism in May 1823, Nimmo's chart of the entrance to Strangford Lough was published and the subsequent criticism of it was made in the January following. Nimmo suspected that the unnamed complainant on that occasion had been motivated by malice; could the criticism, one wonders, have originated from unfinished business around this time?

By the end of 1824, after only eighteen months of activity in the west, Nimmo could boast: 'on the whole there have been made, or are in hands, roads to the extent of 300 Irish miles or 382 British miles in this district, besides assisting in the construction or improvement of 12 harbours or quays, over and above what has been done in that way by the fishery board'.[34] All the fisheries piers were then still under his superintendence, and his contribution was not confined to the twenty piers in the coastal counties of Connaught alone. He built piers and harbours during the 1822 to 1824 period in counties Clare (7), Kerry (3), Cork (3), Waterford (1), Wexford (2), Wicklow (1), Dublin (2) and Down (2). However, it was only in Connaught that he had the opportunity of integrating them with the road network. Settlements formed around some of his harbours, such as Cleggan and Leenane. In two places, Roundstone and Maam, local opposition had threatened to hinder development, so he bought the leases to the land in order that construction could proceed more readily. At Roundstone, he erected a storehouse and workshop near his new pier, and his assistants, including his brother John, lived there for much of their time in the west. This encouraged others to build, and the new village of Roundstone soon arose at the site. At Maam in Joyce's Country he built a pay office/workshop designed specifically to become an inn when the

FIGURE 10.2
Nimmo's office in Maam, now a tavern. The photograph is trimmed to show the building as it may have been before the present wing extensions were added long after Nimmo's time.

roads were completed. This edifice became the Bianconi coach inn a few years after his death and it is still a tavern today, although its structure has been altered from Nimmo's original (figure 10.2). Under the aegis of the resident proprietors, other villages sprang up at, or near, his works in places such as Tullycross, Clifden and Leenane in Galway and at Belmullet, Binghamstown and Mullaghmore elsewhere. All along the lines of his roads, new cabins were built and cultivation of the adjoining land progressed wherever conditions permitted. Shoots of prosperity were beginning to spring up in Connaught for the first time in a long period.

Progress with his unwritten scheme could never have been anything but piecemeal and slow. Each element had to be funded and executed largely in isolation from the whole and with the needs of small localities uppermost; clearly it would take years to come to full completion. His annual reports document the steady progress that was made and the difficulties that accompanied it. Griffith and Killaly also made good progress in their districts, but these were not comparable in scale, in complexity, or in diversity with Nimmo's western district. (Neither of them, for example, had any responsibility for piers and harbours.) Although he did not live to see his grand scheme brought to full fruition, and later circumstances (e.g., the replacement of sail by steam and the eclipsing of the roads by the railways) would alter the strategic importance of some localities, Nimmo's vision for Connaught laid a solid foundation for the subsequent development of the province and shaped its future, as far as it went. No one before him, and very few since, ever took as comprehensive and sanguine a view of the region. The design of its modern communications infrastructure is still largely a product of his innovative and far-sighted approach. However, even in his own day there was not universal approval for what he was trying to achieve. Government parsimony, local opposition, a shortage of indigenous capital and principally Nimmo's own neglectfulness would bring the whole scheme to an early halt.

'... GANG AFT AGLEY'

Nimmo's arrival in county Galway in June 1822 had precipitated a flurry of local activity. Within weeks, men were working on Cleggan pier and road, Clifden and Spiddal piers, the Connemara coast road and the Spiddal to Moycullen road. By the autumn, Costello and Roundstone piers were started, Barna pier was under repair and the Tuam to Galway road was in hand. Galway town in particular benefited from his presence: in June, roads were commenced from Terryland to Menlo, Galway to Clonliffe, and Woodstock to Barna. A new market house, for which Richard Martin of Ross gave a free site, was built to a design most likely Nimmo's;[35] this building, at 43 Eyre Square, now occupied by a bank, is one of the more elegant Georgian buildings in the town. Nimmo's greatest work in Galway – the Slate breakwater – was started in August. It required large amounts of stone from Murrough Quarry and from the Aran Islands, giving increased employment in quarrying and transportation over a much wider area than just the town. By September, reconstruction of the quay and jetties at the Claddagh had commenced.

Applications began flooding in for help with small local projects elsewhere in the county. Many came, for example, from Archbishop Trench, Henry Blake, John D'Arcy, Lord Clonbrock and other proprietors in the eastern side of the county, for sums under £100 each. In all these cases, Nimmo gave money in trust to the applicants who undertook to act as overseers. Nimmo trusted these reputable persons to act in an honourable fashion, although it was rare enough to trust overseers on presentment projects. A major problem was the widespread practice of jobbing: overseers contracted the work to others, who sub-contracted it on, and so on again, until the final contractor had very little to spend on the actual work, which was poorly executed in consequence. To avoid this, Nimmo had been instructed to operate by day labour where possible, paying the workers directly to ensure that the work was properly executed and the benefit went to them and not to middlemen.

As early as November 1822 he came into conflict with one influential overseer, Manus Blake, who was a Galway magistrate. Blake had requested an advance of £72 for work he did as a supervisor prior to Nimmo's arrival. Nimmo considered the work very imperfect and agreed to pay only £60 until it was put in order. Subsequently, he learned that Blake had received 'as much money again for the same work from charitable sources' and therefore he refused payment beyond £60. A contretemps ensued in which Blake 'insulted me in a gross manner by giving me the lie, calling me a scoundrel, saying that I must be in the habit of pocketing the public money when I accused him of so doing and evidently with the intention of provoking an assault'.[36] Nimmo urged Goulburn to enquire into Blake's character and to take whatever steps were necessary to prevent a repetition, 'otherwise it will be impracticable to execute the duties of my office'. Blake defended himself, declaring that Nimmo had been 'the aggressor and that his conduct was so improper and intemperate as to have fully justified me in applying to him much harsher language than any that I used'.[37] The record shows that Nimmo won out on this occasion, and only £60 was ever paid to Blake.[38] The incident is important, not only in illustrating some of the difficulties that Nimmo faced but also as the first instance in Galway that an accusation of improper use of public money was made against him. Perhaps it resonates too with his intemperate behaviour when aroused, remarked on in Inverness Academy many years earlier.

The use of direct day labour made it necessary to employ a large establishment of pay clerks, superintendents and assistant engineers if Nimmo were to keep track of all that was in train. There was also a shortage of skilled men, but no lack of general labourers. Both of these issues caused another early problem for him, when the stonemasons of Galway sent a petition to the Lord-Lieutenant in January 1823. They were, they began, in such a state of wretchedness that they were famishing with hunger and falling victims to fever.[39] They went on:

> Mr Alexander Nimmo the Engineer has a great deal of work laid out in this place and no person will get work from him but such as have been recommended by favourite gentlemen who will put his daily pay into his own pocket, to discharge a prior debt. The Stewards appointed by Mr Nimmo are half pay officers, reduced gentlemen and other strangers who are not acquainted with the distress of the poor of this town. Besides, they wont employ a man except at reduced pay of six pence per day, although said stewards has from a pound to six pounds per week, drinking, eating and sporting the public money and doing nothing, idling about with their horses and servants as will be proved to your Excellency's satisfaction, if you are pleased to direct an enquiry into it.

They named the stewards as Captain Johnston and Mr Beaty from Dublin, S. White and son, Mr Plunket, Mr Jones and unnamed others, strangers to the town, and then continued:

> such masons as will be employed by them must pay themselves a handsome present or give them a weekly entertainment of the best luxury that Galway can produce, by which John Shaughnessy mason and others has permanent employment at a pound per week tho' himself and family are in a flourishing condition. Your Excellency, Sir, will observe that

the Stewards are kept daily in permanent pay, while the masons and labourers are dismissed from work in these short days tho' ready to work at any price, in consequence of the above mentioned Stewards conduct. The work is slowly going on and the money spending and dividing among the above stranger stewards. There is work carried on since the first of August at the Slate point of Galway, two masons work there, eighteen labourers and six or eight Stewards attending said men, kept on permanent pay wasting the money. These stranger Stewards will not employ the poor of Galway.

The petition was signed by John Fitzgerald, Thomas Kelly, Stephen Leary, James Quinn, Frank Davis and others. If true, was a severe indictment of Nimmo's establishment but fortunately for Nimmo, the petitioners exonerated him personally from much of the blame: 'Indeed we exonerate Mr. Nimmo from having any hand into their infamous transactions...this Government is strongly imposed on, behind his back, Mr Nimmo not being in town.' But neither did he get off scot free: 'when he [Nimmo] is in town he overlooks the distress of the people, leaves all to these wicked stranger Stewards who has not the fear of God nor cared if the whole population of the loyal town of Galway starved.'

It was not a good circumstance for him, coming so soon after the business with Blake. He certainly had brought in his own establishment, including the named Michael Plunket (overseer), Simon White and William Johnston (pay clerks) and Samuel Jones (assistant engineer). His brother George, too, was employed in Aran as early as October 1822.[40] Eventually Nimmo would have offices and staff in Galway, Clifden, Roundstone, Maam, Castlebar, Ballina and Carrick-on-Shannon to service the dispersed works. Because of this, his presence on site was not required constantly and he simply made tours of inspection as needed, while he dealt with major issues from the Dublin office or from England. The masons were correct, therefore, in saying he was not always in Galway, nor was he expected to be. Nevertheless, in 1822 he spent a minimum of sixty-five days in the west and thirty-seven days in Dublin.

As regards the employment of masons, there was so much work going on that skilled men, knowledgeable in building piers and arches, were certainly in short supply. This situation was not unique to Nimmo; Griffith was also experiencing a shortage of labour in late 1822.[41] Therefore, Nimmo was obliged to call in assistants from his other commissions elsewhere in Ireland. William England, a stonemason, came from Dunmore to build Raughley harbour, and William O'Hara, assistant engineer, came from Killough to supervise Inishturk pier. Alex McGill from Dunmore, who had twenty years' experience, assisted with Raughley, Ennisturk and Roundstone. The Galway masons may indeed have had a case when they complained that strangers were being employed, but the number involved was miniscule compared with the hundreds of locals engaged. In July 1822, for instance, there were 300 men at work in Cleggan alone, with many more at Roundstone and Costello, as well as those building the coast road.[42] The disappointment of those fortunate enough to get work can be imagined when they received no wages on 16 July, due to the non-arrival of money from Dublin. On this occasion, Nimmo gave an order for £100 to his paymaster, 'that no unpleasant consequence to the poor people may arise from the delay – nine days wages are due'.[42] Delays like this were to become a recurring feature of public works, not just for Nimmo but for Griffith and Killaly also. Griffith, for example, wrote despairingly to Goulburn in June 1825: 'being deeply in debt to the masons, labourers etc. employed by me on the public works in the southern district I have to entreat that … £8,000 be issued to me'.[43]

By August 1822 the authorities had hopes that the worst of the distress was over and that normal funding might soon be restored, namely, the works could proceed thereafter by presentments. When the works were suspended at the end of the season there was no guarantee therefore that they would be taken up again the following spring. Nimmo instructed his staff to reduce those undertakings that were supported entirely by the relief funds and to concentrate on those that came under presentment. He anticipated that presentments would now increase, but he was wrong – to expect the Grand Juries to take up the responsibility of public works just because the immediate crisis appeared to be over was naïve. Already by the end of 1822, Nimmo's outlay had totalled £26,893, Griffith's £17,043 and Killaly's £16,300.[44] Nimmo's figures did not include the sums he disbursed from the fisheries account or those from the Post Office.

Despite a promising end to 1822, the following year was little better and the Government was obliged to continue providing money to keep employment going. Nimmo spent another £28,300 in ongoing activities and commenced more than sixteen new projects in 1823, spent £23,000 in 1824 and almost £25,500 in 1825. Aware now of his penchant for 'great national works', the authorities began to make closer scrutiny of his annual proposals for new projects and they allocated money to him only for specified works of which they approved. This meant that as time went on, some projects were suspended for want of approval, some were abandoned entirely, some funds were diverted from one project to another and little or nothing was available to maintain the completed works in a good state of repair. In 1824, for example, Nimmo was allocated £27,000 for the 1825 season,

> to be applied for the most part to the completion of the several works herein enumerated, it appearing to His Excellency more advisable now that there is no apprehension of any want of provisions in any part of your district, to complete such works as are likely hereafter to be generally useful and not, by expending a small sum upon each work with a view to more general employment, to protract the period at which any of the works can be made useful to the public.[45]

After this instruction, very few entirely new works were initiated, and Nimmo's officers, established at the various locations, concentrated on pursuing those works within their localities that the authorities approved each year. Nimmo made tours of inspection and continued to request new funds and propose new works. In 1825 and 1826 the engineers of all three districts had to make repeated pleas for the release of funds, and it is clear that the authorities were growing impatient with the protracted nature and excessive cost of some of the works. Under the Act 6 Geo IV C. 101, enacted in July 1825, completed roads were put under the management of the Board of Inland Navigation.[46] In the year to March 1828, the first to which records relate, Killaly had placed ninety-seven miles of roads in County Clare and forty-one in County Galway under the board; Griffith had transferred ninety-two miles in County Cork; only thirty-eight miles in the whole of Nimmo's western district had been transferred (twenty-two in County Mayo and sixteen in counties Leitrim and Roscommon). Even that small number from Nimmo was taken on reluctantly by the board. Killaly, who was to become Nimmo's nemesis, happened also to be the board's long-standing engineer. When asked to report on the western district roads, he observed 'those works were in such detached portions and the length of the several divisions in general so trifling that their maintenance … would entail upon the public a very heavy expense'.[47] Naturally, the board did not wish to take these roads on, but they were given no choice: 'The Government intimated their wish that we should at once take the roads in question under our charge, we acted in obedience thereto and entered into contracts for their maintenance. This, we trust, will only be a temporary measure.'[48] Obviously progress was not as great as Nimmo had been reporting, and this discrepancy further alarmed the authorities. Apart from another tranche of about sixty miles of road from Griffith, and eighteen from Nimmo (this latter the Kenmare to Killarney road, not part of his western district at all), no other roads were transferred to the board before it was replaced by the new Office of Public Works in 1831.

When, in 1827, Nimmo submitted his estimate – £30,145 – for the completion of all the works in the 1828 season, the authorities were utterly dismayed. Since 1826 Parliament had voted £26,000, he was informed, for works that he had estimated would cost £32,771. How then, they asked, could the works now require so much more? The Lord-Lieutenant could not submit to Parliament, under any circumstances, a request for more than £12,000. This was followed by the clearest possible admonition:

> His Excellency has commanded me to add that he reserves for consideration, after he shall receive your reply to this communication, whether it may not be more advisable to place that fund at the disposal of some other engineer for the purpose of completing some of the works to which your estimate refers rather than to expose the public to the risk

TABLE 7.

Documented number of days Nimmo spent in different locations in the period 12 June 1822 to 31 December 1831, as estimated from dated documents. Numbers of days in brackets are 'certain', numbers of days without brackets are 'probable' (see text for explanation). The final column gives the percentage of Nimmo's documented time that he spent in Britain; values in brackets are for 'certain' days and values without brackets for 'probable' days.

YEAR	DUBLIN		WEST & CLARE		REST OF IRELAND		BRITAIN		TOTAL DAYS		% TIME IN BRITAIN	
1822	37	(14)	65	(21)	0		0		102	(35)	0	
1823	42	(17)	12	(12)	0		3	(3)	57	(32)	5%	(9%)
1824	46	(32)	74	(27)	20	(8)	48	(32)	188	(99)	26%	(32%)
1825	16	(12)	7	(7)	3	(3)	0		26	(22)	0	
1826	3	(3)	0		7	(6)	44	(28)	54	(37)	81%	(76%)
1827	27	(23)	0		15	(13)	46	(23)	88	(59)	52%	(39%)
1828	6	(6)	0		0		90	(57)	96	(63)	94%	(90%)
1829	0		0		0		9	(9)	9	(9)	100%	
1830	9	(6)	0		3	(3)	60	(37)	72	(46)	83%	(80%)
1831	3	(3)	0		0		125	(58)	128	(61)	98%	(95%)
TOTAL	189		155		48		425		817	(460)	52%	(54%)

of such another miscalculation as that which you have hitherto fallen into. At all events, His Excellency will not consider himself justified in making any further [undeciphered] until some report shall have been had from other competent persons as to the expenditure on the several works during the last year.[49]

Nimmo could hardly have received a clearer warning of the concern the authorities had, and the possibility that someone might be appointed in his place. However, even before the authorities had started to lose patience, Nimmo had already begun to lose interest, and was turning of his own accord from the western district to pastures new.

The year 1823 had seen his fishery charts published and his lecture on the seabed sediments delivered to the Royal Irish Academy. In the summer between these two events he had finished the hydrographic survey of the west coast, commenced work on the Wellesley Bridge in Limerick and cooperated with Telford in investigating the River Mersey in England. His coastal survey (already completed except for the south coast) was officially terminated the following year (1824), which gave him more time to follow new interests. He spent long periods in London during 1824, giving evidence to the Select Committee on the Survey and Valuation in May, and to the House of Lords enquiry on the State of Ireland in June. During whatever time he spent in Ireland, he supervised the Berehaven road with John Killaly on behalf of the Loan Commissioners, and helped Maurice Fitzgerald with his plan for a steamship company. Increasingly his focus was moving away from Connaught, and the personal priority he had previously given to the western district was starting to slip noticeably. This is apparent in the amount of time he spent in the west (and in Ireland generally) from 1826 on, as a close examination of dated letters, documents and reports from 12 June 1822 to the end of December 1831 shows. In all, his location can be determined with a very high degree of certainty for 463 days in that period of 3,490 days (about 13 per cent of the period) or, if certain reasonable assumptions are made, with a lesser but still very high degree of confidence for 820 days (23 per cent of the whole period). In the analyses, his presence at a location is treated as 'certain' on the day of the dated document, plus one day on either side of it. His presence at a location is treated as 'highly probable' by assuming his continued presence there when the interval between dates he was certainly known to be there is short, typically one week or less. For example, if he wrote letters from Dublin on 2, 6 and 14 May he is treated as having been there continuously for fifteen days, from 1 to 15 May inclusive. If, on the other hand, he had written letters from Dublin only on 2 and 14 May he is treated as having been there certainly for only 6 days – 1 to 3 and 13 to 15 May. Table 7 shows the result of this treatment of the documentary sources. Summarising the data, he spent 322 documented days in Ireland and 51 in England between 1822 and 1825 (86 per cent and 14 per cent respectively of the documented days for that period). From 1826 to 1831 the distribution is reversed: he spent only 73 documented days in Ireland and 374 in England (16 per cent and 84 per cent respectively) during that period.

Notwithstanding the relative paucity of the observations, and accepting that we cannot read too much into incomplete data (although they are the best we have), the *prima facie* evidence supports the view that from 1826 onwards he spent much more time in England than in Ireland, and there is no documentary proof that he spent any appreciable time in the western district after 1825! (He may, of course, have been there, but documents are lacking.) Detailed information in later chapters confirms that the focus of his attention and activity moved away from Connaught after 1823, first to projects in Limerick, Courtown, Killarney and elsewhere in Ireland, and then to his work with Telford and to other schemes in England.

One other notable feature of the data is the remarkable shortage of information for the year 1829. In June 1828 he could not travel out of London for a while, 'being too weak to leave town'.[50] By October he was in Harrogate, 'trying to pick up some health to encounter the next campaign'.[51] The winter did not help him and on 6 April 1829 he wrote to Robert Stevenson from London stating: 'I have been very poorly all this winter and have full escaped a fever and cannot much attend to business.'[52] That, in fact, is one of the very few documents of Nimmo's that has surfaced for 1829 and it is probable that he was indisposed for much of that year. However, for most of the next two years, he was back in good health and busy at new ventures mainly in England, which further explain his neglect of duty in the western district. This neglectfulness would precipitate his final severance from the district and lead to the demise of his great scheme for Connaught.

Controversy had never been far from Nimmo. As early as 1824, a Memorial purporting to come from the fishermen of Ennisturk, supported by Earl Spencer, alleged that his pier on the island had destroyed the natural harbour and produced injury rather than advantage to the place. James Donnell, Lieutenant Hutchison of the Royal Navy and John Killaly were asked to examine and report on the matter. When the fishermen, the inhabitants and the coast guard were consulted they unanimously agreed that the pier had served the harbour well, and expressed unease and even alarm at the prospect of its possible removal. Donnell concluded his report: 'I did not meet any person on the island who would admit that he had ever signed or seen the Memorial or believed such a paper ever had been there, although some admitted that they had heard of such a matter at Castlebar.'[53] This effectively exonerated Nimmo. In September 1827, Colonel Bingham wrote to Lamb, the new chief secretary, about the Castlebar to Erris road and its branches on which, he said, Nimmo had done nothing. 'Send down an engineer of the first class, and unconnected with Mr Nimmo, to examine the state of those roads,' he demanded.[54] Nimmo consulted his own resident engineer, Patrick Knight, on the matter before planning a visit to the area himself. Knight wrote back a very positive account of the district[55] in stark contrast to Bingham's tirade, and this allowed Nimmo to defend himself in a letter to Lamb at the end of October.[56]

A complaint against him of a much more serious nature occurred in 1829, when the Commissioners for Auditing the Public Accounts in Ireland published their audit of his accounts for the years 1822 to 1825 inclusive.[57] He had received £80,375 for public works but his accounts failed to balance with that sum, showing a deficit of over £22,000. Even when they allowed for his inexperience of their method of accountancy, they found the inaccuracies were beyond normal explanation. They called his attention to the insufficient and unsatisfactory state of the accounts and to the unnecessary delay in the payment of bills and wages that they found. Although he had explained himself with some effectiveness, they admitted, they nevertheless expressed the hope that his future accounts would 'assume a character of greater method and simplicity'. This was the first official acknowledgement that his accounting was not altogether in good order. Some weeks later, on 6 June, Mr O'Neill rose in the House of Commons with a petition from unnamed persons in Galway, alleging that Nimmo had built roads that no one used, had wasted public money, and built a road to service a place where he had bought land and was developing a settlement.[58] It was a grossly inaccurate allegation, but following so soon on the report of the Audit Commissioners, it needed firm rebuttal. Within days Nimmo, writing from London, had transmitted his report for the year 1828 to Lord Leveson Gower, who was then the chief secretary. The House of Commons ordered the report printed, and on 19 June Leveson Gower rebutted O'Neill's allegation, stating that the charges had been found to be groundless 'and were the more ungracious as they were made without any intimation to Mr Nimmo, the party affected by them'.[59] Maurice Fitzgerald also addressed the House in Nimmo's defence and the matter was officially closed. However, the allegation shows that Nimmo had powerful enemies in the west who would not hesitate to discredit him. Later, in October, Leveson Gower made a tour of the west visiting Galway, Headford, Cong, Renvyle, Clifden, Ballinahinch and Castlemagarrett.[60] From this list of venues it would appear he went along many of Nimmo's new roads, meeting many of those proprietors like Martin (Ballinahinch), D'Arcy (Clifden) Blake (Renvyle) and Browne (Castlemagarrett) who generally supported Nimmo's endeavours. Gower expressed himself highly pleased with what he saw and heard (Plate 11).

The allegations had, however, damaged Nimmo's reputation and he was instructed to prevent further unnecessary expenditure.[61] Priority was to be given to the main lines of road such as the Castlebar to Erris and the central Connemara road; work on branch roads and cross roads was suspended pending completion of these. Samuel Jones was seriously behind in his contract to build the central Connemara road, so Nimmo was instructed to step in and finish it without further delay. This was not quite so simple: legal opinion was that Jones should get proper notice of the intention to remove him before Nimmo could move in.[62] Nimmo was further instructed to prepare an account of all the money he had received up to then for the western district and of all the works he had started since 1822. Concomitantly, the Fisheries Commissioners were asked to account for the state of all the piers built, repaired, improved or in progress under their direction. These inventories, prepared in June 1830 and published in July, give a full account of all that Nimmo had undertaken in the west up till then.[63] By his own reckoning, he had disposed of almost £160,000 (British currency) for public works in the western district and had started about eighty-four separate projects in five counties (see Appendix 3). Unfinished though many of the projects were, it was a remarkable achievement to have undertaken so much, under the exigencies of the time.

The total amount of money that came under Nimmo's administration is a matter that will repay further detailed examination and one that may show the Government in a better light than we might expect. The £160,000 of western district funds was only part of the total. The Fisheries Commissioners, for example, reported thirty-one piers finished and in perfect condition, eighteen in progress, five damaged and seven unfinished; Nimmo had erected most of these with separate fisheries funds (along with associated charitable contributions). He had also expended funds from the Post Office budget (approximately £17,500 in 1822 to 1824),[64] £40,000 in 'extra' funds granted in 1822 'for such measures as the exigency of affairs in Ireland may require',[65] presentment funds, consolidated loan funds, the Lord-Lieutenant's Moiety Act gifts, private and charitable contributions and county Treasurers' cess funds. Up to January 1825 he had spent over £80,000 under the *Employment of the Poor Act* alone and he estimated that a further £57,000 was still needed to complete all the work of the western district. Griffith had received £35,000 for the southern district and Killaly £9,000 for the middle district by that time.[66]

The aggregate of Nimmo's estimates for the years 1825 to 1830 came to £86,000 while those of Griffith came to £21,000 for the same period and Killaly had already finished his tasks before then. It is small wonder, then, that the authorities were growing exasperated with Nimmo's expensive agenda, especially when it did not result in many roads being transferred to the Board of Inland Navigation, unlike in Griffith's and Killaly's cases. The extent of his absences in England cannot have gone unnoticed in Ireland either: almost all his correspondence with the authorities now originated from there rather than from Dublin.

On 23 March 1831, Nimmo wrote to chief secretary Stanley from London enclosing his report for 1830 together with 'Estimates of the expense of repairing and completing these and all the works undertaken and since suspended, also the sum which I would recommend to be laid out during the ensuing seasons, and have added estimates of various other useful lines of road in that district.'[67] The total estimate came to £60,471, being £15,635 for new works, £27,291 to complete ongoing works and £17,544 for works suspended. In 1827 the estimate for completion of everything had been about £30,000; now it was double that, despite the expenditure of £30,000 in the meantime! This account was the straw that broke the camel's back as far as Dublin was concerned, and some action had to be taken urgently to rein in the expenditure. A certain reading of Nimmo's new estimate might suggest that he was actually preparing to withdraw entirely, of his own accord, from the western district and was deliberately making an estimate of all that still needed to be done, in preparation for someone else taking over responsibility. He was already overloaded with work in England and in Dublin with the Kingstown railway; he was directing a new survey of Dublin Bay and another of Valentia harbour, as well as writing his book on piloting directions. On top of all this, his health seems to have taken another turn for the worse and he was once again spending time at the spa in Leamington.

Whatever Nimmo's true intentions, Stanley found a convenient way to put an end to the excessive, unending and seemingly fruitless expenditure in the western district: he initiated the action first threatened in 1827 when Wellesley was Viceroy and Lamb chief secretary. This involved moving Nimmo aside, without actually dismissing him. Killaly had already finished in the middle district and, since he was also the engineer of the Board of Inland Navigation, Stanley commissioned him to report on the state of the roads in Galway and Mayo in order to establish their suitability for transfer to the board. He was clearly seeking a report 'from another competent person as to the expenditure on the several works' – precisely what Wellesley had threatened to do in 1827.

Killaly's letter of appointment, reprinted here, was dated 28 May 1831. However, the twenty-first Report of the Commissioners for Auditing the Public Accounts of Ireland shows that he was employed as 'directing engineer for Mayo and Galway' from 1 April 1831. From then to 31 August he was paid four guineas a day for 153 days, that is, for every single day of the period and his salary from the Board of Inland Navigation was deducted for the 153 days.[68] This means he had taken up the post a full two months before his formal letter of appointment, but only one week after Nimmo's detailed estimate of 23 March.

Killaly may have carried out a preliminary inspection (either by visiting the sites or examining relevant documents) during April and May and reported on what needed to be done. With that information, Stanley was able to draw up the formal letter of appointment of 28 May which placed 'that fund at the disposal of some other engineer for the purpose of completing some of the works' – the main thrust of the Wellesley solution.[69] The detailed instructions in the letter referred to specific roads and bridges, exhibiting an unusually intimate knowledge by Stanley of the then state of the works, probably notified by Killaly, and made clear that the latter's role was to put the roads in fit condition to be taken over by the juries.

The letter of appointment reads as follows:

> Dublin Castle, 28 May 1831.
> Sir,
> His Majesty's Government, having decided to intrust to your superintendence the disposal, upon certain public works in the counties of Mayo and Galway, of a sum of money ... you will proceed forthwith to the district in question for the purpose of putting the works in progress without loss of time, and you will report to me ... the present state of the various works....You will bear in mind that it is not the wish of His Majesty's Government to afford gratuitous assistance upon the commencement of any new works, the intention being to render available to the public those upon which large sums have been already laid out ... and upon which a comparatively small expenditure in future is requisite in order to reap the benefit of that which has already been incurred. Subject to this restriction, the objects which it will be necessary for you to bear in mind, should be 1st the completion of such small portions, now unfinished, of roads already in progress, as form an interruption to the communication, and which if completed would form an unbroken line of intercourse through an extensive district. 2nd the employment of the poor in the parts of the county where the greatest distress prevails and 3rd the placing of such roads in a state of repair in which they may be transferred to the Directors General of Inland Navigation, in order that the repair may hereafter be borne upon the local funds. The sum which it is proposed in the present year amounts to £11,000 ... you will, having reference to the three objects to which I have already alluded, principally direct your attention to the following roads.
> 1st The central or main Erris road from Castlebar to Belmullet and especially to a part of the road near Bangor, where one bridge being broken

down and another in a very bad state of repair, it will probably be more advisable, and more economical to construct a new road short distance [sic] by which both bridges will be avoided.

2nd The road from Westport to Leenane at the head of Killary Harbour. Upon this road a sum of about £2,500 may be expended, in completing the part in progress between Leenane and Glinsk and in repairing and making anew the old road from Glinsk to Westport.

3rd You will pay particular attention to ... the present state of the central road through Cunemara [sic] and more especially that part of it which had been contracted for by Mr Jones ... and you will take the work into your own hands, without loss of time. Upon this road a sum of about £2,000 will probably be applied with advantage ... [Y]our attention is to be immediately and constantly turned to the object of handing over the public roads of the county to be maintained by the county funds ... [Y]ou will also inform me what roads or portions of roads...are in a good state of repair, in order that the Grand Juries of the counties may be informed that it will be necessary for them to provide for their future maintenance.

Signed E. G. Stanley.
To J. Killaly Esq.[70]

Killaly was given the title 'Directing engineer' and his son, Hamilton that of superintending engineer. Hamilton was paid only from 1 June; on 31 August he was made principal engineer, and later he succeeded his father on the latter's death in 1832. Once the Killalys took over, Nimmo informed his own men that their services were no longer needed after 30 June 1831. Naturally enough, this did not make for good relations between Nimmo's and Killaly's men, a situation that was aggravated by arrears in pay that some of the former were allegedly enduring.[71] To the public eye, it appeared that Nimmo had been dismissed with a question mark over his stewardship of public funds. Creditors were alerted by this and began to complain to Dublin Castle that they had not been paid for sundry services.

Returning to 1831, the chronology of events is quite remarkable. The records show that Killaly was engaged from 1 April. On 3 April, Anglesey, who had become Lord-Lieutenant for the second time in December 1830, set out on a tour of the west, arriving in Galway on 8 April. There he cautioned the Grand Jury not to expect too much from Government, telling them 'we live in an age of economy – indeed I might say of parsimony', which he accompanied with threats of force if riot and outrage were not suppressed.[72] Destitution and distress were again rife and once again the poor were being blamed for their own misfortune. Anglesey followed the well-worn visitor route – Ballinahinch, Clifden, Delphi and Westport – before returning to Dublin. Nimmo, unfortunately, was in Leamington at the time and so he lost the opportunity of correcting mischievous comments made to the Viceroy as he progressed through the district. During that progress, he heard allegations that Nimmo had left hundreds of poor people unpaid and on his return to Dublin he started an investigation. Nimmo tried to correct matters in a letter to Stanley on 27 April, explaining the cause of the non-payment and indicating those roads that were ready for immediate transfer to the board.[73] But the damage was already done by then and his explanations were too little, too late. From this chronology, Killaly had been engaged *before* Anglesey's tour commenced, suggesting that Nimmo's downfall might have occurred even if nothing negative was alleged about him during the tour. From here the timing becomes even more remarkable.

Anglesey wrote confidentially to the Prime Minister, Earl Grey, on 16 July (1831), by-passing Stanley, who had recently compromised him in another matter:

> When I made my western tour I received many loud complaints against Mr Nimmo, a Government engineer. Upon enquiry I discovered that hundreds of poor people remained unpaid for work done upon the roads.
>
> Immediately upon my return to Dublin I caused an enquiry to be made respecting Mr Nimmo's accounts and to my amazement I found that they had been passed and that there was no possibility of obtaining justice for these defrauded people. All this led to a closer investigation of the method of settling accounts with public accountants, and the result is the statement that I now send and a prospect [?] of a bill to remedy existing abuses. Of course you will put it in the hands of our chancellor and chief secretary.[74]

Grey replied from 10 Downing Street three days later, on 19 July:

> I have received your letter of the 16th with the accompanying papers. The case of Mr Nimmo is a most crying one, proves the necessity of a general remedy, and, I should add, of the immediate dismissal of Mr Nimmo. I have sent the statement and draft to Stanley and if he and Plunket do not object I will press it forward as much as I can.[75]

By July, Killaly had already been three months in office as directing engineer and Nimmo's role in the western district was over. Did Anglesey not know these facts? If he did, why did he not inform Grey of the action that had already been taken? Had Stanley acted without Anglesey's knowledge and approval (possibly compromising him a second time)? The correspondence proves that Nimmo had not been dismissed by July (or, that neither the Prime Minister nor the Lord-Lieutenant knew anything about it, which amounts to the same thing). Nimmo himself does not appear to have considered himself dismissed.[76] The 1833 audit of his final account would show that his salary for the western district was claimed, *and paid by the auditors*, up to the very day of his death in 1832, confirming that he was never formally dismissed.[77]

By the September after his appointment, John Killaly had completed the up-to-date report requested by Stanley. The Newport to Achill road, the road to Erris and the north Mayo coast road had all been addressed and were expected to be open soon for wheeled carriages. He had had no success in bringing Jones to his duty, nor could he legitimately encroach on that part of the central Connemara road that was within Jones's contract. He repaired sections of the road from Kylemore to Leenane and onwards to Westport. The Ballintubber road was in good enough shape so he had carried out only minor works to it.[78] From the time of their appointment in 1831 to April 1833, the Killalys did about £26,537 of repairs and modifications to Nimmo's works.[79] That compares remarkably closely with Nimmo's estimate of £27,291 for the entire completion of all the works.[80]

Nimmo had continued to press for the payment of arrears by the Government and was eventually successful in September 1831, when his invoice was met in full. When Stanley proposed to pay this account, Anglesey was slow to agree. Under-secretary Gosset informed Stanley: 'There have been so many and apparently well-founded complaints of Mr Nimmo or his agents having neglected to pay for work which had been performed and for which he had been audited by the Government that the Lord-Lieutenant deemed it proper that certain enquiries should be first made before any sum whatever is paid to Mr Nimmo.'[81] The account was paid nevertheless and Nimmo could now finally let his staff go. This effectively severed his active connection with the western district, except for the audit of his final accounts. Memorials and complaints, some genuine and some patently contrived, continued to arrive, making claims on his account. Many more would come in after his death. We will see that in November 1831 the Liverpool and Leeds Railway Company censured him over the large sums he had extracted from that company for his services: obviously his finances, public and private, were little short of chaotic, but he was probably too ill by then, eight weeks prior to his death, to make much of an issue of them.

Earl Grey's comment that Nimmo merited immediate dismissal was a cruel and ignominious remark on which to end a long career of public service. He had certainly neglected the western district in the later years, when his focus had shifted to England. For this he paid dearly. But he had helped significantly to change the face of Connaught and had given to its people hope for better things to come through the improvements he had made to its infrastructure. Taken together, all of his Irish activities truly constituted 'works of great national improvement'.

NOTES

1. A. Nimmo to W. Gregory, 19 June 1822 (NAI, CSORP 1822/861).
2. Sirr, *Memoir of the Last Archbishop of Tuam*.
3. Parliamentary debates, House of Commons, June 17 1822, column 1124.
4. *The Times*, 28 June 1822, p. 5.
5. 'Principles to be attended to in the Distribution of the Money granted to alleviate the Scarcity which

at present prevails in some Districts in Ireland, in Correspondence and Accounts relative to Scarcity in Ireland, 1822 to 1839' (B.P.P.1846 [734], vol. XXXVII, mf 50.311).
6. Instruction written on a copy of the letter from A. Nimmo to W. Gregory, 19 June 1822. (NAI, CSORP 1822/861).
7. W. Gregory to A. Nimmo, 25 June 1822 (NAI, CSORP 1822/861),quoted in K. Villiers-Tuthill, *Alexander Nimmo and the Western District* (Clifden: Connemara Girl Publications, 2006).
8. Plan of proposed road, Killala to Broadhaven, by William Bald 1821 (NAI, OPW5HC/6/0301).
9. This was the name given to the road in a memorial from the magistrates and residents of the Barony of Erris to the Lord-Lieutenant in March 1830 (NAI, CSORP 1830/ 453/1).
10. Expenses shown in various years 1822 to 1825, in Seventeenth Report of the Commissioners for Auditing the Public Accounts in Ireland (B.P.P. 1829 [193], vol. XIII, mf 31.88–9), hereafter referred to as Seventeenth Audit.
11. Plan of the mail coach road between Dublin and Ballina continued from Boyle to Tubbercurry by Alexander Nimmo, 1815 (NAI, OPW 5 HC/6/0108).
12. Calculated from Nimmo's MS accounts for different years (Nat. Archs. UK, Works of A. Nimmo, AO 17/492).
13. A. Nimmo, 'Report on the Western District by Alexander Nimmo, civil engineer', p. 35, in Employment of the Poor, Ireland, Reports by the Civil Engineers employed in superintending the Public Works (B.P.P.1823 [249], Vol. X, mf 25.86–7), hereafter referred to as Report for 1822.
14. A. Nimmo, Report on the progress of the Public Works in the Western District in 1824, appendix 1, p. 10. estimates for the sum that will be required for carrying on certain public works in Ireland during the year ending 5 January 1826 (B.P.P. 1825 [279], vol. XXII, mf 27.124), hereafter referred to as Report for 1824.
15. A. Nimmo, 'Report on the Fishing Stations on the West coast of Ireland from Errishead southwards', appendix 10, p.34, in Eighth Report of the Commissioners for Irish Fisheries (B.P.P. 1826/7 [487], vol. XI, mf 29.89 –9), hereafter referred to as Eighth Fisheries Report.
16. A. Nimmo, Appendix no. 10, p. 32, in Eighth Fisheries Report.
17. Fourth Report of the Commissioners of Irish Fisheries for the year 1822, p.11 (B.P.P. 1823 [383], vol. X, mf 25.85– 6), hereafter referred to as Fourth Fisheries Report.
18. Nimmo, A. Report for 1822, p. 39.
19. Nimmo, A. Accounts for the years 1822 to 1825 (Nat. Archs. UK, Works of A. Nimmo, AO 17/492).
20. *Ballina Impartial* [newspaper], 20 October 1823.
21. *Ballina Impartial*, 7 March 1825, p. 3.
22. Nimmo, A. Evidence, p. 171, in Minutes of Evidence taken before the Select Committee of the House of Lords to examine into the Nature and Extent of the Disturbances which have prevailed in those districts of Ireland which are now subject to the Provisions of the Insurrection Act and to report to the House (B.P.P. 1825 [200], vol.VII, mf 27.60–63).
23. Villiers-Tuthill, *Nimmo and the Western District*.
24. Nimmo, Report for 1822, p. 31.
25. Fourth Report of the Commissioners for the repairing, maintaining and keeping in repair certain roads and bridges in Ireland (B.P.P. 1829 [132], vol. XIII, mf 31.8).
26. Jenkins, *Era of Emancipation*, p. 196.
27. H. Goulburn to Lord Wellesley, 23 May 1823 (BL, Wellesley Papers, Add 37301 fo. 90).
28. Nimmo, Report for 1822, p. 31.
29. Returns of grants and loans for Public Works since the Union (B.P.P. 1839 [540], vol. XLIV, mf 42. 307).
30. A. Nimmo, Report for 1822, p. 31.
31. A. Nimmo, 'Report on the progress of the Public Works in the Western District of Ireland in the year 1829', pp 3–4, in Public Works Ireland (B.P.P. 1830 [199], vol. XXVII, mf 32.194).
32. Nimmo, 'Report on the roads proposed for the colliery district in the west of Lough Allen [July– August 1824]' (NLI, Hibernian Mining Company papers, MS 657).
33. Report of the Select Committee on the Employment of the Poor in Ireland (B.P.P. 1823 [561], vol.VI, mf 25.49–51).
34. Nimmo, Appendix 1, p. 9, in Report for 1824.
35. A. Nimmo to A. Mangin, 26 June 1822 (NAI, CSORP 1822/866).
36. A. Nimmo to H. Goulburn, 16 November 1822 (NAI, CSORP 1822/3270).
37. M. Blake to H. Goulburn, 28 November 1822 (NAI, CSORP 1822/3416).
38. Nimmo, Report for 1822, p. 44.
39. Petition of stonemasons and labourers of Galway to the Lord Lieutenant, 18 January 1823 (NAI, CSORP 1823/5127).
40. G. Nimmo to E. Lees, 12 October 1822 (Library of Peadar O'Dowd, Galway).
41. Report of the Select Committee on the State of the Poor in Ireland and means of improving their condition, p. 38 (B.P.P. 1830 [667], vol. VII, mf 32.47–57).

42. A. Nimmo to H. Townsend, 16 July 1822 (NAI, OPW 8/83 [1]).
43. R. Griffith to H. Goulburn, 5 June 1825 (NAI, OP 972/13).
44. Report for 1822.
45. Draft letter, Lord-Lieutenant's office to A. Nimmo, 4 August 1824 (NAI, CSORP 1824/8855).
46. 6 Geo. IV C. 101 *An Act to provide for the repairing, maintaining and keeping in repair certain roads and bridges in Ireland.* 5 July 1825.
47. Third report of the Commissioners for the repairing, maintaining and keeping in repair, certain roads and bridges in Ireland (B.P.P. 1828 [126] vol. XII, mf 30.82).
48. Ibid.
49. Draft letter to A. Nimmo, undated [1827] (NAI, CSORP 1827/153/11).
50. A. Nimmo to R. Stevenson, 18 June 1828 (NLS, Robert Stevenson and Co. Records. Acc 10706/81/10).
51. A. Nimmo to Lee and Watson, 15 October 1828 (BL, Add 44085 fo. 45).
52. A. Nimmo to R. Stevenson, 6 April 1829 (NLS, MS 19988 fo. 64).
53. J. Donnell, 'Report on the State of the Fishery Piers and Landing Quays on the coast of Ireland', in Eighth Fisheries Report, appendix no. 11, pp. 52, 53
54. D. Bingham to W. Lamb, 29 September 1827 (NAI, OP 972/49/1).
55. P. Knight to A. Nimmo, 19 October 1827 (NAI, OP 972/49/3).
56. A. Nimmo to W. Lamb, 28 October 1827 (NAI, OP 972/49/2).
57. Seventeenth Report of the Commissioners for Auditing the Public Accounts in Ireland (B.P.P. 1829 [193], vol. XIII, mf 31.88–9).
58. *The Times*, 6 June 1829, p. 2
59. *The Times*, 20 June 1829, p.3
60. *Connaught Journal*, 5 October 1829.
61. A. Nimmo to F. Gower, 4 January 1830 (NAI, CSORP 1830/61).
62. The Crown's case against Jones (NAI, CSORP 1830/496).
63. Public Works Ireland. A return of all the public money expended under the direction of Mr Nimmo [...] and of all piers built on the coast [...] 1. Account of the different Public Works in the Western District of Ireland commenced under the direction of Alexander Nimmo in the summer of 1822. (Reproduced here as appendix 4.) 2. An account of all the piers [...] under the direction of the Commissioners of the Irish Fisheries (B.P.P. 1830 [624], vol. XXVII, mf 32.194).
64. Public Works Ireland. Returns of all sums of money advanced in aid of Public Works in Ireland since the Union (B.P.P. 1839 [540], vol. XLIV, mf 42.306–7).
65. Fourteenth Report of the Commissioners for Auditing the Public Accounts in Ireland (B.P.P. 1826 [317], vol. XI, mf 28.67– 9). £200,000 was granted to the Lord Lieutenant by Act 3 Geo IV C. 127 enacted 6 August 1822, 'For such measures as the exigency of affairs in Ireland may require, for the year ending 5 January 1823.' This was separate from the £100,000 granted for the employment of the poor in Ireland by section XVI of the act.
66. Nimmo, Report for 1822.
67. A. Nimmo to E. Stanley, 23 March 1831 (NAI, OP 973/76).
68. J. Killaly, 'Public Works in the Western District', p. 176. In Twenty-first Report of the Commissioners for Auditing the Public Accounts in Ireland (B.P.P. 1833 [102], vol. XVII, mf 36. 124–127).
69. Draft letter to A. Nimmo, undated [1827] (NAI, CSORP 1827/153/11).
70. E. Stanley to J. Killaly, 28 May 1831 (NAI, CSORP 1831/1732).
71. A. Nimmo to E. Stanley, 27 July 1831 (NAI, OP 973).
72. *Connaught Journal*, 11 April 1831.
73. A. Nimmo to E. Stanley, 27 April 1831 (NAI, OP 973/82).
74. Anglesey to Lord Grey, 16 July 1831. (Durham University Library, Grey of Howick Papers, GRE/B3/150).
75. Grey to Lord Anglesey, 19 July 1831. (Durham University Library, Grey of Howick Papers, GRE/B4/155).
76. A. Nimmo to E. Stanley, 27 July 1831 (NAI, OP 973).
77. A. Nimmo, Accounts of A. Nimmo for the years 1831 and 1832 (Nat. Archs. UK, Works of A. Nimmo, AO 17/492). See also the full audit of Nimmo's final accounts in Note 93.
78. J. Killaly to W. Gosset, 30 September 1831 (NAI, OP 973/86).
79. Public Works Ireland. Return to an order of the Honourable House of Commons dated 16 April 1833. No.2. Return showing the detailed expenditure on the principal works in the counties of Mayo and Galway under directions of Messrs. Nimmo and Killaly from the year 1821 to 1833 (B.P.P.1833 [285], vol. XXXV, mf 36.257–8).
80. A. Nimmo to E. Stanley, 23 March 1831 (NAI, OP 973/76).
81. W. Gosset to E. Stanley, 28 September 1831 (NAI, OP 973/84).

CHAPTER 11

Nimmo's Irish Bridges

Since most of Nimmo's roads were in remote and poor regions it is inevitable that his bridges would be more utilitarian than decorative. Where the streams to be crossed were small, he favoured small, sturdy, single semi-circular arch structures. These vernacular bridges are common and largely unremarked along most of his western roads, and are still in use today. He never gave these a name in his reports, so that it is not always possible to confirm his contribution to all of them. For this reason none of his bridges – excepting Sarsfield and Poulaphuca, the only two that have been widely attributed to him until now – have ever appeared in authoritative books on Irish stone bridges, such as those of O'Keeffe and Simington[1] and Cox and Gould.[2] In some instances, his vernacular bridges have been widened in modern times by laying concrete slabs raised on concrete supports alongside them, removing the contiguous parapet, and levelling the deck across the whole, widened surface. This results in functional but inelegant hybrid structures, where one face exhibits the original arched masonry and the other, modern flat concrete. We have already discussed his bridge, modified somewhat like this, on the N70 in Kerry.

A number of his bridges are quite substantial, and from the records it is possible to identify at least thirty with which he was involved with certainty. Table 8 lists them all, indicating their locations and giving references to some of the documents that confirm Nimmo's connection with their design or construction.

The variety exhibited by his bridges suggested to Ruddock that Nimmo wished to be a master of all styles.[3] However, the demands of particular sites are bound to influence the style of bridge that may be suitable in each individual case, and there is no reason why all the bridges along even a single stretch of road should have the same style or structure; indeed, the opposite might often be advisable. For example, the preferred bridge to cross the estuary of a major river within an urban environment will clearly be different to that of one crossing a narrow gorge in a rural setting. Nimmo's Wellesley Bridge in Limerick City and Poulaphuca Bridge in rural county Wicklow are examples of these respec-

NIMMO'S IRISH BRIDGES

TABLE 8.

A list of all the bridges to which Nimmo had an input, as recorded in extant documents. The numbers in the last column gives the source of the information linking him to each structure, as given in the 'notes and references' at the end of the chapter.

No.	Bridge name, Road No., County	River	Location	Type of Bridge	Apprx. Height (feet)	Apprx. Span (feet)	Notes and References
1	Poulaphuca, N81, Wicklow.	Liffey	N 53 07.06 W06 35.11	Single pointed arch.	150	65	See text for references.
2	Shaugh-nessy's, R336, Galway.	Tributary of Foilmore.	N53 29.39 W09 33.51	Single pointed arch.			Now much repaired. 19, 87
3	Glanocally, N59, Mayo.	Glenocally.	N53 37.801 W09 36.66.	Single pointed arch.			20, 76, 87.
4	Tum-neenaun, R345, Mayo.		N53 31.26 W09 23.43.	Single pointed arch		15	22, 76, 88 87, 89.
5	Ardbear New, R341, Galway.	Owenglin	N53 29.189 W10 01.21.	Single pointed arch.			21, 22.
6	Dunmore East, Waterford.	Crossing to small island.	N52 08.56 W06 59.32	Single pointed arch.			Demolished. See text for references.
7	Glenanean, R319, Mayo.	Glenanean river	N53 53.28 W09 48.53.	Single pointed arch			20, 87, 88, 90.
8	Onodoo, `R319, Mayo.	Cartron river	N53 56.14 W09 49.40.	Single elliptical arch.			87, 88.
9	Sarsfield, (Wellesley) N18, Limerick.	River Shannon.	N52 39.60 W08 37.49	Five arches.			See text for references.
10	Youghal, off N25, Cork.	Blackwater.	N51 58.488 W 7 51.297	Thirty timber arches, plus bascule.			Replaced. See text for references.
11	Gowlan, N59, Mayo.	Tributary of Erriff.	N53 39.90 W09 34.06.	Single stilted segmental arch	9	12	76.

cont

12	N59-062-00, N59, Mayo.	Tributary of Erriff.		Stilted, Rounded arch	8	5	
13	Erriff, N59, Mayo.	Owenmore.	N53 39.274 W9 34.401	Single segmental arch.			20, 22, 76, 77, 86, 87, 90, 91.
14	Srahlea, N59, Mayo.	Owenmore.	N53 40.842 W9 33.045	Two arches.			Mostly replaced. 76
15	Cornamona, R345, Mayo.	Dooghta river.	N53 30.864 W9 26.915	Single elliptical arch.			74, 76, 92
16	Toombeola, R341, Galway.	Ballinahinch.	N53 26.09 W09 52.17.	Three stilted round arches.			20, 77, 87, 91.
17	Maam, R336, Galway.	Bealnabrack	N53 30.55 W09 33.31	Single segmental arch			Relaced twice. 20, 74, 77, 87, 90, 92,
18	Tiernakill, R336, Galway.	Foilmore.	N53 30.52 W9 33.923	Single segmental arch.			Mostly covered in concrete. 20, 76, 87, 88, 90.
19	Casla, R 336, Galway.	Fermoyle	N53 16.32 W09 31.52.	?			Replaced. 76, 82, 93, 94,
20	Bundorragha, R335 Mayo.	Bundorragha.	N53 36.29 W09 44.56	?			Replaced. 22, 82, 88, 92, 94, 95,
21	Leenane, N59, Galway.	Lahill river and inlet of the sea.	N53 35.46 W09 41.37.	Three round arches.			Recently swept away 74, 77, 92.
22	Kilhale, R312, Mayo.		N53 54.6 W09 24.66	Single round arch.			
23	Ardbear, R341, Galway	Inlet of the sea.	N53 28.25 W10 01.09	Embankment with two round, stilted arches.			22.
24	Ballinderry, R347, Galway	Grange river	N53 28.20 W08 50.18	Single segmental arch.			22, 82, 74, 93, 96, 94, 97.

cont

25	Annios, RBBB, Mayo.		N53 42.447 W09 12.727	Three stilted round arches.		22, 73, 74, 82, 92, 94, 95.
26	Kilnacarra, Co. Mayo	Manulla River.	N53 53.04 W09 11.27	Three stilted round arches.		82, 94, 95.
27	Shindela, R336, Galway.	Lake Shindela.	N53 27.17 W09 32.31	Single round arch.	12	22, 76.
28	Clydagh, N59, Galway	Loughkip river.	N53 19.569 W9 10.035	Two round arches.		74, 77, 88, 91, 97.
29	Hollymount, R331, Mayo.	Robe River.	N53 39.44 W09 09.37.	Three round arches.		22.
30	Cloongullaun N 26, Mayo.	River Moy.	N53 57.52 W08 59.65	Five round arches		81, 82
31	Barna Bridge, Galway	Inlet of sea	N53 15.17 W09 06.54	One?		Never completed. 20, 22, 73, 74, 76, 77, 87, 92,

tive situations, and they do indeed differ like chalk and cheese. The former is unique in Nimmo's œuvre, but the latter is one of a number of pointed-arch bridges that he designed, usually for ravine sites. Two other general types of bridge worthy of particular note are his timber bridge over the estuary of the River Blackwater near Youghal, County Cork and his iron bridge proposed for Athlunkard in Limerick, but never built. These were more 'artistic' than his usual functional bridges in other places and they merit some extra consideration. The number in brackets given after each bridge refers to its place in table 8.

POULAPHUCA BRIDGE, COUNTY WICKLOW AND OTHER POINTED-ARCH BRIDGES

The pointed-arch bridge at Poulaphuca (1), with its associated smaller convenience bridge nearby, is one of the most famous structures attributed to Nimmo. The main bridge spans the River Liffey where it passes through a deep ravine in a once tumultuous cataract located on the Wicklow–Kildare border, some miles south of Blessington. The river flow is now considerably reduced due to the modern hydroelectric works, and the scene has lost much of its original dramatic impact. Nevertheless, the bridge still adds height, character and ambience to the site. O'Keeffe and Simington give an expert account of the bridge structure, which has a span of about sixty-five feet and a rise from the span line to the arch keystone of almost forty feet. The rise from the river-bed is between 150 and 200 feet, making it one of the highest bridges in Ireland. They also discuss some aspects of its construction history, concluding that their findings 'leave little doubt that Poulaphuca was one of his [Nimmo's] masterpieces'. Yet paradoxically, they surmised that it was built around 1835, after Nimmo's death, from which they say 'it would appear that...he was not involved in the construction'. How can these seemingly contradictory statements be reconciled?

The attribution of Poulaphuca Bridge to Nimmo is, in fact, circumstantial, and Professor de Courcy was the first in modern times to point out emphatically that Nimmo's contribution to it, if any, has never been established in any primary source.[4] The earliest attribution is in Samuel Lewis's *A Topographical Dictionary of Ireland* first published in 1835, just three years after Nimmo's demise.[5] Lewis described it as 'a picturesque bridge of one pointed arch springing from rock to rock, built in an antique style from a design by the late Mr Nimmo Esq., at an expense, including the land arches and approaches of £4,074-15-0'. Published so soon after Nimmo's demise, this account is likely to be accurate and is good but not conclusive circumstantial evidence of his involvement. Since Lewis, most writers have repeated the attribution to Nimmo, so frequently indeed that to question it now can make one seem needlessly incredulous.

The date of its construction can now be established with certainty from a number of primary sources. In the 1820s there was considerable pressure for a mail coach road from Dublin to Waterford through New Ross. If it were to come to fruition it required separate Grand Jury presentments in a number of counties. As a result, the road was constructed in piecemeal fashion, according as the different counties passed the presentments. No one was keener to see the road completed through Wicklow than Lord Downshire through whose property it ran, and Lord Waterford. The latter would gain twice from the road: not only would it ensure a better communication with his southern estates, but also the planned route would, he hoped, pass along the Blessington to Baltinglass road that passed by his Wicklow property. When Downshire saw an advertisement seeking contractors to build the road, he wrote on 29 May 1823 to Sir Edward Lees of the Post Office. The new road, he said, would join the Dublin section of the road to the part already completed in the county of Wicklow 'through Blessington which is my property ... You are no doubt aware that a bridge over the waterfall at Russborough is nearly completed and that the road to it through Blessington is perfectly level and in good order and only requires completion on the Dublin side.'[6] Lees replied confirming that the mail coach would go via Blessington and Baltinglass as soon as ever the new road was open. Downshire's letter, on its own, establishes satisfactorily the date of the bridge at 1823. Also in 1823, the Lord-Lieutenant made a gift of £2,037-7-6 'To make a bridge at Poulafuca' under powers granted by the Moiety Act.[7] However, neither this nor any of the correspondence about the Blessington–Baltinglass road, makes mention of Nimmo or any other engineer.[8]

Coloured architectural drawings of the two bridges – the main bridge and the 'dry' bridge – are conserved in the National Archives of Ireland. The one titled 'Proposed Bridge over the Liffey at the Fall of Poulafuca' is unsigned and undated, showing the pointed-arch bridge drawn to a scale of one inch to ten feet (Plate 8).[9] The second is titled 'Proposed Bridge over the Hollow next the Liffey' (Plate 9), with a manuscript annotation at the top stating 'Dry arch near Poulafuca Bridge'.[10] A table headed 'Estimate of costs' has been almost completely erased from this item. Both drawings seem to be 'of a piece' – same scale, similar colour wash – and subjectively compare well with other drawings that are signed by Nimmo; the inference that they are Nimmo's design drawings for Poulaphuca is almost inescapable. If they are indeed Nimmo's design drawings, why did he not sign them when he was careful to sign almost everything else?

From other documents, Nimmo can be confidently connected with Poulaphuca Bridge, but perhaps not in the way we might have anticipated. He and John Killaly were engineers appointed under the Moiety Act to approve the staged payments to projects funded under that act. When an application was made for the progress of Poulaphuca Bridge to be certified in order to obtain an interim instalment, Nimmo wrote to Secretary Goulburn on 3 July 1823 certifying the progress made. That certificate, signed, dated 30 June 1823, and sealed with Nimmo's seal, approved a payment of £1,018-13-9.[11] Having received Killaly's certificate separately, Goulburn wrote to both engineers on 3 September, pointing out that their certificate needed to be a joint one. Killaly wrote back that he had already reported on 9 August, but enclosing further copies of both his own and Nimmo's certificates.[12] On 30 September Killaly wrote again enclosing another certificate – this time the signing-off certificate – approving a final sum to be paid to the contractors, who were named as Bergin, Woods and MacKenna.[13] They had won the construction contract over the Tassie brothers of Dublin, who complained subsequently to the Lord-Lieutenant that theirs had been the lowest quote and that therefore they should properly have been given the job.[14] A brief investigation showed that the award had been fair, at least by the standards of those days.[15]

As to the cost of the bridges and associated road works, Galloway wrote a letter dated 19 March 1822 to Goulburn informing him that in the Summer Assizes of that year the Dublin Grand Jury had passed a presentment for £973-12-2½ for the road and bridge at Poulaphuca, and the Wicklow Grand Jury had presented for £1,063-15-3½.[16] Adding these sums to the gift of the Lord-Lieutenant in 1823 (£2,037-7-6) the total comes to £4,074-15-0, the exact amount, to the penny, stated by Lewis.

Still, the question remains whether Nimmo's involvement was more than just his Moiety Act supervisory duty? Did he *design* it, and if so, why he did not sign the drawings? The evidence presented here could equally point to Killaly as the designer. During the summer of 1823 Nimmo, who was surveying the

Clare coast for the Fisheries Commissioners, worked closely with Killaly, so much so that they even economised by sharing the same writing paper: on the back of a letter dated 30 September 1823 from Nimmo (in Ennis) to his brother George in Dublin, Killaly had written instructions to Nimmo's Dublin office regarding the Poulaphuca business.[17] One of Killaly's projects in hand was the construction of a road (now the N67) along the coast of Clare and he was working near Miltown Malbay that September. The road crosses the Inagh river by Bealaclugga Bridge, near Quilty. This, too, is a pointed arch structure, similar in design to Poulaphuca Bridge, although the ravine it spans is not nearly as deep. It was built between 1823 and 1825 at a cost of £870.[18] Which of the two engineers designed this bridge, or did they cooperate in the task? Or was Killaly simply recycling the general plan of Poulaphuca at Bealaclugga? The connection between these two bridges may be closer in time, in design and in designer(s) than previously realised.

Two other bridges of similar design built around this time can be attributed with confidence to Nimmo, or to his engineering practice: the bridge on the Maam Cross to Maam road, R336, known as Shaughnessy's Bridge (2) (figure 11.1), and the Glenocally (Glenacally) Bridge (3) on the N59 between Leenane and Westport (figure 11.2). The former was a source of some conflict between Nimmo's office and a contractor, documented in correspondence.[19] The latter, dated 1826 on the upstream keystone, is mentioned by Nimmo in his report for 1828 and elsewhere.[20]

Both these bridges are pointed-arch structures spanning ravines, with the arch springing directly from the rocks on either side. Both are located in Nimmo's western district on roads that he laid out and documented in the reports. On the road between Maam and Cong (R345), well documented as Nimmo's work, there is another pointed-arch bridge spanning a ravine near Carrick. This is Tumneenaun Bridge (4), sometimes called Carrick or Corrib Bridge (figure 11.3). Although not specifically mentioned by name, Nimmo reported that 'The Joyce Country road ... has been gravelled to the extent of eleven miles, the remaining four miles are in hand, as also the bridges', so it is very probably one of Nimmo's design, as the other bridges on this road are known to be. Ardbear New Bridge (5) in Clifden, a slightly pointed, single-arch bridge spanning a chasm on the R341 road as it enters the town from the south, is attributed to Nimmo by the Halls,[21] and is mentioned in one of Nimmo's own reports.[22] At Dunmore East harbour also, the bridge (now demolished) dating from Nimmo's time that crossed the original chasm from the mainland to Eilan na Glioch, had a pointed arch (6). It was mentioned by Coote and Nimmo in the original plan for Dunmore.[23] The earliest picture of it, a photograph by S. D. Goff dated about 1860, is reproduced in an article by June Fennelly.[24] It is also illustrated in photographs in the Lawrence Collection in the National Library as shown in figure 6.2.[25] Since Nimmo is the common thread joining all of these pointed-arch bridges, the inference that Poulaphuca was his design and not that of Killaly is enhanced. It follows that Nimmo may have had some input in the design of Bealaclugga Bridge, perhaps in cooperation with Killaly. The latter was mainly experienced in canal bridges that did not span ravines and no other pointed bridge has ever been attributed to him.

Finally, a bridge of considerable interest and finesse in this context is Glenanean Bridge (7) in Corraun peninsula, some miles north of Mulranny in County Mayo. Carrying the R319 across a deep gorge, it is named on the first edition of the six-inch Ordnance Survey map surveyed in 1838.[26] Nimmo had laid out this road, partly funded by a gift of the Lord-Lieutenant.[27] In the report for the year to January 1829, John Mulloy was paid for the delivery of stone and timber to Glasnanean bridge [sic] and Onandoo bridge (8), which were centred under the supervision of a Mr Smyth.[28] According to Nimmo the whole road into Achill was fully completed by 1830.[29] Although the river Glenanean is small, a high bridge was necessary to span the deep gorge without diverting the road a considerable distance to a more level spot. It is a simple, austere, stilted pointed-arch structure made of local sandstone with a limestone arch ring, about thirty-six feet high from base to keystone. The span is about eight feet and the height from the springing to the keystone about nine feet. Corbels for the support of the centring are still evident in the intrados. The bridge, particularly its upstream face, has a pronounced ecclesiastical feel to it, even when the later buttress is ignored (figure 11.4). Killaly claimed to have overseen the final completion of the Achill road when he was directing engineer in Mayo in 1831, but there is no reason to believe that he had anything to do with this most appealing bridge, an adornment on the road.[30] A little to the north of it, an elliptical arch bridge with a limestone ring and local sandstone spandrels, reminiscent of Errif Bridge except smaller, crosses a much wider river known as the Cartron River at a level place. This is the Onandoo bridge (8) mentioned by Nimmo (figure 11.5); one of the tributaries making up the Cartron River is called the Easandoo River. The limestone rings of both bridges are probably the stones delivered by Mulloy in 1829, since the sandstone used in the spandrels is local.

Returning to Poulaphuca Bridge, it seems, therefore, that O'Keeffe and Simington's paradox can be resolved: Nimmo was certainly connected with, and almost certainly designed it, but it was constructed by Bergin, Woods and MacKenna and its completion to specification was certified by Nimmo and Killaly in September 1823, not 1835. As to why Nimmo did not sign the drawings, we can suggest one possible explanation. Because he was one of the engineers of the Loan and the Moiety Acts whose function was to approve the plans, certify the progress, and confirm the completion of the bridge, prudence – if not propriety – may have dictated that his name should not be too blatant on the original plans when they appeared before the Grand Jury. This explanation is supported by a stricture regarding public works in general made later on by the Select Committee that examined the Irish estimates in

FIGURE 11.1
Downstream face of Shaughnessy's Bridge as drawn by James Mahony and published in *A Tour of Connemara* in 1854.

FIGURE 11.2
Glenocally Bridge. The original cut stone of the arch is evident, contrasting with the modern concrete reinforcement. Note the rubble stone in the soffit.

FIGURE 11.3
Tumneenaun Bridge. An unusual pointed arch bridge on the Killary to Cong road built by Nimmo in 1825–8.

FIGURE 11.4
Glenanean Bridge, upstream face, constructed in 1829.

1828: 'it does not appear to the committee expedient that the engineers employed to superintend the work, to report on the progress, and to certify its completion, should have any pecuniary or other interest in the work, or that they should be made the public accountants.'[31]

WELLESLEY BRIDGE, LIMERICK AND THE PROPOSED BRIDGE AT ATHLUNKARD

By any measure, the posts held by Nimmo under the various acts were substantial sources of engineering work. However, being concerned mainly with remote rural districts, they gave little opportunity for stylish or monumental works, which are more often associated with urban environments. Low cost, minimum size and simplicity of construction were the determinants of all the public works funded by the Government and the Grand Juries in the poorer parts of Ireland. Engineers based in the great cities of England, which were expanding and modernising in the early nineteenth century, had many opportunities to express their talents, and flatter their patrons, by designing exuberant commercial edifices and other monumental works. Limerick was to be the Irish city where such an opportunity would knock on Nimmo's door, and the Wellesley Bridge (9) would provide this (Figure 11.6). It was to be his first and only major Irish urban project where cost would not be the sole determining factor in the design and furtherance of the work. We can be certain that it was designed by Nimmo, but whether 'designed' is quite the correct word for what the experts regard as a reduced-scale copy of a then well-known French bridge, the Pont Neuilly in Paris, designed by Jean Perronet, is a moot point.[32]

Limerick's developing trade, the possibility of a railway linking it with Waterford and the likelihood that steam navigation on the Shannon would open up the hinterland, all pointed towards a bright trading future for the city in the early 1820s. Consequently, the need for improved dock facilities became urgent and the city's connection to the hinterland of Clare and Galway was seriously constrained by the river crossing. In 1822 the merchants conceived a plan to build new floating docks on the south shore with a new bridge that would give access to the northern suburbs and places beyond. The Chamber of Commerce sent an outline proposal to the Loan Commissioners in September 1822, requesting them to send engineers to ascertain the best site for a bridge and docks.[33] A local commission to progress the proposal was formed, headed by Thomas Spring Rice, and Nimmo was commissioned to draw up suitable plans for both the bridge and the floating docks. How this plum commission fell to Nimmo we do not know for certain. Since he was an engineer with the Loan Commissioners he would have been consulted about the proposal, but that does not mean that the design had to be his. It is possible that Spring Rice, who was a convinced admirer of Nimmo, had proposed him for the design work, separate to his official function. However he obtained the commission, Nimmo had drawn up plans and an estimate by January 1823; the whole package was to cost £108,702, quite an enormous sum at the time.

The opportunity that the bridge presented to make a memorable addition to the fabric of the city was not lost on Nimmo and demanded something special from him. Therefore, he drew his design from his knowledge of bridges elsewhere, specifically Perronet's Pont Neuilly. His plan was engraved without delay and published with an elaborate dedication to Richard Colley, Marquis Wellesley, the Lord-Lieutenant.[34] Wellesley, who shared many of Spring Rice's (and Nimmo's) more liberal views on Ireland, would doubtless have appreciated the token of respect that naming so splendid a bridge in his honour represented. Nimmo's coloured drawing of the bridge is shown in Plate 10.

The bridge itself was only one part of the overall scheme, which consisted of five separate structures.[35] First the bridge, which would cost £23,673, would have seven river arches, along with one small arch over an island quay and one draw-arch between the island and the south bank through which tall ships could pass into a floating dock above. Second, there would be a tidal basin below the draw-arch where ships would lie, awaiting the opening of the bridge and upper lock. The shore-side quay wall of this basin would pass under the draw-arch, extending upstream as far as Spaight's Quay. The tidal basin and its associated quays would cost £34,382. Thirdly, two floating docks, estimated at £20,376 and £19,520 respectively, were envisaged, one at Custom-House Quay, the other coming down (along Arthur's Quay today) as far as Spaight's Quay. Finally, less substantial works on the north shore opposite the docks were estimated at £10,750. The docks were designed specifically to accommodate 'sharp vessels', that is, steam ships, and therefore would have been among the most modern docks in Ireland.

The already existing Shannon Navigation canal from Killaloe entered the river at Charlotte Quay near the Custom-House Dock, part of the proposed new floating dock. The following year, Nimmo would also design the Limerick to Waterford Railway, placing its terminus at Charlotte Quay.[36] His overall plan would therefore be a model of transport integration, locating the sea navigation, the inland navigation, the railway communication and the road bridge all at a single central hub, conveniently sited beside the custom-house. No map of his specific dock plan has surfaced, but his map of the proposed railway is extant and shows the whole integration.[37] Nimmo's dock is also shown in Thomas Rhodes's 1833 plan of the Port of Limerick, without the railway and completely dominated by Rhodes's own dock proposal.[38]

A private act for the bridge was steered quickly through Parliament (with Nimmo's help[39]) and passed into law on 17 June 1823.[40] It placed a levy on all shipping using the port that, within ten years, raised £13,500 towards the cost of the docks. The Bridge Commissioners sought a loan from the Consolidated Fund of £60,000, about half the total estimated cost of the whole

FIGURE 11.5
Bridge that Nimmo called Onandoo Bridge, spanning the Cartron River in Corraun, Co. Mayo

FIGURE 11.6
Sarsfield (Wellesley) Bridge, Limerick. The balustraded parapet is modern.

FIGURE 11.7
The underside of one arch of Wellesley Bridge showing the complex arching of the soffit, perhaps the bridge's most sophisticated feature.

plan, mortgaging the tolls of the bridge as security. In due course, the Loan Commissioners approved an advance of £55,384, by far the largest single sum ever awarded under the Loan Act. It was to be repaid in annual instalments of £2,769.[41] The Bridge Commission's engineer, Mr Baker, commenced work in 1824 under Nimmo's direction, and the foundation stone was laid on 25 October of that year.[42] Baker continued as engineer until 1827 when the contract was handed over to Clements and Sons, experienced contractors.

The bridge was, and remains, a light, elegant and graceful structure. Nimmo's original drawing (Plate 10) shows it with seven river arches – five elliptical, with one semi-circular arch abutting each end – but the terminal semi-circular arches were never incorporated into the finished bridge, shown in figure 11.6. Its unique feature, the extraordinarily sophisticated arching in the soffit, is hidden underneath: while the external façades of the arches are segmental on both the upstream and the downstream faces, the soffit is curved to form a double bell-mouth shape, so that the soffit arch is elliptical in the mid-line. This is evident in figure 11.7 and is a feature which, because of its location on the under-surface, is rarely observed and appreciated. Contemporaries of Nimmo, like the mathematician William Rowan Hamilton, noted its mathematical sophistication[43] and even to modern experts, like Ted Ruddock, the bridge remains remarkable:

Nimmo was not content with ordinary *cornes de vâches*; he curved the whole soffit to a double bell-mouth shape ... the masonry of the piers is also dressed to smoothly curved surfaces past which the water flows with the least possible turbulence. Moreover these special forms are not just a proud display of ideal scientific shapes, for the tide rises high on the piers and every tide that is higher than ordinary springs goes part of the way up the streamlined soffits. The design was therefore economical as well as scientific and elegant ... In everything except size Nimmo's bridge is the equal of any of Perronet's and in its basic form at least a little nearer perfection.[44]

How Nimmo might have appreciated this praise!

This sophisticated structure was difficult to build; it was still incomplete at Nimmo's death and the loan funds had already run out. By then the work was in a state of great embarrassment for want of detailed plans and a proper understanding with the contractors. It is hardly surprising that it should have been so; the necessity of using a diving bell to make the foundations, and the sophistication of the centring needed to erect the complex arches, would have made unusual demands on any contractor, no matter how experienced. Nimmo did not spend enough time monitoring and supervising progress, which he should have done for a project that was so technically demanding. One of the first actions of the new Commissioners of Public Works who took it over in late 1831 was to ask Charles Vignoles to assist in arbitrating the claims made by the contractor and to form a plan for the completion of the work. Vignoles executed 'the whole of that very intricate and arduous task' to the entire satisfaction of the board, placing the engineer Mr Grantham in charge of its day-to-day progress. The board – who regarded the bridge as 'an interesting and valuable work' – advanced another £25,000 for the completion of the whole, including the floating docks. The latter were never properly finished, but the bridge was opened on 5 August 1835, having taken eleven years and over £89,000 to complete. It remained a toll bridge for many years, but it never recouped the construction cost, not meeting even the interest charge; the toll was eventually abolished leaving Limerick with one of its finest structural features. It is now named Sarsfield Bridge and, like most of Nimmo's structures, this one – unquestionably his best – has no plaque or any other mark of his contribution to its design. *Sic transit gloria mundi.*

While the bridge was being completed, the merchants of Limerick directed a Memorial to the OPW requesting that the floating docks be progressed and bemoaning the fact that the money collected from the shipping levy had not been applied to the good of shipping. Justice demanded that the levy be discontinued or the dock completed, they complained.[45] Captain Mudge and the engineer Thomas Rhodes (Nimmo's pupil from the Inverness Academy days) inspected the site in 1832 and Rhodes made a proposal for elaborate new docks. In the opinion of some engineers, this would have laid a great part of the city under water at certain times, so it was abandoned in favour of constructing a fine line of quays along the south shore below the bridge.

While engaged in Limerick, Nimmo instructed Ben Meredith to survey a new line of road from the city to Parteen in County Clare. This would cross the Shannon by a new bridge to be erected at Athlunkard. Wellesley Bridge would be the lower-most bridge spanning the Shannon within the city, and Athlunkard would be the uppermost. Stylistic symmetry therefore called for a structure at the latter site equally as striking as that at the former, a demand that Nimmo proposed to fill with his most unusual bridge ever. The crossing at Athlunkard was 450 feet wide and the bridge would require three arches. His proposed structure (figure 11.8) was to be an underslung, chain suspension bridge that was not only structurally unique but probably impossible also. It appears from the engraving that the suspending chains were to extend below the bridge deck, which would attach to the chains from underneath, probably by rigid iron ties. The plan, which is indicative rather than definitive, is signed 'surveyed under the direction of Alexander Nimmo civil engineer by Benj. Meredith.'[46] One can surmise either that Meredith got it wrong, or that Nimmo had some remarkable idea of how such a bridge might be made feasible. The only other remotely like it in design is that by James Dredge at Belleevy in County Down. Unsurprisingly, Nimmo's was not the design chosen when Athlunkard Bridge was eventually built – to the design of the Paine brothers of Limerick. Whatever realistic kind of bridge Nimmo would have eventually designed for Athlunkard, had he been commissioned, would doubtless have complemented Wellesley Bridge. He rarely saw things in isolation, and the symmetry of 'embosoming' Limerick city between two beautiful bridges of his own design would surely have appealed greatly to him.

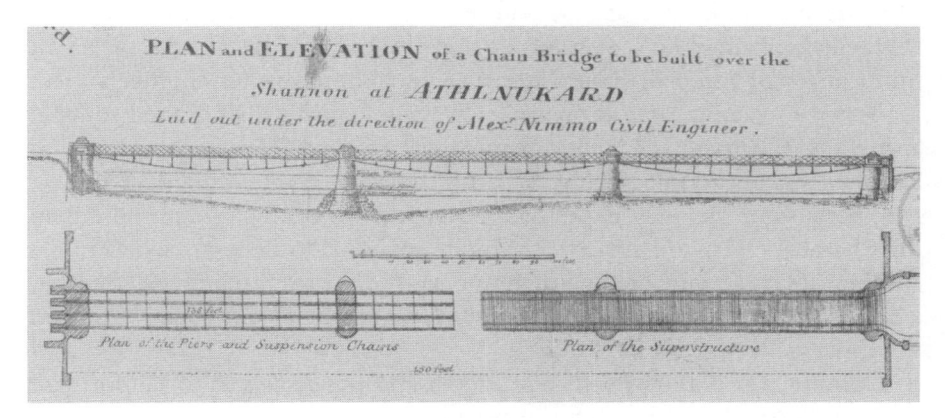

FIGURE 11.8
Nimmo's plan and elevation for a chain bridge over the River Shannon at Athlunkard. Note that the bridge deck is suspended above the arch of the chain. (Detail from ref. 46, reproduced with permission of the National Library of Ireland.)

By happy circumstance, a new footbridge built in the campus of the University of Limerick in 2006 to the design of the architectural practice of Wilkinson and Eyre (in cooperation with Ove Arup and partners) has become the uppermost bridge over the Shannon in Limerick. A suspension bridge, it comprises five underslung arches supporting the overhead deck by solid ties. The historic resonance of this wonderful new bridge, which expresses the unique design feature originally envisaged by Nimmo is remarkable and quite serendipitous: the modern architects do not appear to have been aware of Nimmo's original proposal when they designed this magnificent structure which exhibits a lightness and fluidity of great beauty.

YOUGHAL BRIDGE

The packet service from Milford Haven to Dunmore (to Waterford city after 1825) was the principal mode of carriage of the mails from England to the south and south-west of Ireland. When the packet boat arrived in the late afternoon the mails were off-loaded and sent by road to Waterford. They were stored there overnight and sent off again at 6 a.m. the following morning to Cork and points west.[47] They took ten hours to travel from Waterford to Cork city by way of Dungarvan, Cappoquin, Lismore, Tallow and Youghal. The more direct route from Dungarvan involved a dangerous ferry crossing near Youghal, and it was in order to avoid this that the mail coaches detoured through Cappoquin, crossing the River Blackwater by bridge at Lismore. A suitable bridge at Youghal would knock fourteen miles and almost two hours off the journey, missing out Cappoquin, Lismore and Tallow. It suited the merchants of Youghal to have such a bridge, and as early as 1821 the Chambers of Commerce of Cork and of Waterford made a Memorial for one. It was accompanied by a plan for a bridge at Rincrew, about one and a half miles north of Youghal town. Later that year, the Loan Act engineers were requested to examine the proposed site. Nimmo and Killaly visited Youghal and on 8 February 1822 they made their report.[48] The proposed site at Rincrew was not to their liking, as it might interfere with a good anchorage and, because of the great depth of the water, it would cost much more than the Chambers' plan anticipated. Although not asked to do so, they examined another site half a mile below Rincrew, nearer to the town. Here they thought a bridge might be more safely constructed, where the channel was wider and not quite as deep.

Nothing was done immediately and the proposal seemed to lie dormant for some years, although Nimmo believed that a subscription had been started and funds were accumulating. The proposal was resurrected in 1827, when the Select Committee on the Milford Haven Communication sat.[49] Sir Edward Lees, secretary to the Irish Post Office, was questioned about the detour that the mail coach was obliged to make and while he had some misgivings about bypassing Lismore, he agreed that a bridge at Youghal would speed up the mails considerably.[50] His support was crucial to Youghal's claim, because without it the town could eventually be bypassed by the Cork mail coach. The mails from Waterford to Limerick already went through Cahir, County Tipperary, and there was a reasonable suggestion that the Cork mails might be sent by that route also, diverting to the south from Cahir.

Nimmo gave his evidence to the Select Committee two months later. He explained that he 'had been professionally employed on the particular thing' and agreed that a bridge would shorten the journey. It had not been determined yet what kind of bridge should be erected:

> I have had it surveyed as to the most advisable bridge, but I have laid before them plans for bridges at three different places, of three different descriptions ... I should be very desirous to bring the bridge down to the town of Youghal. If it can be made, unobjectionable to the navigators, in the upper part of the river, about two miles higher up, there is a very favourable situation for a chain bridge; but a timber bridge below, could be got for less expense.[51]

A chain bridge would cost £20,000, a timber bridge nearer Youghal £15,000 to £16,000. He had not yet estimated the cost of another timber bridge even closer to the town, but he claimed 'to have had it all surveyed' already.

By his own account, therefore, Nimmo prepared three bridge designs for Youghal. The first was a chain suspension bridge above Rincrew, where the river narrows. He regarded this, the most up-river site, above where the modern bridge stands today, as very favourable for a chain bridge. No plan of it appears to have survived, but if it had resembled the suspension chain bridge that he had proposed at Athlunkard (see above), it would surely have been an impressive structure, and would have added great refinement to the dramatic, wooded riverscape in the vicinity. Strangely enough, he proposed an iron suspension bridge for the Harbour Trust of Swansea at the same time, and it is tempting to suggest that he may have put forward the same plan for both sites.[52] On the other hand, he was familiar with Lemuel Cox's existing timber bridge across the River Suir at Waterford city, which linked County Waterford with County Kilkenny; the symmetry of having a similar timber bridge linking County Waterford at its opposite end with County Cork would not have been lost on him, nor would the opportunity of designing a structure that would outshine that of the famous American have been unwelcome.

His second plan was for a timber bridge a little further down-river, but still in Rincrew. The plan for this bridge has survived. It shows a masonry and timber arched structure 'at Rincrew, near Youghal' and is signed by George Nimmo.[53] The site was near the location of the present-day bridge, about one-quarter to one-half a mile upstream from Foxhole. The Rincrew (nearer Cork) end of the bridge was to commence with a cut stone abutment having two masonry arches, then about thirty timber arches, then an opening bascule section, followed by another timber arch attaching to a masonry abutment having two arches on the opposite bank of the river. The total width cannot be calculated accurately because the extant plan does not show the full number of arches clearly. However, thirty arches of thirty feet span are clear in elevation, so that the total length, including the masonry arches, must have been more than 900 feet. An undated estimate, archived with this plan and almost certainly relating to the same work, is also signed by George Nimmo. The estimate states that the proposed bridge was to have 'abutments and arches' together with nine pile piers and seventeen piers sunk in timber frames, with 610 linear feet of joisting and flooring and 1,240 linear feet of railings. All this indicates a timber structure about 610 feet long which, with a bascule of eighty feet and two masonry sections of about 210 feet at both ends (as measured from the actual plan), would give a total of 900 feet. The river is actually about this wide at the site. This bridge, too, would have been an impressive structure, had it ever been built. Like the proposed chain bridge, a timber bridge in the wooded riverside setting would have provided a romantic backdrop that would have delighted the eye of the traveller.

Could this plan be the one that was submitted with the Memorial way back in 1822? The cost of the original 1822 plan, according to Killaly and Nimmo's report at the time was £10,500, the exact amount of George's signed, but undated, estimate. It may have suited Alexander to have the plan inscribed with George's name, for the same reason suggested for his failure to sign the Poulaphuca plans, that is, he was the Loan Commissioners' engineer and it would hardly have appeared proper that he adjudicate his own plan. There is no evidence that George Nimmo was sufficiently skilled as an engineer to have designed the bridge all on his own, and it is most likely that Alexander had actually designed it, especially in the light of his own affirmation of the fact. The Post Office drew up a draft Bill for Youghal Bridge in 1827, much to the consternation of the inhabitants of Cappoquin, Lismore and Tallow. In November they addressed an elaborate Memorial to the Lord-Lieutenant pointing out the importance of the boat trade to their towns and their surroundings and claiming that the proposed bridge would materially injure the navigation thereby damaging their trade.[54] When the Post Office informed Nimmo of the objection he wrote to Gregory in rather acerbic terms, pointing out that the site had been recommended by Killaly and himself as long ago as 1822.[55] In proof of this, he enclosed a copy of the report of 1822. Sir Edward Lees then replied to the memorialists on 28 January 1828, deploring the fact that they had not raised their objection at the Select Committee on the Milford Communication and reiterating that a direct route through Youghal was the best for the mails.[56] The Post Office, however, would consider any injury to the navigation shown by any competent and neutral man, he stated, but reminded them that both Killaly and Nimmo had reported on the matter years earlier. The 1822 report was shown to the memorialists, who appear to have been ignorant of its existence, and were therefore taken off-guard by it. The Bill drew no other serious objections and was enacted into law on 23 May 1828 as the local Act 9 Geo 4 cap liii.[57] A Mr Henley wrote subsequently from Tallow complaining that '…the most unjust bill for building a bridge across the mouth of the fine navigable river Blackwater has passed both houses of Parliament on the single information of Mr. Nimmo, notwithstanding the Memorial.'[58]

Given that the 1822 report had been raised and vindicated, it was now essential that the bridge be built at the site recommended in it. That was at Foxhole, about one and a half miles north of Youghal and Nimmo's third, final and successful plan was for a timber bridge at this site (10). From the firm dry land there, he planned a stone embankment through the sand out into the river, as far as low water of spring tide. On the opposite shore, he planned a stone quay at Tinnabinna; the bridge would span the estimated 1,320 feet between that place in County Waterford and the tip of the proposed embankment at Foxhole in County Cork. An elevation of the bridge and its exact location are shown on Nimmo's chart of Youghal Bay (figure 11.9).[59] It had thirty arches, each with a span of thirty-five feet, along with an opening bascule section abutting the Waterford quay. This gave a total bridge length, including the width of the timber piers, sufficient to span the estimated width of the river at that point. As illustrated on the plan, the bridge was an attractively light, elegant and graceful structure. The long embankment on the County Cork side facilitated the enclosure of a tract of ground almost 400 acres in extent, lying between it and the town, and this circumstance was an important consideration. A loan of £10,000, secured on a mortgage of the prospective tolls, was approved for the work in 1829 and advanced out of the Consolidated Fund, to be repaid in annual instalments of £500.[60]

The contract for construction was drawn up on 25 May 1829 between Michael Power of Clashmore and others on the one hand, and Mr Bennett of the City Foundry and George Nimmo on the other. George was to be the contractor, although he was given the title of civil engineer on the contract document.[61] Because of this, and the earlier unused plan inscribed with his name, the design of Youghal Bridge is often erroneously attributed to George. The absence of masonry sections to the bridge that was built makes it clear that it is not the one that George had allegedly designed. The superintending engineer was John E. Jones, one of Nimmo's assistant engineers, and in his words 'It was commenced in the year 1829 under my superintendence and finished in the year 1832.' The title of Jones's published account – *Account and Description of Youghal Bridge designed by Alexander Nimmo* – leaves absolutely no doubt as to who really designed it.[62] Jones never mentioned George by name in this account or anywhere else either, and there may have been little love lost between the two

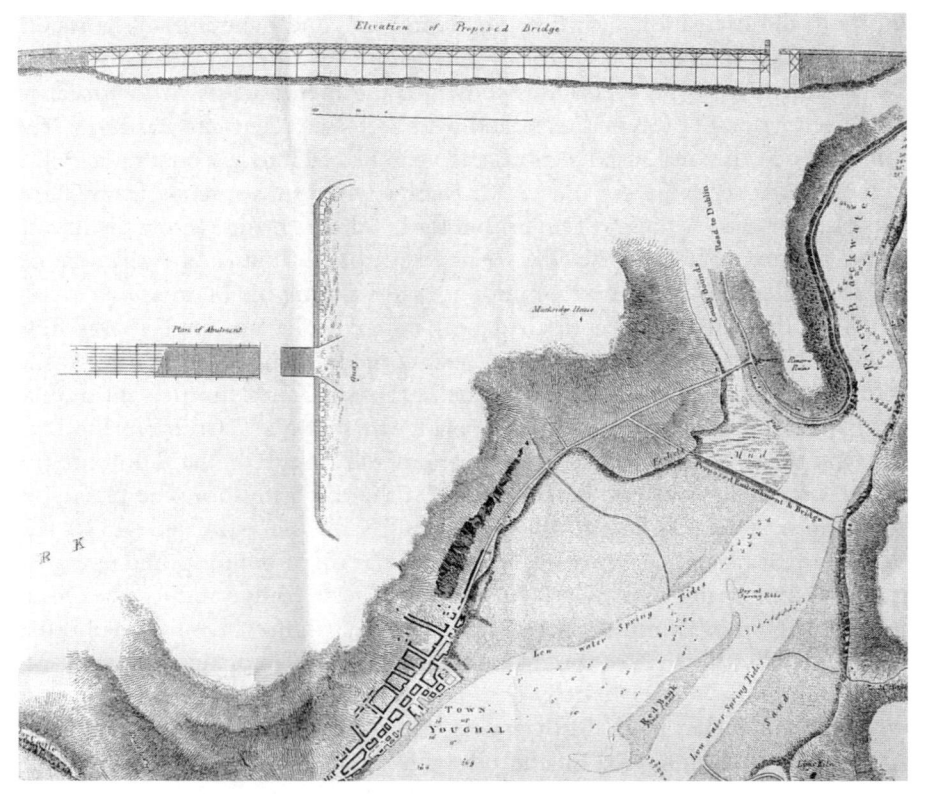

FIGURE 11.9
Detail from Nimmo's plan of Youghal Bay showing the proposed location of the embankment and bridge. Note the elevation of the bridge shown at the top of the plan. (Detail from ref. 59, reproduced with permission of the British Library.)

men. Jones embellished his account with an engraved panoramic view of the new bridge, with the town of Youghal in the background, which was exhibited at the Royal Hibernian Academy in 1828. This must have pleased Jones who had very pronounced artistic leanings and who soon abandoned the engineering profession entirely to become a sculptor, in which role he achieved great renown. He sculpted the bust in the Royal Dublin Society that is the only known extant representation of Nimmo.

The bridge was built of the best timber, Crown-brand Memel. To select and purchase this, one of the contractors – George Nimmo, it would appear from an anecdote recounted by Henry M'Manus[63] – travelled to the Baltic.[64] Jones gave the following linear dimensions for the completed work, which can be relied upon as accurate: the embankment was 1,500 feet long, the timber bridgework 1,542 feet and the quay wall on the Waterford side 200 feet, total 3,242 feet. The timberwork comprised forty-seven bays of thirty-foot span and a bascule section of eighty feet, including the supporting piers. The space occupied by the piles made up the remaining fifty-two feet. Measurements and calculations made from the 1842 six-inch Ordnance Survey map give the following approximate dimensions: embankment 1,540 feet, bridge (including the opening bascule section), 1,760 feet, total 3,300 feet.[65] Actual measurements made recently by GPS indicate that the embankment was 1,540 feet long and the bridge 1,785 feet, a total of 3,325 feet. In all cases the values are only approximations, but the bridge length from all three sources (Jones, Ordnance Survey and GPS) are reasonably close (especially the OS and GPS values), given the different methods of calculation and the problem of the exact root and exact head of the embankment. They prove that the estimated length of the bridge initially given in Nimmo's plan of thirty arches each of 35-foot span was an underestimate.[66] This is understandable, since the full extent of the finished embankment, which would determine the exact length of the timber bridge, was dependant on conditions and circumstances at the time of building. The timber section was nearly twice as long as Cox's timber bridge in Waterford city and was the longest timber bridge ever built in Ireland, features that Nimmo might have relished had he lived to see it completed. As it was, there is evidence that he lost touch with the project early on and never actually saw it under construction.[67] It was not completed until after his death.

The embankment was formed by placing heaps of loose stones about six feet in depth along its proposed length out to low water springs. The stones sank a considerable way into the sand under their own weight. The filling (the hard-core of the eventual mole) was then thrown in and the whole left to settle for twelve months, during which time extra material was added as the earlier material sunk further, until the whole lot became a firm mass. The facing walls were then built of good rubble stone and the head was clad in cut limestone. The embankment rose in height from two to twenty feet along its length; the roadway on top was thirty feet wide with a footpath six feet wide on either side. Where the embankment met low water, the timberworks started.

FIGURE 11.10
View of Youghal Bridge from the north by John E. Jones the engineer who superintended its construction. Note the embankment to the right and the opening bascule section to the left, with a sailing boat approaching. Youghal town is in the background on the right. (Detail from ref 62 reproduced with permission of the Irish Architectural Archive.)

The deck of the bridge was twenty-two feet wide and ten feet above high water. All the beams were fourteen inches square and from forty-four to ninety feet long. The pier piles were driven up to twenty-five feet into the sandy bottom and the upper works fastened to them. The whole structure required 1,132 tons of timber, including piles, beams, joists, flooring, handrails, and so forth, and the final cost of the completed works was £18,000.

Small craft could pass underneath by stepping their masts; larger vessels could pass under the bascule when it was raised. Part of Jones's engraving is reproduced in figure 11.10 which gives a fine impression of the bridge; its lightness and simple elegance are apparent. It achieved its intended purpose immediately, the road over it becoming the main coach route between Cork and Waterford. A toll was payable, with the tollhouse (now demolished) located on the Cork side at the root of the embankment. Another house was built on the apron of the quay on the Waterford side, but that was not for tolls; it is still standing and permanently occupied.

The timber bridge lasted until it was replaced by a shorter iron bridge at the same site in the 1860s; to construct the latter, Nimmo's original embankment at Foxhole was extended out a further 500 feet into the river, reducing the width to be spanned to about 1,285 feet. Today, the original Nimmo embankment remains intact, but because of land reclamation on the south side, it no longer appears to project out into the river. It is best observed by examining its northern face, where the original stonework is clear and forms the retaining wall of the intake. The stonework sharply differentiates Nimmo's work from the later extension for the iron bridge. The original head of Nimmo's embankment can still be identified from the nature and quality of the cut limestone masonry at its seaward end, where the extension for the iron bridge commences (figure 11.11). Most of Nimmo's embankment, with some of the reclaimed land, is now the local municipal dump, but a footpath along the north side to the left of the dump remains open and it is easy to walk along this out to its head and on to 'the iron bridge embankment'. There is no bridge at the site now but on the Tinnabinna (Waterford) side, the limestone quay is still intact and much as it must have been in Nimmo's time.

SOME OTHER NOTEWORTHY BRIDGES

Nimmo laid out and supervised the Leenane to Westport road, the N59 today. The styles of the four main bridges on its course are vastly different: Glenocally (3), discussed above, is a single pointed-arch structure in a rocky, gorge-like site (figure 11.2); Gowlaun (11) is a single-arch segmental bridge (figure 11.12); a short distance further north is a narrow (span approx. five feet), relatively tall (about eight feet from base to keystone) single-arched bridge (12) (figure 11.13), which, although smaller and round-arched,

FIGURE 11.11
Bridge embankment near Youghal. The limestone section to the left is the head of Nimmo's original embankment. The sloping section to the right is the extension erected for the iron bridge that replaced Nimmo's timber bridge. The bridge quay at Tinnabinna can just be seen on the opposite side of the river.

FIGURE 11.12
Gowlaun Bridge. A typical small bridge by Nimmo, with a cut stone stilted round arch and roughly shaped local field stone spandrels and parapet. The coping is modern.

FIGURE 11.13
Bridge N59-062-00 on the Leenane to Westport road.

FIGURE 11.14
Erriff Bridge, an elegant structure on the Leenane to Westport road. Note the ornamentation beside the buttress on the left.

FIGURE 11.15
Cornamona Bridge, its face somewhat disfigured by the pipe.

appears almost transitional between his usual round-arched form and the pointed-arch bridge at Glenanean.

Near it, Erriff Bridge (13) is a large, elegant, single elliptical arch in cut stone, with decorated abutments, spanning the river at a location where the flow is gentle and wide and which Nimmo chose in 1823 specifically to avoid interfering with Erriff Wood.[68] It exhibits an almost urban gracefulness, unexpected in such a rugged location (figure 11.14). Confusingly, Erriff Bridge spans the Owenmore River; where this joins the Owenwee River below the bridge, it forms the Erriff River proper. The bridge was centred in 1827 and the arch closed in 1828.[69] In subsequent years it was raised to the parapets and the river-bed paved; Killaly finished off the work in 1831. Further upstream on the Owenmore River, Srahlea Bridge (14) appears to have had two arches, separated by a central pier with an upstream cut-water, but today it is very much altered. From these examples alone (there are a number of other small, vernacular round-arch bridges along the route also), Ruddock's comment regarding Nimmo's eclectic bridge styles appears amply justified along this sixteen-mile stretch of road.

One of Nimmo's other important roads in Connemara was that from Leenane to Maam (R336), and onwards to Cong (R345). At Cornamona on the R345 in south Co. Mayo, he erected another cut-stone bridge (15), which, although in an isolated location, (now a small village) exhibits an urban ele-

FIGURE 11.16
Toombeola Bridge, upstream face. Note the corbels in the soffits and the generally heavy-set aspect.

FIGURE 11.17
Nimmo's bridge and office at Maam, as drawn by James Mahony and published in *A Tour of Connemara* in 1854.

gance somewhat marred today by a water pipe carried along its downstream face between the parapet and the voussoirs (figure 11.15). Tumneenaun Bridge is situated a few miles further east along this road. The elegant Erriff and Cornamona bridges contrast sharply with the bridge he designed over the estuary at Toombeola (16) in Ballinahinch (figure 11.16). This latter is a rather crudely built, heavy-set bridge of three round arches, lacking ornamentation but with coarse cut-waters of cut stone attached to both the upstream and the downstream faces. It was commenced before 1827, centred by James King in 1828 and finished by Richard Martin under contract in 1831. Corbels in the intrados, used for the original centring, are still evident.

Perhaps the most famous, or infamous, of Nimmo's bridges was that at Maam (17), completed in 1829 and shown here in a drawing from 1854 (figure 11.17). By contemporary accounts, it was not wide enough to release the water that backed up in the Bealnabrack River after heavy rain. Killaly regarded it as a very precarious structure: it was, in fact, swept away later after a heavy flood and was replaced by an iron bridge. That was replaced by the present structure, but Nimmo's masonry abutments and ornamentation were retained and parts are still in place today. Tiernakill Bridge (18), close by on the R336, was also poorly regarded by Killaly, but it has survived and is still in use today. It is a single, high, segmental arch that has been encased in concrete in recent years, so that the original masonry is evident only at the parapets at both ends of the bridge (figure 11.18). Further along the R336 towards Maam Cross is Shaughnessy's Bridge (2) discussed earlier and illustrated in figure 11.19, and Shindela Bridge (27), a single-arch small bridge over the narrows of Shindela Lake at Maam Cross, shown in figure 11.20.

In 2007, Leenane Bridge (21), a structure of three round arches, gave way in extremely wet weather and is currently in course of replacement. This was not the first time that Leenane Bridge failed. When Nimmo first built it in 1825, locals predicted its collapse, alleging that it was located in an insecure spot; according to Maria Edgeworth, it collapsed three times within a couple of years.[70] Nimmo himself recorded that 'the mountain floods and high tides of last year [1828]' had damaged the road and injured the bridge, but not enough to hinder all passage. He repaired it the following year, making the road to Westport passable for carriages again.[71] Such failures were not uncommon in the early nineteenth century, bridges often being improperly built by unskilled labour. The wonder is that so many of them survived in the difficult terrain and under the prevailing constrained conditions. In the case of the Castlebar to Erris road, a mighty storm in the summer of 1826 swept away most of the bridges that had been erected before the road was put in Nimmo's charge. He caused most of them to be rebuilt, but the construction work was poorly carried out and many needed to be repaired by Killaly when he took over. Killaly found that 'parapets and retaining walls are yet wanting to most of those that have been rebuilt, and the backing or approaches thereto are

FIGURE 11.18
Tiernakill Bridge spanning the Foilmore river near Maam in Connemara. The bridge is now completely encased in concrete but retains a certain elegance.

FIGURE 11.19
Downstream face of Shaughnessy's Bridge as it is today. Compare with Mahony's 1854 drawing in figure 11.1.

FIGURE 11.20
Shindela Bridge downstream face. The original arch is still evident.

FIGURE 11.21
Kilhale Bridge. Note the cut stone work in the soffit. The original parapet has been replaced by railings for safety reasons.

quite incomplete'.[72] However, even to an inexpert eye, the masonry of Kilhale Bridge (22) and its buttresses (figure 11.21) is far better than that of the retaining approach walls that were added later.

Obviously, all Nimmo's bridges warrant expert engineering examination for the insight they may give into the history and manner of bridge building in early nineteenth-century Ireland and to make a record of an important feature of our built heritage and transport history. Some of his bridges have already been replaced entirely, such as those at Casla (19) and Bundorragha (20), and one was a complete failure from its very beginning. This was at the mouth of Rusheen Bay near Salthill in Galway. In an attempt to avoid a rocky and steep part of the road near Barna, he proposed to make a new road nearer the sea from Blake's Hill to Silverstrand.[73] This necessitated a bridge to cross the mouth of Rusheen Bay with an embankment on either side of it. By placing a sluice in the bridge he hoped to shut out the sea (but release the waters of the creek at low tide) and reclaim the seventy acres of the bay. He conceived the plan in 1822; nothing had commenced by 1824, probably because of the difficulty of the task involved. In 1825 Patrick McDonagh took it up and was paid £210 for his efforts, which were not really satisfactory.[74] Nimmo believed that McDonagh had delayed excessively and had effectively abandoned the task, so he took the work from him. [75] By then only the bridge was wanting and he was anticipating a successful outcome. After some more delay, John Steel, who was a supervisor at the quarry in Aran, was engaged as supervisor and the work recommenced.[76] He laid a substantial outface for the bridge in hewn stone with four swing doors attached, and the closing of the passage was far advanced by 1828. Nimmo admitted it had been 'a work of some difficulty' but was still positive it would prove successful. He was called on by the Commissioners for Woods and Forests for a plan of the area that would be retained by the embankments.[77] But Nimmo's optimism proved unfounded and there was no evidence of any bridge at the site in 1839.[78]

The remaining extant bridges by Nimmo are less distinctive or less notorious but still worthy of expert evaluation and preservation. Most are small- to medium-sized, sturdy, round-arched (often stilted) structures well suited to their rural locations. The bridge over the River Robe between Levalley (Ballinrobe) and Hollymount (33) is one of the larger, having three arches mounted on massive piers that are supported on broad foundations shaped as cut-waters (figure 11.22). Nimmo began this bridge in 1823, but had to suspend it temporarily for want of funds the following year.[79] The road, which he had surveyed by Michael Philips in 1822, was intended to join Ballinrobe to Claremorris and Ballyhaunis and onwards to the Shannon. The massive bridge foundations resemble those of the single-arched bridge at Ballinderry (24) that Nimmo said were 'entirely artificial' and 'required some careful precautions' to preserve the navigation.[80] This bridge is on the road

FIGURE 11.22
Bridge over the River Robe between Levalley and Hollymount, Co. Mayo. Note the massive foundations and thick piers.

FIGURE 11.23
Bridge of Annios. Two flow arches and one flood arch. Note the absence of a cutwater between the flow arch and the flood arch at the right.

FIGURE 11.24
Bridge at Salt Lake, Ardbear, near Clifden, during ebb tide. Note the widely separated arches

FIGURE 11.25
Canal Bridge on the N59 road near Ballinahinch. Note the ornamentation resembling that at Erriff. Compare this picture with figure 11.27, drawing of 'Benabola Bridge'.

FIGURE 11.26
Downstream face of Weir Bridge, Recess, Co. Galway. Note the complete absence of ornamentation and evidence of multiple repairs.

that connects Tuam with Athenry. In both these cases the foundations are now reinforced with concrete, but the original stonework of the piers and spandrels is still clear. The arch of Ballinderry Bridge was blown up during the civil war era and later repaired, as were two of the five arches of Cloongullane Bridge that crosses the River Moy near Swinford in Mayo. Nimmo had laid out and supervised the construction of the road from Killala to Swinford via Ballina and Foxford that carries this large bridge and had handed the road over to the Board of Inland Navigation by February 1826.[81] It had cost over £4,000 in the years 1822 to 1825.[82] The bridge is therefore most likely Nimmo's or earlier, but its size and the cut-stone arch rings suggest it is an engineered rather than an artisanal bridge. The outermost arches at each side provide overflow channels, dry except during flood conditions.

Another Nimmo bridge with an overflow arch is Annios Bridge (25) on the Ballintubber road running from Westport to Ballyglass, where it joined the mail coach road to Claremorris. Work started in July 1822.[83] The site required an embankment to raise and carry the bridge through the low-lying ground. It is a very well made but rather plain bridge (figure 11. 23), with three arches, two with a permanent channel and one overflow arch. It cost less than £200 and was completed by 1826.[84]

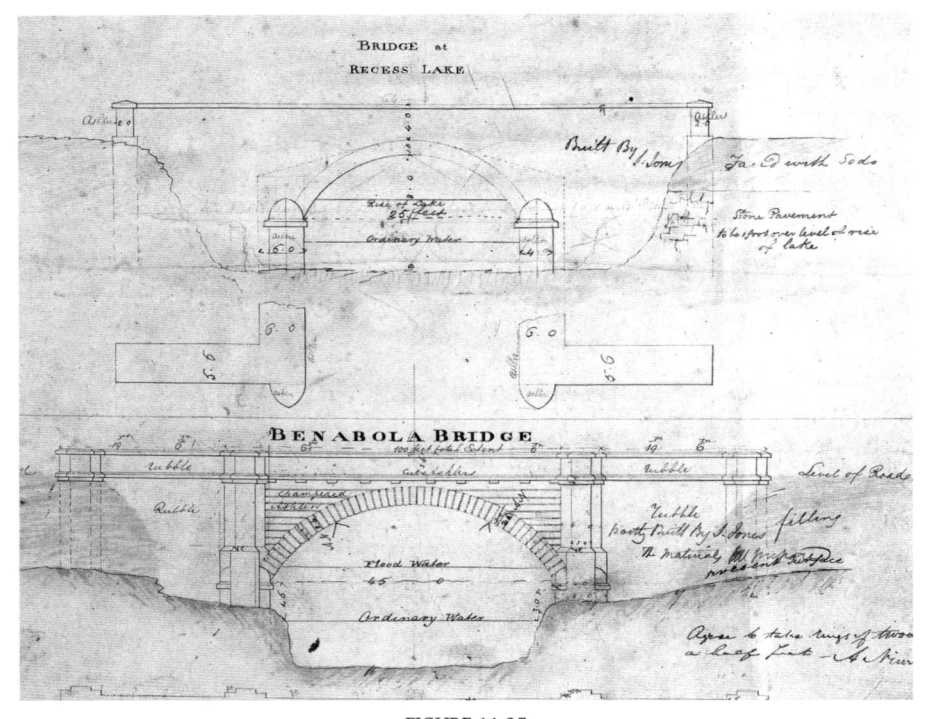

FIGURE 11.27
Original drawings of Bridge at Recess Lake (Weir Bridge) and Benabola Bridge (Canal Bridge) in Connemara, County Galway. (NAI, OPW 5 HC 6/248. Reproduced with permission of the National Archives of Ireland.) The drawing is signed by Nimmo.

Ardbear (Salt Lake) Bridge spans an inlet of the sea near Clifden. Nimmo records in his report for 1824 that '...a new line with two considerable bridges across the creek of Ardbear [this bridge (23)] and river of Clifden [the pointed-arch bridge (5)] is completed into that thriving village.'[85] The Salt Lake bridge is, in fact, a long embankment, faced with random rubble stone, over the rocky ledge where the Salt Lake communicates with the sea in Ardbear Bay. It is pierced by two widely separated, stilted round arches through which the sea flows as rapids when the tide is flowing and ebbing (figure 11.24). Nimmo's comment here is the only one that he made on both these structures and indeed he made little enough comment on any of his bridges in his annual reports; being largely plain and functional, located in isolated places, there was little point in making much ado about them. Because of this, few of the bridges he designed have ever been recorded and confirmation of his contribution to most was only circumstantial until now, and is still not complete. In particular, the contribution of others to their present form needs clarification. Some on the N59 Oughterard to Clifden road such as the Canal Bridge and the Weir Bridge at Recess are almost certainly Nimmo's, as judged by their size, quality, style and original ornamentation. The buttress ornamentation, for example, on the Canal Bridge (figure 11.25) resembles that on Erriff Bridge and on the original Maam Bridge, although the Canal Bridge has been considerably altered on the downstream face. Unfortunately, the original ornamentation on the Weir Bridge has entirely disappeared during the numerous repairs and alterations made over the years (figure 11.26). There is also some good documentary evidence to support the attribution of both these bridges to Nimmo. For example, the drawing reproduced in Figure 11.27 shows the planned elevation of the Canal Bridge (which Samuel Jones called Benbola Bridge) carrying Nimmo's signature and his initials approving modifications that were to be made to it.[86] Other, smaller bridges on that line may be Jones's – he certainly took credit for building them, but to what extent they were Nimmo's designs needs further research. Altogether, there is circumstantial and some documentary evidence to indicate that a further twelve bridges (at least) in the west and the southwest may be attributable to Nimmo after more research; they include bridges between Castlerea and Ballymoe, on the Ballinakill road, on the road from Kylemore to Leenane, on the Erris road and on the Killorglin to Cahirsiveen road. For now, table 8 must suffice as the first and the most complete listing of the bridges to which Nimmo is confirmed to have contributed.

NOTES

1. P. O'Keeffe, and T. Simington, *Irish Stone Bridges: History and Heritage* (Dublin: Irish Academic Press, 1991).
2. R. C. Cox, and M. H. Gould, *Civil Engineering Heritage, Ireland* (London: Thomas Telford, 1998).
3. T. Ruddock, *Arch Bridges and their Builders, 1735 – 1825.* (Cambridge: Cambridge University Press, 1979).
4. S. de Courcy, 'Alexander Nimmo, Engineer. Some Tentative Notes', lecture to the National Library Society of Ireland, November 1981. Privately circulated. In a later, expanded version of this paper, published in *Connemara: Journal of the Clifden and Connemara Heritage Group* (1995), he did not comment on this absence of primary sources and in fact attributed the design of Poulaphuca Bridge to Nimmo.
5. S. Lewis, *A Topographical Dictionary of Ireland* (London: S. Lewis, 1837).
6. Lord Downshire to Sir E. Lees, 29 May 1823 (PRONI, Downshire Papers, D671/C/220/3).
7. 'A particular account of the sums granted by the Lord-Lieutenant, by way of gift, under the Act 1 Geo. 4. C. 81', in Public Works: Return to an Order of the Honourable House of Commons dated the first of March 1824 (B.P.P. 1824 [(278], vol. XXI, mf 26.135).
8. K. Trant, *The Blessington Estate 1667–1908* (Dublin: Anvil Books, 2004).
9. Coloured plan entitled 'Proposed Bridge over the Liffey at the Fall of Poulafouca' (NAI, OPW 5HC/6/0386).
10. Coloured plan entitled 'Proposed Bridge over the Hollow next the Liffey' (NAI, OPW 5HC/6/0387).
11. A. Nimmo to W. Gregory, 3 July 1823 (NAI, CSORP 1823/6474 [3]).
12. J. Killaly to H. Goulburn, 14 September 1823 (NAI, CSORP 1823/6474 [4]).
13. J. Killaly to H. Goulburn, 30 September 1823 (NAI, CSORP 1823/6474 [5]).
14. J. Tassie, and H. Tassie, Memorial to Lord Wellesley, 1822. (NAI, CSORP 1822/2593).
15. H. Arabin to A. Mangin re. Poulaphuca Bridge, 1822 (NAI, CSORP 1822/2593).
16. J. Galloway to H. Goulburn, 19 March 1822 (NAI, CSORP 1823/6474/1).
17. A. Nimmo to G. Nimmo, 30 September 1823 (NAI, OPW8/97/1); J. Killaly to G. Nimmo, 30 September1823 (NAI, OPW8/97/1[rere]).
18. J. Killaly Statement of Receipts and disbursements on Public Works carried out in the Central District in the years 1822 to 1825 (NAI, OP 972 /67).

19. W. Johnston to P. Carney, 15 July 1831 (NAI, CSORP 1831 2092/2).
20. A. Nimmo, Report of Mr Nimmo on the Western District for the year 1828 (B.P.P. 1829 [348], vol. XXII, mf 31.136), hereafter referred to as Nimmo's 1828 report.
21. S. C. Hall, Ireland: Its Scenery, Character, etc. (London: J. How, 1843).
22. A. Nimmo, Report on the Progress of the Public Works of the Western district in 1824 by Alexander Nimmo. Appendix 1, in Estimate of the sum that will be required for carrying on certain Public Works during the year ending 5 January 1826 (B.P.P. 1825 [279], vol. XXII, mf 27.174).
23. C. Coote, and A. Nimmo, 'Messrs Coote and Nimmo's Report on the Practicability of forming a Packet Harbour at Portcullin Cove, near the village of Dunmore, in the county of Waterford, 4 March 1825', in Royal Commission on Tidal Harbours, Second Report, Minutes of Evidence, Appendices, Supplement, p.100 (B.P.P. 1846 [692], vol. XVIII, mf 50.169– 77). (The report is erroneously dated 1825 in the text).
24. J. Fennelly, 'Notes Towards a Maritime History of Dunmore East', *Decies* (1984), pp. 17–21.
25. View of Dunmore Harbour (NLI, Lawrence collection of photographs, no. 2077).
26. Ordnance Survey of Ireland, County Mayo, sheet 66, first edition, six inches to one mile. Surveyed in 1838. The bridge is named 'gothic bridge of Glenanean' on the map of Achill road signed Barth Ellis, 1835(NAI, OPW5 HC 6/1).
27. 'A particular account of the sums granted by the Lord-Lieutenant, by way of gift, under the Act 1 Geo. 4. C. 81', in Public Works: Return to an Order of the Honourable House of Commons dated the first of March 1824 (B.P.P. 1824 [278], vol. XXI, mf 26.135).
28. Works of Alexander Nimmo, year to January 1829 (Nat. Archs., UK, AO 17/492).
29. A. Nimmo to F. L. Gower, 18 September 1828 (NAI, CSORP 1828/1468).
30. J. Killaly to Sir W. Gosset, 30 September 1831 (NAI, OP973/86).
31. Report from the Select Committee on the Irish Estimates, p.26 (B.P.P. 1829 [342], vol. IV, mf 31.23–4).
32. M. Lenihan, Limerick: Its History and Antiquities (Dublin: Hodges, Smith and Co., 1866); O'Keeffe, and Simington, *Irish Stone Bridges*.
33. J. Harvey to H. Goulburn, 23 September 1822 (NAI, CSORP 1822/2014).
34. A. Nimmo, Elevation of the new Bridge of Limerick over the River Shannon by Alexander Nimmo, C.E., F.R.S.E., M.R.I.A. (NLI, maps, 16 H 22 [3]).
35. A. Nimmo, An Estimate of the Expense of constructing a Bridge over the Shannon at Limerick with a Floating Dock and Tide Bason [sic] for the accommodation of Shipping and Draw Arches over the Navigation. Signed and dated January 1823 (NLI, MS 22391).
36. A. Nimmo, 'Report of Alexander Nimmo, Civil Engineer, on the proposed Railway between Limerick and Waterford', *Dublin Philosophical Journal and Scientific Review*, 2, [1826], pp. 1–19.
37. A. Nimmo, and B. Meredith, Map and Section of an intended Railway or Tram Road between the Cities of Limerick and Waterford with Branches to Thurles, Killenaule and Carrick. Designed for the Hibernian Railway Company by Alexander Nimmo, C.E. and Benjamin Meredith, surveyor, 1826 (NLS, SIGNET s.145).
38. T. Rhodes, 'Plan of proposed Docks for the Port of Limerick' (B.P.P. 1834 [592], vol. LI, mf 37.363).
39. A. Nimmo. Evidence, p. 172, in Minutes of Evidence taken before the Select Committee of the House of Lords to examine into the Nature and Extent of the Disturbances which have prevailed in those districts of Ireland which are now subject to the Provisions of the Insurrection Act and to report to the House. (B.P.P. 1825 [200], vol.VII, mf 27.60–3).
40. 4 Geo IV cap xciv. *An Act for the Erection of a Bridge across the River Shannon and of a Floating Dock to accommodate sharp vessels frequenting the Port of Limerick*, 17 June 1823.
41. An Account of all sums of Money advanced on the recommendation of the Irish Loan Commissioners (B.P.P. 1831 [184], vol. XVII, mf 34.128).
42. M. Lenihan, *Limerick: Its History and Antiquities*.
43. W. R. Hamilton to M. Hutton, 25 August 1827, in R. P. Graves, *Life of Sir William Rowan Hamilton*, vol. 1 (Dublin: Hodges Figgis and Co., 1882), pp. 250–2.
44. Ruddock, *Arch Bridges and their Builders*.
45. Memorial from the merchants, traders and inhabitants of Limerick to the Commissioners for the extension and promotion of Public works in Ireland (NLI, MS 22391).
46. Map of a proposed new road and bridge between the city of Limerick and Parteen in the county of Clare. Surveyed under the direction of Alexander Nimmo by Benj. Meredith (NLI, maps, 16 H 22 [4]).
47. Sir E. Lees, Evidence, in First Report from the Select Committee on the Milford Haven Communication (B.P.P.1826/27 [258], Vol. III, mf 29.22–3), hereafter referred to as First Milford Report.
48. J. Killaly, and A. Nimmo, Report of the Engineers on the proposed Bridge near Youghal, 8 February 1822 (NAI, CSORP 1822/2590).
49. The Select Committee on the Milford Haven Communication sat from February to June 1827 and issued two reports (B.P.P. 1826/27 [258] and B.P.P. 1826/27 [472]).
50. Sir E. Lees, Evidence, p. 52 in First Milford Report.
51. A. Nimmo, Evidence, p. 108 in Second Report from the Select Committee on the Milford Haven

Communication (B.P.P.1826/27 [472], vol. III, mf 29.22–3), hereafter referred to as Second Milford Report.
52. A. Nimmo, Evidence, p. 110 in Second Milford Report.
53. G. Nimmo, Estimate, Plan and Elevation of a proposed timber bridge over the Blackwater at Rincrew, near Youghal by George Nimmo (PRONI, Villiers Stuart Papers, T3131/H/10/13/2).
54. Memorial from Cappoquin to the Lord-Lieutenant regarding the proposed Bridge at Youghal (NAI, CSORP 1828/ 52).
55. A. Nimmo to W. Gregory, 18 January 1828 (NAI, CSORP 1828/52).
56. Sir E. Lees to Cappoquin Memorialists, 28 January 1828 (NAI, CSORP 1828/52).
57. 9 Geo IV cap. liii. *An Act for the erection of a Bridge across the river Blackwater, at or near Foxhole and the town of Youghal in the county of Cork, to the opposite side in the county of Waterford, and for making the necessary approaches thereto.* 23 May 1828.
58. M. Henley to W. Gregory, 5 July 1828 (NAI, CSORP 1828/52).
59. A. Nimmo, Bay and Harbour of Youghal surveyed under the direction of Alexander Nimmo C.E. etc., exhibiting the proposed bridge (BL, maps, 11500 [3]).
60. An Account of all sums of Money advanced on the recommendation of the Irish Loan Commissioners (B.P.P. 1831 [184], vol. XVII, mf 34.128).
61. Deed of Contract for the building of Youghal Bridge. Made 25 May 1829 (NLI, MS 7199).
62. J. E. Jones, 'Account and Description of Youghal Bridge designed by Alexander Nimmo', *Transactions of the Institution of Engineers*, 2 [1838], pp. 157–60, plates 13 and 14.
63. H. M'Manus, Quoted in T. Robinson, *Connemara. Listening to the Wind*, p.181 (Dublin: Penguin Ireland, 2002).
64. A. Nimmo to T. C. Harrison, 16 February 1829. Appendix 15 in Twenty-second Report of the Commissioners of Inquiry into the Collection and Management of Revenue in Ireland (Post Office Packet Establishments) (B.P.P.1830 [647], vol.XIV, mf 32.108–16).
65. Ordnance Survey of Ireland, County Cork, sheet 67, first edition, six inches to one mile. Surveyed in 1842.
66. A. Nimmo, Bay and Harbour of Youghal surveyed under the direction of Alexander Nimmo C.E. etc., exhibiting the proposed bridge. (BL, maps, 11500 (3)).
67. Nimmo's letter of 16 February to T. C. Harrrison in reference 64 suggests that he had lost contact with the bridge by then.
68. A. Nimmo, Report on the Progress of the Public Works of the Western District in 1824 by Alexander Nimmo. Appendix 1, p. 9, in Estimate of the sum that will be required for carrying on certain Public Works during the year ending 5 January 1826 (B.P.P. 1825 [279], vol. XXII, mf 27.174).
69. A. Nimmo, Expenditure on Public Works for the Relief of the Poor, Western District, pursuant to 3 Geo IV, cap. 34. Two years to January 1828 (Nat. Archs. UK, Works of Alexander Nimmo, AO 17/492); Nimmo's 1828 report.
70. M. Edgeworth, Tour in Connemara and the Martins of Ballinahinch, ed. H. E. Butler (London: Constable, 1950), p. 75.
71. A. Nimmo, 'Report on the Progress of the Public Works in the Western District of Ireland in the Year 1829' in Public Works Ireland (BPP 1830 [199], vol. XXVII, 32.194).
72. J. Killaly to Sir W. Gosset, 30 September 1831 (NAI, OP973/86).
73. A. Nimmo, 'Report on the Western District by Alexander Nimmo, civil engineer', p. 35, in Employment of the Poor, Ireland, Reports by the Civil Engineers employed in superintending the Public Works (B.P.P.1823 [(249], vol. X, mf 25.86– 7).
74. Nineteenth Report of the Commissioners for auditing the Public Accounts of Ireland (B.P.P. 1831 [12], vol. X, mf 34.75–7).
75. Public Works Ireland. Report of Mr Nimmo on the Western District 1828 (B.P.P. 1829 [348], vol. XXII mf 31.136).
76. A. Nimmo, Expenditure on Public Works for the Relief of the Poor, Western District, pursuant to 3 Geo IV, cap. 34. Two years to January 1828. (Nat. Archs. UK, Works of Alexander Nimmo, AO 17/492).
77. A. Nimmo, 'Report on the progress of the Public Works in the Western District of Ireland in the year 1829', in Public Works Ireland (B.P.P. 1830 [199], vol. XXVII, mf 32.194).
78. Ordnance Survey of Ireland, Galway sheet 93 first edition, surveyed in 1839.
79. A. Nimmo, Report on the Progress of the Public Works of the Western District in 1824, Appendix 1, p.10, in Estimate of the Sums that will be required for carrying on certain Public Works in Ireland during the year ending 5 January 1826 (B.P.P. 1825 [279] vol. XXII, mf 27.174), hereafter referred to as Nimmo's 1824 report.
80. Ibid, p. 8.
81. First Report of the Commissioners for Repairing roads and Bridges. B.P.P. 1826 (102) vol. XI, mf 28.65.
82. Seventeenth Report of the Commissioners for auditing the Public Accounts of Ireland (B.P.P. 1829 [193], vol. XIII, mf 31.88–9).
83. Ibid.
84. Nineteenth Report+ of the Commissioners for auditing the Public Accounts of Ireland (B.P.P. 1831 [12], vol. X, mf 34.75– 7).

85. Nimmo's 1824 report, p. 7.
86. A small portfolio of these bridge drawings is extant (NAI, OPW 5HC/6/0248).
87. A. Nimmo, 'Report on the Public Works of the Western District for the year 1826, furnished 5 January 1827' in Report of the Select Committee on the Irish Miscellaneous Estimates, appendix 28, (C) (B.P.P. 1829 [342], vol. IV, mf 31.222–7).
88. A. Nimmo, Expenditure on Public Works for the Relief of the Poor, Western District, pursuant to 3 Geo IV, cap. 34. One year to January 1829 (Nat. Archs. UK, Works of Alexander Nimmo, AO 17/492).
89. A. Nimmo, Expenditure on Public Works for the Relief of the Poor, Western District, pursuant to 3 Geo IV, cap. 34. One year to January 1830 (Nat. Archs. UK, Works of Alexander Nimmo, AO 17/492).
90. J. Killaly to Sir W. Gossett, 30 September 1831. (NAI, OP973/86).
91. Extract of the Report of Alexander Nimmo on the Western district for 1830. (NAI, OP 973/74).
92. A. Nimmo, Expenditure on Public Works for the Relief of the Poor, Western District, pursuant to 3 Geo IV, cap. 34. One year to January 1826. (Nat. Archs. UK, Works of Alexander Nimmo, AO 17/492).
93. Twenty-first Report of the Commissioners for auditing the Public Accounts of Ireland (B.P.P. 1833 [102], vol. XVII, mf 36.124–7).
94. A. Nimmo, Expenditure on Public Works for the Relief of the Poor, Western District, pursuant to 3 Geo IV, cap. 34. Four years to January 1825 (Nat. Archs. UK, Works of Alexander Nimmo, AO 17/492).
95. A Return of all the Public Money expended under the Direction of Mr Nimmo on Public Works in Ireland; which of these works have been completed, and which left in an unfinished State; and of all the Piers built on the coast, which of them now remaining in a state of perfection, and which in a dilapidated state (B.P.P. 1830 [624], XXVII, mf 32.194), reprinted here as appendix 3.
96. Public Works Ireland. Return to an order of the Honourable House of Commons dated 16 April 1833, no. 2. Return showing the detailed expenditure on the principal works in the counties of Mayo and Galway under directions of Messrs Nimmo and Killaly from the year 1821 to 1833. B.P.P. 1833 (285) vol. XXXV, mf 36.257 –8.
97. A. Nimmo, Expenditure on Public Works for the Relief of the Poor, Western District, pursuant to 3 Geo IV, cap. 34. Two years, 1831 and 1832 (Nat. Archs. UK, Works of Alexander Nimmo, AO 17/492).

CHAPTER

12

Private Commissions in Ireland

BALLINGAULE PIER AND VILLAGE

Nimmo was not exclusively a public works engineer and he took on commissions and consultancies from private sources whenever suitable work presented itself. Getting the public commission for Dunmore harbour had given an important fillip to his private career. Why, if not for his involvement in Dunmore, should the Cork port authorities have felt he could advise them regarding improvements on the River Lee? Before Dunmore, he had no prior experience of harbour engineering of any sort. Arising from the Dunmore and Cork projects others followed. For instance, when travelling along the south coast he had to pass through the Villiers-Stuart estate in County Waterford, whose proprietor wished to build a fishing pier at Ballingaule near Helvick Head. He gave the task of designing it to Nimmo, who completed the task in October 1815 with an estimate of £2,780.[1] His plan was immensely more elaborate than the fishing villages of Tobermory and Pultneytown in Scotland, designed by Telford some years earlier. The site was beside a stream in a long gully leading down to the sea. Fishermen's houses along the cliff, overlooking one another in terraced rows, would communicate by parallel cross streets with the wharf on the shore, where a roofed market-house was to be located.[2] Never one to think small, he envisaged over one hundred substantial dwellings and other buildings, including stores for salt and other items. The various streets were given names, including one called Stuart Street in honour of the proprietor. The pier, 300 feet long and angled in the middle, would extend from the market-house into the bay and have a twelve-foot depth of water at its head at high tide. He envisaged that the area near the mouth of the stream could be used for a scouring basin or a graving dock to be formed later, and eventually a breakwater could be constructed to enclose the harbour completely. The plan, dated 1816 and shown in figure 12.1, proved far too ambitious and indeed it is hard to believe that Nimmo had actually visited the site and examined the topography before designing so elaborate a settlement. For example, he made little of the steep contours, which would have made the exe-

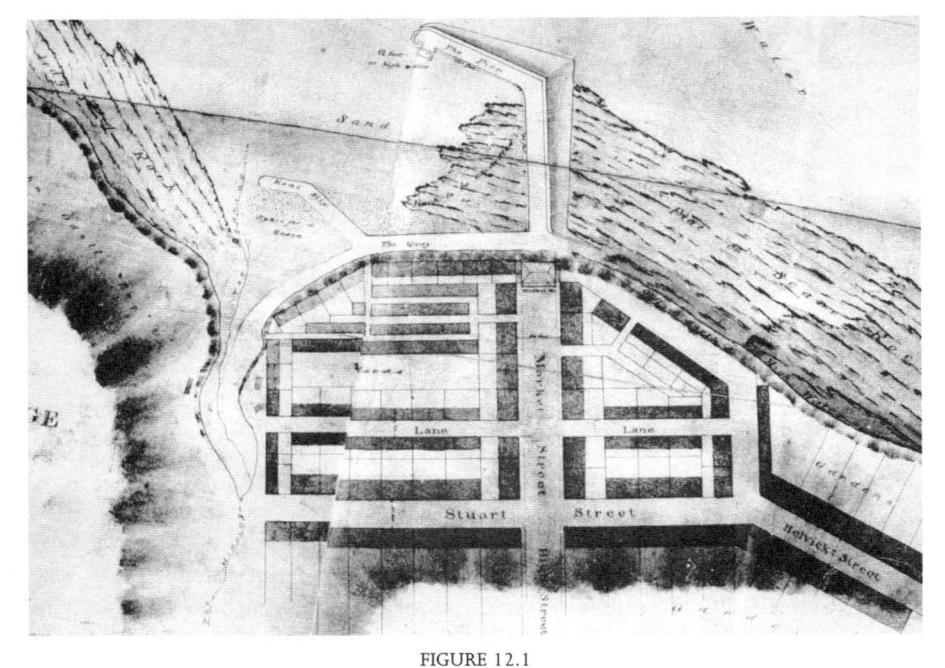

FIGURE 12.1
Plan of Nimmo's proposed harbour and village at Ballingaule, Co. Waterford. (PRONI, T/3131/H/7. Reproduced with permission of the Public Records Office of Northern Ireland.)

cution of the design difficult if not impossible. Had the village been built as he envisaged, it would have been one of his most comprehensive works. A large pier was eventually built there later in the century, and the small village of Ballingaule grew up on the slopes above it. Today, this is quite a charming place, although nothing as elaborate as Nimmo had proposed. Its existence confirms the suitability of the location he had chosen, difficult topography notwithstanding.

COURTOWN HARBOUR

In November 1819, before his appointment by the Fishery Commissioners, Lord Courtown approached Nimmo to design a small harbour at Brenogue, near the village now called Courtown in County Wexford. It was to facilitate the coastal trade and possibly be a packet station for communication with Wales. Nimmo sited the new harbour, comprising a basin enclosed by two piers, in the shelter of Brenogue Hill, near the mouth of the Chapel (Blind) River. First, he had to cut away some of the base of the hill in order to widen the intended basin (figure 12.2).[3] That allowed him to set the northern pier nearly parallel to the shore and reduce the southern pier to just a short breakwater. Behind the harbour, he proposed a backwater, or inner basin, to receive and store water directly from the Chapel River and from a new cut he would make to bring extra water from the Courtown River. The inner basin would be closed off from the outer harbour by sluice gates that could be opened to release the pent-up water to scour the harbour at low tide, and it would double as a useful place where small craft could lie on occasions. The harbour and the new cut would be protected from the open sea by groynes and an embankment that he planned along the sandy shore north of the harbour mouth. Nothing was done to advance the proposal for a number of years, during which the local fishery increased quite considerably and the need for a harbour of refuge increased correspondingly.

Lord Courtown certainly had not forgotten the project. In 1823 he applied for a grant-in-aid under the Moiety Act and Nimmo was asked to re-examine his original plans and make suitable alterations to them. In September 1823 Nimmo wrote from Ennis to his brother George in Dublin requesting him to send the altered design, together with the new estimate (almost £10,000) and report, to John Killaly, who was in Ennis with him at the time.[4] Both men were the engineers of the Moiety Act; unsurprisingly, they approved the works, and almost £5,000 was granted as a 'gift' of the Lord-Lieutenant later in 1823.[5] That sum was one-half of the estimated cost and the other half had now to be found locally by subscription. Lord Fitzwilliam gave £1,000 and Lord Courtown and Viscount Stopford contributed £1,000 between them.[6] The complete sum was raised eventually and a private bill was drawn up and passed on 17 June 1824.[7] Today it appears peculiar that the engineer who designed the harbour should also be the one who approved it for grant aid from the Viceroy, but matters were handled differently in those times and would change due to circumstances like this.

The Courtown Act set up commissioners who started construction under

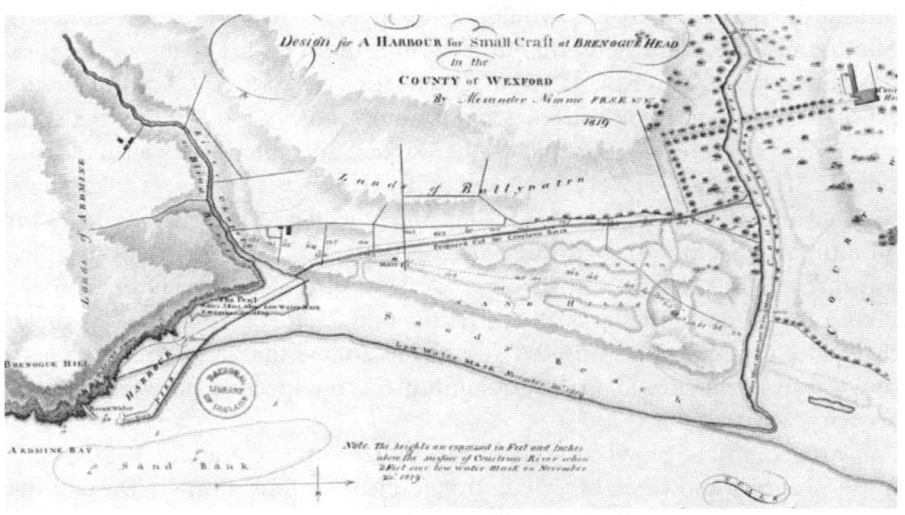

FIGURE 12.2
Nimmo's 1819 plan for a harbour at Courtown, with a cut to the Courtown River. (Detail from ref. 3, with permission of the National Library of Ireland.)

Messrs O'Hara, Simpson and McGill, with the aim of completing the work within two years. O'Hara and McGill worked on many piers and harbours for Nimmo and were among his most trusted assistants. The Harbour Commissioners approached the Fisheries Commissioners offering to exempt all fishing boats of import and harbour exactions if they would make £500 available towards the cost. The Fisheries Commissioners agreed to this: it did not breach their rule of keeping their contribution below half of the total cost and of not exceeding £500 in actual amount for any fishery harbour.[8] As soon as the work started, sand began to accumulate in the outer basin (the backwater was not yet formed) and the prospects for success did not look good. George Halpin of the Dublin Ballast Board was called in for advice in December 1825. He regarded the site as a poor one, very sandy, with poor tidal flow and a low rise of spring tides. He doubted that the flow of the river would ever be strong enough to counteract the action of the sea, so that a bar would most likely be thrown up outside the harbour.[9] Indeed, a sandbank was already obvious (see figure 12.2) in Nimmo's map. Halpin seems to have forgotten that Nimmo had not depended solely on the river alone, but had planned the backwater to provide an increased supply to enhance the scouring. Eight thousand pounds had already been spent and in his, Halpin's, opinion another £8,000 were needed, but the harbour would never earn the amount needed to service these sums.

All this was bad news for the Harbour Commissioners and cannot have been pleasant for Nimmo either. Six months later, on 7 June 1826, he and Thomas Telford visited Courtown to inspect the operations. Telford considered that the harbour would be of great utility and from enquiries he made locally he estimated that it would do about £2,900 of business annually once completed. Approving the utilisation of the river for scouring, he proposed that the harbour be given an outer lock gate so that it would become a floating dock and could have proper landing quays.[10] In its then existing state, the south pier was already well advanced and the north pier had been started. Stone quarries were open locally and there were railways from them to the work site. Telford concluded: 'Having considered the plan made by Mr Nimmo and having gone through the necessary calculations, I am of opinion that £9,000 will be required to complete the plan as now definitively determined upon.' This was much better news for everyone, but it meant that more money was required. The commissioners therefore sought a loan of £6,000 under the Loan Act, assigning the prospective harbour dues as security.[11]

Work continued slowly and Nimmo was back on site again in 1830 examining progress and directing the completion of various works.[12] Despite the backwardness of most of the construction he was sanguine (as ever) about the benefit to be expected of the completed harbour, which he thought was now well calculated to take a steamer. That same year, the Trustees of the Port of

Aberystwyth in Wales had asked him to report on their harbour and make recommendations for its improvement, probably with a view to increasing its trade with Courtown, just across the Irish Sea from it. Nimmo made his report, but had died before the trustees could progress his plan very far.[13] After his death, an engineer named George Bush re-examined Aberystwyth port and substantially supported Nimmo's recommendations. Meantime, the Commissioners of Courtown Harbour were persevering with remediation of their works, calling on George Nimmo for further advice. George visited in April 1833 and recommended they build a timber breakwater some distance from the mouth of the harbour, as a means of diverting the surge of the sea and preventing the sand from entering. This would require fifty-two frames of five piles each, placed five feet apart and braced together, for which his estimate was £3,116. In his estimates he referred to a plan of 1829 that does not appear to have survived and raises the possibility that the breakwater may not have been his own idea. Indeed, it looks very much as if he was recycling the design used in Youghal Bridge, built the previous year.[14] The commissioners had taken the precaution of asking Jacob Owen, an engineer of the Board of Works from whom they were seeking a further loan, to examine the harbour while George was there. Owen agreed that George's breakwater would be useful and he added his own major alterations, costing over £2,700, to the works as they then stood.[15] A sum of £16,000 had already been expended and Owen's modifications, added to George's, would bring the total to almost £22,000. Even then, Owen was not optimistic that the harbour would ever be much good. Moreover, in the previous twelve months the tolls had come to only £20, so the prospective income was unlikely, in his opinion, ever to repay the sums that needed to be expended. The commissioners persisted however, and by 1846, £31,000 had been invested in Courtown harbour, after yet another engineer, Francis Giles, had made recommendations. The history of the harbour since then has been well described by Kinsella. Nimmo would still recognise the inner part of the harbour – what he called the pent – which is now the main basin, and the artificial cut he made from the Courtown River. His projected outer harbour has been replaced by a narrow masonry entrance between new piers. There is no evidence of George's timber breakwater, if it was ever constructed.

NEWRY NAVIGATION

Once Nimmo became engineer to the Fisheries Commission other harbour works flowed in and his reputation as a harbour engineer took off. While his fishery piers and harbours were intentionally small and generally useful only for small-boat purposes, his private clients subscribed to have more elaborate structures designed so that they could be used for general trading purposes. Examples already mentioned are the harbours of Killough, Port Ballintrae and

Bangor. While he was surveying Lough Carlingford on his coastal survey an opportunity arose respecting the Newry Ship Canal. The sea approach to the first lock at Lower Fathom suffered serious and constant silting. Brownrigg and Killaly, engineers with the Board of Inland Navigation, proposed to solve the problem by prolonging the ship canal to Rice's Bay much further down the Lough, at an estimated cost of £80,000.[16] The board, seeking a second opinion on that costly plan, consulted Nimmo. He sounded the inner reaches of the Lough around Warrenpoint, laying down the soundings on a chart, and indicating his preferred solution to the problem.[17] This was to make an entirely new canal from Warrenpoint along the opposite, north, shore through Narrow Water, as far as the deep pool at Upper Fathom. From there, the navigation could continue by a short stretch of the river as far as the existing first lock on the south shore, and thence by the existing canal to Newry. His solution, therefore, involved two separate canals, one on the north shore and the other on the south, linked by a deep section of river. Nothing was done immediately and Sir John Rennie was eventually called in for his advice in 1829. He favoured a variant of Nimmo's plan: he proposed to widen and deepen the existing river channel all the way from Warrenpoint to Upper Fathom, in place of Nimmo's proposed canal. Eventually a mixed solution was adopted: the original ship canal was extended downstream as suggested by Brownrigg and Killaly, but not as far as they had proposed, and the channel below it was widened and deepened, as Rennie had proposed, the extended canal joining the river at a new entrance lock, now called the Victoria lock, located between Upper Fathom and Rice's Bay. These modifications cost almost £90,000.[18]

As we might have anticipated, Nimmo took a broad perspective on the whole Newry navigation and not just on the ship canal below the town. The inland section comprised the existing canal from Newry (the first-ever commercial canal built in Britain or Ireland) to the junction of the upper Bann and Cusher rivers, just south of Portadown. The summit level, between Poyntz Pass and Scarva, was seventy-eight feet above sea level. North of Portadown, the navigation continued by way of the Upper Bann River to Lough Neagh. Several rivers flowed into that lough, but only one, the north-flowing Lower Bann River, drained it. Winter flooding was therefore a constant feature of the lough surrounds and there were calls for over 200 years to do something to mitigate the worst floods. Nimmo saw a way of improving the canal and alleviating the flooding problem in one go:

> it would be a matter of no great expense or difficulty to cut the summit level of the Newry Inland Canal to a depth sufficient to bring the surface level of Lough Neagh through the valley towards Newry, giving a supply of water to the canal and our proposed lower navigation that would be abundant and inexhaustible, saving also the passage of three locks on one

side and four on the other. I would propose to bring the level on to the town … and thus the whole of the inland trade of Newry may be done without lockage at all and the cutt [sic] may be of such dimensions as to admit of steam vessels, so as to excel any kind of land carriage, even in expedition … A slight improvement on the Lower Bann would extend this level navigation to Portglenone, sixty miles, in a straight line, from Newry. We could obtain a command of Lough Neagh such as to run it down some feet in summer and keep its winter rise to the present summer level, thus saving from being flooded six thousand acres of land. We would also bring a water power into Newry of, say, one sixth of the discharge of the lake, with forty-five feet of fall, which I estimate at three thousand horse power, being more than that of all the steam engines of Glasgow.[19]

This novel, enterprising and risky proposal was too much for the Board of Navigation to approve, but it was resurrected again many years later. In 1882 the chairman of the Portadown Town Commissioners suggested that the flow of the Upper Bann could be reversed, thereby relieving Lough Neagh of one-ninth of its inflowing water. The river flow could be diverted into the Newry canal, where it could be made to serve a more useful purpose by dispensing with the locks and making an inland ship canal from Portadown to Newry possible.[20] He estimated this would cost £200,000, a sum sufficient in itself to ensure that local interests would not progress the scheme, even had they thought it feasible. Nimmo's original proposal was never mentioned.

DROGHEDA AND THE RIVER BOYNE

During the 1820s many of the small ports on the east coast had ambitions to increase trade with towns across the Irish Sea. Steam packets were coming into operation that permitted faster, more reliable and more comfortable travel, but they demanded deeper harbours and floating docks than were usual in the days of sail. Drogheda, a town located about four miles from the open sea, lies on the estuary of the River Boyne. Its merchants wanted to increase trade with Liverpool, Glasgow and even, they hoped, to open new trade with America. The River Boyne met the tide about one and a half miles above the town, so that, with the benefit of the river and the tide, adequate depth of water at the town quays could be achieved by excavation. Below the town, the tidal river widened and small weirs of gravel and clay, erected mainly on the north shore confined the flow to a definite channel that ran in a generally easterly direction. These weirs had been adequate to maintain an unobstructed and deep navigation in the channel in earlier days. Below Banktown Ford, the river expanded to a large tidal basin of almost 500 acres, half a mile wide at high tide. The channel ran through this, before making a sudden bend at Crook Point to flow southwards to the open sea between two sandbanks about 120 yards apart. The low North Bull sand bank formed the eastern side of this channel and was overflowed by normal high tide. The last of the flood and the early part of the ebb setting across it made the channel at Crook Point quite dangerous for shipping.

The Harbour Commissioners asked Nimmo to examine the area and make recommendations for its improvement. He set about surveying in autumn 1825 with his customary enthusiasm, expressed with added sentiments that he could reveal to Maurice Fitzgerald, if to no one else:

> At work surveying Dundalk and Drogheda relative to instituting steam passage to Liverpool which is now a great object of the policy of this country and likely to produce effects more important and valuable than all the potential schemes of union and conciliation which the workmen, or rather talking men, of your shop have been at for centuries.[21]

His report traced the changes that had occurred in the channel within the previous fifty years and described the general navigation below the town.[22] He identified the chief obstacles to safe passage to and from Drogheda, and went on to outline solutions to each. First, he proposed to form an entirely new navigation channel from Banktown Ford to Crook Point by cutting through the swash in a direct line (shades of Sligo and Cork); that would make a new channel running almost due east. To rectify the passage beyond Crook Point he proposed to heighten the inner part of the North Bull by raising a bank to high water level that would prevent the tide from setting over it. The material for this could be dredged from the Rock Shoal, thereby removing that obstacle even as the bank was being raised. In a few years, sand would accumulate behind the dredged spoil deposited on the north bank, as indeed it did. He would continue the north bank down to a point opposite the Maiden Tower, thereby turning the whole reach into a sheltered harbour where vessels could wait safely for a fair wind. Below the Maiden Tower, he proposed to cut through the lower extremity of the bar and once again the dredged material could be used to build up the north and south sides of the harbour entrance. Eventually, extra material deposited on the north side could be made into a mole and when brought up to low water level, a stone breakwater could be constructed on it to a level above high water. That way the entrance would be sheltered from northerly gales.

The plan was a very straightforward and economical one, but would require the use of a steam dredger, the first time such a vessel was recommended for use in Ireland. Nimmo estimated the overall cost at £13,344, of which almost £4,000 was for excavating the new direct line through the swash. The plan was acceptable but the cost was too much for the commissioners. By 1827, however, funds had been obtained sufficient to purchase a steam dredger for £5,000.[23] This was the first steam dredger ever used in Ireland. After Nimmo's death, William Bald re-examined the harbour. He revised Nimmo's plan and reduced the estimated cost, much to the satisfaction of the new board of works, who agreed to advance a loan to undertake the works.[24] Bald's modifications added very little that was really new to what Nimmo had already proposed, although he did fill in the detail of the sketchier parts of the latter's proposal. He showed how the direct line from Banktown Ford could be cut progressively and, unlike Nimmo, he proposed to cut away the Carrick shoal entirely. Along the course of the estuary he built small dykes of rubble stone to direct and speed up the flow in the channel, just as the weirs on the north shore had done for generations.

It was John MacNeill (1793–1880), working under Nimmo's direction, who had carried out the original survey, the results of which were presented on Nimmo's large chart of the harbour. Bald delineated his modifications in red ink on a reduced copy of this.[25] It is the only copy of Nimmo's map that has been found. In the years since, the deepening and widening of the channel has progressed, more than 172,000 tons of gravel being removed by steam dredger in 1832 and 1833 alone. By 1835 a very extensive trade was carried on between Drogheda and Nova Scotia and New Brunswick in Canada and there was also a very considerable cross-channel trade, with three steam packets plying to Liverpool and Glasgow.[26] Bald supervised the work for six or seven years, until the money granted by the OPW ran out. He then handed over to Mr Young, who continued the work under the Port Commissioners.

In December 1826, on Nimmo's recommendation, John MacNeill became assistant to Thomas Telford on the London to Shrewsbury part of the Holyhead road and, until Telford's death in 1834, he worked mainly on roads in England.[27] On return to Ireland he contributed to harbour improvements at Dundalk, Drogheda and Arklow and then set about guiding the construction of a large number of Ireland's main-line railways. In 1842 he was appointed to the foundation Chair of Civil Engineering at Trinity College Dublin, becoming another of Nimmo's protégés to enjoy an illustrious career in engineering. He was knighted in 1844.

Today, the port of Drogheda has been considerably enlarged and improved, but apart from the entrance walls the estuary largely retains the general topography that it had in Nimmo's time. The considerable exertions of other engineers, and the application of newer technologies, have done little enough beyond the channelling, deepening and widening actions recommended in Nimmo's original plan.

No information has yet been found on Nimmo's contribution to the port of Dundalk in the same county.

THE HIBERNIAN MINING COMPANY AND NEW COMMERCIAL ENTERPRISES

The mines at Ross Island in Killarney were exploited throughout history for their copper, lead, silver, zinc and cobalt reserves.[28] In the nineteenth century copper was their most important ore, often mined with more zeal than skill.

Because of their proximity to the lake, flooding of the shafts was the greatest hazard and had caused many attempts at mining to be abandoned prematurely. In 1812 they were leased to William White, who introduced a steam engine to drain the flooded workings and get them back into production. The engine proved unequal to the task and Nimmo was called in for advice. On his recommendation, a small mine was opened on Crow Island and secured from flooding. Ore was raised at a cost of two pounds per ton compared with four pounds formerly, and a cargo was shipped to Swansea where it fetched eleven pounds a ton. At the same time, workers repaired the dams of the main mine under Nimmo's direction and brought it back into working order. Activity was suspended due to the early onset of what Nimmo called 'the wet season', and he decided that it should remain suspended until water levels would decline the following summer. By then he was no longer available to continue supervising since his 'engagements at that time led him into distant parts of the kingdom'.[29] The Bogs Commissioners had given him the Connemara district and that became his immediate priority. During his absence, shortage of capital put an end to the enterprise and the lease of the mines reverted to the Earl of Kenmare. It would be twelve more years before the lease would be taken over by new owners and Nimmo would return to the venture.

In 1824 the Hibernian Mining Company, determined on reopening the mines, called on Nimmo to prepare a full account of the resource. Nimmo was nothing if not enthusiastic. Ross Island was, he said, 'one of the richest mines in the world' and would 'afford the adventurers a rich field of unbroken ground'.[29] By his calculations the venture would clear £893 a month for a capital outlay of £20,000. He recommended that half the capital be called up immediately, to be used for the purchase of new steam engines and the repair of the dams. No part of the mines was more than sixty feet deep, which was shallow compared with mines elsewhere. It was only seven miles by land to the best harbour at Kenmare, he said (it is, in fact, more like twenty miles), and 'his new road from Killarney to Kenmare will greatly facilitate the workings of these mines'.[30] (This road was one that received a grant of £6,420 from the Lord-Lieutenant under the Moiety Act.[31]) In reply to further queries, he proposed that a new coffer dam be erected around the main mine at a cost of about £1,266, which he regarded as an economical solution to contain the immediate problem of flooding. Once the mine was drained, a more permanent dam could be erected.[32] Economy being one of his aims, he went on to advise that the fissures and leaks could be rectified for £3,800 and that new steam engines, costing £1,200, were the most suitable sources of power for the operations envisaged. Coal could be imported through Kenmare and brought to Ross Island as a back cargo in return for ore shipped out. The Company took his advice regarding steam engines and by 1827 work at the mine was in a flatteringly satisfactory state, as was the price of the Company's shares. The mine flattered only to deceive, however, and difficulties persisted until the whole operation ceased entirely in 1829. The full story of the Ross Island mine from prehistoric times to today is told in O'Brien's comprehensive study,[28] from which the above account is drawn in part.

Nimmo was not the only engineer consulted by the Hibernian Company. Richard Griffith, among others, was also approached for advice. He was, as usual, cautious and circumspect: whereas Nimmo had said the mine was one of the richest in the world, Griffith had thought it a 'hazardous speculation' – a contrast that neatly encapsulates the difference in approach of these contemporary engineers.[33] Both men also gave advice to the company on other mining ventures. For Nimmo at least, this was to be something of a new beginning. Up to now, most of his work was of a public nature, funded wholly or in part by public funds. The Hibernian Mining Company, on the other hand, was a Joint Stock Company – a PLC – although it had engaged him before it was incorporated by a private Act passed on 17 June 1824.[34]

Nimmo informed the House of Lords in 1824 that he was working with the proprietors of coal mines near Lough Allen to improve the roads in that district: 'we have coal at the head of the Shannon which is now in progress of being worked; and we are opening roads to assist the proprietors in that operation. I refer to Lough Allen at the head of the Shannon, and the Arigna river; that is the coal of the best quality in Ireland, and the most extensive supply.'[35] In July he prepared and delivered a report on these roads to the Hibernian Company.[36] Around this time, he was throwing his weight behind mining ventures with his customary enthusiasm. He had informed the Lords that Arigna coal would probably have a very extensive consumption in Ireland, and be cheaper than turf, provided proper roads were made to the mines. He also proposed a railway from Wicklow to Dublin for the transport of turf and granite to the city and claimed to have raised a subscription of £30,000 towards the venture. As he explained it, 'The railway into the Wicklow mountains is with a view to the extraction of turf and granite, which is a kind of mining, and the [Hibernian] company propose to become miners also themselves, that is, to become turf manufacturers and granite quarriers.'[37] By November 1824, the organisation of the Hibernian Company was complete and it placed advertisements seeking proposals to develop mines anywhere in Ireland.[38] The company's office was located in Marlborough Street, Dublin (possibly in Nimmo's office, then in no. 56), but later it moved to new premises in Sackville Street (1825–27) and then to Eden Quay (1828–29).

Perhaps because of his work with the company, and the need to transport coal cheaply from Lough Allen, he developed a great interest in the improvement of the Erne and Shannon navigations, which he felt should be developed for steamboats. 'I have looked at Lough Erne with a good deal of pains,' he said, 'and my opinion is that they [steamboats] should go from Lough Erne by Lough Melvin and Manor Hamilton to Sligo and there communicate with the head of the Shannon.'[39] Today, this does not sound at all a good route to have recommended. It does, however, indicate that he was thinking of a way to join the great inland navigations of the country – the Neagh, the Erne and the Shannon – a project that had been long hoped for.[40] If Limerick could be linked to Waterford by the railway he envisaged (chapter 14), then a great arc of communication from the south-east to the north-east would be completed. But here his main interest was in shipping coal, slate, sandstone and plaster of Paris from Lough Allen to Dublin and Limerick.[41] He laid out a road beside Lough Allen that he had plans to convert eventually into a railway running northwards to Lough Gill, through which there was an existing navigation to Sligo. Towards the south, the railroad would connect with the Shannon, thereby linking that river to Sligo. By the end of July 1824, he had estimated the cost of extracting Arigna coal at nine pounds per hundred tons, with a delivery cost to Limerick of thirteen shillings a ton. It could be sold there at a profit, but if it were to be trans-shipped for onward delivery to Valentia, the charge would be a further five shillings a ton. This, he believed, would make it uncompetitive in Valentia with seaborne coal from Scotland or Wales.

Next he turned his attention to the relative costs of freightage from Dublin, using information provided by the City of Dublin Steam Packet Company, which was founded by his friend, Charles Wye Williams in 1822 and which operated steam vessels between Dublin and Liverpool. Nimmo calculated that a steamship could run between Dublin and Limerick in forty-eight hours with a cargo of 300 tons at one-third the cost of the existing inland navigation by way of the Grand Canal and the River Shannon. Transport by the latter route could take up to three months to complete! He was convinced that it was much more efficient to go around by steam than to go through 'by trackage'. He and his brother engineers were turning their attention to improving rivers not for trackage but for steaming. With this in mind, he proceeded to outline the alterations that he thought were necessary to improve the Shannon navigation, a work 'I should conceive a great national object.' This included embanking, removing and widening the locks already in place and building quays and wharves, especially on the west bank of the river.[42] It would fall later to his ex-student of Inverness days, Thomas Rhodes, to develop the Shannon navigation in earnest.

At the early stage of its operation, the Hibernian Mining Company was trying to buy extractives such as coal, slate, marble and building stone wherever it could. It sought slate from Valentia, marble from Ballinahinch and coal from Lough Allen, all extractive enterprises with which Nimmo was connected. However, the legislation under which the company acted did not allow it to engage in trading, only in mining. Therefore, a petition for a new bill was introduced in Parliament in March 1825[43] and a new act passed on 22 June,[44] the same day as an act for the new Arigna Iron and Coal Company.[45] This was at the height of an inordinate passion for speculation that gripped Britain, resulting in the formation of numerous joint stock companies of questionable viability.[46] At least four acts were passed setting up mining companies in

Ireland, one setting up a railway and one a transatlantic steamship company. Very often it was the same men of capital who participated in the different Irish enterprises, not simply because so few had capital, but also because only a relatively small number had the development of Irish enterprise at heart. Nimmo, too, was involved with most of these companies, possibly at the behest of Maurice Fitzgerald. But there is very little evidence that he ever invested his own money in any of them; if he had, it might explain why he left so little estate when he died. Robert Mudie, however, believed that Nimmo had invested with William Edgeworth in the Hibernian Mining Company, but that has not yet been corroborated.[47] Neither is Mudie's hint confirmed that they lost money in some other venture. The important feature is that Nimmo was professionally involved with persons who were trying to progress Ireland socially, commercially and economically at a time when this was not a notable feature of Irish life and politics.

VALENTIA AND THE AMERICAN AND COLONIAL STEAM NAVIGATION COMPANY

Maurice Fitzgerald (1772–1849), the eighteenth Knight of Kerry and long-time MP for that county, a man of great energy and enthusiasm, had a poor sense of business and was inclined to let his enthusiasm get the better of his common sense. To such a man, bursting with 'improving' ideas, Nimmo was an ideal colleague, bringing to their relationship a cooler, more technical, head. Sir Peter Fitzgerald, Maurice's son, believed that Nimmo had had huge influence on his father: 'I think it was from intercourse with Nimmo that many of the matters on which my father wrote and worked originated: for instance, packet station, improvement of land, Board of Works, defence of the seaboard of the country etc.'[48] We do not know for certain when they first met, but it may have been during the Bogs Commission days. Nimmo's musings about railways and canals in Iveragh in 1812, however speculative, must have been music to Fitzgerald and would have endeared Nimmo to him. By 1816 the latter was immersed in the Dunmore harbour project, but had not abandoned Kerry altogether, maintaining his involvement in the road from Killorglin to Cahirsiveen that he laid out and that would not be completed for a number of years yet.[49] Fitzgerald, who was trying to develop the mail coach road from Tarbert to Tralee by way of Grand Jury presentments, wrote to Peel requesting that a Government engineer be appointed to the task.[50] Peel was unable to oblige so we do not know for certain, although one has suspicions, whether Fitzgerald was trying to obtain the post for Nimmo. But when Fitzgerald was made a Commissioner for Fisheries in late 1819 it may have given him the opportunity to propose Nimmo as engineer for the coastal survey, even though the latter had admitted that he knew nothing at all of the Irish fisheries. That Fitzgerald (among others) held Nimmo in very high regard we know from a letter of Lord Lansdowne in 1822:

> I was sorry to hear from Kerry that Mr Nimmo, of whom from your account and that of others I had formed a high opinion, is not to be employed in the public works there, as he has acquired more local knowledge as well as skill than any other person. Mr Griffith who is to replace him I have never heard of, but that the Lord-Lieutenant has some good reason for confiding in him.[51]

Nimmo had unrivalled knowledge of Galway and Mayo as well as of Kerry, which is why he was appointed to the larger, more demanding, western district while Richard Griffith was given the Slieve Luachra borders of Cork, Limerick and Kerry.

Fitzgerald worked tirelessly to improve Ireland and the condition of the Irish peasantry by championing the commercial interests of Kerry, and particularly of Valentia. Of his many schemes, his attempted development of Valentia harbour as the first transatlantic steamer port was one of his most ambitious. The potential of the scheme appealed to Nimmo, whose proclivity for grand schemes we have noted already. They must have planned the scheme together, because when it first saw the light of day, at a public meeting held in the City of London Tavern on 25 June 1824, Nimmo was on hand to explain and support the proposal. The idea was to establish a direct steamer communication between the United Kingdom and America. The harbour of Valentia was the nearest suitable point in Europe to the North American continent and they planned to establish a line of steam packets that would ply weekly from there to New York. According to the prospectus, Valentia was 200 miles further west than Falmouth in England and could be reached from London in fifty hours, or from Liverpool in forty. It was well known, the prospectus went on, that 'by far the greater part of the delay and danger of the passage from Falmouth or Liverpool occurs within the Channel and the Bay of Biscay, which, by this line, are avoided'.[52] Valentia harbour was accessible at all tides and wind conditions, even those that closed the western ports of England.

When Fitzgerald addressed the meeting, he was careful to emphasise that it was impossible for any port connected with the American trade to enter into competition with Liverpool because of the extent of that port's existing American trade. However, he had no doubt that passengers, separate from freightage, would welcome the more rapid means of travel that the proposed Valentia steam packets would provide over the Liverpool sailing packets. Although concentrating on passenger traffic, the prospectus nevertheless stated that 'Ballast cargoes may be obtained there [Valentia] in slates, butter, and coarse linen, for the American markets', so some trade in goods as well as passengers was anticipated. Fitzgerald then called on Nimmo to corroborate his assertion of the profitable prospect that the plan held out. Nimmo obliged with an account of recent improvements in steam navigation and an estimate that a vessel of 1,000 tons burden could navigate the Atlantic from Valentia to

Nova Scotia in less than a fortnight using 400 tons of coal. He recommended that only large vessels should make the Atlantic crossing; on arrival in Nova Scotia, smaller vessels could carry the navigation onwards to New York and all parts of Canada. On the Irish side, the smaller vessels that brought the necessary coal to Valentia would convey the merchandise that had arrived from America onwards to other parts of the kingdom. He then submitted an estimate of the likely cost of the whole operation.

Nimmo would criticise many aspects of the scheme later, but at this early stage it succeeded in arousing immediate interest among those in London involved in the American trade. Within three weeks, Robert Day, a relative of Fitzgerald, wrote enthusiastically to him using a formula of words that today might raise eyebrows, or at least give rise to wicked satire: Lord Meath, he said, had told him that 'you and Perkins are getting on swimmingly and that in a few weeks we should have active intercourse of parties of pleasure between Valentia and New York'.[53] Detailed planning commenced and the possibility that coal could be sourced in Arigna, County Leitrim, and shipped down the Shannon to Limerick for onward transport to Valentia was raised with Nimmo. He did not favour the Shannon navigation route, thinking that it would be too expensive to trans-ship coal at Limerick; if it were shipped direct through Sligo, it could be taken straight to Valentia at lower cost.[54] A few months later he did a *volte face*: writing from Boyle, County Roscommon, he was insistent that 'the most important circumstance connected with Valentia is the nature of the Limerick trade, which must be kept prominent in the foreground as affording the supply of fuel on reasonable terms'.[55] One of the great disadvantages of Valentia, the fact that it lacked local supplies of coal to fuel the steam ships, was becoming of greater concern to him as he considered the details of the scheme.

They were buoyed up when the plan was well received at a meeting with some merchants of Bristol.[56] There was also an offer to introduce the plan to Mr Rothschild, whose influence was vast in the city of London, but it seems to have come to nothing.[57] St J. G. Gregory, a London solicitor supporting the scheme, proposed that more attention should be given to Valentia's strategic importance: 'I think the whole point of Valentia harbour by far too important in the national point of view [sic] as a military and naval station and as commanding the whole of the west of Ireland along the line of coast' he wrote, proposing that it should be made available for military purposes.[58] Fitzgerald was impressed by this, indicating to Nimmo that the steam packet might be promoted as a rapid way of connecting military forts on both sides of the Atlantic, in which case Halifax, Nova Scotia would be the obvious North American terminus. Nimmo replied on 4 October 1824 'at the first blush of the project', as he said. He was against the slant that the scheme seemed to be taking: 'Your early military habits make you stress the military benefits of Valentia. I and others don't agree ... you will lose the vantage ground with regard to the commercial view of the project. If Halifax, why not Cork? If Valentia, why not Berehaven?'[59] He thought Cape Canso, his own choice of terminal in Nova Scotia, was much more suitable than Halifax because, in his opinion, 'if Britain means to retain her hold on North American colonies, Halifax must be abandoned as Shelburn was, and the seat of Government and force removed to the Gut of Canoo[?] where it would be inexpurgible'. Nimmo certainly had a point about the commercial consideration, but St J. Gregory's next letter made greater political sense. In it, he urged Fitzgerald to get a charter for the proposed company, emphasising the naval and military grounds 'so that the Duke of Wellington, Duke of York, Lords of the Admiralty etc. will support it'.[60] The conflicting advice threw Fitzgerald into a quandary, so much so that Nimmo had to spur him on to make some decision: 'Take care you are not anticipated by an American company. It is contemplated to have a steamer from New York or Boston to Europe through Cork or some other port in Ireland.'[61]

Early the next year, 1825, the news was more encouraging for Fitzgerald. Wilmot Horton, under-secretary for War and Colonies, wrote from Downing Street that Lord Bathurst, the secretary of the department, was favourably disposed towards 'a steam establishment between the west of Ireland and British North America', especially as it related to the Gulf and the St Lawrence River.[62] With this indication of support, a full prospectus was published in April.[63] Having recounted all the benefits of Valentia once more, it outlined the general plan of the company, which would have a capital base of £600,000 in shares of £100. There would be two steamers a month crossing the Atlantic. One would go to Halifax and after refuelling there, carry on to Boston or New York. The second would go to the West Indies via the Azores, Antigua and Port Royal in Jamaica. This latter, having met local steamers plying to South America, would make the homeward journey to Valentia via Bermuda. Both vessels would be well armed against pirates. The expected revenue from passengers was estimated at half a million pounds per annum and the service would also be useful, the prospectus stated, for the transport of troops and stores, for which the endorsement of military authorities was claimed. Nimmo had clearly lost the debate on the military aspect.

Parliamentary approval was now sought for a limited liability joint stock company to be named the American and Colonial Steam Navigation Company. Sir Edward Lees of the Irish Post Office, wrote on 27 May to congratulate Fitzgerald on the progress he had made in advancing the company, and was effusive in praise of the whole scheme:

> It is replete with everything that is national and important to this country. It holds out the certainty of opening numerous channels for the introduction of British capital and of cementing our connection with England ... My only fear is that it may make Ireland too rapidly great, before its internal feuds have subsided and impress us with notions of separate power that had better be kept in the shade.[64]

Lord Lansdowne introduced a second reading of the bill before the House of Lords on 14 June, saying that he regarded the undertaking as one of great national importance regarding the British colonies and the mother country, in improving their commercial connections and in being calculated to promote the military security of those colonies.[65] The bill succeeded and on 22 June 1825 it passed into law as the private act 6 Geo IV cap. clxvii.[66]

Opposition was not long in coming. Within days, W. Latham of Murray, Latham and Co. wrote to Fitzgerald. He was a shipping agent out of Liverpool for years, he said, and he was convinced that steam packets to America would develop from that port. His advice was that the steam vessels should leave Liverpool with goods and passengers, picking up coal and passengers at Valentia *en route* to New York. Steamers leaving London the same day with goods and passengers for the West Indies could also call at Valentia on their journey. If trans-Atlantic steam packets were to prove successful, this is what would most likely happen he believed, and the competition from the English ports would undermine the new Valentia company.[67] Others, however, were more bullish, congratulating Fitzgerald on the new company and eagerly offering themselves for engagement. Daniel O'Connell, for example, was keen to be a director and subscribed for nineteen shares of £100 each on 30 June, paying 10 per cent deposit.[68] He promised to put 'all my shoulders' to the undertaking, quite an offer from the great man. A Mr Harem made detailed comments on the constitution and composition of the proposed company, offering himself as a possible company secretary.[69] He appears to have become a director, ultimately with dire consequences.

In Ireland, William Edgeworth, who was staying at Glanleam, Fitzgerald's home on Valentia Island, was negotiating the purchase of parcels of land for the company while in Britain the company lost no time in purchasing two steamers.[70] One was the paddle steamer *Calpé*, a timber vessel of 440 tons built by J. H. and J. Duke, Dover, with a steam engine of 100 horsepower made by Messrs Maudley and Field. She was then almost ready for sea. The other was also a timber paddle steamship, named the *United Kingdom*, still under construction by R. Steele and Co. in Greenock. *Calpé* cost over £20,000 to buy and a deposit was placed on the *United Kingdom* pending its completion.[71] At the end of August Nimmo wrote to Fitzgerald from Dublin ostensibly enquiring about the progress of the company, but really concerned at some unwelcome news.[72] He had heard from Glasgow that Mr Pike had bought *Calpé* for the company, giving a considerable premium over the contract price. *Calpé* was much smaller than Nimmo thought prudent for the Atlantic crossing and he was concerned that if bigger vessels were not acquired the whole scheme would stand at great risk. Later it transpired that the premium on *Calpé* was £1,200 and, in addition, the contract price for the larger *United Kingdom* had been inflated by £7,300.[73] As if these problems were not enough, jobbing in the company's shares was rife in London and confidence in the management plummeted. St

J. Gregory wrote to Fitzgerald warning that his 'dictatorial methods' were in danger of splitting the board of directors.[74] Some subscribers complained that they were billed for shares they had not ordered; another complained that he was made a director without his consent and that Fitzgerald acted foolishly because of his 'excessively sanguine temperament'.[75] Meetings were called at which it was made clear that the purchase of the ships had breached the act and had been improperly concluded. Some of the general meetings were tumultuous affairs, and the company was clearly in chaotic disarray with a number of shareholders wishing to pull out entirely. In a private letter to Fitzgerald, the acting company secretary, Mr Wells, blamed Harem for the debacle. Harem had died shortly before, so maybe he was just a convenient scapegoat. 'I should tell him [Harem], was he alive,' Wells wrote to Fitzgerald,

> he had much to answer for in confiding his judgement to people whose names ... he rashly admitted to promote the purchase of the two vessels expressly contrary to the Act of Parliament, without summonsing the board of directors generally, but relying on the judgement and integrity of the solicitor, secretary and two others interested in the purchase of the ships as affording them an opportunity of jobbing in the fitting out and having the command independent of any adventure derived from negotiation in the purchase. This was done without the knowledge of myself and the other directors who were at hand and ought to have been consulted.[76]

There was little for it but to sell the ships and to re-organise the company by obtaining a new act. The directors paid £4,500 to buy their way out of the purchase contract for the *United Kingdom*.[77] They would later meet opposition when this charge was levied on the shareholders, some of whom felt that the directors should shoulder the loss personally, since they had acted illegally in buying the ship at the outset. When launched later in 1826, the *United Kingdom* went into service with the Dublin Steam Packet Company. In 1841 she was bought by the P&O Steam Navigation Company and after a period as a hulk, was eventually sold again, presumably for demolition, in 1849.[78] *Calpé* was sold to the Government of the Netherlands in 1826 at a loss of £9,000.[79] She was renamed *Curacao* and was the first steam vessel in the Royal Netherlands Navy. In April 1827 she made her first crossing of the Atlantic from Holland to Paramaribo, Surinam, becoming the first steamship to cross from Europe to South America. She completed further crossings in 1828 and 1829. When *Savannah* had made the first steamship crossing of the Atlantic in 1821, it was predominantly under sail, with some assistance from the steam engine. The *Curacao* crossings were predominantly under steam power, with some minor assistance from sails. The vessel therefore has a strong claim to fame and some authorities even regard its 1827 voyage as the first true steam crossing of the Atlantic. It leaves us wondering what might have been, had The American and

Colonial Steam Navigation Company only been a success. *Curacao* was eventually broken up in 1850.

As it happened, the company, having paid all its debts and losing £6-14-5 on each share, reorganised itself and obtained a new act in 1826, 7 Geo IV, cap. cxxiv.[80] This allowed the waverers to exit the company, permitted the directors to buy as well as build steamships, and split the value of the shares to make them less risky for investors. Reorganisation continued through 1827 and 1828, a new prospectus proposing that two company steamships would carry one hundred passengers outward and inward each month, one-fifth of that number being sufficient to cover all transport costs.[81] It must have galled the investors to see *United Kingdom* make her maiden voyage, a tour around the British Isles, under the flag of the Dublin Steam Packet Company, and to hear of *Calpé*'s, now *Curacao*'s, historic crossing to South America. Fitzgerald continued to push the Valentia scheme, as did Nimmo, who actively canvassed the Post Office on behalf of the company for a contract to carry the American mails.[82] But in truth, the company would achieve nothing thereafter. Of those who had been most supportive of the whole business, Sir Henry Blackwood would become Commander-in-Chief at the Nore (a naval anchorage in the Thames) in 1827, Admiral Sir Pulteney Malcolm became Commander-in-Chief in the Mediterranean in 1828, Francis Beaufort (later Admiral Sir Francis) became hydrographer to the Navy in 1828 and Fitzgerald was made a Lord of the Treasury in 1827 and a Lord of the Admiralty later. With naval influence of this calibre in early support, the failure of the venture is even more to be wondered at and regretted.

In every sense, the Valentia scheme was truly adventuresome and visionary and Nimmo never lost sight of what the prospect of American trade would do for Ireland. When Fitzgerald later turned his mind to a possible steamer connection between Limerick and London, a much less risky and demanding venture, Nimmo was disappointed. Steam trade to Liverpool, however desirable, lacked the grandeur of the original scheme. He wrote to Fitzgerald, 'The great ultimate object is, as I have always thought, to open a depot at Valentia for the western world: a matter which will do wonders for the whole west of Ireland.'[83] As a business proposition however, trans-Atlantic steaming from Valentia probably had little to recommend it above steam passage from Liverpool or London. Kerry had no indigenous coal, so Valentia would always be dependent on imported coal from England that was cheaper and more readily available there. Fitzgerald had thought that small steamers bringing American imports from Valentia to other ports in Britain would carry coal to Valentia as a back cargo, but what 'back cargo' would the large steamers from America take westwards? The prospect of a ballast trade in butter was unrealistic; as Nimmo warned him, the whole peninsula of Iveragh produced only one hundred tons a year and the whole of Kerry could not provide for a constant trade.[84] Valentia slate was certainly saleable, but Nimmo had tried the

Valentia greenstone at the harbour works at Killough and found it of little value for cut-stone work. English exporters were never likely to agree to tranship goods in Valentia which would be necessary under Fitzgerald's scheme. As for the passengers, how many would cheerfully embrace a steam passage across the Irish Sea, he wondered, followed by a long coach ride through Ireland, before ever setting foot aboard the American-bound steamer?

Sanguine as ever, Fitzgerald now felt that Valentia could be a good base for the American cotton trade and he thought about setting up a settlement there for the manufacture of cotton and silk cloth. Nimmo had to dissuade him gently:

> Your proposal for a cotton or silk colony is a bold one, but you are not aware of the difficulties ... Cheap fuel [turf] you certainly possess, but not as cheap as coal at six shillings a ton. The clothing manufacturer meets cotton in warehouses, not in ships. Manchester is the main centre and it receives cotton from all over, so the manufacturer chooses his raw material there.[85]

Nevertheless, a week later Nimmo was arranging for Fitzgerald to meet persons in the clothing industry in Manchester, Dewsbury and Leeds, presumably to discuss his new cotton plan.[86]

It was in anticipation of the steamer business that a village, now called Knightstown, was planned at the place called the Foot on Valentia Island. As we saw, Edgeworth was buying leases of land there for the company and he kept in touch with Nimmo. One site was for a hotel and another for a coal-yard.[87] Together, these two would be the nucleus of the steamer base and consequently of any prospective village. Nowadays, the layout of Knightstown is commonly attributed to Nimmo, but there is no primary evidence for this. On the contrary, a modern map, drawn by the local historical association and displayed until lately on the external wall of the tourist office in the village, dates many of the buildings. Only three pre-date 1830, and one is dated 1820–40. There was no village shown there in Nimmo's Bogs Commission map of 1813, but a village, laid out much as it is today, is shown on the six-inch Ordnance Survey map of the area, surveyed in 1842.[88] Therefore, the village must have originated between these two dates. Some of the buildings shown on the 1842 map are no longer standing and others there today were erected since then. However, while the present buildings may not be a good guide to earlier structures, the general layout has not changed much in the intervening period and the village has the pronounced appearance of one designed to a definite plan. In Nimmo's hydrographic chart, published in 1832 from a survey he did in 1831, his pier at Valentia is shown along with the slate yard and some scattered houses, distributed in no obvious planned order.[89] The place on the chart is named simply 'The Foot Point'. Had a proper village existed there in 1831, and especially one of his own design, surely Nimmo would have given some greater indication of it on this chart? In summary, Nimmo could have set the layout of Knightstown, but hard evidence is lacking for this; on the contrary, it is likely that he had nothing whatever to do with it. He did, however, design the pier at Knightstown, which was built under his direction by Ben Meredith between 1822 and 1825. It is the one structure in Knightstown that can be assigned incontrovertibly to him. The pier to the north of it is a newer construction, although it looks much older and did not exist in Nimmo's time.

NOTES

1. A. Nimmo to Sir M. Homan, 20 October 1815. (PRONI, Villiers-Stuart Papers, T/3131/G/2/1–145 [MIC/464/reel 9]), hereafter referred to as Villiers-Stuart Papers.
2. A. Nimmo, 'Plan for Ballingaule Pier and Fishing Village by Alexander Nimmo 1816.' (Villiers-Stuart Papers, T/3131/H/10/7/1–72 [MIC 464/22]).
3. A. Nimmo, 'Design for a Harbour for small craft at Breenogue Head in the county of Wexford by Alexander Nimmo' (NLI, maps, 16 J 13 [3]).
4. A. Nimmo to G. Nimmo, 30 September 1823 (NAI, OPW8/97/1).
5. Public Works Ireland. An account of the sums of money advanced out of the Consolidated Fund etc (B.P.P. 1824 [278], vol. XXI, mf 26.135).
6. A. Kinsella, *The Windswept Shore. A History of the Courtown District.* (Dublin: Graphic Services, 1982).
7. 5 Geo IV cap. cxxii. *An Act for completing the Port or Harbour of Courtown at Brenogue Head in the county of Wexford*, 17 June 1824.
8. Correspondence between Fisheries Commissioners and the Commissioners for Courtown Harbour, July to September 1824 (NAI, OPW8/97 [2], [3] & [4]).
9. G. Halpin, 'Report on Courtown Harbour, 12 December 1825' (NAI, OPW 8/97 [7]).
10. T. Telford, 'Report of Thomas Telford on Courtown Harbour, 24 June, 1826' (NAI, OPW 8/97 [9]).
11. A.Kinsella, *Windswept Shore.*
12. A. Nimmo, 'The Report of Alexander Nimmo, engineer, on the Progress of the Works of Courtown Harbour, 28 February, 1830' (NAI, OPW 8/97 [17]).
13. S. Lewis, *A Topographical Dictionary of Wales*, 4th edition (London: S. Lewis, 1844).
14. G. Nimmo, 'An Estimate in detail of a timber pile breakwater in framed work outside the pier heads of Courtown Harbour and at three hundred feet from the present wooden pier in nine feet water of spring tide ebbs expressed on plan of 1829' (NAI, OPW 8/97 [23]).
15. J. Owen (1833), 'Report on Courtown Harbour' (NAI, OPW8/97 [25]).
16. D. Corneille, J. Saurin and H. R. Paine, 'A copy of the Report made by order of the Board of Inland Navigation in Dublin, respecting the Newry Navigation, 15 September 1820' (B.P.P. 1826/1827 [255], vol. XX, mf 29.154).
17. A. Nimmo, Plan and Sections of the Lower Newry Navigation with a design for its improvement by Alexander Nimmo. (NLI, maps, 16 B 22 (4)). Nimmo's chart of Lough Carlingford (NLI, maps, 16 B 22 [1]) is dated 1821, and his canal plan may date from around that time.
18. W. A. McCutcheon, 'The Newry Navigation: The Earliest Inland Canal in the British Isles', *The Geographical Journal*, 129, (1965), pp. 466–80.
19. A. Nimmo, quoted in W. A. McCutcheon, *The Canals of the North of Ireland* (London: David and Charles, 1965), p. 121.
20. McCutcheon, *Canals of the North of Ireland*; Delaney, *Ireland's Inland Waterways.*
21. A. Nimmo to M. Fitzgerald, 7 October 1825 (PRONI, Fitzgerald Papers, MIC/639/11/reel 6/97), hereafter referred to as Fitzgerald Papers.
22. A. Nimmo, *The Report of Alexander Nimmo Engineer on the Improvement of the Harbour of Drogheda* (Drogheda: P. Kelly, 1826).
23. Royal commission on Tidal Harbours. Second Report and Minutes of Evidence, p.58a (B.P.P. 1846 [692], vol. XVIII, mf 50.169–77).
24. Report from the Select Committee on Public Works in Ireland, appendix 2, p. 302 (B.P.P.1835 [537], vol. XX, mf 38.143–8); First Report from the Commissioners for Public Works Ireland (B.P.P. 1833 [75], vol. XVII, mf 36.127–8).
25. W. Bald, Plan No. 24 in Third Report of the Commissioners for Public Works, Ireland (B.P.P. 1835 [76], vol. XXXVI, mf 38.296–8).
26. S. Lewis, *A Topographical Dictionary of Ireland*, pp. 152–3.
27. Cox and Gould, *Civil Engineering Heritage, Ireland.*

28. W. O'Brien, *Ross Island. Mining, Metal and Society in Early Ireland* (Galway: National University of Ireland Press, 2006).
29. A. Nimmo, 'Mr Nimmo's account of the Ross Mines on the Lakes of Killarney' (NLI, Hibernian Mining Company Records, MS 657), hereafter referred to as MS 657.
30. 'Memorandum of a Conversation with Mr Nimmo'(MS 657).
31. 'A particular account of the sums granted by the Lord-Lieutenant, by way of gift, under the Act 1 Geo. 4. C. 81', in Public Works: Return to an Order of the Honourable House of Commons dated the first of March 1824 (B.P.P. 1824 [278], vol. XXI, mf 26.135).
32. A. Nimmo to Hibernian Mining Company, 11 June 1825. (NLI, Hibernian Mining Company Records, MS 658), hereafter referred to as MS 658.
33. R. Griffith to the Hibernian Mining Company, 30 September 1824. (MS 657).
34. 5 Geo IV cap. cxxxvi. *An Act to encourage the working of Mines in Ireland by means of English Capital, and to regulate a Joint Stock Company for that purpose*, 17 June 1824.
35. A. Nimmo, Evidence, p. 170, in Minutes of Evidence taken before the Select Committee of the House of Lords to examine into the Nature and Extent of the Disturbances which have prevailed in those districts of Ireland which are now subject to the Provisions of the Insurrection Act and to report to the House (B.P.P. 1825 [200], vol.VII, mf 27.60–63). Hereafter Lords Enquiry.
36. A. Nimmo, 'Report on the Road proposed for the Colliery District in the West of Lough Allen prepared for the Hibernian Mining Company by A. Nimmo'(MS 657).
37. A. Nimmo, Evidence, p.173, in Lords Enquiry.
38. Advertisement in *The Times*, 30 November 1824.
39. A. Nimmo, Evidence, p.173, in Lords Enquiry.
40. R. Delaney, *Ireland's Inland Waterways*.
41. A. Nimmo to H. Hamill, 31 July 1824 (MS 657).
42. A. Nimmo, Evidence, p.171, in Lords Enquiry.
43. House of Commons (Private Business, Wednesday March 2), in *The Times*, 4 March 1825.
44. 6 Geo IV cap. clxxxii. *An Act to alter, amend and enlarge the Powers of an Act passed in the fifth year of His Present Majesty entitled 'An Act to encourage the working of Mines in Ireland by means of English Capital and to regulate a Joint Stock Company for that purpose'*, 22 June 1825.
45. 6 Geo IV cap. clxxxi. *An Act to encourage the working of Mines in Ireland by means of English Capital and to regulate a Joint Stock Company for that purpose to be called 'The Arigna Iron and Coal Company'*, 22 June 1825.
46. H. W. V. Temperley, 'Great Britain', in A. W. Ward, G. W. Prothero and L. Stanley (eds), *The Cambridge Modern History. Vol. 12, The Restoration* (Cambridge: Cambridge University Press, 1907).
47. Mudie (ed.), *Surveyor, Engineer and Architect*.
48. P. G. Fitzgerald, quoted in A. P. W. Malcolmson, *Catalogue to the Fitzgerald Papers in the Public Records Office of Northern Ireland* (Belfast: PRONI, 2006).
49. Nimmo signed maps of the road from Cahirsiveen to Killarney dated 1818 and 1821 (NAI, OPW5HC 6/0306, 0307).
50. R. Peel to M. Fitzgerald, 5 June 1818 (Fitzgerald Papers, MIC/639/10/reel 5).
51. Lord Lansdowne to M. Fitzgerald, 14 September 1822 (Fitzgerald Papers, MIC/639/11/reel 6/41).
52. 'Steam Communication with America', *The Times*, 28 June 1824.
53. R. Day to M. Fitzgerald, 13 July 1824 (Fitzgerald Papers, MIC/639/11/reel 6).
54. A. Nimmo to H. Hamill, 31 July 1824 (MS 657).
55. A. Nimmo to M. Fitzgerald, 21 November 1824 (Fitzgerald Papers, MIC/639/11/reel 6/56).
56. R. H. Davis to M. Fitzgerald, 26 August 1824 (Fitzgerald Papers, MIC/639/11/reel 6).
57. J. Cottiman to M. Fitzgerald, 4 November 1824 (Fitzgerald Papers, MIC/639/11/reel 6).
58. St J. G. Gregory to M. Fitzgerald, 3 August 1824 (Fitzgerald Papers, MIC/639/11/reel 6).
59. A. Nimmo to M. Fitzgerald, 4 October 1824. (Fitzgerald Papers, MIC/639/reel 6/73).
60. St J. G. Gregory to M. Fitzgerald, 29 December 1824. (Fitzgerald Papers, MIC/639/11/reel 6).
61. A. Nimmo to M. Fitzgerald, 20 December 1824. (Fitzgerald Papers, MIC/639/11/reel 6/78).
62. J. W. Horton to M. Fitzgerald, 17 February 1825. (Fitzgerald Papers, MIC/639/11/reel 6).
63. American and Colonial Steam Navigation Company, *Prospectus of a Joint Stock Company for Steam Navigation from Europe to America and the West Indies*, dated 1825 (RIA, 1350/2).
64. E. Lees to M. Fitzgerald, 25 May 1825 (Fitzgerald Papers, MIC/639/11/reel 6/85).
65. 'American and Colonial Steam Navigation Bill', *The Times*, 15 June 1825.
66. 6 Geo IV cap. clxvii. *An Act to facilitate intercourse between the United Kingdom and the Continent and Islands of America and the West Indies*, 22 June 1825.
67. W. Latham to M. Fitzgerald, 25 June 1825 (Fitzgerald Papers, MIC/639/11/reel 6/87).
68. D. O'Connell to M. Fitzgerald, 30 June 1825 (Fitzgerald Papers, MIC/639/11/reel 6/88).
69. B. Harem to M. Fitzgerald, 31 July 1825 (Fitzgerald Papers, MIC/639/11/reel 6/ 92).
70. W. Edgeworth to M. Fitzgerald, 6 July 1825 (Fitzgerald Papers, MIC/639/11/reel 6/89).
71. 'American and Colonial Steam Navigation Company', *The Times*, 16 March 1826.

72. A. Nimmo to M. Fitzgerald, 31 August 1825 (Fitzgerald Papers, MIC/639/11/reel 6/95).
73. S. E. Rice to M. Fitzgerald, 19 November 1825 (Fitzgerald Papers, MIC/639/12/reel 6/3).
74. S. J. G. Gregory to M. Fitzgerald, 11 July 1825 (Fitzgerald Papers, MIC/639/11/reel 6/90).
75. O. Williams to M. Fitzgerald, 1 November 1825; S. E. Rice to M. Fitzgerald, 19 November 1825 (Fitzgerald Papers, MIC/639/12/reel 6/4).
76. J. Wells to M. Fitzgerald, 10 November 1825 (Fitzgerald Papers, MIC/639/12/reel 6/10).
77. J. Wells to M. Fitzgerald, 31 December 1825 (Fitzgerald Papers, MIC/ 639/12/reel 6/16).
78. Ted Finch, personal communication.
79. R. Bell to M. Fitzgerald, 29 August 1826 (Fitzgerald Papers, MIC/639/12/reel 6/27).
80. 7 Geo. IV cap. cxxiv. *An Act to amend an Act of the last session of Parliament for facilitating intercourse by steam navigation between the United Kingdom and the continent and islands of America and the West Indies*, 26 May 1826.
81. American and Colonial Steam Navigation Company, Prospectus of a Joint Stock Company for Steam Navigation from Europe to America and the West Indies, dated 1827 (Fitzgerald Papers, MIC/639/6 [82]).
82. A. Nimmo to T. C. Harrison, 16 February 1829, in Commissioners of Inquiry into the Collection and Management of Revenue in Ireland, Twenty-Second Report (Post Office Packet Establishments), appendix 15. B.P.P. 1830 [647], vol.XIV, mf 32.108–16.
83. A. Nimmo to M. Fitzgerald, 21 September 1828 (Fitzgerald Papers, MIC/639/11/reel 6/105).
84. A. Nimmo to M. Fitzgerald, 7 October 1825 (Fitzgerald Papers, MIC/639/11/reel 6/97).
85. A. Nimmo to M. Fitzgerald, 12 October 1829 (Fitzgerald Papers, MIC/639/11/reel 6/109).
86. A. Nimmo to M. Fitzgerald, 20 October 1828 (Fitzgerald Papers, MIC/639/13/reel 7/2).
87. W. Edgeworth to M. Fitzgerald, 6 July 1825 (Fitzgerald Papers, MIC/639/11/reel 6/89).
88. Ordnance Survey of Ireland, County Kerry, sheet 79, 1st edition, six inches to one mile. Surveyed in 1842.
89. A. Nimmo, 'The Harbour of Valentia by Alexander Nimmo civil engineer dated 1831' (London: Hydrographical Office of the Admiralty, 1832).

CHAPTER

13

The Wallasey Pool Plan and other Works in Britain

Nimmo's career in Ireland reached its peak in 1823/24 and thereafter he turned increasingly to the wider engineering scene in Britain. He was already a regular traveller to London, where he often stayed at the Salopian Coffee House, Telford's lodgings. He was firmly one of the circle of engineers that surrounded that great man and in fact many of Nimmo's English activities that we know of were carried out with Telford. In some instances their paths would cross simply in the normal course of business, when both were engaged as independent consultants, as occurred, for instance, in April 1826. It was proposed to make an artificial cut from the River Yare to the River Waveney, known as the Haddiscoe cut, which would permit the passage of boats between the inland town of Norwich and the port of Lowestoft. The intended bill was examined by a parliamentary committee and both Nimmo and Telford were called in to support the opinions of Mr Cubitt, the proposers' civil engineer. Nimmo's contribution would hardly merit notice except in so far as it revealed him in a new, more strident light. He stated that he had constructed altogether about forty or fifty harbours, and had surveyed 'a great many other harbours on the continent'.[1] He gave his opinion regarding back-waters and their usefulness for scouring and keeping harbours free of silt. Cross-examined by Mr Turner, who prefaced his questioning by noting that Nimmo had 'only' fourteen years experience, Nimmo displayed a distinct tetchiness: 'I have constructed harbours in far heavier seas than are seen upon the coast of England, and in far deeper water; the harbour at Dunmore is in 24 feet of water and upon the Atlantic sea.' Asked whether any lasting benefit had accrued from the harbours he had constructed, he replied testily, 'I think there is hardly any harbour I have constructed, or been concerned in, that I have any doubt of in the result, if they go and complete the thing.' Now that he had Nimmo riled, Turner went on to query his calculations on the scouring effect of water and the movement of sand in suspension, showing a grasp of hydraulics unexpected in a lawyer who was not a trained engineer. Nimmo was quite discomforted by the keen cross-examination. When pressed on his assertion that he had examined the port of Lowestoft, he had to admit that he had remained only one hour in that town and had done little more than walk along the pier, although in fairness to him he did have one of his assistants working on the harbour. The cross-questioning punctured a certain aura of arrogance that was apparent in his main evidence, and taught him a lesson regarding his own preparedness that he would put to good use later. A minor highlight of his appearance was the following humorously testy exchange. Asked about the cost of the sluices at Ostend harbour, he had replied 'I do not recollect it, but they were built in the reign of George II; and rebuilt since we blew them up.' Questioned again on '[The last rebuilding]: what was the cost of it?,' he retorted, 'A great deal less than blowing them up cost!' Telford's evidence was independently supportive of Cubitt and Nimmo, but need not detain us here.

The Haddiscoe Cut was a minor matter compared with another enquiry that was ongoing around that time and to which both Nimmo and Telford contributed. This was the enquiry of the Select Committee on the Milford Haven Communication in 1827. Packet boats between Dunmore and Wales were a convenient means of communication for passengers and for general trade, along with the mails. However, since no improvements had yet been made at Milford Haven, the packet boats had to lie there at their moorings and passengers and goods had to embark and disembark using open boats, a dangerous and inconvenient transfer.[2] In fact, that was only one of three major obstacles on the route from London to Ireland identified by the Select Committee: these were the Severn crossing, the river crossings at Neath and Swansea, and the state of Milford harbour. Giving evidence between February and June 1827, Nimmo was questioned on all three matters.

Mails from London arriving at Bristol might cross the Severn by ferry, or detour to Gloucester to cross by bridge, proceeding onwards through the mountains by Hereford and Brecon on the mediaeval roadway to the harbour at Milford. If they crossed further south near Chepstow, where there were two ferries, the mail coaches could travel along the South Wales lowland corridor through Cardiff, Neath and Swansea, avoiding the mountains and shortening the journey to the port. The Aust to Beachley ferry, also known as the 'Old Passage', had operated since the twelfth century. The river was one mile wide and the tides fast and dangerous at this place. The Pilning to Sudbrook crossing, called the 'New Passage' and which Nimmo regarded as 'the worst ferry in His Majesty's dominions', was two miles further downstream. He said he could improve it considerably, at little expense, by constructing proper slips on both banks, the plans for which he handed in to the Committee.[3] That on the Monmouthshire side should be paved in stone, he proposed, at a cost of £3,000; the Gloucestershire slip needed only a cut through the marl to low water, with a gravelled path alongside it, costing £2,000. He thought it best to let the high tide flow over the slips rather than raise them above the tide level so that ferry-boats could come alongside at any state of the tide. In neither case did he envisage a high pier or any masonry works like those that the Duke of Beaufort was erecting at the Old Passage, under the engineering supervision of Henry Price, Nimmo's late apprentice and colleague on the diving bell. The duke's plan seems to have been the better on this occasion: by 1830, the mail coaches from Bristol were bypassing the New Passage ferry and diverting to the Old Passage, because it had better piers and faster steamboats.

Nimmo had novel ideas for the New Passage ferry, but these were less practical at the time. One was to lay a warping chain across the bed of the river so that the boats could pick it up and warp their way across. He had seen a warping cable in operation on the river Maes in Holland, where the distance to be crossed was similar. When it was pointed out to him that the swell was much greater in the Severn, he replied weakly that it was worth a try at least. More innovatively, he suggested that a cable suspension bridge, what he called a *pont volant*, might be considered at the site, but it would be expensive.[4] Over 120 years later, the first and second Severn Bridges were erected across the Old and New Passages respectively, putting both ferries permanently out of operation, but bringing to reality Nimmo's plan for a *pont volant*.

The river crossings at Neath and Swansea were the next obstacles on the way to Milford. The Harbour Trust of Swansea asked Nimmo and Telford to examine the possibility of bridges at both sites and they consulted together on the matter. Nimmo's plan for Swansea was to cross the river with a lofty, ornamental, iron arched bridge estimated at £10,000, or a little less 'if they use less ornament'.[5] Telford's plan was to cross the river with a drawbridge near the centre of the town. There were serious objections to that, because of the navigation, but eventually they agreed on a bridge at the lower end of Swansea, provided they could 'make still water and float the harbour'. They both prepared and gave in separate plans, although Nimmo was concerned that the expense would be more than Swansea could bear. When questioned, he indicated that a floating harbour would cost not less than £30,000–£40,000. On the other hand, a tunnel under the river Tawe would be a less expensive alternative in his opinion, and would not be a difficult undertaking, although its excavation would cause a considerable temporary interruption to trade. As to the river crossing at Neath, he favoured a timber bridge, but agreed with Telford that a chain bridge at Briton Ferry was preferable.

Finally, the packet station at Milford left much to be desired and Nimmo submitted a plan and estimate of £8,566 for a new station at Hobbs' Point near Pembroke town.[6] His plan, in effect, was to move the packet station from Milford to what is now Pembroke Dock. He proposed a jetty, having 200 feet of quay, out to a rock just off the point. From its base it would be prolonged along the shore in one to two fathoms of water at low springs. Construction would require the diving bell, which he could get from Dunmore. He proposed to make part of the road along Lanyon Bay into a coal wharf, but the actual coal stores would be located at the main quay, close to the steamer berth. Associated with the packet station he planned a hotel for the comfort of intending passengers, his estimate of the total cost for everything being £12,980.

In 1828 the Government voted a sum of £7,000 towards the erection and completion of the pier.[7] Construction started on 16 December 1829 to a final plan by Colonel Fanshaw under the superintendence of Captain W. J. Savage of the Royal Engineers, and the whole was completed in 1834.[8] The front line of the wharf was 200 feet long, fifty feet on the level and 150 feet inclined, so that steamers could lie alongside and discharge easily at all states of the tide. The inclined section differs from Nimmo's indicative plan, but it is the same feature as he had recommended for the Severn ferry crossing (and, indeed, at Ballinacourty in County Galway). The whole of the wharf was built on solid rock, excavated to eighteen feet below the sea-bed, requiring the use of four diving bells. One of these was almost certainly the Dunmore diving bell, which Nimmo had offered for the task. The dimensions given by Savage match closely those of Nimmo's bell, given in chapter 6. A hotel was erected at Government expense a few years later;[9] although a smaller building, it bore some external resemblance to the market-house erected in Galway in 1822. It has since been demolished.

Long before the Milford enquiry, Nimmo and Telford had worked successfully together on the Mersey navigation at Liverpool. The port of Liverpool came under the control of the corporation of the city and the floating docks were managed by a board of trustees, acting through a docks committee, whose members included representatives of the corporation as well as elected merchants of the city. Despite this, the two bodies – the Docks Trustees and the Liverpool Corporation – were separate legal entities. Vessels using the port paid dues to the Liverpool Corporation; those entering into the floating docks paid charges to the trustees. Neither entity had any real interest in the Wirral shore of the Mersey at the time. Access by sea to Liverpool was hazardous at the best of times, with shoals and sandbanks occluding the estuary and extending out into the Irish Sea for a distance of almost five miles. The passage could not be safely sailed at night; local pilots were needed during the day to navigate the known channels in varying tide and wind conditions. From 1819 onwards Liverpool employed numerous engineers, among them the Rennies, Whidbey, Chapman and Telford, to survey the river and sound the approaches in an effort to overcome the problem of access; none of them was able to solve the problem entirely.

While Liverpool was growing rapidly, so too was Manchester, and easy communication between these two great cities was becoming crucial to their mutual progress. The Mersey and Irwell Navigation Company (M&I) provided a navigation to facilitate transport between them, using both the River Mersey itself and short canals cut off it to avoid the long bends that characterise the river's sinuous course above the tideway. The company's lowest cut formed the Lachford to Runcorn canal and in 1823 the M&I engaged Telford to advise them on improving the entrance at Runcorn. When he recommended the formation of a tidal holding basin near the entrance, the company sent his report to (Sir) John Rennie and to William Chapman for their observations. Rennie made some comments that prompted Telford to call on Nimmo for assistance. Together they produced a new joint report that explained the scheme in greater detail.[10] They pointed out that ships had to wait for long periods outside the canal entrance and it was necessary to construct a harbour for them immediately below the tide lock on the Cheshire side of the river, in a well-sheltered bight that extended up to Runcorn. They proposed to wall this off from the river, converting about four acres into a tidal basin fifteen feet deep. They considered this would improve matters by reducing the number of boats waiting in the roads. Rennie and Chapman took a different view, saying that it would deprive the river of a much needed backwater.[11] While agreeing that a holding basin was necessary, they objected to the way it projected into the river, fearing it would deflect the current. The company should indeed be facilitated, they wrote, due to the great increase in trade that was being experienced, but not at the expense of the river navigation. Telford and Nimmo's joint report had made clear that the tidal basin was to be emptied every tide and this made the scheme more acceptable to Rennie in particular. In truth, though, Rennie's objections do not seem to have been very trenchantly held or expressed.[12]

In 1827, Telford, Nimmo and Nicholas Brown were again called in on the side of the company.[13] This time, Liverpool Corporation had alleged that by abstracting water out of the Mersey at Woolaston [sic] weir and diverting it to their canal, the M&I had caused the currents to be greatly weakened and 'several navigable parts of the river were choaked [sic] up, narrowed and silted up and that divers sandbanks have been formed in the said navigable parts of the river viz. between Runcorn and Bank Quay [Warrington]'. The engineers examined the area, calculating the amount of water abstracted and the likely effect this had on the scouring and general discharge of the river. Far from inflicting damage to the public interest it was manifest, they said, that the company had made an important improvement to the navigation at no public expense and this could be easily judged by the vast extent of business that was now done on it. Nothing the company had done, they maintained, departed from the ordinary modes by which rivers elsewhere had been improved for the purpose of inland navigation. Unconvinced, the Corporation took the matter to the courts.

The engineers' report bolstered the Company's defence when the case came before the Assize in Lancaster on 12 September 1827.[14] Nimmo was present in person and on this occasion he was well prepared. Cross-examined by Mr Brougham, he acquitted himself with such confidence and mathematical erudition that his performance was applauded and long remembered by the members of his profession. He had learned his lesson well at the hands of Mr Turner the year before, and it certainly stood to him when in the hands of Brougham, one of England's foremost counsel at the time. At the end of the case, the jury took no time at all in finding the company not guilty of damaging the navigation. In truth, the Mersey did suffer from low flows, especially in summer, and the diversion of water from the river to the canal cannot have improved matters. Even Nimmo had to admit, as the Corporation had argued, that the resultant low flow meant that material in suspension was deposited on the river-bed increasing siltation, rather than the stream having its otherwise natural scouring effect. Furthermore, the engineers' observation that the company's works were an important improvement to the navigation achieved at no expense to the public was true enough, seen from the company's perspective. It was the company, after all, which reaped the benefit through its charges on vessels using the canal, and the more silted the river became, the more essential the canal became. This very success was probably what most irked Liverpool Corporation, rather than the silting-up of the river above Runcorn *per se*: Liverpool could hardly have been expected to view with equanimity the by-passing of its own docks by vessels making direct for Manchester.

The following year would see the Corporation again defend its docks stoutly, this time by making a pre-emptive strike against a scheme devised for Wallasey Pool on the Wirral shore of the river. Nimmo and Telford were deeply involved together in the scheme, the eventual outcome of which may have caused a rift between them late in their friendship, which never healed. Their professional relationship, along with a tantalising foretaste of the affair, is best described by Rickman in his introduction to Telford's autobiography:

> Mr Telford [was] always friendly to [Mr Nimmo], more than once going to Ireland on his behalf. Mr Nimmo often visited Liverpool, in his journeys between London and Dublin, and observing the various shoals and sand-banks which obstruct direct access to the estuary of the Mersey, he formed a grand scheme for a broad and deep ship-canal, extending about seven miles from the south side of that estuary to the estuary of the Dee, near Helbrè Isle, where it was found that a safe entrance to the canal might be constructed, and thus a fair way from the sea to Liverpool secured, with the advantage of such unexampled wharfage accommodation as the increasing commerce of Liverpool seemed to require, in assuming the character of the Western Commercial Capital of Great Britain. The project was magnificent, and in the year 1827 Mr Telford

> was called in to sustain, by his authority, the proposal of Mr Nimmo; and he entered upon the investigation with the utmost zeal; but suddenly that affair was terminated on the payment of a large sum of money, by the Corporation of Liverpool, to a person who had secured right of pre-emption of the only possible position of the northern entrance of the intended canal. Mr Telford, extremely disgusted in being made the instrument of a transaction which in his opinion was premeditated and collusive, seems to have destroyed all his documents relative to this grand scheme, and never spoke of the cause of failure without singular indignation; so that this short notice of the affair, as regarding Mr Telford, must suffice.[15]

That is all that Rickman had to say of the affair. However, it is not too strained to suggest that he was pointing an accusatory finger, however circumspectly, in Nimmo's direction. Both Nimmo and Telford were already dead by the time Rickman wrote this somewhat ambivalent account. On a positive note, the notice confirms that the two men were well acquainted and friendly, as we might have gathered already from events in other chapters. It also outlines succinctly the difficulties of the sea approach to Liverpool in the early nineteenth century.

In 1823 William Laird, a shipbuilder in Liverpool, moved his business across the Mersey to Wallasey Pool on the Wirral peninsula where he planned to erect two docks that would be no real threat to Liverpool's extensive floating docks. He shared his ideas with Sir John Tobin, a wealthy Liverpool merchant and ship owner. Once a privateer and master of a slave ship, elected mayor of Liverpool in 1819 in an election said to be one of the most corrupt ever held in the most corrupt city of England, Tobin was very influential, even if there was a lingering whiff of gun smoke about him still.[16] In March 1824 he bought land at Wallasey Pool jointly with Laird and in March 1826 Laird bought another parcel, half of which he immediately conveyed to Sir John. Later, Sir John was to say that they were honour-bound to each other and one would not sell without the other's agreement. The following December, Laird sold small parts of his holdings to Messrs Lawton and Brassey.[17]

Laird, Sir John and their friends saw these purchases as good speculations and were prepared to wait for them to mature. Others, too, saw the potential of the Pool. Duncan Gibb, a timber merchant, claimed in 1848 that he had proposed a dock at Wallasey Pool to Laird and that it was on the strength of that proposal that Laird and Sir John first bought land there.[18] Francis Jordan said that he had suggested a canal development there to the Mayor in 1824 or 1825, but the Mayor had treated the idea lightly. Jordan claimed he had then gone directly to Sir John with the idea and had received no encouragement, 'but he made a particular request of me that I would not mention the subject to anyone else'. This statement elicited laughter in the court, as did his parting shot: 'I met Sir John the other day [in 1833] and told him that I thought he ought to have made me a present of £5,000 for my information.' However, neither Gibb nor Jordan ever had any systematic plan or scheme in mind for the Pool; theirs were just vague ideas on its usability.

Laird moved on the idea of building a floating dock at Wallasey in 1827. In December he wrote to Robert Stevenson, the lighthouse engineer enquiring whether he (Stevenson) was free to carry out a survey 'with a view to construct a private dock and docks there for the accommodation of the individual [?] shipping that resort [sic] to this port, particularly those in the timber trade'.[19] Sir John Tobin would join in, he wrote, and the Lord of the Manor was willing to pay half the cost of any bill in Parliament that would be necessary. They were also intent on inviting Telford to join them. Stevenson was delighted to help, naturally enough, provided the proposed plan would not interfere with the general navigation in the Mersey.[20] He was already engaged professionally with Liverpool Corporation and did not want to seem to act contrary to the Corporation's interest. Laird knew that the Corporation would not be well pleased with the idea of competing docks on the Wirral side: 'We expect warm opposition from the Corporation as a body,' he wrote, 'but we will be supported by many influential merchants and ship owners in town.'[21] Sir John meantime asked Nimmo to join them and then approached Telford with the same invitation.[22] Initially Telford was hesitant, although he thought the docks plan appeared suitable, commenting somewhat archly, 'Although neither my time nor health will permit me at present to take up this business, yet when the able engineers you refer to have examined the premises, formed an outline and made a report respecting the practicability of the plan, I shall be ready to examine the documents and if necessary visit the place, and give my opinion of the propriety of the design.'[23] Stevenson was happy to work with either Telford or Nimmo, or both, and he proposed that they carry out a 'perambulatory survey' prior to giving directions for a ground plan.[24] Nimmo had already contacted him with some ideas, and had sent one of his assistants to Wallasey to scout out the land. Stevenson, eager to hear what Nimmo's assistant would report, delayed pursuing his own survey 'until I should hear from you, or of your matters'.[25] Up to this point, neither Laird, nor Stevenson, nor Tobin had ever made mention of any development other than floating docks at the Pool.

Nimmo was already well familiar with the difficulties of the sea approach and the shortage of dock space in Liverpool, and he saw in the nascent Wallasey Pool scheme a potential solution to both problems. He therefore expanded Laird's initial dock scheme into the magnificent project described by Rickman. The proposed new docks in the Pool would meet the need for extra accommodation in the Mersey; a new ship canal across the Wirral to Helbrè in the Dee estuary would avoid the hazards of the final sea approach by replacing it with an 'inland navigation'; a new port at Helbrè Island would be large enough to admit safely even the largest ships that he could envisage. It was, as Rickman stated, a truly magnificent conception, far exceeding in scale and ambition the initial dock proposal of Laird.

On 10 February 1828 Nimmo, Stevenson, Laird and Tobin met at Woodside (Birkenhead) to walk the land.[26] Over the next week they examined all the area from Wallasey north-west to Leasowe, along Mockbeggar wharf, then westwards as far as the Dee estuary, to Hoylake and Helbrè. They directed certain surveys to be made and levels to be taken, in order to confirm the suitability of the area for the enhanced scheme. Within a week, Stevenson and Nimmo had issued their first report entitled 'Preliminary Report of Robert Stevenson and Alexander Nimmo, civil engineers, on the proposed improvements at Wallasey Pool', addressing it 'To the subscribers for the proposed wet docks at Wallasey Pool' and dated 23 February 1828.[27] It described the proposal in all its full-blown glory, reading like a company prospectus (it was not called such), setting out the objectives and broad scope of the scheme. In their view the Pool afforded, beyond any doubt, 'the most favourable position in the vicinity of Liverpool for an extension of the accommodation of the shipping of the port, at a very moderate expense'. The 250 acres would create a half-tide basin where ships could lie-up while waiting entry to docks either on the Liverpool side or in the Pool itself. The ground was level, the soil watertight and easy to excavate, so that there would be no difficulty in forming docks of any desired extent on its shores. Since these would be out of the main seaway and out of the stream of the tide, they could be constructed with great economy using good building stones readily available nearby. The truly novel part of the proposal was revolutionary: the level ground made it practical to open a direct passage for ships by inland canal across the Wirral from Wallasey to a new tidal harbour they would form at Helbrè on the estuary of the Dee that was only five miles from the outer edge of the sandbanks. The intricate sea passage to Liverpool would therefore be replaced by seven miles of inland navigation, and the Mersey would be opened to the navigation of the Dee estuary.

Beyond this preliminary report, they were unwilling to go until they had met their eminent friend and colleague Mr Telford, who, they were satisfied, would agree with them on the likely, but still unmentioned, cost. Meantime, they recommended 'to the thriving and enlightened community of Liverpool to weigh well the advantages alluded to and the benefit of now extending their operations to the Cheshire shore'.[28] Liverpool Corporation and the Docks Trustees could hardly claim not to have been put on the qui vive by this first account of the scheme.

By mid-March, George Nimmo and Stevenson's assistant James Ritson were busy taking levels and soundings around Helbrè. Laird wrote to Stevenson that Telford was highly pleased with the whole scheme, and when the preliminary report had been read to the council of Liverpool it had excited a great sensation.[29] Nimmo arrived at Hoylake on 25 March; much surveying remained to be completed, but four lines of level had been marked out between Wallasey and Hoylake for the canal, the summit level of which would be only six or seven feet over high water at Laird's property.[30] Informed of the progress, Stevenson wished to walk over the debated ground once more with Nimmo, George Nimmo and Telford, whenever the latter could find the time.[31] He wrote to Telford explaining that Nimmo was already at Hoylake and that he (Stevenson) looked forward with pleasure to meeting him (Telford) when the latter would visit that place around 10 April. He concluded, 'I heard with great pleasure that you approve generally of the views of Sir John Tobin and those who co-operate with him in this measure connected with Wallasea [sic] Pool.'[32] As it happened, Telford arrived in Liverpool unexpectedly, and together with Sir Henry Parnell, Sir John Tobin, Laird and Nimmo walked over the proposed line of the canal.[33] As this was only two weeks after he had first read it, the preliminary report had clearly excited Telford sufficiently that his time and health were suddenly obstacles no longer to his participation in the grand scheme. Arrangements were made for Stevenson, Nimmo and Telford to meet in London to finalise the plan. Meantime, Ritson continued taking soundings in Helbrè Swash, the direct channel from Helbrè Isle to the floating light outside the banks. Nimmo was on the Lancaster shore of the Mersey 'on the Duchy business chiefly', having been given a warrant to survey the strand on that side.[34] He expected to be engaged on that for some considerable time.[35]

Stevenson had business in London in late April and early May. Nimmo and Telford were there at the same time (on 12 May they wrote a joint report for William Merritt in America regarding the Welland Canal[36]) and Stevenson had his much delayed meeting with Telford. The outcome came rapidly: on 16 May the three engineers issued from London their detailed account of the scheme under their joint signatures.[37] Tobin was positively delighted – so much so that he paid them an advance of £500 on account, before the month was out.[38]

The full proposal started with an historical account of the navigation in the Dee and Mersey estuaries. This bore all the hallmarks of an essay by Nimmo. It continued with an account of the Pool and the line of the proposed canal, emphasising the general lie of the land and its suitability for their purposes. Since the ground was favourable as far as the estuary of the Dee just below the hill of Grange, it proposed to extend the canal there and join it to Helbrè Isle by embankments built through the sands. Its sea locks would be protected from the prevailing westerlies by the island itself and from northerlies by the bank of East Hoyle. The channel known as the Helbrè Swash ran in the ebb stream of the Dee directly past the site, which would therefore remain deep and navigable at all tides. The proposed embankments would enclose a huge pond that would fill at spring tides to provide a water source to scour the new harbour. From this reservoir, the ship canal would run the seven miles to Wallasey. It would not be just a canal: made suitably wide, it would double as a floating dock seven miles long that could be lined with warehouses and

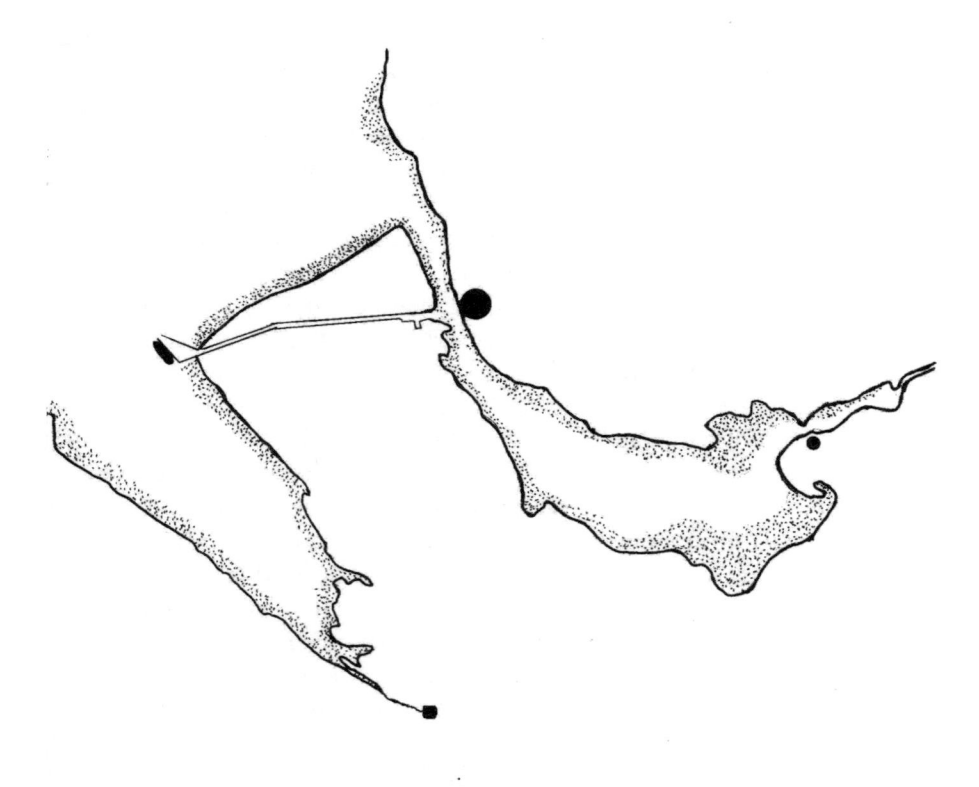

FIGURE 13.1
Approximate location of the proposed Wallasey docks, canal and new harbour at Helbrè island in the Dee estuary.

other services (figure 13.1).

At the Mersey end, the Pool would be quayed on both sides for 400 yards and have a large tidal basin with two wet docks, one for large ships and one for barges. The entrance to the canal would lie beside the ship dock, but separated from it. The south wall of the canal basin would be straight, parallel to and abutting the ship dock; the north wall would curve outwards to increase the water space so that ships could pass each other with ease. The ship dock, 400 yards long and 100 wide, would be an extension of the ship canal and would communicate with it at its inner end, with locks to the Pool at its seaward end. It was to be lined with warehouses behind which, on the side opposite the basin, would be a narrow parallel canal suitable for flats (flats were shallow canal boats). The barge dock was entirely separate and did not connect directly with the ship canal. Figure 13.2 gives an indication of the proposed layout as attributed, possibly mistakenly, to Telford.[39]

The canal was to be 124 feet wide near Wallasey, widening to 163 feet as it proceeded from there. When within half a mile of the shore near Leasowe, it would turn more to the west, entering the Dee estuary between Hoylake and West Kirby. From there it would cross the strand between the proposed new embankments, the southern one of which would run along the island and its rocks to a pier head 300 feet in length. A quay wall of hewn stone would run from the pier head back along the island for 600 yards to the tide lock. The north jetty would be just a rough stone barrier with an opening of 300 feet between it and the southern pier. Further embankment near Kirby church would form a pond 640 acres in extent and nine feet deep containing sufficient water both for scouring and for lockage purposes. The estimate for the whole scheme, which was nothing short of visionary, was £1,400,948, quite a staggering amount, but one the engineers justified by concluding:

> For the above sum a floating harbour will be obtained of seven miles in length capable of indefinite enlargement, with extensive warehouse accommodation, and with a sea port at either end on the two separate estuaries. That this is not too great for the wants of the country will be at once admitted by those who consider the vast extent of shipping usually moored in the Thames, notwithstanding all its docks; the total inapplicability of the rivers Mersey and Dee to such a purpose; and the confined space which even the docks of Liverpool can afford for the accommodation of a trade now hardly inferior to that of the metropolis, and certainly and rapidly increasing.[40]

Laird was keen to embark on the plan immediately, but Nimmo, for whatever reason, was slow to make the drawings available.[41] Perhaps it was just as well, because Tobin was encountering difficulty in enticing further subscribers to the scheme, which is hardly surprising given the breadth of its vision and the enormity of its cost. Within weeks, he and Laird were already considering how to trim costs. Laird wrote to Stevenson on 12 June: 'Sir John Tobin agrees with me that the report in its present state would ruin our project if made public on account of the expense. It will therefore be necessary that we should meet you in order to get a plan on a more limited scale at a less expense.'[42] Nimmo had already informed Stevenson of the need to cut back and Stevenson had replied on 5 June, 'I think we may lop off Wallasea [sic] Pool dock and keep to the Helbrè Harbour and canal.'[43] Laird and Tobin were now becoming impatient to have a meeting with the engineers, 'to see to reduce the estimate otherwise the scheme made public must fall to the ground', and Stevenson tried, unsuccessfully, to have the meeting in Edinburgh.[44] Nimmo wrote to him again on 18 June outlining certain alterations he had proposed to Sir John:

> I wrote to Sir John Tobin asking him to forward you a copy of what appears to be practicable in the way of reducing the works. Giving up the docks for the present and making use of the canal, enlarging the barge bason [sic] and not building warehouses would save £500,000,

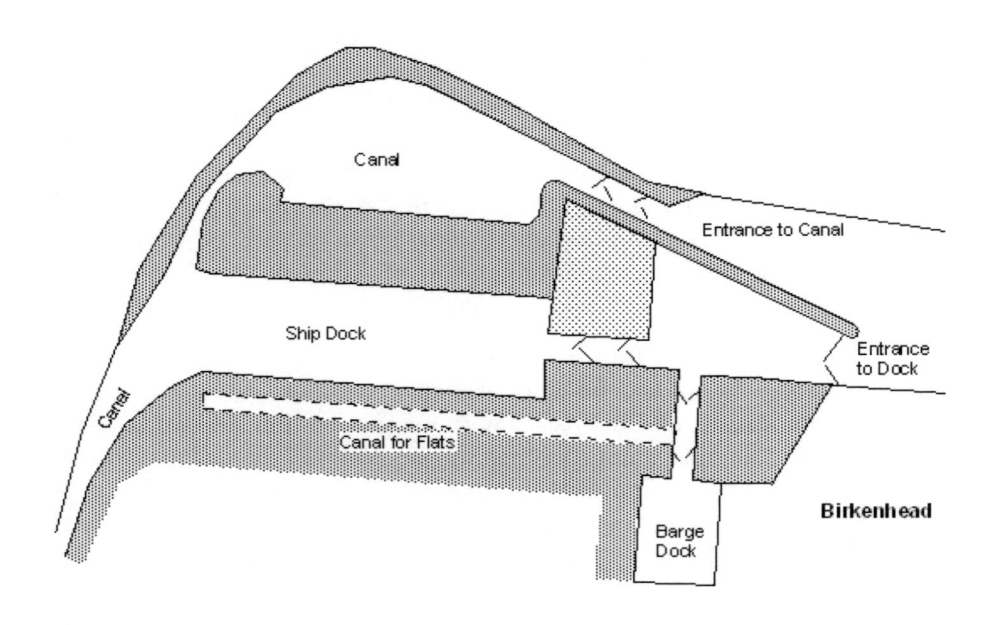

FIGURE 13.2
Plan of the proposed docks and canal at Wallasey Pool. Incoming ships from the canal (at left) could either enter the ship dock and unload into warehouses alongside, or pass directly through the widened part of the canal and lock down to Wallasey Pool and on to Liverpool. Ships exiting Liverpool could lock into the canal directly, bypassing the ship dock and go directly by the canal to the sea at Helbrè. (Based on Nimmo's MS plan in ref. 45 and the plan attributed to Telford in ref. 39.)

still keeping the canal and that at Helbrè, giving up some of the locks etc. This will bring it to Sir John's arrangements.[45]

He sketched the new plan on the back of this letter, which seems to be the only original manuscript drawing of the Wallasey scheme still extant. Figure 13.2 is based on this plan, which pre-dates that attributed to Telford. Some days later Nimmo indicated that without the docks at Wallasey he had been able to reduce the estimate to £800,000, Sir John's idea of what could be raised.[46] Nimmo favoured Chester for their meeting and must have got his way, because a revised plan and estimate were issued from that town by the three engineers on 14 July.[47]

Acknowledging that not all the works needed to proceed immediately, they proposed to make the canal from Wallasey to Helbrè wide enough for three great ships, so that it was usable as a floating harbour, still leaving room for navigation. If the trade should increase as they anticipated, a parallel canal could be built later, the bank between the two then serving as a common towpath, leaving the whole of the opposite banks for berthage and commercial establishments. They dispensed with many of the features at Wallasey Pool, including docks and quays, but recommended 'it will be highly proper to secure a sufficient quantity of land to enable all these improvements to be undertaken at some future period'.[48] They did not deem it advisable to give up the enlargement and deepening of the entrance to the Pool on which the whole scheme depended, nor did they abandon the proposed harbour at Helbrè. The revised estimate came to a more reasonable £734,163.

By now Liverpool was agog with rumour and speculation.[49] Gossip had it that there would be no public issue of shares in the venture, but that a London house would put up all the necessary finance.[50] Alarm bells began to ring in the office of the Mayor, Thomas Colley Porter and that of Alderman George Case, chairman of the finance committee. The Common Council met on 6 August and the Mayor informed them a bill for the scheme was about to be sought in Parliament. A week later he laid before them a letter from Laird dated 9 August enclosing a summary of the plan and a copy of the engineer's [sic] report for their inspection.[51] The council referred these to the engineers Giles and Walker for their opinion and also informed the docks committee. Laird was not too happy with this turn of events, but Stevenson had little difficulty with it, provided Nimmo and Telford agreed. It struck Stevenson that 'the Corporation can be neither for nor against the measure until they obtain a professional report'.[52] He had met Giles recently in Carlisle, he wrote, and the latter 'spoke in *a far off way* [Stevenson's emphasis] on the subject'.

The Mayor, however, took matters into his own hands immediately and, having first ascertained 'with the utmost precision and at the same time with the utmost caution the situation of the various interests in the land', he commenced, with Mr Case, to purchase land on the south shore of the Pool on behalf of the Corporation. When he reported this on 5 September, the council requested that he continue to acquire as much as possible of the frontage in the Pool on their behalf. At their meeting of 15 September he was happy to report that 'he, together with Mr Case, had through various private channels and agencies, been able to negociate [sic] and enter into agreements for the purchase of the whole of the frontage land on the south side of the Pool, as appeared important to them to possess; and of such a portion of the frontage land on the north side of the Pool as will enable the Corporation to command any objects which may be desirable.'[53] The council applauded their diligence and zeal; nothing was said of the high price that the land had commanded due to the interest of the Corporation in it. For example, Mr J. Mallabey, a solicitor and later clerk to the Commissioners for Birkenhead, bought land at £2,250 that had previously sold for £505.

Matters may not have been quite as straightforward as the September report of the Mayor might suggest. On 28 August, Laird had written a letter to Stevenson that was a real bombshell:

> The weather has been favourable for some days for taking soundings at the entrance to Helbrè Swash and I have been engaged in doing so accompanied by Mr Giles and Mr Walker and I am sorry to say that George Nimmo has made a most egregious mistake in the depth of water, there is not more than six feet on the bar. I have written Alexander Nimmo and Sir John Tobin has written Mr Telford and he wishes you to write him also as it puts us all in an unknown situation and at a loss how to proceed ... [undecipherable] ... or dump the whole scheme, particularly with our friends here who had promised us support. In fact it causes distrust of all the other parts of the plan. I should be glad to hear from you on this subject as soon as possible.[54]

He followed this up with another letter on 15 September, the day of the Mayor's boast of buying up the whole of the frontage on the south side of the Pool:

> Sir Richard Stanley declines giving up any of the frontage to the Company and all our subscribers have withdrawn their names in consequence of the error in the soundings. In consequence of this, I have accepted a liberal offer from the Corporation for my land and I have abandoned the scheme altogether. I shall write to you more fully next week.[55]

Laird could hardly have withdrawn more peremptorily.

Nimmo realised straight away that this scuttled the scheme. He wrote to Maurice Fitzgerald on 21 September 'The Corporation of Liverpool have taken alarm at our projects and have purchased up the frontage on the Mersey side which throws us out. We have to thank the cupidity of the chief landholders on the line who will now feel themselves left in the lurch. Your landlords always want to grasp at the goose in the egg and Vyners in this case was for such terms as disgusted our men *in toto*.'[56] He made no mention of Laird.

Porter and Case continued to purchase such tracts as they considered appropriate to safeguard the interests of Liverpool and its port. Sir John Tobin was a member of the finance committee of the council and his advice was that they should buy land *en bloc* rather than by small piecemeal purchases. As regards buying the land owned by himself and Laird, he said there would not be much difficulty provided the Corporation 'would give their solemn engagement not to let the grand project of a canal drop, but become the leaders themselves, employing Mr. Telford and the other engineers who had made the report'.[57] The committee had serious reservations about this advice, and perhaps about Sir John's role in the whole affair. Knowing already that the scheme was doomed because of Laird's agreement to sell out, Sir John said he could no longer have stood alone 'now that the beauty of the scheme was destroyed' and had reluctantly agreed to sell his own land. The sale records show that Laird sold twenty-nine acres, closing in October 1828, for £21,126; Francis Price sold forty-two acres for £17,105 and Sir John sold eighty-three acres for £60,531, closing in December.[58] What excellent advice he had given the finance committee to buy *en bloc*! Messrs Askew, Brassey and Lawton, among others, sold smaller parcels, so that in aggregate the Corporation spent £157,925 buying land at Wallasey Pool. Joseph Mallabey was in no doubt of the reason behind the purchase. Asked directly if the purchase by the Corporation was intended to prevent the execution of the scheme for a ship canal, he replied immediately: 'Oh yes, there is no doubt of that.'[59] John Taylor complained that by its actions the Corporation had deprived the traders of a rival port that would have kept the Liverpool port dues lower.

Francis Price wished to place a condition on the sale of his land, namely that the council would give a pledge that they would use it for a dock or other commercial purpose. At a special meeting of the council on 15 October it was resolved that the Pool would be appropriated to the purposes of trade connected with the interests of the port of Liverpool. The finance committee was instructed to draw up plans for the best uses to which the area could be applied, having regard to the interests of the Dock Trustees and the Corporation. It was further authorised to give notice of its intention to apply to Parliament to introduce a bill in the next session giving effect to any matters that might seem advisable, following consultation with the docks committee.[60] The finance committee drew up the notice one week later, making it clear that the proposed Wallasey docks 'are, and are intended to, communicate with the river Mersey in the said Pool in the port of Liverpool', putting an end to any lingering hope that a canal might be made through to the estuary of the Dee. In reality, however, the potential threat of such development had been forestalled by the land purchases and there was little real enthusiasm for unforced new developments on the Wirral side. The docks committee could therefore afford the luxury of the following resolution:

> it is the opinion of this committee that the present dock accommodation, consisting of forty-seven acres and 1035 yards together with the new docks now constructing [Clarence dock and Half-tide basin] will be amply sufficient to supply the wants of this port for many years to come and that, therefore, this Committee do not concur in the propriety of giving the proposed notice for an Act of Parliament for making docks in Wallasey Pool.[61]

There was no way that Liverpool Corporation could progress the necessary bill without the agreement of the Dock Trustees. As well as that, the interests of the two bodies did not necessarily coincide. For example, it was alleged at the time that conflict could arise between them over Wallasey: if the Dock Trustees were to need land for docks on the Lancashire side they would be obliged to buy or lease it from the Corporation. Should that prove too expensive, the trustees might have sought to develop at Wallasey on their own account. But the purchases by the Corporation now effectively closed off that possibility to them.[62] On the other hand, the trustees' own docks committee had not been idle onlookers in the whole affair. They had quietly bought out the lease of Helbrè Isle in July 1828 for the establishment of a semaphore communication station so that there was no way that the Corporation, or anyone else for that matter, could build a harbour or docks on that property without their consent.[63]

If the council's purchase of the land at Wallasey was the death knell of the magnificent venture, the docks committee's purchase of the lease of Helbrè was the final nail in its coffin. Years later, Stevenson's son, David, would write about the scheme: 'it is almost unnecessary to tell professional readers that after a lapse of nearly a quarter of a century the embryo but comprehensive proposal of Telford, Stevenson and Nimmo resulted in the modified but still large Birkenhead Dock scheme of J. M. Rendel.'[64] The canal and the harbour at Helbrè were never built and even the original drawings for the full project have not survived, in part for the reason given by Rickman. But the origin of the development of modern Birkenhead lies clearly in the plans of the three engineers.

The collapse of the scheme left Stevenson quite bewildered in Edinburgh. On 28 November 1828 he wrote to Telford, ostensibly to thank him for placing his (Stevenson's) son on the Birmingham Canal works, and then taking up the Wallasey matter: 'I have not heard from our friends at Liverpool I believe since the month of August or September when Mr Laird mentioned the intention of a meeting with the Corporation. In the public I have since heard of a bargain for £60,000. This is a large sum but if it had finally been arranged I think Mr Laird would have written to me.'[65] Perhaps sensing that trouble lay ahead, he took care to get himself out of the line of fire: 'In the course of these arrangements at Liverpool something will be said about accounts from the engineers. *You will take the lead in this* [italics added here] and let us know as to the propriety of a joint or separate accounts. I shall try Nimmo with another letter, on Liverpool business.' He wrote the same day and in the same vein to Nimmo, angling for information: 'indistinctly I have since heard that of [sic] a bargain for £60,000 but I do not as yet know the particulars distinctly'.[66]

Three days later, Stevenson received a letter from Laird which he promptly copied to Telford and Nimmo:

> Sir John Tobin and I are anxious to get the expenses connected with our late scheme settled as soon as possible, and I will feel much obliged if you will communicate with Mr Telford and Mr Nimmo and let us know the amount. I hope shortly to see you in Edinburgh and will then explain the cause of not writing you for some time. With best wishes etc., W. Laird.[67]

Laird's son, who was attending classes at Edinburgh University, fell dangerously ill in January 1829, so Laird went north to be with him. He met Stevenson on several occasions but nothing new passed between them. Stevenson was aware of Laird's family burden and for that reason 'that sort of intercourse that I should have had with him while here, is in great measure cut off'.[68] It would have been interesting to hear Laird's version of events.

Nimmo now settled his account and wrote to Stevenson on 6 April:

> I received yours mentioning you had sent your account to Sir John Tobin which I have also done and received payment. The total of my charge ... is about £830 and this ends our connection with the Wallasey business, a work which promised to be so important, and which will still, I think, be taken up some day. My only satisfaction is on account of our worthy friend Laird, whose speculations I fear were beginning to be rather ticklish on account of want of funds. I have been very poorly all this winter and have just escaped a fever [?] and cannot attend much to business.[69]

When Stevenson was again in Liverpool in July, Laird and Tobin paid him in full, but Telford had still not made out his account by November of the year. His destruction of all his papers relating to the affair has not helped in elucidating exactly what transpired between all the participants.

The affair may have been over by July 1829, but the fall-out from it had yet to be faced. Liverpool was left with a bill of almost £158,000 from the land purchases made in the Mayor's pre-emptive strike. In 1833 a court of enquiry into the state of the Corporation reviewed, among other matters, the Wallasey Pool purchases and concluded:

> It is not very easy to comprehend the plausibility of a plan which at the cost of a million and a half was to transfer the trade of Liverpool from the Mersey to the Dee for the mere sake of avoiding some difficulties in the navigation of the mouth of the river which had been found not of sufficient magnitude to arrest the hitherto unrivalled progressive importance of her commercial state; but giving all latitude to their faith in engineers the Common Council might have been placed on their guard by Mr Laird's abandonment of his cherished scheme, the cause of which remains unexplained to this day. And therefore it is at least not uncharitable to represent the Corporation in the character of dupes throughout this transaction. It is, however, proper to state that no charge of collusion or corrupt practice was openly made against the Corporation in respect of this, or any other of their investments ... and it would seem that the Common Council themselves seriously entertained, at that time, the project of forming under the authority of Parliament, new docks at Wallasey Pool.[70]

If the Corporation of Liverpool had indeed been duped, who exactly duped whom? Rickman seems to suggest that Nimmo duped Telford into an arrangement that was pre-meditated and collusive. This seems harsh, implying that Nimmo was a knave and Telford a fool. Telford may have been included to give weight and credibility to the project, but that was hardly an unusual or uncommon arrangement, and in any case it was Tobin, not Nimmo, who first approached him. Of the principals, Tobin felt duped by Laird's selling the land and thereby collapsing the scheme without prior agreement. But Tobin himself did very well from the affair, having made a huge profit on the sale of his land – he had bought it for about £80 an acre and sold for £729 an acre. Since he was an active member of the Corporation, its finance committee, and the docks committee throughout the whole proceedings, the propriety of his behaviour may be a matter for some censure. Laird did correspondingly well, but with a smaller holding. Nimmo seemed to suggest that Laird's precarious finances were eased by the sale, and that it was others who had showed cupidity. But none of the engineers had any personal pecuniary interest beyond their professional fees, and they are exonerated.

Certain features of the affair are still mysterious even today. Laird and his friends initiated the scheme as a floating dock. Its expansion into a canal and harbour at Helbrè on the Dee seems to have come from Nimmo. Why, then, was it that when it needed to be scaled back, it was the initial Wallasey dock part that was cut out, not the later canal proposal? Why did Nimmo and Stevenson, whose assistant Ritson was actually involved in the soundings, never refer to the supposed 'egregious mistake' of George Nimmo? Was the mistake, if there was one, all that bad, and could it not have been overcome by dredging? Why did Stevenson continue to express ignorance of the alleged cause of the collapse, despite Laird's clear, unambiguous letters to him regarding the 'mistake'? Was the court of enquiry in 1833 not told of the alleged mistake? It was never mentioned in correspondence by any of the protagonists (except the two letters to Stevenson in which Laird raised it), or by any enquiry or investigation, then or since. Nimmo never mentioned it. In fact, now is the first time that an explanation has been discovered for Laird's peremptory, premature withdrawal that collapsed the scheme.

Most peculiarly, the role played by the engineers Giles and Walker is obscure and merits greater examination. If they had surveyed Helbrè Swash with Laird in August 1828, as Laird said they had, and found George Nimmo's alleged mistake in the soundings, why did they not advise the Mayor or Corporation who had commissioned them to investigate the scheme? How could they have stood by and let the Corporation buy the land at the Pool in September and October (at greatly inflated prices) if they already knew in August that the scheme was doomed because of the 'mistake'? For that matter, the same charge can be levelled at Sir John Tobin, but his personal land holdings at the Pool and their potential sale value probably explain – but certainly fail to justify – his silence in his official roles. What was Laird doing 'taking soundings at Helbrè Swash' with the Corporation's engineers anyway? Was the scheme indeed a conspiracy of some sort? If it was, then Laird seems most obviously the chief conspirator.

A close reading of the documents suggests that there could indeed have been an attempt to manoeuvre the Corporation into buying the land at Wallasey, even though it had no interest there. The first proposal of 23 February 1828 had put Liverpool on notice that a scheme was in preparation that would develop the Wirral shore of the Mersey and join it to a new harbour proposed for the Dee. Not only that, the proposal was grounded firmly on the perceived shortage of dock space in Liverpool, something that should have alerted the port authorities straight away. When the full proposal of 16 May appeared, there could no longer have been any uncertainty as to what was afoot. The preamble to the plan was ambivalent, to say the least, regarding the sister estuaries of the Mersey and the Dee: 'In one or other of these must always continue to be the great port of the north-west of England,' it stated. In Saxon times the River Dee was an important navigation, and the port of Chester the great port of the west, while the Mersey was little known and Liverpool only a fishing village. Over time, the movement of the sandbanks in the Dee estuary had closed Chester to the sea and caused the trade to transfer to the Mersey, making Liverpool the great port that it now was. Such a reminder of Liverpool's insignificant past, and of its continued vulnerability to the shifting sands, would hardly have fallen comfortably on the ears of the Liverpool city fathers.

The substantive plan could have stood on its own engineering and mercantile merits and the engineers might prudently have left it at that, a scheme aimed primarily at improving access to Liverpool and adding much-needed accommodation to the Mersey docks. Why then, one wonders, did it contain the following paragraph:

> An important advantage obtained by this plan is, that the proposed entrance at Helbrè is within the jurisdiction of the port of Chester, of which it is recorded as a creek in Sir Matthew Hale's *De portibus maris*; and business done there or upon its waters, even as far as Wallasey Pool, being within the port of Chester, will have to pay the dues at that port; and unless ships and goods lock in to the Mersey they are exempted from the dues of Liverpool.[71]

Were they really trying to win over the Liverpool city and docks authorities to this idea? Did they not anticipate the alarm that mention of port dues being paid to the rival city Chester, rather than to Liverpool, was certain to raise? Could they not realise the threat that the proposed new docks and warehouses along the canal (within the jurisdiction of Chester) represented to Liverpool's interests? Or was there another, intended, purpose in raising these issues?

When the *Further Report* of 14 July appeared, it seemed to carry another indication of antipathy to Liverpool's interests: it severely curtailed the original docks proposed for Wallasey Pool in the Mersey, while retaining the new harbour in the Dee and the connecting canal. That had the effect of confining *all* the proposed works to the jurisdiction of Chester, and *all* of the anticipated revenue would therefore accrue to that port. To add salt to the wound, the *Further Report* was (albeit by coincidence?) issued from Chester!

No city Mayor, then or now, who had his duty towards his city at heart and in his head, could have failed to detect an implicit threat in sentiments such as these. Naturally, he would have been alarmed at the prospect of such competing developments planned close to, but outwith, his official jurisdiction. In reacting as he did, Mayor Porter was being prudent, not paranoid. It is for others to evaluate this reading of the documents, but there is here cause for legitimate suspicion that the proposal, in its intention, may not have been as benign towards Liverpool as has been previously thought. The Mayor and Corporation may not have been 'duped' so much as set up: they had to respond to the perceived threat, either by taking some expensive pre-emptive action to stop it, or by taking over the project themselves, with all the expense that would entail. Either way, the Corporation would be obliged to buy the land at a price dictated largely by the scheme's landowning proposers. To have let the project go forward as a private speculation – and it was regarded locally as a good speculation 'with great advantages for ship building and other mercantile purposes, affording the prospect of great pecuniary advantage to any person embarking in the project' – was simply too much of a risk to the city's, and its docks', interests.

The proposal documents give every appearance of having been researched and written by Nimmo – especially the critical introduction to the main proposal quoted above – so we can ask whether he was a knowing and willing party to a deception at least, if not to a fully premeditated and collusive conspiracy? The potential benefit to him of the implementation of the scheme was vastly greater than that of its abandonment. It was indeed a magnificent project, showing again his ability to conceive things on a truly grand scale. Strangely, it was not altogether dissimilar to his ship canal and floating docks proposed for Cork thirteen years earlier; either scheme would have established his reputation as one of the great docks engineers of his day and, as it turned out, aspects of both schemes (but not their canals) would come to be implemented long after him. If, then, it was not his intention to influence the Corporation improperly, how do we explain the thrust of the written proposal? Any ambiguity or ambivalence in it, if not purely accidental, must be ascribed to that political naïvety that was common enough in his writings. He wrote of things as he knew and understood them, and made his detailed calculations in an open fashion. Stevenson knew this when he said of him, 'there is no bringing him [Nimmo] to rule of thumb work'[72] and he had even written to

Nimmo himself 'there is no bringing you to rule, or to a condescendence'.[73] In the event, perhaps it was Nimmo's frankness regarding the historical and administrative position of Chester, and his poor anticipation of how this would be perceived politically in Liverpool, that principally ignited alarm at the proposal.

Apart from a pronounced disinclination to condone the behaviour of Sir John Tobin, who ran with the hare and chased with the hounds, and Laird, with his 'ticklish speculations', there is really no evidence that the motives of any of the principals, including Nimmo, were other than genuine and serious at all times. It was a good idea at the time, irrespective of who was to carry it into effect. Its implementation would have been a wonderful achievement to grace Nimmo's life in engineering, which even then was advancing unknowingly to a premature close. He remained friends with Sir John and they would work together on railway interests until Nimmo's death, as we shall see in the next chapter. He also appears to have been sympathetic to Laird after the debacle ended. If George Nimmo was traduced by the allegation of an 'egregious mistake', there was no one who defended him and he remains a forgotten victim of the affair. Ironically, fifteen years after the event, the Corporation sold the land to John Laird, son of William, and to new developers who went on to create the docks of Birkenhead. Whether or not Telford remained friendly towards Nimmo is not clear. Rickman's account seems to imply a breakdown in relations between them. When Telford was asked to comment on Nimmo's Dublin to Kingstown railway proposal in 1831 he refused, on the grounds of his own ill-health. Was this, one wonders, a diplomatic excuse? Both men died between 1832 and 1834, bringing to an end a professional relationship that had endured a quarter of a century since the days of the Inverness Survey in 1807.

NOTES

1. A. Nimmo, Evidence to the Committee on a Bill for making and maintaining a Navigable Communication for Ships between Norwich and the Sea at Lowestoft (B.P.P. 1826 (369) vol. IV, mf 28.24–6).
2. Select Committee on the Milford Haven Communication, First Report and Minutes of Evidence(B.P.P. 1826/1827 [258], vol. III, mf 29.21).
3. Nimmo's plans for the slips at New Ferry are given on pp.110–13, and illustrated in Plans 6 and 7 in Select Committee on the Milford Haven Communication, Second Report and Minutes of Evidence (B.P.P. 1826/1827 [472], vol. III, mf 29.22) (Nimmo's evidence on 14 May, 22 May and 7 June 1827), hereafter referred to as Second Report, Milford Haven.
4. Second Report, Milford Haven, p. 114.
5. Second Report, Milford Haven, p. 112.
6. The plan for the jetty and coal store at Pembroke is shown in plan 5, Second report, Milford Haven.
7. 9 Geo IV C. 95. *An Act to apply the sums of money therein mentioned for the service of the year 1828 and to appropriate the sums granted in this session of Parliament,* Section 10.
8. W. J. Savage, 'Description of the landing wharf erected at Hobbs' Point, Milford Haven for the accommodation of His Majesty's Post Office Steam-Packet Establishment at that Station, and of the diving bells and machinery used in the Erection. Written May 1836', in *Papers on subjects connected with the duties of the Corps of Royal Engineers,* vol. 1, 2nd edition. (London: John Weale, 1844) (BL., P.P. 4050.i); Fanshaw, E., Evidence, pp 122–3 in Report from the Select Committee on Post Office Communication with Ireland (B.P.P. 1831/2 [716], vol. XVII, mf 35.136–140).
9. S. Lewis, *A Topographical Dictionary of Wales.* (London: S. Lewis, 1844).

10. T. Telford, and A. Nimmo, 'Report of Thomas Telford and Alexander Nimmo, 17 June 1823' (Institute of Civil Engineers London, Library cat. 627.4 [427.1]).
11. W. Chapman to Liverpool Corporation, 18 July 1823, 'Mersey and Irwell Navigation' (ICE, library cat. 627.4 [427.1]).
12. J. Rennie to Liverpool Corporation, 28 July 1823. (ICE, Library cat. 627.4 [427.1]).
13. The Report of Thomas Telford, Nicholas Brown and Alexander Nimmo, civil engineers, on the lower part of the Mersey and Irwell Navigation, 19 May 1827 (ICE, library cat. 627.4 [427.2]).
14. 'The King against Grimshaw and others, Summer Assizes Lancaster, 10 September 1827'. *The Times*, 14 September 1827, p. 2.
15. J. Rickman, (ed.), *Life of Thomas Telford, Civil Engineer, Written by Himself* (London: Hansard and Sons, 1838).
16. M. Lynn, 'Sir John Tobin', in Matthew and Harrison (eds), *Oxford Dictionary of National Biography*.
17. G. H. Wilkinson, and T. J. Hogg, *A Report of the Proceedings of a Court of Inquiry into the Existing State of the Corporation of Liverpool*. (Liverpool: J. and J. Mawdsley, 1833). (Merseyside Maritime Museum, 710 LIV/R/OS).
18. T. Webster, *Minutes of Evidence and Proceedings of the Liverpool and Birkenhead Dock bills for the sessions 1848, 1850, 1851 and 1852* (London: John Weale, 1853).
19. W. Laird to R. Stevenson, 4 December 1827. (NLS, Stevenson Records, incoming letters 1827–1828, p.101. Acc 10706/81), hereafter referred to as Stevenson Records 10706/81.
20. R. Stevenson to W. Laird, 6 December 1827. (NLS, Stevenson Records, outgoing letters no. 9, 1827–1831. Acc 10706/14), hereafter referred to as Stevenson Records 10706/14.
21. W. Laird to R. Stevenson, 13 December 1827. Stevenson Records 10706/81.
22. Index to Report Books, 1811–1834, p.174. (NLS, Stevenson Records. Acc 10706/104).
23. T. Telford to R. Stevenson, 12 January 1828. Stevenson Records 10706/81.
24. R. Stevenson to W. Laird, 2 January 1828, Stevenson Records 10706/14.
25. R. Stevenson to A. Nimmo, 2 January 1828, Stevenson Records 10706/14.
26. D. Stevenson, *Life of Robert Stevenson* (Edinburgh: A. and C. Black, 1878), p. 132.
27. R. Stevenson, and A. Nimmo, 'Preliminary Report of Robert Stevenson and Alexander Nimmo, civil engineers, on the proposed improvements at Wallasey Pool, 23 February 1828', in Stevenson, *Life of Robert Stevenson*, p.132.
28. Stevenson, *Life of Robert Stevenson*, p.135.
29. W. Laird to R. Stevenson, 19 March 1828. (NLS, Stevenson Records, incoming letters, 1827/8, No 9, p.150. Acc 10706/8), hereafter referred to as Stevenson Records 10706/8.
30. A. Nimmo to R. Stevenson, 25 March 1828, Stevenson Records 10706/8.
31. R. Stevenson to A. Nimmo, 28 March 1828, Stevenson Records 10706/14.
32. R. Stevenson to T. Telford, 29 March 1828, Stevenson Records 10706/14.
33. W. Laird to R. Stevenson, 9 April 1828, Stevenson Records 10706/81.
34. R. Stevenson to A. Nimmo, 11 April 1828, Stevenson Records 10706/14.
35. W. Laird to R. Stevenson, 19 March 1828, Stevenson Records 10706/8.
36. T. Telford, and A. Nimmo to W. H. Merritt, 12 May 1828 (BL, Add 38756 fo. 121).
37. T. Telford, R. Stevenson and A. Nimmo, 'The Report of Thomas Telford, Robert Stevenson and Alexander Nimmo, civil engineers recommending two extensive new sea ports etc. etc. in the rivers Dee and Mersey adjacent to Liverpool with a Floating Harbour or Ship Canal to connect them, 16 May 1828' (Northumberland Record Office, WAT/2/3 pp. 11–39), later published in Stevenson, *Life of Robert Stevenson*, pp. 135–148.
38. R. Stevenson to A. Nimmo, 31 May 1828. Stevenson Records 10706/14.
39. Webster, *Minutes of Evidence*.
40. Stevenson, *Life of Robert Stevenson*, p.148.
41. R. Stevenson to A. Nimmo, 31 May 1828. Stevenson Records 10706/14.
42. W. Laird to R. Stevenson, 12 June 1828. Stevenson Records 10706/81.
43. R. Stevenson to A. Nimmo, 5 June 1828. Stevenson Records 10706/14.
44. R. Stevenson to A. Nimmo, 14 June 1828. Stevenson Records 10706/14.
45. A. Nimmo to R. Stevenson, 18 June 1828. Stevenson Records 10706/81.
46. R. Stevenson to T. Telford, 24 June 1828. Stevenson Records 10706/14, pp. 186–7.
47. T. Telford, R. Stevenson and A. Nimmo, 'Further report respecting the proposed two new ports etc., on the Rivers Dee and Mersey, adjacent to Liverpool, July 14, 1828', in Stevenson, *Life of Robert Stevenson*.
48. D. Stevenson, *Life of Robert Stevenson*, p. 149.
49. W. S. McIntyre, 'The first scheme for docks at Birkenhead and the proposed canal across the Wirral', *Transactions of the Historic Society of Lancashire and Chester*, 124 (1972), pp. 108–27.
50. Wilkinson, and Hogg, *A Report of the Proceedings of a Court of Inquiry into the Existing State of the Corporation of Liverpool*.
51. W. Laird to the Mayor of Liverpool, 9 August 1828, printed in J. A. Picton, *Municipal Archives and Records from A.D. 1700 to the passing of the Municipal Reform Act 1835* (Liverpool: Edward Howell, 1907), p. 342.
52. R. Stevenson to W. Laird, 218 August 1828. Stevenson Records 10706/14.

53. Webster, *Minutes of Evidence*.
54. W. Laird to R. Stevenson, 28 August 1828, Stevenson Records 10706/81.
55. W. Laird to R. Stevenson, 15 September 1828, Stevenson Records 10706/81.
56. A. Nimmo to M. Fitzgerald, 21 September 1828 (PRONI, Fitzgerald Papers, MIC/639/11/reel 6/105).
57. Sir J. Tobin, 'Evidence to the Court of Enquiry', in Wilkinson, G. H. & Hogg, T. J., *A Report of the Proceedings of a Court of Inquiry into the Existing State of the Corporation of Liverpool*, p. 216.
58. Wilkinson, and Hogg, *Report into the Existing State of the Corporation of Liverpool*, pp. 453–8.
59. J. Mallabey, 'Evidence to the Court of Enquiry', in Wilkinson, G. H. & Hogg, T. J., *A Report into the Existing State of the Corporation of Liverpool*, pp. 453 ff.
60. Special Meeting of the Council of Liverpool Corporation, 15 October 1828 (Liverpool Record Office, 352 MIN DOC 1 1/5), hereafter referred to as 352 MIN DOC 1 1/5.
61. Meeting of the Liverpool docks committee, 28 October 1828, pp. 115 ff (352 MIN DOC 1 1/5).
62. J. Hornby, 'Evidence to the Court of Enquiry', in Wilkinson and Hogg, *Report of the Proceedings into the Existing State of the Corporation of Liverpool*.
63. Meeting of the Liverpool Docks Committee, 1 July 1828, p. 44. (352 MIN DOC 1 1/5).
64. Stevenson, *Life of Robert Stevenson*, p. 151.
65. R. Stevenson to T. Telford, 18 November 1828, Stevenson Records 10706/14.
66. R. Stevenson, to A. Nimmo, 18 November 1828, Stevenson Records 10706/14.
67. W. Laird to R. Stevenson, 20 November 1828, Stevenson Records 10706/81.
68. R. Stevenson to T. Telford, 31 January 1829, Stevenson Records 10706/14.
69. A. Nimmo to R. Stevenson, 6 April 1829, Stevenson Records 10706/14.
70. Quoted in McIntyre, 'First scheme for docks at Birkenhead'.
71. T. Telford, R. Stevenson and A. Nimmo, 'The Report of Thomas Telford, Robert Stevenson and Alexander Nimmo, civil engineers recommending two extensive new sea ports etc. etc. in the rivers Dee and Mersey adjacent to Liverpool with a Floating Harbour or Ship Canal to connect them, 16 May 1828', in Stevenson, *Life of Robert Stevenson*, pp. 135–48.
72. R. Stevenson to T. Telford, 18 November 1828, Stevenson Records 10706/14.
73. R. Stevenson to A. Nimmo, 18 November, 1828, Stevenson Records 10706/14.

CHAPTER

14

Making the Way Permanent: Nimmo and Railways

Although he had not countenanced the construction of railways in Iveragh in 1811, Nimmo was nevertheless fully acquainted with their usefulness and did not hesitate to use them when constructing Dunmore and Courtown. By 1825 his interest was sufficiently lively for him to contribute an article 'On Railways' to the *Dublin Philosophical Journal*,[1] giving the general reader an account of their peculiar advantages over canals, and explaining the revolutionary implications of steam power for land transport. Increased velocity was the most significant advance that the railways offered, he wrote, so that they would assist in the speedy conveyance not only of bulky articles 'but much of what now passes by coaches and vans, and especially passengers, and perishable articles for the supply of the markets of great towns'. He recognised that the collieries of Newcastle, where steam engines and railways were in use the most, had the particular advantage of abundant coal, at almost no cost. But in the bogs of Ireland fuel was also cheap, so the benefit of steam-powered railways to this country could not be discounted. Although many of the railways then under consideration in Britain 'originate only in the excessive zeal for speculation which characterises the present period', the benefit of railways was, in his opinion, undeniable. Even independent of rail tracks, the consequence to mankind of steam power in land transport might, in his view, be as great as the invention of gunpowder or the use of the compass in navigation.

Later that same year, he had an opportunity to stoke up the 'zeal for speculation' in Ireland by laying down a line of railway from Limerick to Waterford. On 11 January 1825, an advertisement appeared in *The Times* regarding a new company that had been formed, called the Limerick and Waterford Railway Company (L&W).[2] Its object was to facilitate a direct, cheap and expeditious railway or tramway communication between the two cities, thereby avoiding the hazards of the sea passage around the coasts of Kerry and Cork. The company would have £300,000 capital in 6,000 shares of fifty pounds. Twenty-one subscribers were named in the act that they were seeking, eighteen to be directors, twelve of whom were gentlemen of the first eminence in the counties of Limerick,

Tipperary and Waterford. Three days later an amended advertisement appeared; in this, a branch of the railway 'through Ennis by Gort to the town and bay of Galway' was included in the plan.[3]

About two weeks later, on 27 January, another proposal for a different Irish railway company was brought to the attention of the public.[4] A group meeting in the City of London Tavern a week earlier (on 19 January) had decided to set up this second company. Subsequent events suggest that this may have been a reaction by Irish canal interests to the earlier notice about the L&W Railway Company. The new group met again on 26 January with the Earl of Glengall in the chair. They unanimously resolved to form a joint stock general company, the Hibernian General Railway Company, with capital of one million pounds divided into 20,000 shares of fifty pounds. Tellingly, the meeting went on to resolve:

> That it shall be a leading principle in the operation of this company not to interfere with any canal or inland navigation already formed in Ireland, the general object of the promoters of this measure being to give increased facilities to the commercial intercourse of the country, and develop its resources, without injury to any existing interests.

It further resolved

> That the immediate supporters of the railway from Waterford to Limerick, having concurred in the suggestion made to them of the expediency of their forming a junction with the promoters of the objects of this company, and as such proposed line is calculated to render important service to that district without prejudice to any other, it is considered expedient to give, in the first instance, the attention of the Hibernian General Railway Company to the construction of the railway from Limerick to Waterford, it being however the wish of the Company to keep in view at all times the lasting benefit of the whole kingdom in preference to any local interest.

These resolutions suggest a take-over of the original company, even before the latter got properly off the ground, by interests concerned to ensure that railways would not supersede canals in Ireland. A board consisting of a president, vice-presidents and directors was nominated for the new Hibernian Company: Glengall was to be chairman and W. H. Hyett company secretary. Prominent Irish directors would include Richard Martin, MP, Colonel Trench, MP and Richard Wellesley, MP. Four of those named in the original L&W Company – the Earl of Belfast, T. H. Burke, H. H. Hunt and J. C. Mitchell – were also named in the Hibernian. Elmes and Hollingworth were appointed engineers. A portion of the shares was to be 'reserved for Ireland' and 'an extended time granted for applications from that country', a gesture the irony of which hardly needs pointing out.

Not everyone regarded the new Hibernian company as a white knight: the day after the last mentioned meeting, a notice appeared in the *The Times* headed 'Royal Hibernian Railroad Company'.[5] This gave notice that

> A prospectus of this Company, intended to combine great national and political benefits, with the full advantage of commercial enterprise, will be published in a few days. Such prospectus will be best authenticated by the extracts it will comprise from letters received from many distinguished noblemen and gentlemen, large landed proprietors in Ireland, expressive of their opinions in favour of this plan and of their zealous desire to cooperate in it. Many eminent capitalists, merchants and bankers of this metropolis have also been consulted on the occasion, and active measures are pursuing under a committee of nomination in making and giving full effect to the measures with a view to constitute an extensive Company on the solid basis of public and individual good, by the combination of the proprietors of the soil with the great mercantile interests of both Kingdoms.

A rider followed: 'Any precipitate or practical measure, elicited by a knowledge of and in anticipation of this noble object can have no other effect than to stimulate the parties promoting the present plan to increased exertion in bringing it to completion.' Further particulars were supposedly available from Rogers and Co. (bankers), Freshfield (solicitors) and William Tooke (solicitor), all names that, with hindsight, sound suspicious. Had the company name not been incorrectly given, the notice might well be dismissed as the unbridled enthusiasm of a prospective investor. But Glengall, in his capacity as chairman of the Hibernian General Railway Company, put out an immediate disclaimer saying that the notice was 'of an insidious nature' and declaring its observations to be totally false and unfounded.[6] Had he not done this, the notice could be dismissed; it is his response to it that encourages a jaundiced reading, which suggests that the effective take-over of the original Limerick to Waterford project had been unwelcome, that the Hibernian Company was a conspiracy to frustrate local Irish enterprise and that the original supporters of the L&W line were suffering intimidation.

Only two days after the Hibernian Company's meeting of 26 January 1825, Daniel O'Connell wrote to Maurice Fitzgerald: 'We are determined to have railways as well as our neighbours and the utility of them seems to me so obvious that I have consented to become a director of the Northern line free of all objection because it does not interfere with property already sunk in the canals. Whatever be our policy of protecting canals, it is preferable at least in point of good feeling not to interfere with them.'[7] His sentiment echoed uncannily the resolution of the Hibernian meeting of 26 January, and suggests also that it was the Hibernian company that petitioned the House of Commons in February 1825 for a railway from Dublin to Belfast (O'Connell's

Northern line?), which never in fact went ahead.⁸ On 5 March the Hibernian Company placed another advertisement in *The Times*, listing its full board of governors and directors, now comprising over fifty named persons, twenty-four of them titled and ten Members of Parliament. Glengall was chairman, and the chairman of the Dublin committee was one Arthur Guinness. Richard Martin, MP was no longer included, but many bright lights of Irish affairs were newly named, for example, George Beresford, MP, William Vesey Fitzgerald, MP, Peter LaTouche, Standish O'Grady, MP, James Pim (of the Grand Canal Company) and various others. The Earl of Belfast was the only one of the original L&W subscribers who appeared on the board. The solicitors in Dublin were the law practice of P. and D. Mahony and there was no mention of company engineers in Ireland. By May the committee of management (as the company was not yet incorporated it could not properly call them a board of directors) had selected the intended line for the track from Limerick to Waterford.⁹

The Hibernian Company commissioned Nimmo to survey and lay out the line, which he did, assisted by Ben Meredith. His plan was to lay a permanent iron railway between the two cities, passing through the towns of Tipperary, Cahir, Clonmel and Carrick-on-Suir, with branches to other places of importance.¹⁰ Nowhere along the line was the elevation found to exceed 280 feet, and the greatest declivity occurred between Cahir and Clonmel, where the line descended 1 foot in 150 feet. With suitable cuttings through minor elevations, and embankment over various hollows, he concluded that it would not be necessary to use stationary engines to overcome ascents or descents. Horses would be used to pull the train of carriages, also locomotive engines, the latter allowing any degree of speed that was desired. The total length of line was eighty miles, longer than any track then in existence. Iron rails were normally made of flat cast iron bars laid on edge; for his track, Nimmo wanted bars of malleable iron, rolled thicker on top to give greater strength and to lessen wear, laid on a continuous curb of stone. The wheels of the wagons were to be three inches broad with a one-inch groove in the middle that could take the rail. Wagons with such wheels could be used on the ordinary roads whenever required, unlike wagons with flanged wheels such as were in use in England. He estimated that a horse could pull a load of ten tons on such a track, making the railway more than six times more efficient than the ordinary road. But it was the power of steam engines, especially locomotives, that most caught his attention. '[E]ngines have been invented,' he wrote, 'capable of travelling along a railway on wheels, and which draw trains of heavy loaded wagons, each containing 53 cwt, at the rate of six or seven miles an hour; nay, some late experiments have shown that this can be done with a speed of even ten or twelve miles an hour upon straight rail roads.'

The map (never published) describing the route illustrates even better than his report the ideas that occupied Nimmo's mind.¹¹ The Limerick terminus was located alongside his proposed floating dock beside the Wellesley Bridge. The bridge was very far from completion at the time, but he showed it on the map nevertheless. Clearly, he intended that the railway would link with the dock, for easy transfer of goods between the water-borne trade and the rail trade, with additional easy road access to the north across the new bridge. The railway was to leave Limerick city and pass by Caherconlish and Gurrane (Pallas Green today) to Tipperary town and from there run towards Cahir. At Barnalaha (Barnalough today?), about half way between Tipperary and Cahir, the first branch line would go northwards to Cashel, passing to the east of Golden, and terminate ultimately at Thurles. From this branch, a spur line would pass eastwards, connecting, by an inclined plane, with the coal mines at Killenaule. This spur line was meant to facilitate the delivery of turf, stone coal and culm from Killenaule for use as fuel both for trade and for the locomotives. The main line was to proceed from Cahir to Clonmel, passing to the north of that town. Leaving Clonmel, it would go to Clochnacarraigea (Cloghcarrigeen, Kilsheelin today), then through the parishes of Whitechurch and Owning, Fiddown and Kilmacow on a line north of the present N24 road, before crossing the River Suir near Granny Castle, to enter Waterford city along the south side of the river. At Clochnacarraigea, which is about four miles east of Clonmel, it would give off a second branch, to Duffhill (Dovehill today?) and Carrick-on-Suir. This meant that goods from Clonmel could either go directly to Waterford by the main rail line or to Carrick by this branch and then onwards by barge down the Suir.

Once again, Nimmo took his usual broad view in this scheme. While the Limerick terminus linked to the dock there, Waterford was its chosen eastern terminus in order to take advantage of the steam packets giving access to the English markets. By then the packets were by-passing Dunmore harbour and steaming direct to Waterford city, where the new railway would meet them. Since Limerick was linked to the interior and the western counties by means of the River Shannon, the new railway had the potential to service a very wide area. Its reach would be made even greater by Nimmo's intention to link the head of the Shannon with Sligo by inland navigation and, through Lough Melvin, with Lough Erne.[12] This would have completed an orbital communication from the north-west to the south-east (albeit a hybrid railway/waterway one) that joined the packet service to Milford Haven and onwards to London. The proposed extension of the railway from Limerick to Galway Bay through Ennis and Gort, advertised on 14 January 1825, sounds like it may have been Nimmo's idea; he was still engineer of the western district at the time.

Although not shown on the map, Nimmo stated in his report that he had plans to send another branch from Cahir along the south-east side of the Galtee Mountains – along the general route of the present N8 road – to Mitchelstown that eventually could be extended to Cork. Indeed, he suggested that 'Should the railway system be extended, these branches [to Thurles and to

Mitchelstown] may become a portion of a main line from Dublin and the Grand Canal to Cork etc. and all the south of Ireland.'[13] His wider plan, therefore, encompassed much more than just a connection between Limerick and Waterford. His vision was one of a communication network serving the whole southern and western region of the country, carrying passengers as well as goods. Using estimates of the amount of internal trade, together with information on existing imports and exports through Waterford and Limerick, he calculated that six and a half million tons of goods would travel the line annually, which, at three half pence a ton, would produce a 'moderate' revenue of over £40,000 each year.

On 31 May 1826, Parliament passed the first Irish Railway Act, the local Act, 7 Geo. IV cap. cxxxix, *An Act for making and maintaining a Railway or Tramway from the City of Limerick to the town of Carrick in the county of Tipperary with several branches therefrom in the county of Tipperary aforesaid and in the county of the City of Waterford.*[14] Secure with this and Nimmo's plan, the Hibernian Company applied to the Lord-Lieutenant for a loan of £20,000 from the Consolidated Fund. Telford submitted a positive report on the scheme on behalf of the Exchequer and Nimmo did the same on behalf of the company. John Killaly, acting for the Loan Commissioners, was asked to scrutinise the application.[15] 'Both these gentlemen [T.T. and A.N.] anticipate the most flattering results as to the profits that are to arise out of this undertaking,' he reported, 'founded on premises, and in great manner laid down by themselves and in the extent of which, I regret to say, I am not prepared to concur ... I consider part of the project highly deserving of public aid.' He preferred Clonmel as the eastern terminus of the line, because of the large amount of fixed capital already invested in the Barrow, Suir and Nore navigations. He estimated that a canal from Limerick to Clonmel would cost £35,000 as against £87,000 for a railway. In any case, he considered a loan of £20,000 would not be significant support for a railroad that had only eight to ten thousand pounds in called-up capital and that would find it difficult to raise any more because of the shortage of capital in Ireland. The company should, in his view, put up one-fifth at least of the full cost before being supported by Government. Killaly, who was engineer to the Board of Inland Navigation and had been chief engineer to the Grand Canal Company, had a deep and life-long commitment to canals that seriously biased his better judgement. It is no wonder, therefore, that he took the attitude he did, which brooked no interference with the interests of the Barrow, Suir and Nore navigations, for which he had worked extensively. Because of his myopia, compounded by Nimmo's excessively sanguine expectations of success, the commissioners refused the loan, the company withdrew and Ireland lost the possibility of having the world's first commercial passenger and goods railway.

A meeting of the defunct L&W Railway Company was called for 24 March 1829 in the London offices of the Hibernian General Railway Company in an effort to resurrect the original proposal.[16] Five months later the company advertised for contractors willing to execute the line or parts of it, but nothing further was heard of it.[17] In 1838 the Railway Commissioners favoured a new railway line from Limerick to Waterford that bypassed both Tipperary town and Cahir. A Memorial from interested parties argued that Nimmo's original line was far better, passing through those two important towns and shortening the whole distance by five miles.[18] The commission relented and adopted the proposal to reinstate Nimmo's line. It was 1845 before a new act was granted to an entirely new company – this one called the Waterford to Limerick Railway Company – to progress the line.[19] That company also owned steam vessels plying the River Shannon, so that Nimmo's vision of a communication from Waterford to Sligo was implemented in large part: it became possible to send goods by rail and inland navigation from Waterford as far as Leitrim and Roscommon via Limerick and the Shannon. The company would eventually merge with another railway, so that it acquired a complete direct rail connection from Waterford to Sligo via Limerick, finally implementing Nimmo's dream of a full orbital communication constituting a western transport corridor.

In 1827 Nimmo was again involved in Irish railways. He had been asked to prepare plans and an estimate for a harbour at Wicklow for Lord Fitzwilliam.[20] It required good stone, and the nearest suitable quarry he knew of was at Carrick mountain, which necessitated a railway to transport the quartz blocks to the harbour. The line could be extended as far as Rathdrum as a commercial venture, carrying coal, corn, timber and copper, he said, for about £8,000. Nothing permanent came of this proposal either, but it does serve to show that the introduction of rail transport to Ireland was prominent in his mind. Railways were also uppermost in the minds of a group of merchants and traders of Dublin since early 1825. They had petitioned Parliament in February of that year for a railway or tram road from the harbour at Kingstown (Dún Laoghaire) to the city, but supporters of a ship canal, a hoary old chestnut even then, won the day and the bill failed. The passing of the Limerick and Waterford Railway Act must have galled the Dublin petitioners beyond belief: it meant that both the Dublin to Belfast and the Dublin to Kingstown lines had been rejected by Parliament, while the longer, more complex Limerick to Waterford line planned by Nimmo had been approved. Things do not often happen this way in Ireland.

In October 1829 the railway industry in Britain made one of its major advances when the Rainhill trials for steam locomotives were held near Liverpool and Stevenson's *Rocket* made its historic debut. In those early days, steam engines were relatively inefficient and needed to sustain a very fierce heat in the furnace in order to generate the necessary pressure in the boiler. The second favourite at Rainhill, the locomotive *Novelty*, had retired due to boiler and bellows problems and its sponsors proceeded to alter its design. By

May of the following year (1830) they were carrying on static tests with a new 'low pressure boiler on the exhausting principle'.[21] An evacuating pump was rigged so that it sucked the flue gases through the flue pipes in the boiler, greatly aiding the heating process and consuming the smoke, thereby preventing the appearance of soot in the exhaust gases. Nimmo and Charles Vignoles were the engineers who undertook the tests, which were performed in Laird's iron works at Birkenhead (Wallasey Pool) before an audience of interested parties. The engineers were very satisfied with the outcome and, according to their report, the gases exiting the exhaustion chamber were 'so far cooled that the hand and arm might be placed with impunity down the [exhaust] tube, the temperature not exceeding 180 degrees of Fahrenheit. Not the slightest smoke was perceptible.' They estimated that the boiler was a 'forty horse boiler' and from their calculations they estimated that the expense of fuelling it with coke would 'probably not exceed one shilling per hour'. Vignoles immediately sent a copy of the report, dated 29 May 1830 and signed by Nimmo and himself, to the Institution of Civil Engineers in London to be read at its next meeting.

Nimmo's participation confirms that his interests embraced topics that we would today call mechanical engineering. Indeed earlier in 1817 he had made alterations to the steam packet boat that was to serve the mail run between Ireland and Britain, mainly involving the machinery attached to the engine, rather than the steam engine itself.[22] He also made modifications to the steam valves to prevent them sticking in place and allowing them to be opened easily whenever necessary. He reported his results to the select committee on the prevention of boiler explosions in 1817.[23] He tested the modified vessel in the Bristol Channel and actually travelled in it 120 miles down to Tenby. As we saw earlier, he had also modified the air supply of the Dunmore diving bell, and according to one report he had invented a new clinometer for use in surveying, which was for sale by Jones of Charing Cross.[24] His interests therefore were not confined to civil engineering and it is little wonder that he was a member of the scientific council of the British Invention and Discovery Company.[25]

By now he was moving in the circle of persons in Liverpool deeply involved in steam ships and locomotives, such as Sir John Tobin, William Laird and Charles Wye Williams. Tobin and some of his friends formed an enterprise that would become the Liverpool and Leeds Railway Company, initially called the Liverpool and Lancashire Northern Railway Company. It held its first meeting in Liverpool on 20 September 1830, when a provisional committee of management was appointed with Sir John in the chair.[26] They immediately appointed Nimmo engineer to the company, directing him to survey the intended line of railway with all possible expedition. From the outset they intended to build a number of railways in Lancashire and Nimmo was to survey, select and lay out the various lines. The main line would go from Liverpool to Manchester, not across Chat Moss as the Manchester and

Liverpool Railway Company did, but on a more northerly line through Ashton, Clifton and Pendlebury, and would then continue on to Leeds.

Within days Nimmo had examined almost the whole line to Manchester, which he found to be, in general, most promising.²⁷ If the railway only went as far as Ashton, he estimated that it would yield about £30,000 a year, quite a handsome return. He believed that the line would reduce the price of coal in Liverpool by three shillings a ton, which in turn would result in a saving of £10,000 a year in Dublin. Here Nimmo made an observation that might not have occurred to someone whose sole focus was on England: the projected saving in Dublin coal prices would be favoured by Irish Members of Parliament who, for that reason, could be expected to support the measure in the House, and thus neutralise 'any private interest on this side'. At the time, no railway could be built without an enabling Act of Parliament; gathering the necessary support to place a bill before the House and steer it successfully to enactment was no small task. But it was one that Nimmo was experienced in and very good at. As he had told the House of Lords some years earlier:

> I suggested to the people of Cork and those of Waterford, and those of Limerick, the local Acts by which their works are now carried on … and we are coming next year to Parliament to get a local Act to improve their [Ballina] harbour; perhaps most of what has been done of that kind has been through my suggestions.²⁸

He had also helped the Courtown Harbour Bill and the Dunmore Harbour Bill through the process, probably also the Limerick and Waterford Railway Bill. So, like any good lobbyist, he was keenly attuned to the nuances that influenced a bill's acceptability and for this alone his value to the new company was significant.

His proposed line to Manchester was to pass through the land of the Earl of Derby and the company agreed that Sir John, Nimmo and four others should attend the earl to make whatever arrangements were necessary to secure property in the line. The deputation seems to have been successful because there were no serious objections to the line subsequently. They next resolved to build a railway from Bury to Todmorden, 'possibly forming a junction with the Manchester, Bolton and Bury (MBB) canal'. Things were now moving with great expedition as railway mania took hold: on 1 October Nimmo, who was only eleven days in the company's employment, reported that he had already examined sites in Manchester, finding the extremity of Oldfield Lane exceedingly well adapted for a possible passenger entrance to their railway. A handsome portico, completely insolated, might be made nearly opposite Salford church. Two thousand pounds would buy the site, in his opinion.²⁹ His other good news was that the rise from Clifton to Pendlebury was not as great as he had originally anticipated, so it was unlikely to create an insurmountable problem on the chosen line.

In this report he revealed a major and revolutionary feature of the company's plan: he proposed to convert the existing MBB canal into a railway. That was quite an original idea, requiring no new traverse of anyone's land – provided, of course, that the canal company could be bought out (a simple enough task, requiring only sufficient money), and the rises at the locks overcome (a far greater problem). Bolton was to be an important centre in their scheme: according to Nimmo, the quantity of coal sent into Manchester annually from west of the River Ribble was 150,000 tons which, at one shilling a ton, was £7,500 in revenue. Branch lines from the collieries would ensure that the coal was delivered to Bolton for onward transport to Manchester, thus winning the support of the coal proprietors of the area. On his own initiative Nimmo met the committee of the Old Quay Company, a haulage and canal boat company, in the hope of persuading them to cooperate in building a railway bridge over the River Irwell beside their warehouses. 'A splendid entrance may be made into Manchester this way,' he wrote.[30] Here again, as with the company's proposed passenger terminus at Oldfield Lane, we see Nimmo's sense of grandeur and his extended vision in action.

Meeting on 9 October 1830, the committee agreed to purchase the shares in the MBB Canal offered by William Hutchinson at £155 per share.[31] Nimmo was present, so the meeting went on to consider his most recent report. The main line from Liverpool was turning out well, although a small part near Ashton was yet to be levelled. He was now ready, he informed them, to go to Parliament to secure a bill for the line as far as Manchester, but the line onwards into Yorkshire could not possibly be ready by then. They nevertheless instructed him to prepare a small map of the line all the way to Leeds that could accompany their prospectus.[32] He was later instructed to undertake surveys for some other branches in case it would become advisable some time in the future to make other lines.[33] In preparation for the intended bill, Nimmo and Radcliffe, the company solicitor, drew up notices of the company's intentions that appeared in the Manchester papers on 2 October followed by a draft Petition to Parliament that was approved by the committee on 20 December.[34] Both men were instructed to attend at their respective offices to 'receive possession of evidence in support of the bill': Parliament would need assurance that the rights and interests of all affected parties were properly protected or catered for by way of their written assent before considering any bill. Gathering such evidence suited solicitors, hardly engineers, and Nimmo proved difficult to pin down; he had to be requested a second time to attend at Radcliffe's for the taking of evidences.[35]

Some weeks earlier he had submitted a report and valuation of the MBB canal purchase and now he and the solicitor were instructed to examine drafts of a second bill to convert it into a railway. A new act would be required and a new company formed named the MBB Canal Railway Company. Opposition was expected from Earl Sefton whose land the projected line would cross, but, although he had his own plan for it, he did not oppose the company. He asked that Nimmo wait on him the next day (27 December) to explain the committee's plan. Nimmo did so, reporting back on 29 December that Sefton had agreed generally to the company's plans and wished to attend the company's office to settle a price. He would not haggle, Nimmo said, as long as the company made him a liberal offer.[36] He had agreed to certain roads being altered so as not to cross and recross the railway, but felt that some bridges of accommodation might be necessary and some fences 'to keep children and stragglers out'. Nimmo had been more successful with him than anyone could have hoped, proof that Nimmo did not exasperate everyone he had to deal with.

He was now directed to take levels around Halifax with a view to pressing on with the Yorkshire part of the line. Meantime, the committee engaged Charles Vignoles 'to examine the line taken by Mr Nimmo and prepare himself to give evidence as to the general correctness of it, and its comparative advantages, and if any improvements should suggest themselves to him that he be requested to communicate them to Mr Nimmo'.[37] Nothing negative came of this and the next we know is that Vignoles submitted his account for £363 in May 1831.[38] In a classic example of the kind of incestuous professional connections between engineers of the time, the company referred Vignoles's bill to Nimmo to report on its accuracy! In late March and early April 1831 Nimmo was again in London. The bill was not going well and he recommended that the company seek the opinion of the Speaker of the House. On his way back to Liverpool he stopped off at the spa in Leamington, as he was now becoming quite debilitated.[39] When the committee received the draft of the Bill along with the recommendations of the Speaker, they sent all the engineering clauses to Nimmo for his opinion on each, especially those that subjected the projected Canal Railway Company to penalties. On 11 June Nimmo reported his observations and in July and August he was back again in London on the company's behalf. The revised bill was in serious danger of failing because of problems that no engineer could have predicted and for which no allowances had been made in the draft. In the event, the bill was modified successfully and finally passed as a local act in August 1831.[40] It made no provision for the main line from Liverpool – a major, unexplained omission – and did not meet the ambitions of the committee in other respects either. Quite clearly, the whole venture needed to be re-examined and the committee ordered that all plans, maps and documents of the engineer and the solicitors be deposited in the company offices and all outstanding bills be prepared for payment.[41] A sub-committee was set up to consider how best to proceed from there on.

Sir John Tobin reported from the sub-committee on 28 September with a new, slimmed-down plan, which he placed before the meeting: the canal would remain, with a new railway alongside it; of the main line, only the part from Liverpool to Clifton would be proceeded with; and the line from Bury would be continued to Rochdale.[42] The committee refused to adopt this plan, voting three for, five against, but Tobin was not quite satisfied that his proposal was entirely lost.

A general meeting of shareholders was held on 3 October with Nimmo in attendance and 'The Report of the Engineer on the Manchester, Bolton and Bury canal railway, the main line from Bury to Rochdale and the main line from Liverpool to Clifton, with a branch to Newton from about Rainsford Moss' – in essence, Nimmo's endorsement of Tobin's defeated plan – was read. After discussion, the shareholders rejected it.[43] That seemed to put paid to Tobin's plan and the meeting then adjourned. One month later, on 2 November, the committee received an offer from Ben Meredith, Nimmo's colleague on the Limerick to Waterford railway and the Kerry roads, to execute the work on the supposedly rejected MBB railway line.[44] Nimmo must surely have alerted him that this opportunity was about to come up: only three days later the committee resolved 'that the plan for a railway of two lines along or near to the MBB canal reported by the engineer as the best between Manchester, Bolton and Bury for passengers and goods conformable to the Act, on the principle of the canal navigation being retained, be submitted to the adjourned General Meeting along with the engineer's estimates of the expense of the execution.'[45] At the adjourned general meeting on 9 November, it was agreed without controversy to seek a new bill for the Liverpool to Clifton line.[46] This, along with the earlier act that remained in effect, would allow Tobin's whole grand plan to be implemented. How Tobin engineered this apparent *volte face* by the shareholders is not known, but he was not an altogether savoury character and allegations of bribery were not entirely alien to him.[47]

If it appears that Nimmo lived a frenetic life in the fourteen months since his appointment – not just doing the surveying and engineering work, but also some of the legal work, writing reports, lobbying at Westminster, drafting and promoting bills, taking evidences and keeping contacts with landowners and the public – it certainly was the case, and it had entirely predictable consequences for the company. He had an office in Liverpool where he employed a number of surveyors and other assistants, among whom were John, later Sir John, Hawkshaw (who would become a famous engineer and President of the Institution of Civil Engineers) and Noblett St Leger, who had worked with Nimmo in the western district (later to become county surveyor of Sligo). His establishment costs were therefore high, as were the costs of the actual surveys done, not to mention his own professional fees. From the beginning he had been requested to submit his accounts on a monthly basis and to attend the committee in person or by report every Saturday. Nimmo had done neither, submitting his accounts only at irregular intervals. By the adjourned general meeting of 11 November 1831, he had already been paid £4,250. At a meeting only two weeks later the committee resolved to pay him another £500 on account of his work on the canal railway line, in addition to the balance of his account — £356-3-8 — for work on the main line from Liverpool.[48] This was the straw that broke the camel's back: the finance committee met within days and made the payments as instructed, but adding a memorandum:

> The finance committee, on a review of the large sums of money paid under the head of the surveying department, consider it highly important to call the attention of the general committee to the subject: and more especially to that portion of the charge that refers to the recent work done on the canal line. The accounts having hitherto been paid without regard to the engineer's mode of charge for his own time or that of his assistants, and without any control as to the number of assistants to be employed, induced the finance committee to resubmit, that immediate measures be adopted for better regulating and limiting the expenditure in this department in the future.[49]

In response, the general committee asked Nimmo to make appropriate adjustments to the costs of the surveying department. He made new arrangements that he forwarded to the committee meeting of 14 December, but did not attend in person. St Leger, misfortunately for him, did attend to inform them that Nimmo wanted to send Hawkshaw out to gather assents to the proposed new bill. This gave the committee an opportunity to put Nimmo and St Leger firmly in their place: it moved 'that Mr St Leger be directed not to communicate with Mr Nimmo at the expense of this company without the previous authority of this committee'.[50]

Robert Stevenson once wrote to Telford that 'engineers did everything more willingly than sit down to their money transactions'.[51] But Nimmo's neglect in this regard was beyond all excuse. After his death, his accounts were in a notoriously bad state, prompting a question at a select committee: 'Did you ever hear that he [Nimmo] was a person who kept his accounts in a very slovenly manner and has left his affairs in a very complicated state?' Although the respondent, H. Lambert, replied in the negative the question implied a situation hardly to be desired.[52] Deficient accounting had marred his tenure in the western district, and here in this instance, the accounts of his private practice were clearly just as deficient. The company's solicitors had been much better aware of the mood in the company; on 23 November 1831 they had resigned 'there being a just feeling among the subscribers against the employment of so many solicitors', but indicating their willingness to act again at some future time. Nimmo had always lacked the kind of political acumen that this action displayed.

The reprimand from the company's finance committee came only weeks before his death, so maybe he was not as attentive to worldly affairs as he otherwise might have been. Nevertheless, he had received almost £5,000 for fifteen months' work for the company, more than he had earned for his many years with the Fisheries Commissioners. During those fifteen months he had continued as engineer of the western district and with his other commissions in Ireland, however inadequately. Maybe all of his company earnings went to cover the expenses of his Liverpool office and its staff, but still he can hardly have been impecunious when approaching his final days.

In the end, his involvement with the Liverpool and Leeds Company was, perhaps, much ado about nothing. It does, however, serve to show just how enthusiastically he had entered into the new engineering based on steam and railways. To dismiss his contribution to the company may be to ignore the reality that his role was that of engineer only, not that of director or manager. It was not his fault that the company floundered, although his high expenditure cannot have helped. He was extremely well paid, and in that sense his consultancy was a success. In contrast, Vignoles was appointed engineer to the St Helens Railway and Canal Company in 1830 at the much lower salary of £650 a year, with the obligation to provide his own assistants.[53] The objective of the latter company was to make and maintain a railway to Runcorn Gap and to construct a dock there.[54] Vignoles drew up a plan for this and the directors instructed him to get Nimmo's opinion on it. Nimmo attended the meeting of the directors on 15 October, supporting Vignoles's estimate of £36,000 for the whole project. Vignoles was then instructed to proceed with the assistance of Nimmo.[55] While the work was in progress the following year, Nimmo proposed certain alterations, explaining his observations in person to a meeting of the directors on 31 August 1831.[56] They agreed to meet him at the dock site within a few days, and meantime requested Vignoles to make certain alterations to the works. After the site meeting, Nimmo was requested to prepare a report, which was read at a meeting on 10 October 1831 along with Vignoles's revised estimate of £119,000 to complete the project.[57] Nimmo did not come cheap, either in what he charged or in what he advised! He also contributed to the Preston and Wigan Railway Company, to which Vignoles had been appointed engineer.[58] In this case, his remuneration was a trivial £199-16-6, much to everyone's relief, no doubt.[59]

His railway work in England gave Nimmo a growing profile in this new business that would stand to him in Ireland. While the frenzy of railway mania was at its height, dissident members of the Grand Canal Company in Dublin, including James Pim, were resurrecting the idea of making a railway from the city to the harbour at Kingstown. Pim had been a director of the Hibernian Railway Company in 1825 when it swallowed up the first L&W Railway Company, adding to the suspicion that the Hibernian Company had really been a stratagem to protect canal interests. In 1825 and again in 1827 the Grand Canal Company had opposed the idea of a railway from Dublin to Kingstown, dismissing it in a parliamentary petition as 'an insignificant railroad' that would not solve the problem of access from the sea to Dublin city.[60]

The difficulty of the sea approach to Dublin resulted from the bar at the mouth of the Liffey, and great engineers like Jessop and Rennie had previously proposed that a canal communication between the city quays and a deep water harbour, either to the north (Howth) or south (Kingstown), was the best solution. At Pim's request in 1826, John Killaly had produced yet another plan to build a canal across the strand from Kingstown to Merrion, and thence to the Grand Canal docks near Ringsend.[61] The strand would need to be embanked and puddled to maintain the canal, leaving a 'pool of tide water between its bank and the shore for the convenience of the bathing establishments along that side of the Bay'.

By 1830 Pim had come to realise that railways would inevitably supersede canals, so he abandoned the canal idea entirely:

> I turned my attention to the construction of a railway. Knowing that Killaly's (I do not like to call them prejudices) were strongly in favour of canals as compared with railways, I employed the late Mr Nimmo to make the necessary surveys for a railway between Kingstown Harbour and Dublin: I made myself personally responsible to him for the expenses attending such survey.[62]

He also set up a committee of like-minded colleagues to further the proposal to form a new railway company, while Nimmo lost no time in laying down a suitable line for the track. Pim, Nimmo and Mahony (solicitor; late solicitor to the Hibernian Railway Company) submitted Nimmo's report and plan to the Lord-Lieutenant at the end of January 1831, with a view to promoting a bill in Parliament.[63]

To ensure that a canal proposal would not prevail over the new railway once again, Nimmo had prepared a written assessment of the relative merits of a ship canal versus a railway to accompany the submission to the Lord-Lieutenant on 3 February.[64] He did not doubt the feasibility of Killaly's canal, but he felt that there would not be sufficient lockage water to maintain the navigation, nor had any allowance been made by Killaly for the release of sewage from the city. Furthermore, it would be useful only for the carriage of goods, since the canal passage between the city and harbour would take three hours. The railway, on the other hand, was designed for passenger transport, with the carriage of goods as 'a collateral advantage', and the journey would take only twenty minutes. It may have pleased Nimmo that, at last, he was able in this way to trump Killaly's earlier (canal-based) opposition to the Limerick to Waterford Railway. Nimmo's arguments prevailed and in February 1831 Pim, Nimmo and colleagues received approval to present a petition to Parliament stating:

> That the petitioners, with several others, have agreed ... to form a company under the authority of Parliament, to be called the Dublin and Kingstown Railway Company, for the purpose of making and maintaining

> a railway or railways, with proper works and conveniences connected therewith, for the passage of wagons and other carriages from, at or near Trinity College, in the city of Dublin, to the pier at Kingstown, in the county of Dublin, as set forth in the plans and estimates therefor prepared by Alexander Nimmo, Esquire ... and praying that leave be given to bring in a Bill for the same.[65]

By naming Nimmo in the petition, the proposers were emphasising the seriousness and technical feasibility of their proposal, without which the petition might never have left Ireland with credibility (given the continuing pressure for a ship canal) and almost certainly would have met more trenchant opposition in Westminster. By this time Nimmo was well regarded, not only in Ireland but around Westminster also, for his experience with railways and railway bills. For example, he was assisting the bill for the MBB canal railway through the parliamentary committee stage at the very same time as the Dublin petition, and eventually bills for both railways would be enacted into law almost simultaneously in 1831 – the MBB as private act number lx[66] and the Kingstown railway as Act number lxix.[67]

But, full of ideas as ever, Nimmo contributed to the lingering ambition for a ship canal by submitting a new indicative proposal to the Lord-Lieutenant on 25 March 1831 – barely two months after the railway proposal. As we might expect, his plan was one of broad scope, befitting the final scheme he would ever propose in Ireland:

> By the construction of an open channel from the Liffey at Ringsend, directly across the South Bull, to the deeper part of Kingstown Harbour, turning the Liffey above the Custom House behind the docks, and through the north lots for *sewerage* only, joining the North and South Walls at Ringsend, with a lock into the Poolbeg channel, we are enabled to throw the whole of the trading part of the river, the Custom House Docks, the Grand and Royal Canal Docks, into one magnificent floating harbour not liable to be encumbered as the present river is, and of such depth as may be thought proper ...
>
> It will be sufficiently protected on the sea-side by a stout breakwater of stone, which may unite with the present South wall at the Pigeonhouse ...
>
> As the upper level of the Royal Canal will be brought forward over the New Liffey Channel, both navigations participate in the benefit of the floating harbour, and the lockage water of both, as well as the flow of the Dodder, is applicable to the tide-locks at Kingstown. The South Bull Sand will become a lake on which a town might be built, with the conveniences of universal water-carriage, like Venice or Stockholm, or the Dutch towns, and, for many purposes of manufacture, this situation would be singularly convenient ... The New Channel will be, in fact, the real port of Dublin.[68]

This was a far more ambitious vision for Dublin than has ever been mooted, before or since. Unsurprisingly, it was dismissed peremptorily, one comment being: 'A knowledge of the localities will show any professional man the utter inadequacy of this sum (the estimated cost was £500,000) for so extensive a project.'[69] Perhaps this was exactly the response that the proposal was meant to elicit, in order to bolster the railway plan? Yet, this ambitious plan for Dublin exhibited a grandeur and vision of a magnitude that only Nimmo could have conceived and that was typical of him.

From its inception, the planned Dublin to Kingstown railway envisaged the carriage of passengers primarily. The reason was that, unlike Lancashire, there could be no transport of coals from Dublin to balance the incoming trade in goods, but there was considerable passenger communication from the city to the salubrious southern suburbs and back, giving a reliable two-way operation. This was confirmed by a census carried out by the company, which showed that, in the nine months from February to October 1831, there passed '29,256 private carriages, 5,999 hackney coaches, 113,495 private jaunting cars, 149,754 public cars, 20,070 gigs, 40,485 saddle horses and 58,297 carts' along the road at Blackrock.[70] In November the company held a meeting at which the hope was expressed that 'Kingstown will become a spot to which all classes will be attracted by the opportunity for the enjoyment of healthy exercise amidst a pure atmosphere and beautiful and romantic surroundings.'[71]

Dublin was about to become the first capital city with a passenger railway, years before London would enjoy that advantage. But Nimmo was not fated to see his plan implemented. Construction only began in April 1833, a year after his death, by which time Vignoles had been appointed in his place. The latter altered Nimmo's plan, as we might have anticipated, and it is Vignoles's name, along with that of William Dargan who actually built the line, that is associated with Ireland's first railway. The virtual erasure of Nimmo's name from Ireland's railway history, both in regard to Ireland's first Railway Act of 1825, the visionary Limerick to Waterford railway, and the successful Dublin to Kingstown railway, is as unworthy as its erasure from our geological and hydrographic history.

NOTES

1. A. Nimmo, 'On Railways', *Dublin Philosophical Journal and Scientific Review*, 1 (1825), pp. 221–6.
2. Limerick and Waterford Railway Company, advertisement in *The Times*, 11 January 1825.
3. Limerick and Waterford Railway Company, advertisement in *The Times*, 14 January 1825.
4. Hibernian General Railway Company, advertisement in *The Times*, 27 January 1825.
5. Royal Hibernian General Railroad Company, advertisement in *The Times*, 28 January 1825.
6. Hibernian General Railway Company. Advertisement in *The Times*, 31 January 1825.
7. D. O'Connell to M. Fitzgerald, 28 January 1825 (PRONI, Fitzgerald Papers. MIC/639/11/reel 6/), hereafter referred to as Fitzgerald Papers.
8. 'House of Commons (Report of 11 February'). *The Times*, 14 February 1825.
9. Hibernian General Railway Company, advertisement in *The Times*, 28 May 1825.
10. A. Nimmo, 'Report of Alexander Nimmo, Civil Engineer, on the proposed Railway between Limerick and Waterford', *Dublin Philosophical Journal and Scientific Review*, 2, (1826), pp. 1–19, hereafter referred to as Nimmo's L.& W. Railway report.

11. A. Nimmo and B. Meredith, Map and Section of an intended Railway or Tram Road between the Cities of Limerick and Waterford with Branches to Thurles, Killenaule and Carrick. Designed for the Hibernian Railway Company by Alexander Nimmo, C.E. and Benjamin Meredith, surveyor, 1826 (NLS, SIGNET, s.145).
12. A. Nimmo, Evidence, p. 173, in Minutes of Evidence taken before the Select Committee of the House of Lords to examine into the Nature and Extent of the Disturbances which have prevailed in those districts of Ireland which are now subject to the Provisions of the Insurrection Act and to report to the House (B.P.P. 1825 [200], vol.VII, mf 27.60–3), hereafter referred to as Lords Enquiry.
13. Nimmo's L & W Railway report, p. 9.
14. 7 Geo IV cap. cxxxix. *An Act for making and maintaining a Railway or Tramway from the City of Limerick to the town of Carrick in the county of Tipperary with several branches therefrom in the county of Tipperary aforesaid and in the county of the City of Waterford,*. 31 May 1826.
15. J. Killaly to W. Gregory, 14 June 1827 (NAI, CSORP 1827 935).
16. Limerick and Waterford Railway Company, advertisement in *The Times*, 4 March 1829.
17. Limerick and Waterford Railway Company, advertisement in *The Times*, 10 August 1829.
18. Memorial of Land Proprietors etc., in Tipperary, December 1838. In: Resolutions and Memorials respecting Railroads in Ireland (B.P.P.1839 [154], vol. XLVI, mf 42.317–8).
19. C. E. J. Fryer, *The Waterford and Limerick Railway*. (Usk, Monmouthshire: Oakwood Press,2000).
20. A. Nimmo to R. Chaloner, 20 February 1827 (Sheffield Archive, wwm/F64/185 [wwm mf 44]).
21. C. B.Vignoles, and A. Nimmo, 'Memorandum relative to the experiments made at Mr Laird's works at North Birkenhead (opposite Liverpool) with the new Low Pressure Boiler on the exhausting principle of Messrs Braithwaite and Ericsson, 29 May, 1830' (Institution of Civil Engineers, London, O.C. 64).
22. A. Nimmo to R. Peel, 1 April 1817 (BL, Add 40264 fo. 103).
23. A. Nimmo Evidence, pp.38–9. In Report of the Select Committee on means of preventing Explosions on Steamboats, 24 June 1817 (B.P.P.1817 [422], vol. VI, mf 18.30).
24. N. St Leger to N. Colthurst, 4 February 1834 (UCL, Greenough Papers, 162).
25. He is listed as a member of the scientific council of the British Invention and Discovery Company in an advertisement in the *Edinburgh Advertiser*, 1825.
26. Minutes of Meeting of the Liverpool and Lancashire Northern Railway, 20 September 1830 (Nat. Archs. UK, RAIL 370/1), hereafter referred to as RAIL 370/1.
27. Meeting of 25 September 1830. RAIL 370/1.
28. Lords Enquiry, p.172.
29. Meeting of 4 October 1830, RAIL 370/1.
30. Meeting of 7 October 1830, RAIL 370/1.
31. Meeting of 9 October 1830, RAIL 370/1.
32. Meeting of 20 October 1830, RAIL 370/1.
33. Meeting of 2 November 1830, RAIL 370/1.
34. Meeting of 20 December 1830, RAIL 370/1.
35. Meeting of 22 December 1830, RAIL 370/1.
36. Meeting of 29 December 1830, RAIL 370/1.
37. Meeting of 5 January 1831, RAIL 370/1.
38. Meeting of 25 May 1831, RAIL 370/1.
39. Meeting of 4 April 1831, RAIL 370/1.
40. 1&2 Wm IV cap. lx. *An Act to enable the company of proprietors of the canal navigation from Manchester to Bolton and to Bury to make and maintain a railway from Manchester to Bolton and to Bury in the county Palatinate of Lancaster upon or near the line of the said canal navigation and to make and maintain a collateral branch to communicate therewith.* 1831.
41. Meeting of 17 August 1831, RAIL 370/1.
42. Meeting of 28 September 1831, RAIL 370/1.
43. General Meeting of 3 October 1831. RAIL 370/1.
44. Meeting of 2 November 1831, RAIL 370/1.
45. Meeting of 5 November 1831, RAIL 370/1.
46. Adjourned General Meeting of 9 November 1831, RAIL 370/1.
47. M. Lynn, 'Sir John Tobin', in Matthew and Harrison (eds) *Oxford Dictionary of National Biography*.
48. Meeting of 23 November 1831, RAIL 370/1.
49. Meeting of the finance committee, 26 November 1831, RAIL 370/1.
50. Meeting of 14 December 1831, RAIL 370/1.
51. R. Stevenson to T. Telford, 31 January 1829 (NLS, Stevenson Records, outgoing letters no. 9, 1827–1831. Acc 10706/14).
52. Query 4100, p. 234, in Report from the Select Committee on Post Office Communication with Ireland. (B.P.P. 1832 [716], vol. XVII, mf 35. 136–40).
53. Minutes of the first meeting of the Directors of the St Helens Railway and Canal Company, [September] 1830. (Nat. Archs. UK, RAIL 593/1), hereafter referred to as RAIL 593/1.

54. St Helens Railway and Canal Company, Joint Stock Registration Form, 1844 (Nat. Archs. UK, BT 41/906/5326).
55. Meeting of [October] 1830, RAIL 593/1.
56. Meeting of 31 August 1831, RAIL 593/1.
57. Meeting of 10 October 1831, RAIL 593/1.
58. Minutes of the first Meeting of the Directors of the Preston and Wigan Railway Company, 12 May 1831 (Nat. Archs. UK, RAIL 534/4), hereafter referred to as RAIL 534/4.
59. Meeting of 14 June 1831. RAIL 534/4.
60. Quoted in K. A. Murray, *Ireland's First Railway* (Dublin: Irish Railway Record Society, 1981), pp. 13–4.
61. Memorials and Maps relating to the Dublin to Kingstown Ship Canal (B.P.P. 1833 [603], vol. XXXV, mf 36.252–3).
62. J. Pim, Evidence, p. 117, in Select Committee on the Dublin and Kingstown Ship Canal, Report, Minutes of Evidence and Appendix (B.P.P. 1833 [591], vol. XVI, mf 36.121–3).
63. P. O'Mahony to W. Gosset, 3 February 1831 (NAI, CSORP 1831 2119).
64. A. Nimmo, Considerations on the proposed Ship Canal from Dublin to Kingstown as compared with the Railway, 2 February 1831 (NAI, CSORP 1831 2119).
65. Quoted in Murray, K. A., *Ireland's First Railway*, p. 15.
66. 1 & 2 Wm IV cap. lx. *An Act to enable the company of proprietors of the canal navigation from Manchester to Bolton and to Bury to make and maintain a railway from Manchester to Bolton and to Bury in the county Palatinate of Lancaster upon or near the line of the said canal navigation and to make and maintain a collateral branch to communicate therewith.* 1831.
67. 1 & 2 Wm IV cap. lxix. *An Act for making and maintaining a Railroad from Westland Row in the city of Dublin to the head of the western pier of the Royal Harbour of Kingstown in the county of Dublin with branches to communicate therewith.* 1831.
68. 'Dublin and Kingstown Railway. Mr Nimmo's letter to the Marquis of Anglesey dated 25 March [1831]: together with his observations on the proposed ship canal' (Royal Irish Academy, 3 B 53.56/411).
69. Negative response to A N's ship canal.
70. Proceedings of a meeting of the proprietors, 25 November 1831, quoted in Murray, *Ireland's First Railway*, p. 15.
71. Ibid.

CHAPTER 15

A Most Humane and Cultivated Man

For all his public works, Nimmo was essentially a very private person and after his death it was difficult to find anyone who knew much about him that was personal. Even his friend Robert Mudie remarked on this when he wrote, 'we rarely meet with any loophole, save his reports and his works, through which to contemplate his real character as an engineer or a man'.[1] In 1834, George Colthurst of Cork wrote to Noblett St Leger for information about him. St Leger forwarded the query to Samson Carter, who responded with a copy of Charles Wye Williams's obituary of him, while recommending that St Leger contact John Nimmo, who was still alive and living in Connemara.[2] Later, in 1838, J. P. Nichols wrote to Thomas Crofton Croker, clerk at the Admiralty, also seeking information on Nimmo. (This Croker was no relation of J. Wilson Croker, also of the Admiralty, who had clashed with Nimmo over the coastal survey.) Crofton Croker referred the query to Sir Francis Beaufort, who in turn suggested that St Leger, or William Cubitt the engineer, were the best ones to contact.[3] None of them, apparently, had known Nimmo intimately, except for Charles Wye Williams. The summary biographies that have been published until now, for example those in *The Annual Biography and Obituary* (1833) and Connolly's *Eminent Men of Fife* (1866), right up to modern times, are repetitions of, or extracts from, the one that first appeared in the *Gentleman's Magazine* in 1832.[4] The main exceptions are Ted Ruddock's extended essay in Skempton's *Lives of the Engineers*,[5] and Paxton's essay in the *Oxford Dictionary of National Biography*,[6] which provide the most accurate, although still very incomplete, accounts of him with partial listings of his works. The late Professor Sean de Courcy wrote the only modern Irish biographical notes on him which, as the title suggests, are mostly sketchy and incomplete.[7]

Although none of his engineering contemporaries wrote an obituary, and none of them commented in public on his life, his work, or the loss occasioned by his demise, some contributed anecdotes about him. Most of these were either garbled or thoroughly inaccurate versions of occurrences and events whose veracity, or lack thereof, has now been verified in the archives. Their stories did Nimmo no great service, and indicated just how little they knew of a man with whom they had previously had many professional and social interactions. Samson Carter, for example, recounted the following anecdote:

> Of what lead to his receiving such a superior education I have often heard his eldest brother John (when in one of <u>his merry moods</u> [Carter's emphasis]) relate as nearly as follows: '... his father kept a hardware shop in the town of Kirkcaldy and Alexander, when between 10 and 11, was frequently left to watch it. His principal amusement at such times was piling up the marbles in conical and pyramidical [sic] heaps, measuring them and calculating their contents, and from the diameter or circumference of a single marble, computing the quantity in each pile. One day in performing this feat he unfortunately overturned and strewed the marbles about the shop and his father, imagining it was but mischievous idleness, corrected him very severely. While young Nimmo was suffering under the punishment, the minister of the town happened to be passing and coming in and being informed of the truth, he immediately made it a request of his father (and in fact represented the sin it would be to confine a boy of such talents to the mere drudgery of a shop) that he would send him to a school and give him a university education. The father had good sense enough to profit from the advice and dispatched him immediately to Aberdeen. Upon being represented to the Professor as a self-taught child (for he was then no man) he requested he might be brought to him and upon asking him had he read some trifling elementary work on geometry, young Nimmo answered "yes, long ago" and then enumerated a list of all he had read, among which was Muller's *Fortification*. All this he had mastered completely and the Professor was so much delighted with him that he patted him on the head saying "you and I must be better acquainted" and from that time forth he was an especial favourite.'[8]

From the records it is incontrovertible that Nimmo went to St Andrews University, not to Aberdeen. Nevertheless, the anecdote carries a certain aura of plausibility, and may not be that far from the truth, if we imagine it was Playfair (who was a minister until then), or his patron Ferguson, who entered the shop and came upon the calculating Nimmo. At the very least the story is more informative than Beaufort's brief and inaccurate comment that Nimmo's circumstances in early life 'were very confined, that he was a self-taught mathematician and actually learned his profession with the tools of his hand'.[9] Some of his brief early biographies say that he attended Kirkcaldy Grammar School, but that was at a time when that school had not yet come into existence. However, he could have attended some local school before going to St Andrews. All accounts acknowledge that he was naturally gifted from a very early age, and throughout his life people remarked on his genius and erudition.

A voracious appetite for reading accompanied his native genius and the two together must have made him rather bookish in his early years. When travelling on the Continent in 1814 he sought out old maps, books and documents wherever he could, not least among the old bookshops of Amsterdam, which suggests that his curiosity had a certain antiquarian flavour beyond his immediate engineering interests.[10] He certainly exhibited a penchant for quoting arcane sources at every opportunity. Yet he seems to have had the skills of a really good teacher. Maria Edgeworth, some sixteen years his senior, tells us that she came under his pedagogic spell: 'further and further, and higher and higher, Nimmo and my brother William raised my curiosity and deepened my interest about that country'.[11] Maria also informs us that Nimmo taught a 'course of engineering' to Mary Martin, daughter of Thomas Martin of Ballinahinch and, on that tenuous evidence, some commentators elevate her to the role of first woman engineer in the country. In fact, Mary also reputedly underwent 'a course of tactics' (whatever Maria Edgeworth meant by that!) from a French army officer, and learned French from (the same?) officer, M. du Bois, who was one of Bonaparte's legion of honour. To what military rank, one wonders, would the over-zealous commentators elevate Mary on that account? One suspects that neither course can have been much more than some desultory conversations on the interests of the respective visitors to Ballinahinch. Much more realistically, we may ask what young woman could fail to be flattered with such attention from older, experienced and worldly-wise men? Nimmo at least is unlikely to have harboured romantic sentiments towards her, as she was then little more than a child.

That is not to imply that he was a misogynist or without female admirers, even if he did not encourage their attention. The mathematician William Rowan Hamilton, describing how he was persuaded to accompany Nimmo on a journey to the south of Ireland, recounts the following incident concerning him.[12] Nimmo made the invitation while they breakfasted together in the home of Dr Robinson, the astronomer and director of Armagh Observatory in 1827. Robinson expressed a strong desire for Hamilton to remain on at Armagh, but Nimmo pressed him so much with offers of anticipated scientific and touristic delights that Hamilton gave in, agreeing to the proposed travel arrangements. A young female cousin of Robinson's was present at the exchange and Hamilton describes what happened next. 'Miss Hewison, a cousin of Dr Robinson, after Mr Nimmo had left the breakfast-room, on this arrangement being thus suddenly settled, could not refrain from giving vent to expressions of astonishment. "Well," said she to me, "I wish I had but half your powers of attraction!"' Could it be she was wishful of some greater attention, romantic or otherwise, from Nimmo?

There is no evidence that Nimmo ever married. If he did not do so while a teacher (when he had a good house, a fair salary and nine years in which to consider the prospect), it is difficult to see when he would have had the opportunity later in his career, or where he could have hoped to find a suitably compliant partner who would tolerate his constant absences. As he told the House of Lords: 'in the pursuit of my profession in Ireland I have been employed in almost every part of the kingdom and I have generally made from two to three journeys *per annum* round the whole kingdom.'[13] Truly, he was a person of no fixed abode and his various office premises were his only homes at one time or another. In 1824, for example, he retired to Clifden on 20 December to draft his fisheries report, the accounts for the Clifden office that year including £18 for rent, turf and candles.[14] At the time, Clifden was not the attractive town it is today and his Christmas, although not lacking in heat and light, can hardly have been one of great comfort and joy.

For all his dedication to work, he was not a killjoy. According to St Leger: 'He was of a very cheerful disposition and would join in a merry party and sing a bacchanalian song with as much glee as any youngster of the party.'[15] Mitchell recorded that he was 'a social man' while still rector at Inverness.[16] One of the best indications of his demeanour that we have is the account by Hamilton of his journeys with Nimmo in 1827, the genesis of which is recounted above. In enticing Hamilton, Nimmo had regaled him with the opportunities he would have 'of seeing various processes in practical astronomy, making observations, determining latitudes and longitudes, and seeing the Lakes of Killarney'. True to his word, he made Hamilton's trip memorable in more ways than one. Leaving Dublin on 9 August, their first overnight stop was in Limerick, where Nimmo's Wellesley Bridge was in progress. 'The interior surface of the intended bridge interested me very much,' Hamilton wrote to his aunt, 'for it is built on a new plan, and affords a curious illustration of some mathematical principles.'[17] The bridge indeed remains probably Nimmo's most sophisticated and best-designed structure. But Nimmo had other excitements in store for his erudite young companion: together, they descended in the diving bell to the foundations of the bridge beneath the Shannon. This was one of the Dunmore bells, now brought to Limerick for the work there. Hamilton was fascinated: 'the descent, too, in the diving bell served as an experiment to illustrate various theorems in hydrostatics and pneumatics, particularly the great condensation of the air by the increased pressure of the water, which we felt in a very painful manner.' Back on dry land, they took observations of the sun's altitude, in order to establish the latitude of Limerick, before moving on to Killarney. There, Nimmo took him on a short tour of the new Killarney to Kenmare road that he was supervising. Later, they went down into the mines at Ross Island, much to Hamilton's excitement.

Nimmo was due in Hereford within some days, so they soon left Killarney, travelling to Dunmore, where they boarded the Waterford to Bristol steamer out in the estuary. Eventually, after visiting Hereford and Cheltenham, they arrived at Dudley. From the battlements of the ruined castle they observed the numerous fires coming from the chimneys of the various iron-smelting works and from abandoned heaps of burning small coal and slag scattered about. They retired for the night with Hamilton's mind filled with fiery images and disturbing thoughts. Rising early the following morning they went out and penetrated into the underground mines through a deep cave by a stream, with only a young boy for a guide. As they went deeper, Hamilton became more uneasy: his disturbing thoughts and images of the previous night were becoming real, the deeper in they went. The path was slippery and uneven, the cavern pitch black, and the unsteady flame of the boy's rude torch cast dark, flittering shadows on the walls. The air was full of the rumbling sounds of the unseen barrows on the subterranean railway far ahead, and 'the strange and solemn sound' of intermittent explosions of gunpowder, with which the miners blasted the rocks, added to the eeriness. Stalwart Nimmo forged ahead in the van; Hamilton nervously (he had to ask the boy for a small piece of the torch so as not to be entirely in the dark) brought up the rear. For him it was an awesome experience, conjuring up visions of the Styx or the Acheron, a true descent to the house of Hades.[18] For Nimmo, it was all grist to his exuberant mill and a welcome diversion from his impending appearance at Hereford, where he was about to give evidence at Assizes. They came at last upon the miners, and Hamilton's much lightened feelings were once again ones of amazement – at the size of the working chamber and the appearance of the men.

Hamilton was then 22 years old and about to take up the posts of astronomer at Dunsink Observatory and professor at Trinity College, Dublin. In his letters he appears innocent and unworldly beyond modern belief, his adventures with Nimmo being among the most exciting he had ever experienced. Within days they parted company temporarily and Hamilton went on to Liverpool, from where he wrote to Nimmo on 31 August: 'My visit to Robinson and my trip with you have contributed to call forth a taste for practical knowledge, in which before I was very deficient. I open my eyes more, and instead of being content with knowing a little of the mathematical theory of an operation I find myself asking, *could I do this myself?*'[19]

Once again Nimmo's natural gifts as a teacher had made themselves felt and his contribution to Hamilton's practical education is not without importance.

By mid-September Nimmo had rejoined him in the Lake District, as had the Reverend Caesar Otway. Nimmo and Otway were going on presently to Edinburgh and the Scottish Highlands, and Hamilton decided to tag along. But first, the group made an ascent of Helvellyn, which greatly thrilled Hamilton. The rest of the company, too, was most congenial, as he records: 'Mr Otway and Mr Nimmo have a good deal of the poetic spirit, and Mr Jones (an apprentice of Mr Nimmo) is an artist – draws, paints models etc., besides singing both pathetic and comical songs, and telling stories so humorously, that being in bodily fear for my sides I was sometimes obliged to cry out *quarter*.'[20] The merry bunch returned to Ambleside, where they waited by invitation on the poet William Wordsworth for tea. Hamilton and Wordsworth struck up an instant empathy. Nimmo and Otway also made an impression on the great poet: in a letter to Hamilton on 24 September he wrote: 'I was much pleased with him [Otway] and with your fellow-traveller Mr Nimmo, as I should have been with no doubt with the young Irishman [Jones], had not our conversation taken so serious a turn.'[21] Robert Southey, the poet, also spoke of Nimmo at that time, but his remarks were more ambivalent: 'Nimmo the engineer was with them [Hamilton and Otway], but he indeed is a Scotchman, and a young pupil of his, Jones by name, all very pleasant and original men ... Isaac Weld called on me one day – a clever man – but not to be liked like these; for there is more of this world about him, and less of the other.'[22]

Nimmo returned to Dublin via Armagh in early October and filled Dr Robinson in on the adventures of the journey. The scientists Herschel and Babbage were visiting Robinson at the time and it is likely that Nimmo met them while there. The diversity and eminence of those with whom he was on friendly terms and in whose company he was at ease – poets, artists, writers, engineers, architects, scientists, clergymen, politicians and businessmen – is remarkable, and confirms the observation of Charles Wye Williams: 'he knew how to mix the *utile* and the *dulce* in the cup of life.'[23]

Languages, both classical and modern, were among Nimmo's *fortes*, and reports indicate that he was skilled in French, Dutch and Italian. In both his Scottish and Irish reports he showed a keen interest in the meanings of Gaelic words and place names, and he would doubtless have relished the fanciful, alliterative story about his breakwater in Galway called Nimmo's Pier, in Irish *Céibh Nimmo*. The story goes that a local wag in Connemara, never having known of its designer, was convinced the pier was the place where boats from the Aran Islands landed and embarked cattle for the Galway market. For this reason, he told everyone, the pier was called *Céibh na mbó*, which in Irish means the Quay of the Cows! [24]

It is understandable that this bright young man of *petit bourgeois* background should have been attracted to the teaching profession at first. The experience helped to mature his obvious talents, and he never lost a certain schoolmasterish way of presenting his observations and writing his reports. His pursuance of the Inverness survey, however short, followed by his much longer commitment to the Bogs Commission in Ireland, suggest that he possessed a self-reliance and a capacity to embrace the outdoor life in remote locations that are unexpected in one from his town background. All his early professional travel was done on foot or on horseback. There were no roads and consequently few, if any, inns in which to stay in the course of his surveys. He spent much of his time in deeply mountainous terrain under inclement conditions, with few companions, or none, for company, so that he cannot have had much comfort or camaraderie. That did not faze him, for he was indeed very self-sufficient. Mudie, who had been introduced to him by Playfair and who remained a friend for all of Nimmo's life, described him as follows:

> Mr Nimmo was naturally well-formed; his body was as firmly knit, and strong in proportion to its size, as his mind was vigorous. There was, indeed, no possibility of fatiguing him, either in the bodily or mental sense of the word. He was capable of enduring the hardest labour for the whole day and the greater part of the night, and this for many days in succession; and we have traversed moors and hedges, climbed precipices, jumped over ravines, and waded rivers with him the livelong day; and when at last we came after midnight to the sheltered covert, the lonely hut, or the superior mansion, as any of them chanced to be the place of rest, Nimmo landed in full vigour.[25]

From his earliest years, however, 'there was something very peculiar in his manner', Mudie went on, and while yet a boy he 'was regarded as a genius, though a peculiar and eccentric one'. This aura of peculiarity had its ugly consequences in later life: 'The superior talent which Nimmo showed caused him to be disliked. There was no vanity in his composition, but still there was a range and correctness in his knowledge of things, and indeed of persons too, which raised him so high, as to excite the envy and ensure the hatred of by far the greater number of those by whom he was surrounded.' In spite of this, he endeared himself to Telford, with whom he shared a congeniality of feeling. When in London, Nimmo – said to be as delightful in private as he was splendid in public – was a welcome dinner guest in Telford's house in Abingdon Street, sometimes with disastrously eccentric consequences. On one occasion, he 'button-holed' Dr Olynthus Gregory in the drawing-room and refused to let him into dinner, so deeply engaged was he in their (largely one-sided) disputation. Dinner grew cold and the other guests concerned, because Telford was scheduled to leave early. In the end, Telford had to leave the table and prize Gregory away so that both Gregory and Nimmo could be shepherded at last to table, where Telford seated Nimmo on his right. When Telford eventually went off to his engagement he called Nimmo to act as his deputy, and 'a glorious chairman he made'.[26]

Nimmo's peculiarity or eccentricity shines through many of his official writings, and it is easy to see why he so often raised passions among those from whom he differed. Nowhere was this more obvious than in the western district where opposition to him resulted, as we have seen, in complaints from a number of sources, yet where he did more for the people, well-off and otherwise, than anywhere else. Colonel Bingham, Manus Blake, J. Wilson Croker, Denis Browne and others were at various times exasperated by him, but by and large he was welcome wherever he went.

During his travels around Ireland he had unrivalled opportunity to observe the character and the living conditions of the people, his factual, objective description of their condition being stark: 'I conceive the peasantry of Ireland to be in general in almost the lowest possible state of existence. Their cabins are in the most miserable condition and their food is potatoes with water, frequently even without salt.'[27] This, however, did not hinder his social interaction with ordinary people. 'I have sometimes slept in their cabins and had frequent intercourse with them in the south and west of Ireland,' he claimed.[28] When it was put to him by the House of Lords that the Irish were indolent and idle of habit and would rely excessively on the poor law, he was spirited in their defence:

> Really I am not inclined to think the Irish are an indolent people. I think that, as far as spirit is concerned, I would look with more confidence to the spirit of the Irish people in maintaining their independence, than perhaps I should look to the population of either England or Scotland ... I have a far higher opinion of the spirit of independence of the people than perhaps many persons who are immediately concerned with the country.[29]

It took others almost another hundred years to come to this same understanding.

His comments on Irish poverty are entirely free of the pious moralising and the self-servingly patronising tone of many contemporary observers. 'It is unquestionable that the great cause of the present miserable condition of the people of Ireland, and of their disturbances, is the management of the land ... There is no other means of employment for an Irish peasant nor no certainty that he has an existence for another year, nor even for another day, but by getting possession of a portion of land on which he can plant potatoes ... Anything that will give the peasant property, independent of the produce of the land alone, would be useful to him,' he told the House of Lords, before going on to outline his solution for the immediate problem: 'It is my opinion that nothing will ameliorate the condition of the people of Ireland, but some measure that will extend employment generally over the whole kingdom by making it in the interest of local associations, small communities such as parishes or towns or landed proprietors to have spare labour made productive.'[30] This could be done, he said, by introducing that form of the English Poor Law which made the support of the poor the responsibility of the parish. What he wanted was paid work, not poorhouses, which had the disadvantage 'that they do not provide for the early education of the youth in habits of industry; they are rather modes of punishment than modes of education'.[31] His was hardly a diplomatic stance to espouse within the House of Lords, despite its sagacity. Asked had he ever heard of a pauper in England being grateful for any relief afforded to him, Nimmo compounded his unrestrained candour:

> I have no reason to believe that a pauper, on account of the relief that has been given him in England, ever showed any gratitude, nor did I ever

expect him to show it ... What I contend for is the general principle, that the attention of the public, and especially of the proprietors of land, should be turned by some statute to the necessity of funding employment of the poor, and preserving them from being, as at present, an unprofitable burthen to the community.[32]

This philosophy was not at all what their lordships wished to hear, but it was Nimmo at his straight-talking best. A final barbed question was directed to him, as much a rebuke as a rhetorical question: 'Have you never heard that the inhabitants of the districts where the poor laws have been introduced deeply lament their introduction into those districts?' His reply was equally barbed: 'The people who pay lament, I dare say, not those who receive.'[33]

On a daily, practical level, his personal sentiments and behaviour towards his workers was one of humanity and compassion. When, in 1824, he was obliged to suspend work on Liscannor pier for want of funds an employee, now without work and with nowhere to go, wrote to the Fishery Commissioners, who consulted Nimmo. 'Though I have doubt whether this man be legally entitled to pay for the time he remained at that place since the work was suspended and have constantly warned him not to expect it,' Nimmo replied, 'yet I think it would be reasonable and fair of the Board to give him some remuneration for the care of the tools and work – say eight to ten shillings per week. I understand he is a person of good character, with a large family in great distress.'[34] This was not the only instance of his care for the workers, as he went on 'there are some other persons in a similar situation in Aran, unable to leave the place on account of the debts they owe. In other parts of Connaught when we had strangers I was able, by giving them other employment, to prevent the accumulating claims of this kind.' In 1830, when Government funds were slow in coming, Nimmo 'commenced employment giving at my own risk notes for potatoes and meal but am of course quite unable to do so to any great extent'.[35] Piteous appeals to him for help did not go unanswered. 'I have frequently had occasion,' he told the House of Lords, 'to meet persons that begged me on their knees for the love of God to give them some promise of employment that from the credit of that they might get the means of supporting themselves for a few months until I could employ them.'[36] He knew that they would make good on their promise to work when it arose: 'Nothing gives the Irish peasant more happiness than to have some prospect of employment and we find it uniformly causes the giving us no reason to complain of them in any way ... the value they set on their employment makes them industrious, sober and honest.'[37] Indeed, Maurice Fitzgerald went even further about them: 'When able to obtain labour by contract, or by task as it is called, the peasantry are frequently known to overwork themselves, in a manner injurious to their health.'[38]

Because of circumstances in the west of Ireland, Nimmo preferred to pay for work by the measure, 'so that the man, his wife, and children may all have something to do together in breaking stones, and covering roads with clay; but in general there are two persons in a family of six persons, a man and one of his sons, or if the man be old, two of his sons'. In August 1829 he paid for medical attendance for labourers Darby O'Brien and James Walsh, both injured by an explosion. Later, Hugh O'Dwyer received twelve weeks' allowance of four pounds and one shilling for injuries that he sustained at work, and John Hughes received one pound, one shilling and eight pence for five days' injury allowance.[40] These, surely, are early, voluntary examples of workmen's compensation that might not have been expected at that time.

At heart, therefore, Nimmo was a compassionate, humane man but also a practical one, familiar with the ideas and debates in political economy not least those of his late fellow-townsman Adam Smith and many others of the day. His championing of the Poor Laws did not always endear him to the landed class, on whom he relied for commissions. For instance, when Denis Browne, MP, a wealthy landowner in Mayo, was asked whether Nimmo had not become unpopular in Ireland 'for his known attachment to the system of poor rate', he replied: 'He spoke to me about a system of poor's [sic] rates, and that did not make him popular with me.'[40] For a disinterested Government engineer to become unpopular because of his espousal of the cause of the poor in Ireland, speaks well of his humane sentiments. Yet for all that, he was at home with the gentry, at ease in their great houses, and not without some vanity in their company. Maria Edgeworth describes him riding welcomely up the drive of her home in Edgeworthstown 'in a fine German barouche', quite the gentleman!deroga[41] He was made just as welcome by the Martins in Ballinahinch, where even his large retinue of assistants was received hospitably. One of these, Samuel Jones, was a friend of William Edgeworth, from whom he had learned the art of surveying before joining Nimmo's staff. Jones eventually married into the D'Arcy family of Clifden Castle and became master of that household in due course. He became a contractor on the Clifden to Oughterard road and ultimately county surveyor of Tipperary.

Maurice Fitzgerald was a friend with whom Nimmo felt he could be open and frank. In 1824 Nimmo wrote a testimonial in Fitzgerald's defence in which he attributed the measures then in operation for the relief of the poor by public works to Fitzgerald's active participation in the 1819 Select Committee on the State of the Poor.[42] That committee marked a critical turning point in the response of Government to the plight of the Irish poor. In 1828, after the election that saw Daniel O'Connell elected, Nimmo wrote to Fitzgerald asking, 'What do you think of your neighbour and friend O'Connell? Are we to have a regular "jacquerie" in Ireland?'[43] (A jacquerie was a continental faction of peasants in revolt who attacked the castles of their rich overlords in the fourteenth century.) Nimmo was sure that O'Connell would sit in Parliament, but was unsure of the likely response of the Catholic

Association to its success. 'On the whole, I look upon it as the nearest approach to a revolution of anything since I have known Ireland and the most important in its probable results, totally distinct from the religious question,' he wrote. A degree of intimacy with Fitzgerald is evident in these overtly political comments, and especially in his calling the members of the House of Commons, of whom Fitzgerald was one, 'the workmen, or rather talking men, of your shop'.[44]

Henry Petty-Fitzmaurice, Lord Lansdowne, a neighbour of Fitzgerald, was also impressed with Nimmo, although not so much when Nimmo misread plans for a fair in Kenmare that delayed Lansdowne's proposed development of the town.[45] However, this did not stop him from getting Nimmo to design a pier near Kenmare later on. Thomas Spring Rice, MP for Limerick, later Lord Monteagle, was a lifelong admirer of him. These political men, liberal Whigs all, became very influential in the Government after 1830. Had Nimmo lived longer he might reasonably have anticipated some personal advancement through their friendship and influence, but only after the difficulties with his public accounts were cleared up, as they were in 1833.

Nimmo's attitude to money was one that was problematic from his earliest years in the Inverness Academy, and at his death his accounts were in a notoriously confused state. Maria Edgeworth said he died in debt, an unlikely claim given the amount of money he earned in the last decade of his life. His salary from the western district came to £8,149 and from the Loan Commissioners he received probably an absolute minimum of £1,000. His salary from the Fisheries Commissioners was £4,576.[46] He was engaged by them from 14 September 1822 until 1 July 1827 at the latest (this was the date of his last payment). Assuming he charged his usual three guineas a day for six days a week, every week of that time – 1,500 weekdays – his salary would have amounted to that exact sum. All these payments from the public purse, totalling almost £14,000 over the decade 1822 to 1832, excluded office, travel, staff and other expenses, for which he was reimbursed separately. Nor has his remuneration for earlier public works for the Bogs Commission (£1720), the Post Office (about £700)[47] or work at Dunmore harbour (minimum £720)[48] been included in this total.

His private works must also have given him a considerable recompense – we saw, for example, that he drew down more than £5,000 in fifteen months from the Liverpool and Leeds Railway Company, and smaller sums from other railway companies – and surely in cases such as the Wellesley Bridge, Drogheda harbour, Youghal Bridge, Ross mines, Sligo harbour, Courtown harbour and so on he must have received some financial recompense for his efforts. However, in his private works he had office expenses and assistants to pay so we should, perhaps, ignore these. Even so, it is difficult to see how he would have died in debt, as Maria Edgeworth claims. Wye Williams said of him: 'He entertained an indifference almost bordering on contempt for pecuniary considerations …

and thus he ever failed in turning to his own advantage the opportunities he enjoyed and the many works in which he was engaged, and the little property he left behind him may be said rather to have been the result of accident than otherwise.'[49] Mudie, too, remarked that he 'disregarded the world and its possessions, and, we have heard, allowed great part of the earnings of his most scientific and valuable industry to be squandered by those who could do him no honour'.[50] These sentiments echo the comments of Noblett St Leger, who had written: 'As to his [Alexander's] private character he was what few Scotchmen are, generous, particularly to his brothers who from their ignorance and presumption were a source of constant annoyance to him.'[51] He certainly settled an annuity of £50 per annum on his niece Mary, in the event that the Reverend Brabazon Ellis married her, but only one payment was made under the settlement before the grant lapsed on Nimmo's death. This charitable bequest led to a celebrated case in Chancery, *Ellis v. Nimmo*, when Ellis attempted to enforce a promise connected with it.[52] Alexander died intestate, his estate, as far as can be estimated, consisting of a house and store in Roundstone, the lease at Maam and his office in Marlborough Street, Dublin, no mean legacy. No evidence has been adduced that he left any other wealth of a pecuniary nature. All we know of his personal estate is that it was sold in Dublin soon after his death.[53]

While Peter LaTouche hinted in 1829 that Nimmo should not be trusted with managing public funds, there is no imputation that this reflected cupidity or malice on Nimmo's part.[54] His apparent disdain for money, and his sometimes dismissive attitude to the execution of his own works, exhibited some of that 'otherworldliness' that Southey detected in him. His nonchalance must have exasperated immensely those who paid him, and those who worked with him; it certainly did his reputation no good. This demeanour extended to other aspects of life also, as an anecdote recounted by Samson Carter indicates:

> His utter disregard of all form or ceremony was truly ludicrous and Mr Barrington tells a story of him which exemplifies this trait in his disposition most laughably. Mr Nimmo either having to visit or being sent for by Mr Peel (now Sir Robert), left his hotel accompanied by Barrington and though in deep and earnest conversation while crossing St James's Park Mr Barrington could not help remarking the continually increasing crowd who were following and hooting. After attentively examining both their persons to see if there were any thing extraordinary thereon, Mr B. at last discovered Mr N. had on his red morning slippers and upon mentioning it to Nimmo and saying he must go back and change them as he surely could not go to Mr Peel in that state, the only reply he received from Mr N. was, after looking at him very indignantly, 'Hoot, mun, d'ye think it was to look at my <u>sleepers</u> he sent for me?' and proceeded on.[55]

Idiosyncrasy of this kind can of course often appear as disdain if not downright arrogance, however understandable. According to Wye Williams,

> he was peculiarly irritable under the infliction of what he called the 'preaching' of the superficial and ignorant pretender. He was indeed that painful pre-eminence which under the influence of his great and varied talents and experience saw clearly and judged too correctly to bear with patience the young and mushroom race of would-be engineers and Parliamentary squires with whom his professional avocations brought him in collision.[56]

We saw instances of this in his appearances in the House of Lords and at the Yare Navigation hearing. Still, his predominant mien in private life was 'convivial, instructive and entertaining, with a high relish for poetry and works of imagination ... so as to be at once the philosopher and the man of the world, alternately imparting the profoundest truths, or enlivening by his sallies the friends about him.' He was, in short, an individual with the virtues and the faults we recognise in ourselves and in others: a delight to his friends in private, an irritant to those who opposed him in public. That he was erudite beyond the norm was obvious to all, yet Wye Williams was at pains to point out 'Mr Nimmo was <u>without exception the most unobtrusive</u> [emphasis added here] yet most erudite man of the age in all that appertained to the profession of a civil engineer.' In Ireland he won the respect of persons of all classes, so that Mudie could say of him:

> though some of the Irish disliked him at the outset, he very soon became a general favourite among all classes, and, perhaps, there has not been a man, within the last 100 years, over whose bier the people of Ireland sorrowed so much as they did that of Alexander Nimmo.[57]

The *Galway Advertiser*, on hearing of his death, eulogised him with similar sentiments:

> His professional distinctions ... were but dust weighed in the balance when compared with the sterling talent and intrinsic merit of this excellent and lamented individual. Eulogium is unnecessary as the word *Ireland* alone will be both his most merited monument and suitable epithet ... [I]t may safely be said 'the British Empire in general has sustained an almost irreparable loss.'[58]

To place Nimmo in the top rank of engineers of his period – in the company of men like Smeaton, Brindley, the Rennies, Telford, Watt, the Stevensons and Stephensons – seems excessive, *if we consider only the physical structures he left behind*. Yet, to dismiss him from high rank would be harsh. After all, his contemporaries, including Telford and Stevenson, did hold him in high regard. David Stevenson remarked, when writing of his father the great Robert

Stevenson: 'Acting as he did with Rennie, Telford, Nimmo and afterwards with Walker, George Rennie and Cubitt, with all of whom he remained in friendly intercourse' – a comment that serves to place Nimmo in the illustrious company of the engineering greats of his time, Irish focus and residence notwithstanding.[59] So, too, does the comment of a writer to *The Surveyor, Engineer and Architect* ten years after his death: 'No one can reflect upon the gigantic and successful efforts of a Watt, a Telford, a Rennie and a Nimmo, in the construction of roads, bridges, canals and harbours...and say that we lingered in the grand march of improvements.'[60] Yet, while the structural legacies of the great engineers are many, left in stone and iron to last into our own day and probably well beyond, that of Nimmo seems altogether slighter, the Wellesley Bridge, Dunmore harbour and Galway breakwater being his principal structural or monumental works that are comparable in scale, style, visual impact and longevity with the monuments of the greats. The bulk of his remaining structural legacy is, perhaps, less dramatic than these, although the style and presence of Glenanean and Erriff bridges, Raughly harbour, and the Killarney to Kenmare road are not to be discounted.

Essentially, he was a man of ideas and theory, of great plans and grand schemes, a scientist in outlook, perhaps even more than an engineer. This was no mere accident: he was one the first to enter the engineering profession from an entirely academic background, rather than through pupilage or apprenticeship as his predecessors and contemporaries did. Because of this, he was largely lacking in the practical knowledge and skills that come from the drudgery of long hours of work as a stonemason or millwright, miner or journeyman carpenter, such as Telford and the other early engineers possessed, and it is no surprise that he showed little interest in the mundane aspects of even his own structural output. As Wye Williams said of him:'All mind [Williams' emphasis], and mind of a superior cast, it was almost natural that he should be indolent in following up the dull drudgery of carrying into execution the works and offspring even of his own genius.'[61] The consequences of leaving the implementation of his schemes to assistants, or to men of lesser vision, did not appear to dawn on him, although he was driven once to remark in frustration, if not exasperation: 'I think there is hardly a harbour that I have constructed, or been concerned in, that I have any doubt of the result, if they go on and complete the thing.'[62] He had, too, an overwhelming belief in the primacy of mathematics in the solution of engineering problems – 'we highly value the sublime geometry' he had remarked in his article on the theory of bridges[63] – approaching all his endeavours in a calculating and quantifying manner. But far too often he failed to calculate the likely response of others to his plans, or to quantify sufficiently their opinions and objections, not from disdain, or malice, or arrogance, but from the certainty of the correctness of his mathematical analyses.

To judge the early engineers only by the persistence of their structures into our own time diminishes the importance we should rightly give to the diverse contributions they made to the contemporary society in which they functioned. If we take this wider approach, Nimmo's importance is enhanced beyond just the physical structures that he left behind. To say, for example, that he alone precipitated the geological description of Ireland, or the Ordnance Survey, or the Office of Public Works, or the Hydrographic Survey, would be grossly incorrect. But to say that these institutions owe nothing at all to his labours would be even more perverse. To forget the men he trained, to neglect his fine maps, to belittle his grand plans, to ridicule the mathematical rigour he brought to the simplest task, to deny his concern for the poor, to ignore the hope he brought to starving Connaught and to overlook the respect shown to him by his illustrious contemporaries would be foolish beyond credibility.

The emergence of the named public institutions early in the nineteenth century more than anything else signalled the dawning of a new, technological Ireland. Some people, native Irish and British alike, perceived in the Union the potential for economic progress and social improvement and were willing to work towards its realisation in Ireland. Among such men were Thomas Spring Rice, MP, Henry Petty-Fitzmaurice (Lord Lansdowne), MP, Sir John Newport, MP, Maurice Fitzgerald, MP and many others. These forward-looking men needed technical experts to devise schemes and implement developments that they could promote in political and business circles. In Ireland's situation, a technical expert needed to be at once cartographer, surveyor, geologist, hydrographer, agronomist, mineralogist, architect and contractor. If such an array of skills resided in any one person, it was in Nimmo; the Bogs Commissioners had no way of knowing when they engaged him, that they were bringing to Ireland a man who, more than any other engineer, would be one on whom forward-looking men would come to rely for such schemes and expertise. He had in abundance the necessary scientific and engineering skills, the broad vision for development, the appreciation of natural resources and an indomitable disposition that was not dismayed by even the most distressing conditions of poverty. In order to advance in the new Union, Ireland needed men with these attributes, and needed the necessary capital to deploy them effectively; of these requirements, the capital proved most wanting.

To say these things of Nimmo, and to place him in this important role, is not to demean the contributions or the personages of his better-known contemporaries, but rather to restore Nimmo's name and reputation to its rightful place in the ranks of those who worked to change, rather than die for, Ireland. His was a broad and comprehensive vision for his adopted country, where all that he proposed and worked for had its ordered place and role. Nowhere did he express this vision with greater clarity and succinctness than towards the end of his life, when he wrote:

> It is to be feared that many of the works ... have been undertaken ... without that unity of design which would be required to make the several works portions of one great whole, slowly, perhaps, but regularly tending to the attainment of the great ultimate object.[64]

He wrote this regarding Dublin, but it expresses his approach to all his Irish works, its main sentiment giving context to his penchant for 'great national works'. His was a radical, integrated and unifying vision for Ireland that would be absent from relief works during the great famine fifteen years later. It was rekindled briefly in the enterprise of the Congested Districts Board towards the close of the century, and is only now making real progress in Irish infrastructural development.

Nimmo's health began to deteriorate in 1827. In February he was unable through indisposition to keep up with his correspondence. The following year he was worse and we find him spending time in Harrogate 'trying to pick up some health'.[65] That was the start of a bad winter for him and by April 1829 he was writing to Robert Stevenson from London, 'I have been very poorly all this winter and have full escaped a fever and cannot much attend to business.'[66] In late 1831 he was again 'taking the waters' regularly at Leamington Spa, as he was becoming quite debilitated. He was still applying himself to work as much as he could, but was obviously neglecting many aspects of it. One of his last letters from England was yet another request that the authorities in Dublin pay the sums owing to his staff of the western district.[67] He returned to Ireland to finalise his chart and prepare his book, *New Piloting Directions for St George's Channel and the Coast of Ireland* for publication.[68] News was abroad in Dublin that he was approaching the end, even as the Board of Works was busy gathering as much as possible of his notes, plans and documents relating to the works they were taking over. Their haste was as unseemly as their concern was unnecessary – it is not as if Nimmo could take the materials with him. At 7.30 in the evening of 20 January 1832 he passed away in his home 'over the shop' in 78 Marlborough Street Dublin, ending an eventful and productive life journey that had started over the watchmaker's shop in Kirkcaldy. *The Times* recorded his passing with the short and wickedly mordant observation, 'he had long been an invalid from rheumatic pains: a dropsy terminated his suffering.'[69] George Halpin Jnr, his erstwhile apprentice, finished off the *Sailing Directions* and steered it through the press to publication later that year.

Nimmo's two brothers, George and John, joined Alexander's Irish engineering practice in 1822 and 1825 respectively. Samson Carter called them 'his good-for-nothing family' and there is reason to think that they were less of a help and more of a hindrance to him, although Alexander never commented on them publicly. Carter and St Leger worked for a long time with him in the western district and would have known his brothers well. Having started in the Aran Islands, George served in a variety of posts, such as pay clerk, paymaster and assistant engineer at different times, but spent most of his time in the Dublin office. The salary he is credited with receiving, just £748 over ten years, suggests that he must have had another source of income, and some reports state that he was a contractor in his own right, having built the house at 78 Marlborough Street that was Nimmo's office. He appeared as an engineer on one of the plans for Youghal Bridge. After Alexander's death, he tried to keep the Nimmo engineering practice in Dublin functioning, although most of its main projects had been taken over by the Board of Works and the staff had scattered. His work at Courtown harbour in 1833 was the last engineering project he engaged in. Nimmo's Dublin practice appeared in a directory of Dublin businesses as 'George and John Nimmo, civil engineers' in 1833, but it was not mentioned subsequently.

The next we hear of George, he was staying in Alexander's house at Maam, which had become an inn by then, run as Maria Edgeworth tells us, 'by his [Alexander's] Scotch servant who used to come with him to Edgeworthstown'.[70] The Revd Caesar Otway, Nimmo's friend, stayed there during his tour of Connemara in the 1830s and confirmed that the inn was run by a servant of Nimmo's, 'who gives universal satisfaction by his urbanity'.[71] Otway also met George there whom he described as 'a valetudinarian whose state of health had forced him to retire from active business. I found in him much of the information and ability of his brother and he seemed intimately acquainted with this western district and quite alive to its great capabilities.' Henry M'Manus, an Ulster Presbyterian clergyman met George in Roundstone in 1840 and found him 'advanced in years, yet loving wild scenery and ... of a kind disposition'.[72] Apparently he regaled M'Manus with stories of foreign parts, where he claimed to have travelled to buy timber. This indicates that he may have had another job (maybe his alleged contracting work) that might explain his limited income from his employment with Alexander.

A curious printed item from 1838, archived in the USA, may refer to George. It is a pamphlet in the Firestone Library of Princeton University, catalogued under 'Nimmo, George and Monteagle, Thomas Spring Rice 1838', entitled *Observations on the distillery Laws of the United Kingdom*.[73] Even though the duty on spirits in England was greater than that elsewhere in the United Kingdom, the pamphlet argued that the revenues from spirits of Ireland and Scotland were disadvantaged relative to England because of the method used to assess the duty and the different modes of bonding in the different countries. The writer wanted something done about this perceived injustice. Could this pamphlet be by 'our' George Nimmo? It is signed 'George Nimmo' with an address at 5, Fenchurch Buildings, London. George had earlier been on a delegation to London appealing against the duties on spirits and this pamphlet may have been a statement of his case and position.[74] Spring Rice was then Chancellor of the Exchequer, which accounts for his name appearing in the bibliographic reference.

George does not seem to have married, but John was already married and had a son Alexander and a daughter Mary by the time he came to Ireland. He lived in Nimmo's house in Roundstone with his wife and family. From October 1825 to January 1832 he earned £1,395 as a pay clerk, assistant engineer and general inspector of works at several of Alexander's sites. Apart from this, he featured little enough in Alexander's work, maybe for the reason identified by Carter in his letter to St Leger: 'The best way to secure [stories about Alexander's life] is to come over to this country, seize on John Nimmo and keep him in a state of inebriety (to which he will be nothing loath) for a fortnight or three weeks.'[75] Obviously John's interest in the distillers' art differed in kind from George's and it was tragic, but perhaps understandable, that he drowned when he fell from a small boat that was accidentally upset in Roundstone Bay around 1849.[76]

John's daughter, Mary, married the Revd Philip Brabazon Ellis, the Church of Ireland curate at Roundstone in April 1831. Ellis had been appointed curate in 1830 and found the post remote and difficult, with no school or church and a community of only seventy souls.[77] He was not an easy character to deal with, even for his archbishop, Le Poer Trench. On his marriage he became anxious to leave Roundstone (maybe to remove himself and his young wife from John's influence or behaviour) and, exchanging curacies with the Revd Mr Foster, curate of Moyris (Renvyle), he took up residence in Clifden. He became joint secretary with the Revd Foster of a society they formed for affording instruction and protection 'on the colonising system' to poor Protestants, especially converts from Popery. Hyacinth D'Arcy of Clifden Castle (who also took orders subsequently) was the society's treasurer. One of its aims was to buy land in order to set up smallholdings at low rent for willing Protestants (especially converts), a scheme perhaps not too far removed from Nimmo's entirely secular venture in setting up Roundstone village. Maurice Fitzgerald had proposed something similar in 1829 for the relief of the poorest of the poor.[78] This aspect of the society's work came to nothing in the end, and after many tribulations in Clifden, Ellis was appointed to a curacy in Turlough, County Mayo. Later, he took up the post of Rector of St Paul's Church of England in Burslem, Staffordshire in 1840, and the family moved there.

Ellis and Mary had a number of children. Their first, Mary Frances, was born in 1836 and died an infant; she is buried in Turlough. They had eleven more, born either in Mayo or in Staffordshire. One was named Alexander Nimmo Ellis, continuing the name of Mary's illustrious uncle and another child, interestingly, was named Hyacinth Darcy Ellis (he was born many years after the family had left Ireland). Mary died in 1876 and is buried in Burslem with her husband, who pre-deceased her in 1861. Many of their children migrated eventually to Australia, where descendants are still living today.[79]

John's son, Alexander junior, commenced work in his uncle's practice in

1828. Up to April 1831 he had earned £150, hardly a handsome reward. He emigrated to Canada around then, where he worked on the Welland Canal, but his exile did not last long. Returning to Ireland soon after his uncle Alexander's death, he obtained employment in the Dublin office of Charles Vignoles and attempted to don the mantle of his illustrious uncle, with whom he is sometimes confused. He gave evidence to the Select Committee on the Connaught Lakes in 1834.[80] Subsequently, he published a report that he had prepared for the committee of the Lough Corrib Improvement Company in 1837, illustrating a new canal he proposed to join the Lough to the sea.[81] He also laid out a line of railroad from Dublin to Mullingar that was well regarded by Vignoles. While he appears to have lacked the ability of his uncle, he might nevertheless have looked forward to a productive career in railway engineering. However, like his uncle, he was fated to die young in 1839 before he was 40 years of age. With the death of his father, John, in 1849 the Nimmo family came to an end in Ireland.

NOTES

1. Mudie (ed.), Surveyor, Engineer and Architect, p. 5.
2. S. Carter to N. St Leger, no date [1834] (University College London, Greenough Papers), hereafter referred to as Greenough Papers.
3. T. Crofton Croker to J. P. Nichols, 2 April 1838 (BL, Add 37021 fo. 43).
4. Annual Biography and Obituary, vol. 17 (London: Longman, Rees and Co., 1833); M. F. Connolly, Biographical Dictionary of Eminent Men of Fife, p. 349 (Edinburgh: Inglis and Jack, 1866), p. 349; S. Urban, Gentleman's Magazine, 102 (1832), pt 1, pp. 370–1.
5. Skempton, Biographical Dictionary of Civil Engineers.
6. R. Paxton, 'Alexander Nimmo', in Matthew and Harrison, Oxford Dictionary of National Biography.
7. S. De Courcy, 'Alexander Nimmo: Engineer Extraordinary'.
8. S. Carter to N. St Leger, no date [1834]. Greenough Papers.
9. T. Crofton Croker to J. P. Nichols, 2 April 1838 (BL, Add 37021 fo. 43).
10. A. Nimmo, Evidence, p. 76, in Report from the Select Committee on the Survey and Valuation of Ireland (B.P.P. 1824 [445], vol.VIII, mf 26.47–51).
11. M. Edgeworth, Tour in Connemara and the Martins of Ballinahinch (London: Constable, 1950), p. 4.
12. W. R. Hamilton to S. Hamilton, 2 August 1827, in R. P. Graves, Life of Sir William Rowan Hamilton (Dublin: Hodges and Figgis, 1882), vol. 3, pp. 249–50.
13. A. Nimmo, Evidence, p.161, in Minutes of Evidence taken before the Select Committee of the House of Lords to examine into the Nature and Extent of the Disturbances which have prevailed in those districts of Ireland which are now subject to the Provisions of the Insurrection Act and to report to the House (B.P.P. 1825 [200], vol.VII. mf 27.60–3), hereafter referred to as Lords Enquiry.
14. A. Nimmo, 'Expenditure on Public Works for Relief of the Poor, Western District.' (Nat. Archs. UK, Works of A. Nimmo. AO 17/492), hereafter referred to as AO 17/492.
15. N. St Leger to Sir N. Colthurst, 4 February 1834, Greenough Papers.
16. J. Mitchell, Reminiscences of my Life in the Highlands (London: privately printed, 1883), vol. 1, pp. 27ff..
17. W. R. Hamilton to M. Hutton, 25 August 1827, in Graves, Life of Sir William Rowan Hamilton, vol. 3, pp. 250–3.
18. W. R. Hamilton to his sister Eliza, 30 August 1827, in ibid., pp. 253–7.
19. W. R. Hamilton to A. Nimmo, 31 August 1827, in ibid., pp. 258–9.
20. W. R. Hamilton to his sister Eliza, 16 September 1827, in ibid., pp. 261–6.
21. W. Wordsworth to W. R. Hamilton, 24 September 1827, in ibid., pp. 266–8.
22. R. Southey to T. D. LaTouche, 9 December 1827, in ibid., pp. 270– 1.
23. C. Wye Williams, Obituary of Alexander Nimmo. Greenough Papers.
24. T. Collins, personal communication.
25. R. Mudie, ed., Surveyor, Enginee and Architect, pp. 4–5.
26. Anonymous ('C'), 'Anecdote of the late Mr Nimmo', in Mudie, ed., Surveyor, Engineer and Architect, pp 47–8.
27. Nimmo, Evidence, p. 131, in Lords Enquiry.

28. Nimmo, ibid., p. 130.
29. Nimmo, ibid., p. 179.
30. Nimmo, ibid., pp. 131–50.
31. Nimmo, ibid, p. 179.
32. Nimmo, ibid., pp. 174–5.
33. Nimmo, ibid., p. 180.
34. A. Nimmo to the Fisheries Commissioners, 22 September 1824 (NAI, OPW8/236 1 [25]).
35. A. Nimmo to Lord Leveson Gower, 4 June 1830 (NAI, CSORP 1830/5039).
36. Nimmo, Evidence, p. 131, in Lords Enquiry.
37. Ibid., p.160.
38. M. Fitzgerald, Evidence, p.158, in Report from the Select Committee on the Employment of the Poor in Ireland (B.P.P. 1823 [561], vol. VI. mf 25.49–51).
39. A. Nimmo, 'Sums for Public Works for the Relief of the Poor in the Western District. Year to January 1830', AO17/492.
40. D. Browne, Evidence, p. 53, in Report from the Select Committee on the Employment of the Poor in Ireland (B.P.P. 1823 [561], vol. VI, mf 25.49–51).
41. Maria Edgeworth, quoted in J. W. de Courcy, 'Alexander Nimmo, Engineer. Some Tentative Notes'. Lecture to the National Library of Ireland Society, November 1981.
42. A. Nimmo to unknown recipient, 31 August 1824 (PRONI, Fitzgerald Papers, MIC/639/11/reel 6/71), hereafter referred to as Fitzgerald Papers.
43. A. Nimmo to M. Fitzgerald, 21 September 1828, Fitzgerald Papers, MIC/639/12/reel 6/105.
44. A. Nimmo to M. Fitzgerald, 7 October 1825, Fitzgerald Papers, MIC/639/12/reel 6/105.
45. Lord Lansdowne to M. Fitzgerald, 31 December 1826, Fitzgerald Papers, MIC/639/12/reel 6/32.
46. Calculated by addition of the various sums recorded in the annual Reports of the Fisheries Commissioners.
47. Sixth Report of the Commissioners for Auditing the Public Accounts of Ireland (B.P.P. 1818 [154], vol. X, mf 19.54–6).
48. He received £210 in 1823 (Eleventh Report of the Commissioners for Auditing the Public Accounts of Ireland, B.P.P. 1823 [199]) and his executors claimed and received £515 on his behalf in 1832 (Report of the Select Committee on Post Office Communication with Ireland, p. 352,B.P.P. 1831/32 [716], vol. XVII, mf 35.136–40).
49. Wye Williams, Obituary of Alexander Nimmo, Greenough Papers.
50. Mudie (ed,), *Surveyor, Engineer and Architect*, p. 5.
51. N. St Leger to Sir N. Colthurst, 4 February 1834. Greenough Papers.
52. Cases in Chancery, Ellis v. Nimmo, p. 333
53. Frederick Bourne claimed to have purchased a map of Dublin Bay belonging to Nimmo at the latter's sale. Select Committee on the Dublin to Kingstown Ship Canal, p. 14 (B.P.P. 1833 [591], vol. XVI, mf 36.121–3).
54. P. LaTouche to M. Fitzgerald, 27 December 1829, Fitzgerald Papers, MIC/639/13/reel 7 (28).
55. N. St Leger to Sir N. Colthurst, 4 February 1834, Greenough Papers.
56. Wye Williams, Obituary of Alexander Nimmo, Greenough Papers.
57. Mudie (ed.), *Surveyor, Engineer and Architect*, p. 6.
58. *Galway Advertiser*.
59. Stevenson, *Life of Robert Stevenson*.
60. Anonymous ('A Friend to all Improvements'), 'Drainage and Cultivation of Waste Land', in Mudie (ed.), *Surveyor, Engineer and Architect for the year 1842*, pp.277–8.
61. Wye Williams, Obituary of Alexander Nimmo, Greenough Papers.
62. A. Nimmo, Evidence, p. 407, in Minutes of Evidence before the Committee on the Norwich and Lowestoft Navigation Bill (B.P.P. 1826 [396], vol. IV, mf 28.24–6).
63. A. Nimmo, 'Bridge', in *The Edinburgh Encyclopaedia* (Edinburgh: Blackwood, 1819), vol. 4, part 2.
64. A. Nimmo to Marquis of Anglesey, 25 March 1831 (Royal Irish Academy, 3 B 53.56/411).
65. A. Nimmo to Lee and Calder, 15 October 1828 (BL, Add 44085 fo. 45).
66. A. Nimmo to R. Stevenson, 6 April 1829 (NLS, MS 19988 f 64).
67. A. Nimmo to E. G. Stanley, 20 September 1831 (NAI, CSORP 1831 2952).
68. A. Nimmo, *New Piloting Directions for St George's Channel and the Coast of Ireland. Written to accompany the Chart of St George's Channel and the Coast of Ireland, drawn for the Corporation for Improving the Port of Dublin etc.* (Dublin: A. Thom, 1832).
69. *The Times*, 26 January 1832, p. 4.
70. Edgeworth, *Tour in Connemara and the Martins of Ballinahinch*.
71. C. Otway, *A Tour in Connaught* (Dublin: W. Curry, 1839), pp. 231–2.
72. Quoted in T. Robinson, *Connemara. Listening to the Wind* (Dublin: Penguin Ireland, 2002), p. 181.
73. G. Nimmo, *Observations on the Distillery Laws of the United Kingdom* (London: Darley and Sons, 1838).

74. Sixth and Seventh Reports of the Commissioners of Inquiry into Excise Establishments (B.P.P. 1834 [6], & [7], vols. XXIV and XXV, mf 37.173–9).
75. S. Carter to N. St Leger, no date [1834]. Greenough Papers.
76. T. Robinson, *Connemara. Listening to the Wind*, p.181.
77. Sirr, *Memoir of the Last Archbishop of Tuam*.
78. M. Fitzgerald to Duke of Wellington, June 1829, regarding Public Works in Ireland. Report of the Select Committee on Public Works (Ireland), Appendix 3, pp. 319–28 (B.P.P. 1835 [537], vol. XX, mf 38.143–8).
79. Personal Communication from Mrs Morrison, Australia.
80. Report of the Select Committee on navigable inland communication through the Connaught Lakes (B.P.P. 1835 [354], vol. XX, mf 38.14).
81. A. Nimmo, Jnr, 'Report to the Committee of the Lough Corrib Improvement Company, Palace Yard Westminster, London, 11 April 1937' (Princeton University Library, 9179.331.999 v. 1).

Appendix 1
A Gazetteer of Nimmo's Piers and Harbours

It is common in Galway to hear that all the piers of Connemara, sometimes all the piers of the west of Ireland, were built by Nimmo. He certainly was involved with many of them, but there were many more, before and since, with which he had no connection. After all, he was engaged on fishery piers and harbours only in the decade 1820 to 1830, and then not for all of the time. His employment with the Fisheries Commission lasted only from 1820 to 1826 and he had started designing and constructing piers for the western district only from 1822. The commissioners appointed James Donnell as harbour engineer in 1824, when Nimmo was dealing with more than thirty-eight piers and harbours as well as with roads and other projects all over the country. Nimmo alone, or even with his team of young men, was clearly unable to keep up with all the work, and from the time of Donnell's appointment his involvement with fishery piers quickly tapered off. Donnell made numerous visits to the sites where work was in progress and he made recommendations for alterations, modifications and sometimes even abandonment, as he thought appropriate. From the very earliest years, then, Nimmo's contribution was in some danger of being progressively obscured, even obliterated. Many times later in the century the Government initiated schemes of pier building and improvement that resulted in the wealth of stone piers and harbours that we now have, some quite massive and dramatic, even in the remotest parts of the country. Some of these were entirely new; others were enlargements of Nimmo's originals. In the twentieth century many of the earliest piers and harbours were further enlarged as the needs of shipping and the fishing industry changed. With some honourable exceptions, enlargement and improvement have meant encasing the early stonework in concrete, obliterating the original and converting some into places where maritime paraphernalia are dumped and derelict boats abandoned.

What, then, can have survived that we can still see of Nimmo's piers? Perhaps the most telling feature of his work was the quality of his site selection. In each case, he found a convincing fishery, agricultural or commercial reason for the precise locations he selected, and topographically he had an unerring eye for the exact site within a locality that best suited the tides, the winds and the need for shelter. For that reason, there remain today piers and harbours at all but one of his original sites. Very few of them have been entirely abandoned, and if today they serve mainly the small-boat fisheries, that is exactly the purpose for which they were first intended in the 1820s. As ColonelBurgoyne of the OPW acknowledged in 1835: 'I think the small fishery piers have been of great advantage, those small harbours erected by Mr Nimmo were exceedingly beneficial to the lower class of people and in that respect they are deserving of great encouragement.'

Many of them had cost relatively trivial sums to erect and most were built with a view to providing public works for the relief of the poor, rather than with permanence in mind. We must remember too, that they were built in the age of sail (the age of rowboats and curraghs, where Ireland was concerned), when boats lay grounded rather than afloat when not in passage; Nimmo's piers were rarely meant as floating docks. That anything should still remain of them after 200 years is itself a tribute to his skill and to the efforts and achievements of the starving and distressed poor who built them. Some are rude or crude structures; others exhibit a quality of stonemasonry that would do credit to craftsmen of any generation.

The commissioners reviewed the status of all the piers and harbours in 1829, whether finished or still in progress.[1] That review, along with Nimmo's[2] and Donnell's[3] earlier reports, allow us to trace Nimmo's contribution to many of them. By reference to all the published reports, to Nimmo's unpublished letters, plans and maps, and to the western district reports, along with actual visits to all the sites, this gazetteer identifies the precise location of the piers and harbours to which he contributed, including those not funded by the Fisheries Commissioners and therefore not appearing in their annual reports. It summarises their history and progress from 1820 to 1829, and indicates where relevant detailed information may be accessed. This will be a convenient starting point for anyone wishing to research any of the piers or harbours in question. The general state of each structure today, as judged visually, is also indicated briefly. Other connected matters, such as aspects of the events, personages and difficulties at the various sites, will be found elsewhere in the text.

The gazetteer starts at Mullaghmore harbour in County Sligo and works around the coast in a counter-clockwise direction.

COUNTY SLIGO.

Mullaghmore Harbour

Location: N 54 27.952, W 8 26.828

Travelling south on the N15 Ballyshannon to Sligo road, take the minor road to the right at Cliffony village, signposted for Mullaghmore and continue to its end.

FIGURE APP.1.1
Mullaghmore Harbour. The pier on the left is Nimmo's, now much enlarged. That on the right is Lord Palmerston's later pier.

Nimmo regarded this site as the only suitable place of shelter from westerly winds between Sligo and the Donegal coast. It was home to twenty or thirty boats that fished turbot on the local banks. He planned to improve the small natural cove by building a new quay ending in a pier along the rocks on its east side, with a rough mole breakwater running across to it from the west side, leaving an opening of seventy feet between the two. He gave a small, indicative plan in the Fourth Report of the Commissioners.[4] By 1824, after the commissioners had given £545 for the work already under way, Lord Palmerston took over responsibility for it and requested Nimmo to extend the pier beyond what was already envisaged. Nimmo responded with a new plan for a much larger structure costing an additional £1,500, shown in the Sixth Report of the Commissioners.[5] Stone was quarried locally and brought to the harbour on a railway laid down specifically for that purpose. Nimmo expressed the hope of extending the railway to Lough Erne as a commercial venture, but that was too much for Palmerston at the time. Donnell reported that the pier was well in progress in 1826 and he prepared a modified plan to extend both the pier and the mole breakwater even further.[6] Although the harbour was completed to Nimmo's plan by 1828, Lord Palmerston decided to build a new, much larger pier south of Nimmo's to make the harbour even bigger. This was completed in 1844 and the harbour has remained largely unchanged since then. In modern times a new breakwater has been completed outside Nimmo's old pier, which itself has been raised and strengthened (figure App. 1.1). Except for some large stones scattered on the landward side, there is no evidence today of Nimmo's rough cross mole. A new concrete slipway has been added at the innermost, oldest part of the harbour, which therefore bears little enough resemblance to what Nimmo originally designed.

Raughley Harbour

Location: N54 19.368, W08 38.556

Travel south on the N15 towards Sligo, almost as far as Drumcliff. Just before Drumcliff church, watch for a junction to the right, signposted Raughley, beside a large tavern. Travel down that side road through Carney, past Lissadell and on to Raughley at its end.

The potential of Raughley as a fishing station was obvious from its position near the outer edge of Sligo Bay and by the concentration of fishermen in the village (there were thirty to forty rowboats and three sailing boats stationed there in 1822). They used to launch from the west or the east side of the peninsula, depending on the wind, but they lacked a safe place to lie when drawn up. Nimmo proposed to build a quay and pier just inside the point of the peninsula, cutting inwards through the beach to an existing pond that could be converted into a small inner harbour of about seven acres. He illustrated his ideas, which would cost £1,152, with an indicative plan in the Fourth Report of the Commissioners, who granted £720 for the work, half coming from the Dublin Castle and London funds.[7]

Construction started in 1822. Costs increased rapidly and the work could not be continued until Nimmo was permitted to apply funds from the western district towards its completion. By 1824 the works were substantially complete according to Nimmo, who presented a sketch plan of the harbour, shown in the Sixth Report of the Commissioners.[8] It consisted of a stone pier extending nearly 200 feet from the top of the beach as far as low water, with a cant of forty feet at its head, all finished in hammered limestone. Opposite the main pier there was a groyne pier, seventy-four feet in length, sheltering the opening. The inner harbour was one acre in extent, stone-lined, and there was a reservoir closed by a sluice providing a 'backwater' of about six acres. These and other features are illustrated in Nimmo's plan. It had cost a total of £1,606.

When Donnell visited in 1828 he found the works incomplete and needing a further £200 to secure them. At the commissioners' request, he altered aspects of Nimmo's scheme, presenting a revised plan of his own that received the commissioners' approval. He raised the height of Nimmo's pier and redesigned the inner harbour, reducing its extensive quays and embanking the backwater so that the final area of the harbour was considerably less than Nimmo proposal. Donnell's plan – the clearest one – is shown in the Tenth Report of the Commissioners.[9] Today, the harbour is essentially as Nimmo and

FIGURE APP. 1.2
Raughley Harbour. The pier is still unpaved, as originally completed. The gravel bank is thrown up after severe weather.

Donnell designed and completed it (figure App. 1.2). A gravel bar is thrown up at its mouth after stormy weather and the inner dock is now partly filled with sand. A road passes over the embankment enclosing the backwater, which communicates with the dock by a sluice. Raughley has about it a quiet, almost rustic ambience, probably not much changed since Nimmo's time. It is well worthy of conservation.

Pollagheeny (Pollaheeny) Harbour

Location: N 54 16.093, W 09 03.652

Going south on the R297 from Easkey towards Enniscrone, pass through the village of Rathlee. After about a mile watch for the signpost for Pollaheeny harbour and follow the signs to the shingle beach and modern slip.

This small anchorage, a little to the north of Enniscrone in County Sligo, was granted £300 by the commissioners in 1822 for improvement works. Its main advantage was that it lay outside the bar of Killala and the river Moy, and boats could lie up awaiting entry to those ports. A great number of yawls were kept there, drawn up on the shingle beach. There were two coves, separated by a stony spit leading to a low rocky islet. Nimmo proposed to raise the level of the spit to form a higher barrier between the two coves. The southern cove was deeper and he proposed to increase its shelter by building a breakwater inside the rocky islet. In that way, the southern cove would be sheltered on its north by the raised spit and on the west by the new breakwater, making it a much safer anchorage; the outline plan is shown in the Sixth Report of the Commissioners.[10] Although his plan was misunderstood at first when Mr Ham of Ballina was in charge, Nimmo rectified matters in 1823 and progressed it vigorously as far as funds permitted.[11] Eventually the low breakwater was completed at 400 feet long and forty-five feet wide. Although topped by high tide, it nevertheless provided improved shelter within the cove, especially when the spit had been raised with local blocks. Nimmo diverted funds from a road he was building nearby (under contract to Mr Ham) in order to bring the harbour to completion; it would have required an additional £1,600 to make it fully secure, for example by

FIGURE APP. 1.3
Pollagheeny harbour. The sea is breaking along the line of stones laid down by Nimmo just inside the islet in the background.

cladding and quaying the breakwater. Donnell regarded it as very insecure and unfinished in 1826 and he would not recommend any further expenditure so that the commissioners effectively abandoned the site in 1829. For all their supposed inadequacy, the remnants of the works still provide, even today, much needed shelter to fishing boats working in Killala Bay, and local fishermen are well familiar with the benefit of the low walls that were completed. They can be seen when the sea breaks over them, or seen and examined when the tide is low, but it is not generally realised that they are man-made, not natural structures (figure App. 1.3).

COUNTY MAYO

River Moy

Location: N 54 07.055, W 09 08.277

Cross the River Moy from east to west in Ballina town by the bridge near the cathedral. Turn right into Nally Street beside the civic offices and follow along to its end at the public park.

Nimmo had been commissioned by the businessmen of Ballina to improve the navigation up to that town separate to his fisheries responsibilities; he could not, therefore, allocate fishery funds to it. The clergy, traders and inhabitants of Ballina responded by sending a memorial to the Lord-Lieutenant in January 1823 seeking Government support 'for the works in navigation being under the able superintendence of Alexander Nimmo'.[12] Three weeks later Nimmo wrote a letter of support to Secretary Goulburn explaining that he had already spent £1,123 of western district funds on the Moy works in 1822.[13] This had provided a new quay at Belleek, a new road to it and cuts through the rocky bar that obstructed boat passage upriver. The cuts were meant to open a passage to the town through which boats could be warped past the shallows in the main stem. By 1823 boats could work up to his quay at Belleek, and when the cuts would be finished they could, in theory, go as far as the town itself. Anticipation of success was so high in the beginning that the merchants held a gala evening for Nimmo in a local hotel where speech-

FIGURE APP. 1.4
Works on the River Moy. The artificial channel cut by Nimmo is on the left, inside the islet formed by the work. Note the riffle on the rocky bar in the main stem of the river, to the right. Photo taken from the site of Nimmo's quay at Belleek, now a leisure area.

making and singing went on until the late hours.[14] When further funds were not made available to mature the works in the following years, enthusiasm waned and nothing more was done to complete his Moy works, despite further attempts to garner support. When a new commercial quay was eventually erected in 1836, it was located on the opposite (east) bank of the river a little further downstream. Nimmo's quay site at Belleek was abandoned and is now part of a public park, a jumble of stones being all that remains of the quay itself; the cuts through the rocky bar are still clear, located to the north and the south of the quay site and separated from the main course of the river by the small islets that were formed when the cuts were made (figure App. 1.4). There is little local awareness of the origin and nature of these islets in the river.

Nimmo's coloured map (scale of 1 to 528) of the planned improvements survives in the National Archive of Ireland, and a detail is reproduced in Plate 6.[15]

Killala Harbour

Location of pier: N 54 13.021, W 09 12.842

On entering Killala on the R314 road from Ballina, take the turn to the right marked for heavy vehicles. Turn left at the foot of the short hill and follow the road around to the right as far as the pier. From this road the low embankment wall can best be seen as a straight line across the strand to the right at half tide.

Killala was granted a total of £150, half from the commissioners and half from the Dublin Castle and London funds, for work on the existing pier in 1822. Nimmo added funds from the western district to these in order to employ a greater number of destitute people during the prevailing distress. He commenced to extend a new quay wall and make a cut through the sandbanks that obstructed direct access to the old pier. The latter had been erected by the merchants of the town, funded by a levy of one penny on each sack of oats exported, and had proved of great value, although it had no great depth of water. To increase the depth along the new quay, Nimmo undertook an interesting experiment. He built an embankment across the strand from the Green Isle to the mainland, effectively enclosing an area of about fifty acres at high water. It comprised a dry-stone wall, three feet thick on each side with puddle in between, rising only two and a half feet over the strand level, and was covered with five feet of water at high tide. His idea was to force the whole of the early flood and the latter half of the ebb of the fifty acres to flow along the quay, thereby scouring it of sand and ensuring it maintained a good depth alongside. The quay improvements were not fully finished, but the embankment was, when the famine receded and the relief works were peremptorily stopped. Nevertheless, Nimmo felt that the embankment and cut had materially improved and deepened the port. The plan of the proposed works, shown here in figure App. 1.5, was reproduced in the

APPENDIX 1 A GAZETTEER OF NIMMO'S PIERS AND HARBOURS

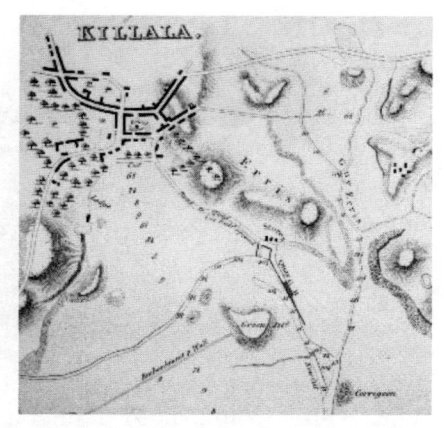

FIGURE APP. 1.5
Nimmo's map of Killala Harbour showing the embankment wall extending out to Green Island as a diagonal straight line from the bottom left of the drawing. (Reproduced from ref. 16.)

Report for 1824 of the commissioners, who had contributed only to the extension of the quay, not the embankment.[16]

Donnell considered the finished pier and quay only as a commercial facility and of little benefit to the fisheries. The fishermen preferred to lie alongside the embanked road to the town, not alongside the quay, so he recommended only some minor work at the site. By 1829 mooring posts and rings had been added to the pier but little else was done up to then. Today the pier has been considerably extended and the direct cut from the pier through the sandbanks has been preserved with appropriate timber baulks. Nimmo's embankment wall is still visible at low tide (or at mid-tide by the growth of seaweed along it, see figure App. 1.6). Many locals seem not to realise what it is, or how it came to be there.

Belmullet canal (Broadhaven)

Location: N 54 13.548, W 09 59.302

At the west end of Belmullet town, the canal is crossed by a fixed bridge on the R313 road.

Neither Nimmo nor Donnell was enthusiastic about Belmullet as a fishery port, but both agreed that a proposal by the owner, Mr Carter to build a pier there would be valuable for trade and for supplying the needs of the fleet.

FIGURE APP. 1.6
Nimmo's embankment at Killala visible as a straight line of seaweed at half tide.

FIGURE APP. 1.7
The canal at Belmullet, looking south into Blacksod Bay.

When Carter commenced a pier in Blacksod Bay south of the town, he requested financial support from the commissioners and Donnell drew up plans for a proper structure. Nimmo had recommended piers at Rinrua (Rinroe point) and Inver in Broadhaven Bay, north of the town, but his main proposal for the area concerned the Mullet isthmus. In 1824 he proposed that a canal be cut through the isthmus in the town, linking both bays.[17] This, along with Achill Sound, would give 'a kind of inland navigation', sheltered from the stormy Atlantic, to boats plying between Westport and Donegal Bay. In total, the canal would be almost 700 yards long, costing about £5,000, and would permit the passage in most tides of 'such vessels as use the Forth and Clyde canal'. There would be little difficulty in making the cut, he thought, and it could be preserved by jetties or gravel lining. Since the tide in Blacksod rose earlier and higher than in Broadhaven, he envisaged that the current through the canal would help keep it scoured. He annexed a plan of the proposal to his report, but the commissioners did not publish it, possibly because they did not regard it as aiding the fishery and therefore had no intention of funding it. The merchants argued that boats could enter and sail out of one bay with the same wind that opposed them in the other, by which they justified piers in both bays, and the canal would help to maximise the benefit of this. The commissioners, however, had other more pressing demands on their funds than this mercantile object, so they refused a grant. A plan of Belmullet 'showing the line of the cut proposed in 1824' was published in 1836 by Patrick Knight, who was Nimmo's assistant in Mayo from 1822 to 1831.[18] This is almost certainly Nimmo's original line, with a floating dock located at its Blacksod Bay entrance. A canal, over a quarter of a mile long, was eventually built along this line in the 1880s, allowing safe passage for small vessels from Clifden to Donegal Bay via Achill Sound and the canal, exactly as envisaged by Nimmo. Although still in use by small boats and crossed by a small bridge (figure App. 1.7), the canal today has a somewhat neglected air. The permanent road bridge (R313) that crosses it now means it is no longer suitable for any but the smallest craft.

APPENDIX 1 A GAZETTEER OF NIMMO'S PIERS AND HARBOURS

Saleen (Ely Erris) Harbour

Location: N 54 11.628, W 10 02.322

Take the R313 road from Belmullet towards Blacksod. About half a mile south of *An Geata Mór* (Binghamstown) village the road makes a right-angled turn to the right at the site of the harbour.

The Mansion House fund, the Dublin Castle fund and the commissioners supported this pier and harbour with £300 in 1822, and work commenced under William Bald's supervision. A quay had already been built of rough stone along the side of a small promontory and it was proposed to continue this outwards as a straight pier projecting 140 feet from land. A short groyne pier was built opposite this to form an entrance into an inner dock shown in the Sixth Report of the Commissioners.[19] The pier and harbour were not principally for shelter, since Saleen was already well sheltered, but to provide a landing place convenient to the new village of Binghamstown then under construction. By 1824 work had made some good progress, almost one hundred feet of the pier being completed, but at great expense, since Bald had faced it on both sides with expensive hewn granite blocks drawn from Tarmon Point. Nimmo furnished a new specification and plan for its improvement, adding a cant or return, raising the height and excavating the inner part of the basin at an estimated cost of £378.[20] Work was still under way in 1829 when Donnell visited, but was progressing very slowly. By then the growth of Belmullet had greatly exceeded that of Binghamstown and the importance of the latter was already in decline. Work on the pier continued to completion under the new Board of Works after 1831.

The hewn granite gives the Saleen piers a certain solidity uncommon in structures of their rather small size (figure App. 1.8). Despite this, a section of the main pier has collapsed into the entrance channel very recently, possibly because a boat may have collided with it. The inner basin has been lined with concrete slabs, strengthening the small neat harbour. Overall, this harbour is reminiscent of Raughley in County Sligo. Both are good examples of the small, vernacular harbours that Nimmo envisaged and designed, although the modern concrete lining of the inner basin at Saleen detracts from its period

FIGURE APP. 1.8
The granite stone work of William Bald at Saleen pier.

layout. Both Saleen and Raughley contrast strikingly with the more massive structures erected later in the century under the Congested Districts Board and other schemes.

Blacksod (Tarmon) Pier

Location: N 54 05.929, W 10 03.604

Continue southwards along the R313 through the Mullet peninsula. Blacksod is at the very end of the road.

This pier is on the south-eastern tip of the Mullet peninsula on a site chosen by Nimmo and supported by the London and Dublin Castle funds. A small landing pier was started in 1822 under Bald's supervision.[21] Excellent blocks of granite were available locally and, like Saleen, were used to face the quay in hewn stone, adding so much to its cost that all the money was expended before the pier was made of much use. Nimmo was not at all happy with this state of affairs and he recommended that part of Bald's work be taken down and rebuilt to a new plan, shown in the Eighth Report of the Commissioners.[22] This involved building an embankment from the land to a small stone and gravel island offshore, and building a jetty pier from the end of the island so as to enclose a small harbour. Material cut from the inner edge of the island could be used to raise the embankment and the rest of the island to pier height, so that the whole formed a curved mole from the land to the pier head. This was then coped with stone taken from Bald's original structure, and paved. Nimmo closed the harbour with a rough cross mole, twenty-one feet broad, extending from the mainland towards the pier head, enclosing a space, then mainly a bog, which was to be cleared to form a dock.

The whole place was in an unfinished state when Donnell visited in 1826, so he arranged to have it completed by contract with new funds from the commissioners. The new contractor, like the previous one, eventually failed to complete the work. A new government-funded road to the pier (now the R313) was completed under Nimmo by 1829 and the pier itself was finally completed by then. Today the area at the base of the pier has been largely filled in and built upon, Blacksod lighthouse occupying the site. The pier base,

FIGURE APP. 1.9
The large granite blocks forming the root of the pier at Tarmon (Blacksod), laid down by William Bald working under Nimmo's direction.

largely Bald's work, has been extended in concrete, doubling its original length and halving its original beauty; its innermost, original section, made of well-cut granite blocks, has an aspect of rugged solidity, as shown in figure App. 1.9.

Achill Sound Pier

Location: N 53 52.30, W 09 56.54

Take the road to the left in the middle of the village of Achill Sound and continue south on this until you reach the tip of the island. The pier is down a short road to the left.

At its southern tip, Achill Island is joined to a smaller island, Achill Beg, by a spit that dries at low water. East of the spit was a natural anchorage that was one of the sites chosen by the Fisheries Commissioners for financial support of £400 in 1822, half of which came from the Mansion House and Dublin Castle funds. Nimmo prepared an indicative plan for a landing pier, 150 feet long, located on the Achill side and shown in the Eighth Report of the Commissioners.[23] Construction was carried out under Bald's supervision and was soon completed to near low water mark.[24] The location, having no road to it, proved of little use for the fisheries but the pier served the locals well for landing kelp and for commerce with Westport. Donnell thought it was poorly built and was not surprised when it was damaged by the sea before 1825.[25] He recommended repairs costing £120 to avoid further dilapidation, but because the site was given low priority nothing substantial was done by 1829. Today, the pier has been considerably extended but the original masonry laid by Bald is still visible at its landward extremity, with the original steps still in place on the eastern face. The head of the original pier just beyond these steps is still easily identifiable; the modern pier extends outwards from it (figure App. 1.10). Although somewhat shaken, Bald's dry stonework is in reasonable condition and, although less fine, is reminiscent of his other pier work at Tarmon and Saleen. Despite Donnell's jaundiced view of its utility, the pier remains today in active use by boats servicing the local salmon farming industry.

FIGURE APP. 1.10
The modern pier at Achill Beg extends outwards from the head of Nimmo's pier that was built by William Bald. The picture shows the original head and its granite construction.

Piers on Clare Island

Locations: Old Castle: N 53 47.59. W 09 57.05; Port A' Chuille: N 53 47.26, W 09 59.08

A pier that had been built at Old Castle on the east side of Clare Island by the O'Malley family was still in use when Nimmo visited. He proposed to excavate some rocks within and without the pier, clearing the channel and forming a small harbour, as shown on his plan in the Eighth Report of the commissioners. His MS plan, illustrated in Plate 12, shows the existing pier in dark grey extending to the north-west. The excavated rock could be used to form a mole outside the pier that could eventually be made into a breakwater extending towards the north-east and shown in pink on the plan. This would give full shelter to the inner pier. A small jetty, shown in dark grey, could be built from the opposite side towards the inner pier; an eighty-foot opening between this and the pier could be an entrance into the inner, excavated area that would now become an inner dock. These improvements were to cost about £3,660, a sum he realised would be difficult to obtain.[26] In January 1824 Sir Samuel O' Malley was granted £276 to implement Nimmo's plan, hardly enthusiastic support. Nevertheless, work started as best it could with so little money. Donnell refused to confirm that the work was completed when payment was demanded in 1826, but everything was completed to his satisfaction by 1829. The harbour today bears little resemblance to Nimmo's plan, which must have proved too expensive to implement fully.

Nimmo also gave a plan for a small harbour on the south side of Clare Island at a place called Port A'Chuille. This is illustrated in the Eighth Report of the Commissioners and could be completed for about £250.[27] They granted Sir Samuel £92 for this work on condition that he contributed a like sum. He engaged Bald for the task and the latter changed Nimmo's plan after he had come under pressure to do so from the local fishermen. They felt that the proposed north pier would produce a reflux against the main south pier to the detriment of the harbour. Because of the disagreement, work never got started and Donnell felt that the plans could, in any case, never be completed properly for the sum available. Nothing further was done and the commissioners revoked their grant some years later.

Old Head Pier

Location: N 53 47.040, W 009 41.357

From Westport, take the R335 road towards Louisburgh. About three miles beyond Lecanvey village watch for a turn to the right signposted for Old Head. Follow that road (L1827) to the car-park at its end close to the pier.

This locality was approved by the Mansion House Committee for a pier in 1822 and received a total of £461, Nimmo's full estimated cost of construction. It was begun immediately and made good progress until the funds ran out and it was left unfinished. The facing was made in hewn sandstone ash-

APPENDIX 1 A GAZETTEER OF NIMMO'S PIERS AND HARBOURS

FIGURE APP. 1.11
Old Head Pier (detail). The original head of Nimmo's pier was at the kink between the two chains seen hanging in this photo. Note the difference in the stone work on either side of the kink.

lars brought from the north side of Clew Bay (the same sandstone as was later used for Glenanean Bridge), which had contributed to the high cost.[28] Nimmo requested further support to raise the pier another course, pave it, and give it a new parapet and perhaps a return head, or cant, forty feet long. He saw no benefit in extending it a further one hundred feet, as some had proposed. When Donnell visited in February 1825 he found the work still unfinished, estimating that it needed a further £1,340 to complete to his own plan. He also devised an alternative plan on a smaller scale costing half that amount. The Marquis of Sligo insisted on retaining and extending the original, and agreed to contribute £417 to its furtherance if the Commissioners would add £927 more, which they did. The final cost came to £1,806 and it was almost fully completed in 1829, when Donnell again reported. Thereafter the pier was dropped from further mention in the Fisheries Commissioners' reports.

The phases of its construction can still be seen in the pier today. The smallish sandstone ashlars at its root mark Nimmo's original section: they extend outwards in fair, unmortered courses (topped with a layer of larger limestone blocks) for about 150 feet to a protrusion in the pier face that probably marks the head of the original structure (figure App. 1.11). Beyond the protrusion, the pier extends another seventy-five feet approximately before turning right in a cant of about thirty feet. The stone blocks in the extended section are mixed and larger than in the previous, laid in tighter courses and are far better cut and worked. The whole modern surface is paved in concrete and the parapet is made of quite different stone from the pier face. Repairs have been effected near the root and a slip has been built to the right of the pier. As usual, concrete has been liberally used to 'improve' the general area. Nimmo's original indicative plan is shown in the Eighth Report of the Commissioners.[29] As it stands, the structure is worthy of further expert study and conservation.

Inishturk Pier

Location: N 53 42.157, W 10 05. 224

Inishturk is an island about six miles off the coast of Co. Mayo. Take the ferry from Roonagh Quay direct to the island.

Inishturk had no real harbour in 1822, its only place of anchorage being a small bay with two inner coves on the east side of the island. The usual landing place was in one of the coves and even that was not safe in easterly winds or during storms. For this reason, only two or three yawls were stationed there, despite its proximity to some of the best fishing banks along this coast. A Mr S. Knight, who surveyed the island on behalf of Lord Lucan, impressed on Nimmo, in the latter's capacity as engineer of the western district, the need for a pier. Nimmo accordingly proposed to the commissioners that whenever they would come to fund a second group of piers, Inishturk should be on the list. Later on, he changed his mind for some unknown reason and wanted the money for Inishturk to be given to Inishbofin, but it was too late to change. The eventual allocation was £382, half coming from the London and Dublin Castle funds. He visited the island in 1823 with Lieutenant Dundas, RN and examined the landing cove. Together they concluded that the swell within it could be largely stilled by constructing a breakwater mole from the south shore out to a rock in mid-channel, closing off part of the entrance that was not navigable in any case. In that way, part of the swell would be shut out and would spend itself on the rocks outside. The plan is shown in the Eighth Report of the Commissioners.[30]

It was, in truth, a minimalist solution and William Bald was left to make the mole. Hearing later that he had spent most of the money on unnecessary stonemasons and stonecutters with little result to show for it, Nimmo sent an experienced assistant, Alex McGill from Dunmore harbour, to prosecute the work. Later he sent another assistant, Mr O'Hara from Killough harbour, to assist and in due course they had the mole completed. The entire fisheries grant was exhausted by then, but at least the scheme seemed to work and the inner part of the cove was much quietened, even before any other structures were started. This was the fourth time that Nimmo had engaged Bald as supervisor and on each occasion there was a lack of satisfaction, due largely to Bald's profligate expenditure.

James Donnell visited in November 1824 when the breakwater extended half way across the mouth of the cove.[31] He reported that the work had indeed quietened the harbour but some more needed to be done. Despite his approval, a memorial was allegedly sent to the Lord-Lieutenant in December 1824 complaining of supposed damage to the cove inflicted by the works; this was promptly denied, as we saw in chapter 10. The breakwater was finished and things were in a generally satisfactory state when Donnell next reported in 1827. He prepared specifications and plans, shown in the Tenth Report of the Commissioners, for a parapet to be put in place, steps to be made at the root of the pier and various rocks to be blasted to make the approach cleaner.[32]

Bundurra (Bundorragha) Quay and Pier

Location: N 53 36.350, W 09 45.162

On the R335 road from Louisburgh to Leenane turn right to the pier at the bridge where the road approaches Killary harbour.

APPENDIX 1 A GAZETTEER OF NIMMO'S PIERS AND HARBOURS

The Marquis of Sligo wrote to the Fisheries Commissioners in May 1822 offering £300 towards the cost if they would erect a pier at this site, located at the mouth of the Bundorragha River where it enters Killary Harbour.[33] He was one of the earliest to make such an offer and he also financed the pier at Old Head elsewhere on his vast estate. Nimmo had already surveyed the locality and drawn up a suitable design (probably in response to a private request from the Marquis), which the Fishery Commissioners duly forwarded to his Lordship for his well-anticipated approval. The works were constructed under a contract dated 12 August 1824, Nimmo's assistant William O'Hara being the successful contractor at an estimated cost of £600. A quay 152 feet long, was projected along the shore on the west side of the river, with a short jetty-pier of seventy-five feet projecting out from it, having stone steps on its east face. Fishermen's cottages were planned behind the quay, some of which were built before 1829. The articles of agreement, with Nimmo's plan of the site, including the proposed small fishermen's village, have survived.[34] The plan is reproduced here in Plate 13. By November 1824 the work was sufficiently advanced for O'Hara to receive a staged payment, and everything was completed by January of 1825, a very rapid outcome.[35]

The site was protected from the open Killary Harbour by a spit of land on the east so that boats had good shelter for the winter. At the time there was no road from there to Leenane so the location of this pier across from Leenane pier made it quite useful for general trade as well as the fisheries. When all the agreed structures were almost completed at the end of 1824 the Marquis requested a new slip be built to accommodate the drawing-up of salmon draft-net boats. Nimmo estimated that this would add another £25 to £30 to the cost, a supplement the Marquis was understandably reluctant to pay. However, he relented in the end and agreed to undertake the work to the specification of James Donnell, using his own men, but only after the Commissioners agreed in 1827 to give him £15 towards the cost.36 Today the works are substantially as designed by Nimmo (Figure App. 1.12) except for the recently applied concrete cap that almost ubiquitously disfigures Irish piers of the nineteenth century.

FIGURE APP. 1.12
Nimmo's quay at Bundorragha, with the Marquis of Sligo's slip in the foreground. Picture taken from the jetty projecting from the quay, now covered in concrete.

FIGURE APP. 1.13
Nimmo's pier at Leenane. Note the rubble stonework in rude courses, the steps and the unpaved surface.

COUNTY GALWAY

Leenane pier

Location: N 53 35.757, W 09 42.383

On the right-hand side of the R335 road going from Leenane towards Kylemore, just beyond Leenane Hotel.

Nimmo made all the roads leading to Leenane. Two were from Westport and Cong. The third was a development of the old bridle road that ran along the south shore of Killary harbour from Kylemore. While making an embankment along the shore to support this road he took the opportunity to erect a rough jetty-pier near the village. In his own words, 'this work, although rough, answers as an excellent boat quay and is of considerable value in the time of the fishery'.[37] It was used for trading with Galway, rather than an aid to the fisheries, its location near the junction of the three roads leading into Leenane making it particularly useful for a wide catchment. Receiving no support from the Fishery Commissioners, Nimmo built it entirely with western district funds, so that it was never mentioned in fisheries reports. Rough it may have been, but it still stands today and serves the fish-farming industry that now occupies Killary harbour (figure App. 1.13). It is well protected by a later pier built alongside it. Together, both piers form a small harbour at the site. The vernacular nature of the old pier's construction, with its rude stones laid in uneven courses and rough-cut steps, is reminiscent of Cloonisle but even more rustic and, like Cloonisle and Raughley, its unpaved, grassy surface lends it an air of pastoral simplicity, contrasting markedly with the finished appearance of its partner alongside. Like Cloonisle, it well merits conservation as an outstanding example of its *genre*.

Derryinver Pier

Location: N 53 34.297, W 09 58.864

Take the N59 road into Letterfrack. At the village, take the road signposted for Renvyle. Take the left fork on this road, following the coast loop until you see the pier on your left just beyond Derryinver village.

APPENDIX 1 A GAZETTEER OF NIMMO'S PIERS AND HARBOURS

Nimmo built a small harbour here with western district funds, on the estate of the Blakes of Renvyle in 1822 or 1823. It consisted of a pier built along the rocks, then turning through a right angle to enclose a basin of about one hundred feet square, shown in the Eighth Report of the Commissioners, although it was not funded by them.[38] Donnell never mentioned it and it did not feature in the commissioners' list of piers in 1829. Today it has been considerably altered by restoration work but the original stonework is visible in places and the general lay out is unchanged from Nimmo's time (figure App. 1.14).

Cleggan Pier

Location: N53 33.433, W10 06.674

On the N59 road from Letterfrack to Clifden, look for the wide turn to the right at Streamstown, signposted for Cleggan. Carry on to the end of that road at the village of Cleggan.

Nimmo found that Cleggan Bay had five to ten fathoms of water and a clean bottom, making it an acceptable harbour for boats of all sizes. Its location made it suitable for communication between Inishbofin and the mainland and it was also close to the fishing banks. For these reasons he chose it as one of the twelve sites to be funded by the Mansion House committee in 1822 and it was granted £461, a quarter coming from the Mansion House and a quarter from the Dublin Castle fund. He selected a small bight on the south shore of the bay that had a bog hollow inside it, located on property owned by his friend Thomas Martin of Ballinahinch. A cut into the bog hollow would give an inner dock, the entrance to which could be protected by a short pier; a plan is shown in the Eighth Report of the Commissioners.[39] Operations were placed in the hands of Alexander Hay, 'a young architect of talent and good character, though not of great experience in such works'. Initial progress was good, but a gale in October caused considerable damage and filled the entrance with gravel. Nimmo sent two experienced men to do what they could to fix the dilapidation, but by then the money had run out and all work beyond necessary repairs had to be suspended. Already the pier was home to over one hundred boats engaged in the herring fishery, and the resultant cart traffic was cutting up the surfaces on the new Connemara roads.[40] A further grant of £200 for the pier

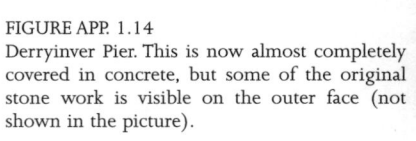

FIGURE APP. 1.14
Derryinver Pier. This is now almost completely covered in concrete, but some of the original stone work is visible on the outer face (not shown in the picture).

was made in January 1824 and Nimmo brought in Alexander McGill from Dunmore to take over. McGill had twenty years' experience in pier building and Nimmo had great confidence in him. It was due to his skill and exertions that the works at Cleggan were put in a secure state and eventually taken on almost to completion. The pier was straight, 200 feet long and fourteen feet high at the head. It had a parapet wall and a sloping pavement to seaward that was rough but substantial. According to Nimmo, it needed a cant, or return jetty, at the head, and the entrance to the dock needed to be deepened and protected. McGill therefore commenced to build a pier on the opposite side of the entrance. The inner basin excavated from the bog was one hundred feet by sixty feet, without any proper quays, although some masonry work was started early on and would only be completed much later.

Donnell inspected Cleggan in October 1824 and again in 1825.[41] He proposed some modifications and alterations that would cost £378 in the long term, but he decided to do just £55 worth of holding repairs in the 1825 season in order to check the progress of dilapidation. He expressed concern that Nimmo had charged £150 extra to the Cleggan account (possibly the wages and other requirements of McGill) and he, Donnell, did not know where that money had come from. Perhaps it was just as well that he did not know that Nimmo had 300 men employed on the road to the pier in 1822![42] Donnell appears not to have appreciated the 'relief' nature of much of the work undertaken by Nimmo, or the fact that Nimmo had authority to utilise western district funds quite distinct from the Fisheries Commissioners' funds.

A colour drawing of the works in 1826, archived in the National Library, is shown in Plate 7. In January 1829 the commissioners published Donnell's plan for Cleggan in their Tenth Report.[43] It included major alterations to McGill's wall alongside the entrance, making a proper passage into the inner dock. Failure of local interests to contribute to the cost meant that little was done immediately, but later on a new jetty was added opposite the main pier. The harbour was further altered by public works in 1884; Robinson says it was altered again in 1909.[44] Today there are small outer and inner docks at the site and both are extensively covered in concrete. Otherwise, the general layout of the site is recognisable from Nimmo's and Donnell's time except for the landward corner of the inner dock, which was once 'Cribben's garden'.

Clifden Quay and Doughbeg Wall

Locations: N 53 29.15, W 10 01.7810(Quay); N 53 29.13, W 10 03.01 (Wall)

Leave Clifden by the only road leading to the quay, less than half a mile distant. Then continue along that road to its end near Clifden Lifeboat premises, which are located where the Doughbeg wall joined the mainland to the small island off this point.

Clifden did not exist as a proper village before 1822. Having no pier or landing place all that the local fishermen could do was draw their boats up on a small bank of gravel at Doughbeg on the north side of the sea arm leading to the site of the eventual village, but over a mile from it. John D'Arcy, the proprietor, had a small storehouse on the shore there, near his newly erected castle. These were the earliest structures of what would become Clifden. In 1821 D'Arcy requested the Fisheries Commissioners to help with the provision of a pier, and a contract was soon agreed with him to start construction with a grant of £495. He did nothing immediately and in 1822 the commissioners authorised Nimmo to take the matter in hand without further delay as a means of employing the poor of the district. Nimmo first built a stone wall, backed with gravel, to join the Doughbeg gravel bank to the mainland, thereby making an immediate improvement to the landing place.[45] This created a sheltered lay-by in the outer harbour and was used by some boats as a landing wharf. More importantly, he surveyed the whole location and determined on a site that was suitable for a proper stone quay further inland, the site of the present quays.[46] The commissioners granted £225 for the work, including the cost of the Doughbeg wall, and D'Arcy was to provide the rest. Nimmo's plan, shown in the Eighth Report of the Commissioners, called for a longshore quay of 600 feet with an elbow bend in the middle.[47] The upper section was intended as a boat quay, where small craft could lie alongside on the mud; the lower part was meant as a ship quay, with deeper water where larger boats could lie afloat. D'Arcy began work in 1822 as required by the original contract, using Nimmo's plan and employing Samuel Jones as supervisor. Unsurprisingly, the money soon ran out, even though Nimmo added an extra £342 from western district funds. When Donnell reported in 1825, only 175 feet of the ship quay (below the elbow) had already been constructed and 314 feet, effectively all, of the boat quay (above the elbow) had been raised to within two feet of its planned height. The upper boat quay was made as deep as the lower, a modification that Nimmo felt, in hindsight, was justified since trading boat traffic had increased rapidly. Donnell complained that the whole should have been completed by August 1823, but had not been, and he drew up specifications to prolong and secure the ship quay.[48] However, far from being seriously behind schedule, it is remarkable that so much had been achieved in the

FIGURE APP. 1.15
The quay at Clifden. The boat quay is the section above the elbow; the ship quay is in the foreground below the elbow.

prevailing circumstances. By January 1827 Donnell was again making recommendations for repairs and modifications to facilitate fisheries use only, not for commercial trade. By 1831, however, the quays were of more value in general trade than in the fisheries, thereby making Clifden an important port in the region. Today, the Doughbeg wall is gone, its position being marked by some large blocks and stones scattered along the low causeway joining the land beside the lifeboat premises to the small grassy island that was once the sandbank. At Clifden, the quays still provide mooring and landing facilities for small craft (figure App. 1.15).

Roundstone Harbour

Location: N 53 23.77, W 09 55.05

In the centre of the village of Roundstone, Connemara.

This is another of the piers that received relief funds (£400) in 1822, half from the Commissioners and £91 each from the Mansion House and the Dublin Castle committees. There was no village there when Nimmo selected the site and constructed a wharf 150 feet long with a jetty originally sixty feet long. The indicative plan is shown in the Eighth Report of the commissioners.[49] Work was begun by day labour in 1822 under the management of the young architect Alexander Hay. The tenant of the farm on which the pier was erected alleged that he had suffered damages during the construction and was eager to be compensated. At his own expense, Nimmo bought out the man's interest in the lease as the best way to settle the claim. Thereafter Nimmo held it by lease under Thomas Martin of Ballinahinch. He built an office and store that he later leased to the Fisheries Commissioners and subsequently used as an office for his western district staff. By winter 1822 the harbour works had been completed to spring tide level using rough granite blocks brought in specially, not for any aesthetic effect but simply to foster employment in the wider district. The harbour had an immediate effect in stimulating trade and fisheries, being the last port of shelter before rounding Slyne Head, and having five feet depth of water at the pier head. By 1824 there was some

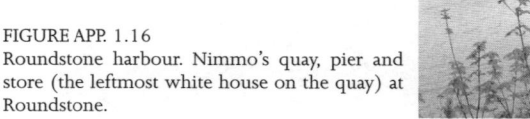

FIGURE APP. 1.16
Roundstone harbour. Nimmo's quay, pier and store (the leftmost white house on the quay) at Roundstone.

dilapidation, so another of Nimmo's assistants, Alex McGill, was called on to rectify matters.[50]

Once started, the works were carried well beyond the original plan so that when Donnell visited in 1825 the wharf was 217 feet long and the pier 150 feet. Nimmo had allocated £150 from the western district funds to it, something he had done elsewhere as well, much to Donnell's stated disapproval. Donnell proposed to re-cope and raise the work, putting a parapet on the jetty, work that was carried out in and after 1825. By 1829 it only needed some mooring posts to be made perfect.[51] By then a new village was growing rapidly and the pier (figure App. 1.16) was in use for trade, for fisheries and even for careening the gun-boat named *Plumper*.

Later in the century, a second pier was built opposite Nimmo's, enclosing the small harbour that is Roundstone today. Nimmo's pier is the more southerly of the two, where his much-altered storehouse is now a residence occupied by the cartographer Tim Robinson, making an interesting connection between the two great historiographers and cartographers of Connemara.

Cloonisle Quay

Location: N 53 26.03, W 09 51.36

The quay forms part of the road R342 from Ballinahinch to Cashel, about a half-mile beyond Toombeola bridge.

It was Nimmo's policy to make a quay whenever any road he was building tracked along the edge of the shore. At the head of Roundstone Bay his new Connemara coast road crossed the Ballinahinch River at Toombeola. Near there, the road passes through Cloonisle, where it skirts the shore beside some steep rocky ground. He took the opportunity of forming a small quay as part of the road, which he said 'may yet be of importance to the district, especially for the shipment of the beautiful green marble from the Twelve Pins mountains'.[52] Because it was never intended for fishery use it received no funds from the Commissioners and neither they nor Donnell ever alluded to it at any time. Nimmo provided the necessary finance from the western district funds. The quay and jetty are still intact today, largely as Nimmo left them, although

FIGURE APP. 1.17
Cloonisle is an attractive simple road-side quay and jetty built by Nimmo in a vernacular style with stones laid in rude courses. It is now in need of some attention and its conservation is much to be desired.

they are somewhat distressed and in need of repair. Formed of rough-cut stones laid in rude dry courses, with the jetty at one end, they constitute a fine example of a simple, vernacular quay and jetty of the time, well worthy of conservation (figure App. 1.17).

Gorumna Pier, Greatman's Bay

Location: N 53 16.772, W 09 39.156

Take the R374 road through *Ceanntar na n-Oileán*. As you enter Gorumna Island at Tír an Fhiadh village the pier is visible down to the left. A small road leads down to it.

Although communication in Greatman's Bay and throughout *Ceanntar na n-Oileán* was almost entirely by water in 1822, the region had not a single quay or pier for unloading, or where boats could be safely laid up. When local interests presented a memorial seeking a harbour in 1822, Nimmo had no hesitation in identifying a suitable site near the village and started construction of a pier by way of relief work; his indicative plan is shown in the Eighth Report of the Commissioners.[53] The famine of 1822 had passed before the work was fully completed. By 1826 the pier was nearly 500 feet in length; half of it was walled but not filled in behind, and the outermost part was still little more than a rough breakwater.[54] It was designed to have a return head of one hundred feet, but that was never built. An application was made to the commissioners to complete the works in 1828, but since the undertaking had been initiated with western district funds rather than fisheries funds, they refused support. The pier continued to go unrecognised by the commissioners and their engineer James Donnell never inspected it. The shortened pier was finished off in stone, but the mole that was to make the remainder of the pier and the return head were left unfinished. Today the rough mole can still be seen, extending straight outwards from the pier head and then turning sharply to the left. When viewed from the beach along its right hand side, as in figure App. 1.18, the pier exhibits quite a graceful, almost feminine, sinuosity, not common in harbour works.

FIGURE APP. 1.18
Gorumna pier at *Tír an Fhiadh*, showing its curved form.

APPENDIX 1 A GAZETTEER OF NIMMO'S PIERS AND HARBOURS

An tSean Céibh Pier, Rossaveal, Casla (Costello) Bay

Location: N 53 16.02, W 09 33.58

From the R336 Screeb to Galway road take the R372 road to the right at Casla Bridge and follow the signs for the Aran ferries. Nimmo's quay is on your right as you enter the ferry area.

The turn-off from the R336 road to the R372 occurs at Casla Bridge, originally built by Nimmo, but now replaced entirely with a new structure. Nimmo's pier is part of the ferry port, the ferries departing from its head. The original stonework can be seen along part of the right-hand side of the pier; to the left-hand side, behind the pier wall, the inlet is now filled in and covered over, making Nimmo's part of the pier a quay rather than a jetty, as it was originally.

Located on the eastern side of Casla Bay, this pier was supported in 1822 with funds from the commissioners, the Mansion House committee and the Dublin Castle committee. Nimmo's plan for it is shown in the eighth report of the Commissioners.[55] The work was superintended in the first instance by Lieutenant Dundas, RN, a local resident, but was soon given on contract to a Galway builder.[56] It was increased in size from the original plan because of the high rise of the tide in Casla Bay, and in the course of this alteration the head was turned from east to west. When finished, the pier was almost 300 feet long and twenty-one feet wide, quay-faced on both sides and coped with hewn limestone from Aran. It had four feet of water at its head at low tide, so that it was accessible at all tides to sailing boats, whether traders or involved in the fisheries. Nimmo contributed £375 of western district funds to it and to a spur road (now the R372) connecting it to his new road to Galway (now the R336). In 1824 Donnell submitted estimates to repair some defects and these were rectified, not with complete satisfaction, the following year.[57] By then the pier had proved its usefulness both for the fisheries and for the coastal trade and by 1828 it was in a very satisfactory condition.

This quay is known today as 'an tSeancéibh' (the old quay) and was the original quay at this well-known place – Rossaveal – now a major centre for the west coast fishery fleet and the ferry terminal for the Aran Islands. The

FIGURE APP. 1.19
Nimmo's quay, An tSean Céibh, at Rossaveal. In parts of this quay the original walls can still be seen, but the original jetty has been converted into a large concreted apron.

363

bight to the west of the pier has been reclaimed and the west face of 'an tSeancéibh' has been completely built over. Part of the east face has been concreted but some of the original stonework is still visible (figure App. 1.19).

Spiddle (An Spidéal) Pier

Location: N 53 14.57, W 09 18.267

Take the turning towards the sea at the Galway end of Spiddle village (not the crossroad at the village centre). This leads directly to the pier.

Only the ruins of an earlier pier remained here when Nimmo examined the site in 1822. It was allocated £150 by the commissioners, including £75 from the combined Dublin Castle and London funds. Local proprietors subscribed £125 and Nimmo contributed £175 of western district funds. He began the work with day labour (including some men from Dunmore harbour) under the management of Mr Morris, a local proprietor. The original indicative plan is shown in the Eighth Report of the Commissioners.[58] A further £262 was needed from the Government and £100 from Sir Robert Staples in order to continue construction in 1824.[59] The quay was made of rough granite, protected on its seaward side by existing rocks (figure App. 1.20). The cove sheltered by the pier, about one and a half acres in extent, was clean and sandy, but it dried at low tide and had only ten feet at high water. The pier itself extended 216 feet and the inner shore of the cove was walled as a quay, but in a very inferior way. In 1824 James Donnell specified some necessary but rather trivial improvements, which were carried out in 1826.[60] The whole harbour was in good condition in 1828, but its shallow nature severely limited its value. Today it is much as Nimmo designed it, but has been extensively strengthened and the inner margins of the harbour have been re-quayed in recent times. The pier is known as 'an tSeancéibh' and should not be confused with the much larger and later pier west of it, which is Spiddle pier to most people.

FIGURE APP. 1.20
Nimmo's pier (an tSean Céibh) at Spiddle.

APPENDIX 1 A GAZETTEER OF NIMMO'S PIERS AND HARBOURS

Barna Pier

Location: N 53 14.938, W 09 08.934

Travelling from Spiddle towards Galway on the R336, turn right at the crossroads (traffic lights) in Barna village.

Prior to Nimmo's time there was a small village named Freeport and a harbour here, built at the private expense of the Lynch family. The pier was 470 feet in extent, with a lighthouse at its head. Inside, the harbour was walled with quays to an extent of 620 feet. The pier was demolished in one night by a storm many years before Nimmo, and the remainder of the structure had suffered severe damage after that, being a virtual ruin in 1822. The inhabitants were then in great distress, so Nimmo gave £150 from the western district funds to General Elly in order to commence repairs by way of relief work, and the commissioners gave £276, which included £145 from the London and Dublin Castle committees. When Elly left (he was there, according to Nimmo, only for the sea-bathing!) the work was carried forward by day labour.[61] The pier was substantially rebuilt to an extent of 300 feet, about as much as was really necessary, but the shortage of funds did not permit any work on the inner quays or the parapet, and the pier head was left unfinished. Nimmo drew up a further plan and specifications for its completion and later, in October 1824, Donnell drew up more specifications for rebuilding at a cost of a further £587, half to be raised locally.[62] He was of the opinion that Galway Bay had already received more than its fair share of public money and if local subscriptions were not forthcoming, Barna pier should simply be tidied up and left. No local contribution was made and the commissioners duly instructed him just to make the works secure. A plan of the pier as it was in 1826 is shown in the Eighth Report of the Commissioners.[63] Its state in 1829, with the pier head still incomplete, is shown in their tenth report.[64] Nothing more was done immediately, although the pier head and inner wall were completed some years later. Today the pier is a substantial structure that protects a small harbour that dries out considerably on low tide (figure App. 1.21).

FIGURE APP. 1.21
Barna pier as it is today.

Harbour of Galway, Slate breakwater (Nimmo's pier)

Location: N 53 16.01, W 009 02.48

At the entrance to the harbour, Claddagh, Galway city.

The harbour of Galway was very defective in 1822, with only one small dock, now known as the Mud dock, and two short, narrow jetties on the Claddagh shore. It was a difficult harbour to enter: the outflow of the River Corrib was too great to permit sailing boats to enter safely at low tide. They had to wait outside in the roads for high tide, and even then entry could be tricky when the wind was strong or contrary. The merchants of the town petitioned the Lord-Lieutenant in 1822 to make improvements and he forwarded their memorial to Nimmo on his arrival.

Nimmo saw immediately what needed to be done: the Claddagh fishermen needed a quay and a proper place to draw up their boats in safety; the whole harbour needed better protection from the west and south-west, together with a new floating dock; and the town needed a canal joining the harbour to Lough Corrib. At the time, the boatmen of Menlo used to haul their boats across the town to the sea when the herring fishery commenced.[65] He put men to work immediately, deepening the Claddagh beach, raising the jetties and making a rough quay wall along the shore, his plan for which is shown in the Eighth Report of the Fisheries Commission.[66] This was only a temporary expedient to give much-needed employment and Donnell would give specifications for the extension and improvement of the works in 1828.[67]

Nimmo's main priority was to increase the shelter of the harbour by erecting a breakwater along the rocky outcrop known as the Slate at the west of the entrance. It was to be 500 feet long with a return of forty feet at its head, constructed of hewn limestone from Aran, with a sloping pavement to seaward. The work started in 1822 and was completed by 1827, giving the harbour considerable shelter and contributing greatly to its general usefulness (figure App. 1.22). The breakwater was detached from the land but was continuous with a long spit of gravel that stretched westwards from it along the open shore. To get to the new breakwater it was necessary to cross a creek and marsh

FIGURE APP. 1.22
The Slate pier (Nimmo's pier) Galway, dwarfed by the modern Japanese fishing boat as it enters Galway port.

of about twenty acres that was overrun by high tides, located in the area between the gravel spit and the present Claddagh. Nimmo proposed to build an embanked road 300 yards long across the marsh and creek, joining the breakwater to Claddagh village and providing more quay space for boats. In this he planned an opening, crossed by a stone-arched or a swivel bridge, leading into the creek and marsh, transforming them into a five-acre dock for fishing boats.[68] The gravel spit could be easily raised above high tide level, giving extra protection to the new dock. This was the same idea as he would propose later for St Kitt's (Killeenaran) pier and for Kilbaha and Ballinacourty piers; it would have cost over £1,000 to carry out in the Claddagh. The marsh and creek are shown in pictures of the place dating from 1888.[69] Altogether his proposal would have made a tidy small harbour for the Claddagh, but it was never pursued. The spit was raised and the marsh subsequently filled in, the site of Nimmo's proposed harbour becoming a municipal dump in the twentieth century, now converted into a leisure park (South Park).

West of the Slate breakwater, he envisaged raising the low natural causeway that joined Mutton Island to the mainland to a level above the tide, thereby increasing the shelter in the roadstead and providing, on its eastern face, extra wharfage that boats could use in preference to going all the way to the inner harbour. On the island itself he proposed a small floating dock in the sandy bight at its eastern end. None of these proposals met with approval at the time. In recent years the natural causeway has been raised as he proposed, with a service road running its length, but it has not been quayed. The gravel spit linking the causeway to the Slate breakwater has also been raised and paved and is now rarely overflowed by the tide. The Slate breakwater was joined to the Claddagh, as he proposed, by an embanked road some time later in the century. The whole embankment and pier, 600 yards long and all now paved, are today called Nimmo's pier, but the pier proper only commences at the end of the parapet and sloping sea-wall on the right-hand side, about 400 yards from the Claddagh roadway. The sloping cut of the sea pavement limestone ashlars behind the parapet is quite a remarkable feature of the pier, which overall is a monument worthy of Nimmo (figure App. 1.23).

Galway Floating Harbour and Corrib Canal
Location: N 53 16.15, W 009 03.03

The docks are part of the central area of Galway city.

Once the Slate breakwater was completed the harbour was much more sheltered and easy to access. The merchants therefore approached Nimmo to design floating docks and to pursue the necessary parliamentary approval that such a venture necessitated.[70] They also wanted a canal joining the docks to Lough Corrib.[71] Nimmo set to work and on 17 June 1830 a local act was passed, 11 Geo IV & 1 Wm. IV, cap. cxxii approving the venture.[72] The act was obtained principally on the strength of Nimmo's plan for a canal, whose route

FIGURE APP. 1.23
The finely cut, sloping ashlars of the sea face of Nimmo's pier, Galway.

was to be from Woodquay, along today's Eglington Street (approximately), through today's Corrib shopping centre to Victoria Place and into the proposed new dock basin.[73] It was the route he had suggested almost eighteen years previously.[74] The Act set up harbour commissioners with the power to borrow £50,000 towards the venture. The commission formed a deputation to take Nimmo's plan to the Loan Commissioners, who sent it to Killaly, their engineer. The latter thought the plan too expensive and proposed a smaller, less costly scheme based essentially on Nimmo's scheme but omitting the canal. The harbour act therefore had to be replaced by a new act in August 1831, 1 & 2 Wm. IV, cap. liv, which reduced the commission's borrowing power to £40,000.[75] The new board of works approved Killaly's less expensive plan (although the Galway commissioners preferred Nimmo's); it was cheaper because it omitted Nimmo's canal.[76] Nimmo's own drawings for the harbour have not come to light, but his proposed canal is shown in Killaly's drawing in the report of the OPW and shown here in figure App. 1.24.[77]

Construction of the docks commenced in 1832, under the supervision of Killaly and his son, Hamilton, who were by then engineers of the OPW. It was

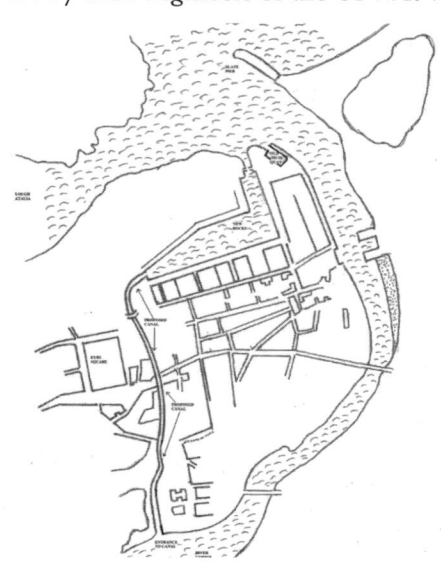

FIGURE APP. 1.24
Killaly drawing of Galway Docks. Note the canal proposed by Nimmo shown running vertically though the sketch.

one of the last Irish works for both men: John Killaly died later in 1832 and Hamilton left Ireland for Canada, where he made a very great reputation for himself in engineering, becoming eventually President of the Canadian Institute of Engineers, among other honours. While Galway docks were not built to Nimmo's exact design, nevertheless it was his initial work that informed the enabling legislation making docks possible, and his Slate breakwater that made the place secure for them. The canal to Lough Corrib that he championed would eventually be made (the present Eglinton canal), but not along the route he had favoured since 1813.

FIGURE APP. 1.25
New Harbour, Rinville. Different phases of construction are evident in the stone work, culminating in modern concrete.

New Harbour (Rinville)

Location: N 53 14.40, W 008 57.56

In Oranmore village, take the road signposted for Maree. Then follow the signs for the Galway Bay Yacht Club, which is located at the pier.

This harbour traditionally provided a safe refuge for small boats in westerly winds. The commissioners granted £377 for a pier, including £200 from the Dublin Castle and London committees in 1822, with an extra £147 in 1823, making a total of £535. As planned, it comprised a wharf 150 feet long extending sixty feet from the shore with a jetty, twenty-three feet long, projecting at one end, all faced with hewn limestone. An indicative plan is shown in the Eighth Report of the Fisheries Commissioners.[78] There was no road to the site, which was used mainly for careening and wintering, as it still is today. In October 1824 construction was in progress by contract when Donnell visited and recommended improvements to the standard of the work, indicating some necessary repairs.[79] In January 1827 he recommended further repairs involving the use of iron bolts to strengthen the structure where the stonework was loose and shaken.[80] This was completed by 1828, when the structure was in good repair. In modern times the whole has been increased in size and capped with concrete (figure App. 1.25). Today it is mainly used for yachting and leisure purposes.

Ballinacourty Pier

Location: N 53 12.582, W 08 57.515

On entering Clarinbridge village from Galway, take the first road to the right. Take the left fork on this road and carry on to the T-junction. At the T- junction, go left and continue along the road past the graveyard on your left until you come to old farm buildings, also on your left. Turn left down the track at the old thatched house and stop at the pier.

The commissioners granted £276, half from the Dublin Castle and London funds, for a pier at Ballinacourty in 1822. Nimmo made a cut across a narrow gravel bank into a pool behind, which was to serve as a lying-up harbour. The cut was quayed on the east side, forming a wharf 154 feet long clad with hewn limestone. From the outer end of this, he projected an extension eighty feet in length, declining from half-tide level down to low water neaps. That gave added shelter as a boat quay and a landing place at half flood; boats could move up to the main wharf at high tide. The inclined section could be built up to high water level and a return head put on it at a future date should the harbour develop, a modification that was later carried out. Nimmo's plan is shown in the Eighth Report of the Commissioners.[81] A coloured plan signed by him is also extant in the National Archives of Ireland.[82] Work was very slow in starting and in March 1824 he drew new specifications for the work, which he forwarded to the commissioners.[83] A further allocation of funds was approved and two local contractors, White and Coen, tendered successfully for the work at £389.[84] However, in September 1824 Nimmo's assistant Alex Hay reported that little progress had been made, although by then the total expenditure had exceeded £785.[85]

Donnell inspected the site later that year; he confirmed that work was then at a standstill and all the funds had been spent. Twenty-five pounds would be needed for repairs to secure the work from damage by the sea, and £500 would be needed to perfect and complete the harbour. By June 1825 Hay reported that the pier was completed in a workmanlike manner agreeable to Nimmo's specifications. When Donnell next inspected it, in January 1827, he detailed some minor repairs that were needed, but agreed that the pier was then substantially complete.[86] Later it had fallen into disrepair and had to be extensively restored, when the inclined section was built up and a return head, facing upstream, was added. This gave the small harbour much greater shelter and usefulness. In modern times, the pier has been considerably extended and concreted; a plaque on it names it Lynch's pier, and it should not be confused with Dooras pier, on which there is a contemporary memorial plaque indicating that it was constructed by Patrick Lynch of Parkmore in 1823.

St Kitt's (Killeenaran) Pier

Location: N 53 11.84, W 08 56.45

From the N67 Kilcolgan to Kinvarra road take the road to the right in the centre

APPENDIX 1 A GAZETTEER OF NIMMO'S PIERS AND HARBOURS

FIGURE APP. 1.26
St. Kitt's pier, Killeenaran, Co, Galway. The arch leads through the pier into a small pond that Nimmo intended as a harbour for small craft.

of Ballinderreen village. Where this road divides take the right fork and continue on until the T- junction at its end (Killeenaran). Go left there and follow the road to the pier at its end.

St Kitts pier, otherwise known as Killeenaran pier, received £232, approximately half from the Dublin Castle and London committees in 1822 and a further £230 a year later. According to Nimmo, the village had 300 inhabitants in 1822, making it quite a settlement, and it was much frequented by Galway fishermen during the herring season. Nimmo started work by day labour and completed it by contract. The pier was 120 feet long with a return head of twenty feet, all clad in hewn limestone.[87] Towards the landward end, a small opening covered by a bridge pierced it; rowboats could pass through the opening into the pool beyond, where they could lie in safety separated from the open sea by a bank of gravel. The pier, of which a plan appears in the Eighth Report of the Commissioners, could be easily approached at high water but was dry at low water.[88] Work progressed until early 1824, when the pier had grown to 160 feet. It was then left unfinished, probably because it had already cost over £464 and the money had simply run out. In March 1824 Nimmo revised the specifications for its completion and the contractors, White and Coen, offered to finish the task for £326.[89] When Donnell inspected it the following October, it appeared to be progressing well but once again it was left in an unfinished state. It needed to be raised above high spring tide level and about fifty feet at the landward end (between the arched opening and the approach causeway) remained to be finished. He estimated that it needed another £250, an expenditure he was not prepared to recommend since the site was used very little for the fisheries; it was mainly used for landing turf.[90] Nimmo was not quite so negative: when writing to the commissioners on the state of the Galway piers in April 1825, he had begged them to get St Kitt's completed in the same style as other piers in the locality (Ballinacourty and Burrin New Quay). The exasperated reply of the commissioners is noted on the rear of Nimmo's letter: 'Remind him [Nimmo] that we requested a plan from him on 8/1/24 and 1/7/24.'[91] The Commissioners must have forgotten that he had sent the relevant specifications to them in April 1824.⁹² The works were subsequently completed satisfactorily and the pier continued to be useful for turf and other goods, but never really for the fisheries.

The pier is faced on its inner face with cut limestone; its sea wall of rough field stones is much ruder and more poorly finished. Today the arched opening is still mostly unobstructed, but the lying-up area behind is very silted and not suitable to shelter boats, which lie by preference within the shelter of the pier (figure App. 1.26). Nimmo had proposed, but not built, similar small lying-up harbours at Old Head and Claddagh. St Kitt's was the only one actually constructed, and for yhis reason its conservation is well merited on this ground.

Duras (Dooras) Pier

Location: N 53 10.17, W 09 00.32

Travel west from Kinvara on the N67. Take the turn to the right signposted for Dooras Hostel. Continue until you see the Travellers Inn tavern. Take the left turn just before the tavern and continue on that road, following the signs for Dooras church and school. Where the road swings to the left, signposted for the church, a small unpaved road continues straight on to the pier.

The proprietor of this place, Patrick Lynch, joined with the commissioners in funding the pier that he built in 1823 to a plan of Nimmo's, as shown in the Eighth Report of the Commissioners.⁹³ Built of hewn limestone with a substantial parapet, it was 120 feet long with a cant of forty feet. There was a depth of only ten feet alongside at high tide, and it was completely dry at low water. It was, in fact, sited far too high on the shore, so that it was of relatively limited usefulness, as both Nimmo and Donnell acknowledged. Lynch must have been very proud of it, however: he had a plaque inserted in the parapet wall recording that he had erected it in 1823. The commissioners had given him £156 towards it, together with the loan of some machinery during its construction, but refused to give any more once they realised its main usefulness was for trading with Galway, rather than an aid to the fisheries.94 Neither were they at all happy that no mention was made on the memorial plaque of their contribution to its construction. In Nimmo's words, it was 'neatly built', and Donnell said it was

FIGURE APP. 1.27
Dooras pier stands almost fortress-like, high up the shore.

'neatly and substantially executed'. Today it is substantially unaltered from the original. Up close it is indeed an imposing structure, made of finely cut limestone blocks laid with great precision and well meriting the engineers' complimentary words (figure App. 1.27). The parapet wall is currently breached in two places, but Lynch's memorial plaque is still in situ, hidden by weeds. To a casual observer, the whole structure is patently located too far from the water, even appearing slightly incongruous in its position when the tide is out. The sheltered harbour is almost occluded with gravel and stones and is never used nowadays.

Killany (Killeany) Pier (Aran Islands)

Location: N 53 06.18, W 009 39.37

Killeany pier is on Inish Mór, the largest of the Aran Islands. Ferries ply from Nimmo's old pier at Rossaveal to the Islands a number of times each day. Distress was so severe in the Aran Islands in 1822 that Inish Mór was one of the very first places where Nimmo took action. Immediately on appointment he engaged men to quarry and cut limestone ashlars for use in public works on the mainland. When the Fishery Commissioners granted £250 (half from the Dublin Castle and London committees) to build a pier near the village of Killeany, he laid out a plan for a harbour there, shown in the Eighth Report of the Commissioners.[95] Its execution was, in his own words 'not perfectly understood, at least not exactly followed',[96] which is rather strange, since his brother George was the supervisor there and he at least should have understood Alexander's plan, or been able to advise on it. This was the first time that George was mentioned in his brother's service; he was to stay with him until Alexander's death.[97] Although only twenty-six men were employed with a total weekly pay bill of less than seven pounds, the funds were exhausted before the work was completed. In January 1825 a further £461 was granted and the work was taken up again on contract under Donnell. Donnell inspected the work in October that year and revised the specifications for its completion.[98] After some further delays and repairs, the whole was completed to his satisfaction by 1826 and two years later it was proving to be of immense value, with a greatly increased fishing fleet attending there. The pier was located at the upper part of the inlet, but was dry at low tide. It was 245 feet long with a cant, completed in hewn local limestone. The road from the village was quayed as well, forming a useful summer wharf 326 feet in length. Up to a hundred boats could lie safely at Killeany but even that was not sufficient to accommodate all the boats that lay there during the herring season.

COUNTY CLARE

Burrin New Quay

Location: N 53 09.38, W 09 04.56

Travel west on the N57 from Kinvara towards Ballyvaughan. After about five

FIGURE APP. 1.28
The quay and jetty at Burrin New Quay.

miles, take the small road to the right signposted for New Quay. Continue on this road until you see the pier, which is located behind Linnane's tavern.

Before Nimmo, two attempts had been made to build a quay here. Only ruins of these remained when he arrived and he doubted that either of the earlier quays had ever been of any use. The commissioners granted £250 for a quay, half from the Dublin Castle and London committees. He designed the structure in hewn limestone near the innermost of the old ruined jetties, leaving an experienced man from Killough harbour to carry on the work by day labour.[99] As usual, the funds proved inadequate and the work was suspended for two seasons. When further funds were allocated, it was taken up once again. Costing £413 altogether, it consisted of a longshore quay 105 feet in length, projecting seventy-two feet from the land. At one end, a jetty pier seventy feet long and fifteen feet high, with a good parapet, extended outwards from it to low water. This was to become an important quay for fisheries and general trade, with a summer packet service to Galway. Apart from requiring mooring posts and rings, Donnell approved of it as a general utility quay in 1828, complaining only that it was 'sometimes too small to contain the numerous craft seeking shelter in it'.[100]

FIGURE APP. 1.29
Steps in the jetty at Burrin New Quay.

APPENDIX 1 A GAZETTEER OF NIMMO'S PIERS AND HARBOURS

Today, the jetty has been extended and capped with concrete, but the original dry cut stone east face is extant and visible (figure App. 1.28). The quality of the stonework at the steps is quite remarkable (figure App. 1.29).

Liscannor pier

Location: N 52 56.17, W 009 22.60

The R478 road from the Cliffs of Moher to Lahinch passes through Liscannor village. The pier is close to the main road in the village.

Built at the behest of General Sir Augustine Fitzgerald the local proprietor, this was the first of the fishery piers undertaken by Nimmo.[101] Within one week of being instructed by the commissioners, he had provided a plan and specifications, dated 24 May 1822, and work commenced immediately.[102] His plan was for a quay along the side of a small bight, with a jetty projecting 600 feet from it. Further in from the quay he proposed to excavate an inner basin where boats could lie in shelter. The site was a particularly difficult one due to the ferocity of the seas around Hag's Head, which accounts for his original intention to excavate inwards rather than to build outwards. However, initial trials revealed the difficulty and expense of excavation, so by April 1824 he was seeking an extra £700 to extend the jetty outwards a further one hundred feet.[103] That exposed the works to the swell and it became quickly apparent that rough seas would overtop the rather low pier that he envisaged. When Nimmo later increased the estimated cost yet again, the work was stopped temporarily while the commissioners considered how to proceed. They sent Donnell, their new harbour engineer, to examine the site and he presented a new plan in April 1825, proposing to extend Nimmo's jetty a further 200 feet, ending it in a forty-foot return head.[104] By building a new breakwater of 120 feet on the shore opposite this main pier, the small bight could be made fully sheltered and, crucially, it could all be done for less than Nimmo's original estimate. Donnell modified his scheme in December 1825 and tenders for the work were submitted in April 1826, a Mr McDonagh winning the contract for £1720.[105]

FIGURE APP. 1.30
Innermost part of Liscannor pier showing section built by Nimmo.

The work was recommenced in 1827, when the ferocity of the sea was again a problem leading to disputes between Donnell and Fitzgerald.[106] The latter had agreed to contribute £150, Major McNamara £50 and sundry other gentlemen £377, in addition to the £923 already allocated by the commissioners, so they had some right to be listened to. John Madden, overseer of the work, wrote to the commission explaining the problem: the initial weight of the storm surge fell on the angle of the pier head, dividing there and then rolling along the face of the quay up to the bog garden at the innermost part; the recoil of the surge around the new head covered the pier two feet deep and rose as far as Nimmo's old works.[107]

With hindsight, the rationale of Nimmo's plan to excavate inwards rather than build outwards in such an exposed place was obvious. Construction of quays and piers in such exposed sites was not a precise art and modifications had to be made to accommodate both the location and the skill of the builders. As a pier progressed, the flow of the tide around it and the deposition of beach material behind it would alter, and this would influence the structure being built. Donnell's plan was no less subject to these constraints than Nimmo's original, and to address the ongoing difficulties the commissioners called in Killaly. He visited Liscannor and proposed modification to the angle of the return head and an increase in the sea pavement. A composite plan of the proposals of Nimmo, Donnell and Killaly is conserved in the National Archives of Ireland, with detailed sketches illustrating the various proposed modifications in separate paper overlays.[108] These documents fully explain the evolution of the harbour. By 1828 when Donnell visited again, the works were approaching completion, but even the structure as then finalised required major alterations up to and beyond 1885.

Today the pier and harbour are much as Donnell had proposed, but with newer extensions made since his time. Nimmo's contribution to the present structure can be identified by examining the quay wall alongside the modern slipway (figure App. 1.30). The innermost part is Nimmo's original, constructed of rough-hewn stones and flags in rude courses; outside this, the next section is in well-cut stone laid in regular courses and is the part designed by Donnell and built by McDonagh; both these sections are topped with stone laid on edge. Beyond these again, the stonework is much more regular and square-cut, topped out with cut limestone. The harbour is still in use for fishery and leisure purposes.

Seafield Pier
Location: N 52 48.34, W 09 29.17
Travel south from Miltown Malbay on the N67. At the north entrance to the village of Quilty take the right fork signposted for Seafield. Continue along this road until the signposted turn-off to the right leading to the pier.

When the Fisheries Commissioners had first sent Nimmo to Liscannor they specified that 'he be further directed to continue his general survey on that line

APPENDIX 1 A GAZETTEER OF NIMMO'S PIERS AND HARBOURS

of the coast and report to the Board'.[109] He therefore did a rapid tour of the Clare coast, noting sites that seemed suitable for piers and making indicative drawings.[110] Seafield was one of those sites and a small, coloured indicative sketch of the location and proposed pier, most probably done by Nimmo, is extant.[111]

His original pier, on the Marquis of Conyngham estate, was 330 feet long, commenced in 1822 at a cost of £461, including £115 from the Mansion House Committee.112 In January 1824, when it was virtually completed, a further £184 was granted in order to cut a channel through the rock making access easier for boats. When Donnell reported its damaged condition in 1824 he was granted another £200 to finish the work to the specifications he drew up. The pier, however, remained essentially as Nimmo designed it. After some initial difficulty in finding a competent contractor, the work was finished by 1826 and proved useful to the fisheries.[113] Donnell specified final details of repairs in 1827, and his imposing plan is shown in the Tenth Report of the commissioners.[114]

Kilbaha Pier

Location: N 52 34.13, W 009 51.45.

From Kilkee take the R478 road to Loop Head. Kilbaha is the last village on the road.

Nimmo planned to make a small harbour at this site by cutting through the beach into the bog behind, where he would excavate a sheltered inner basin and dress it with a stone quay wall.[115] The cut across the beach was supported by two groyne piers, each 140 feet long, set forty-five feet apart. The commission granted £461 for the works in 1822, which included £115 each from the Mansion House and the Dublin Castle Committees.[116] In 1824, further sums were granted, bringing the total to £823. The works were in a ruinous state when Donnell visited in late 1824, so a further allocation of £114 was advanced for repairs and refurbishment. Even that was not enough, and the fol-

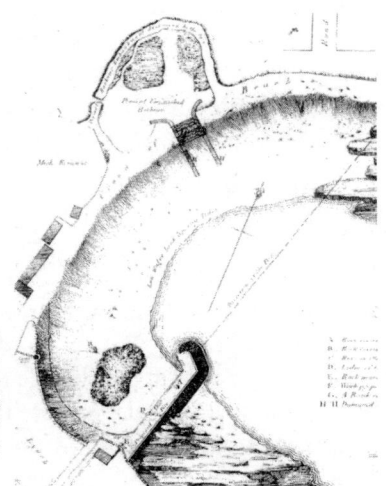

FIGURE APP. 1.31
Nimmo's plan for Kilbaha Harbour (upper part of picture) and Donnell's pier at this location (lower part of picture). Nimmo planned an entrance between two groyne piers leading into an excavated harbour backed by an embankment that could be quayed. Nimmo's groyne piers were undermined by the sea and eventually removed entirely to provide stone for Donnell's structure. (From ref. 118.)

FIGURE APP. 1.32
The site of Nimmo's harbour at Kilbaha. The pond on the left is the inner harbour, now cut off from the sea by the modern road.

lowing year Donnell estimated that £775 more would be needed to finish Nimmo's plan. Altogether, he expressed himself embarrassed at the state of the works: sand was being eroded from the back of the groynes, exposing them to dilapidation by the sea. Given his choice, he would abandon Nimmo's plan, although it had already absorbed over £930 pounds, and build a pier in a different part of the beach. His advice was taken and the pier he designed is located exactly where he suggested, across the bight from Nimmo's site.[117]

Fortunately Donnell made a drawing of Nimmo's planned harbour, in its then dilapidated state, in his report in 1827.[118] It is reproduced here in figure App. 1.31. The groynes faced directly into the stroke of the sea and altogether the precise site appears to have been a remarkably poor one for the entrance to a small harbour, most untypical of Nimmo. The stones of Nimmo's groyne piers were removed and used to build Donnell's pier nearby, which was nearing completion when he reported in 1827. Today, there is no evidence of any man-made structure on the beach at Nimmo's precise site. Without Donnell's plan, Nimmo's harbour might never have been recorded and its exact site never been known. Today the R487 road separates the beach from Nimmo's excavated inner harbour, which is still visible as a pond, along with part of its embankment, on the landward side of the road (figure App. 1.32). It is the only part of Nimmo's works still evident at the location, the single example of an unsuitable site selected by him for a harbour. Today, nothing seems to be known locally about the history of this interesting pond. Nearby, Donnell's pier still stands and remains in use.

Carrigaholt Pier

Location: N 52 36.146, W 009 42.269

After leaving Kilbaha on the R487, watch for a minor road to the right, signposted for Carrigaholt. Pass through the village, exiting on the R488. The pier is on the right.

There was a small pier beside the creek when Nimmo visited. He planned to extend that to 150 feet, excavate the bog and build a new quay almost 300 feet long on the opposite side of the creek. His coloured plan is conserved in the National Library and is reproduced here as Plate 14.[119] The works received £369

APPENDIX 1 A GAZETTEER OF NIMMO'S PIERS AND HARBOURS

FIGURE APP. 1.33
The back of the pier at Carrigaholt County Clare.

in 1822, £100 of which was from the London Committee. When Nimmo came to approve the final expenditure in 1824 he claimed that the work had not been completed either to the dimensions of his plan or to the exact measurements of his assistants made in 1823.120 Despite his misgivings, he agreed to pay the balance. The older part of the works fell within a year and had to be rebuilt. Donnell had to spend a further £87 later to put everything in full order and by then 'it was a very useful work, in good order, and has facilitated the fishery, and promoted the industry, of the district'.121 By 1833 the old section had fallen again, due allegedly to the very small stones that had originally been used in its construction.122 The whole works had to be rebuilt in 1884, retaining Nimmo's original plan. Today the pier is a substantial structure, but with a pronounced recreational ambience, its commercial usefulness having been largely superseded by a much later pier built south of the village (figure App. 1.33).

COUNTY KERRY

The vast bulk of Nimmo's public works were in Connaught and county Clare, largely because of his western district duties. But his duties as engineer to the Fisheries Commission meant he had responsibility for piers in all the other counties also. Six sites were selected in County Kerry for inclusion on the 1822 list of piers (see table 6), and Nimmo's work in that county was further increased by his own abiding interest in it, dating from his Bogs Commission days.

Barrow Harbour

Location: N 52 17.57, W 009 51.05

This was a location listed (but misspelt Bana) among the first twelve piers to be funded in 1822, receiving a total of £231 which included £58 from the Mansion House committee. A coloured drawing by Nimmo of the site is conserved in the National Library.123 A landing quay, thirty-eight feet in length was designed and built; Nimmo reported it was in good shape in February 1824.124 However, it was never fully completed thereafter, even though a further £231

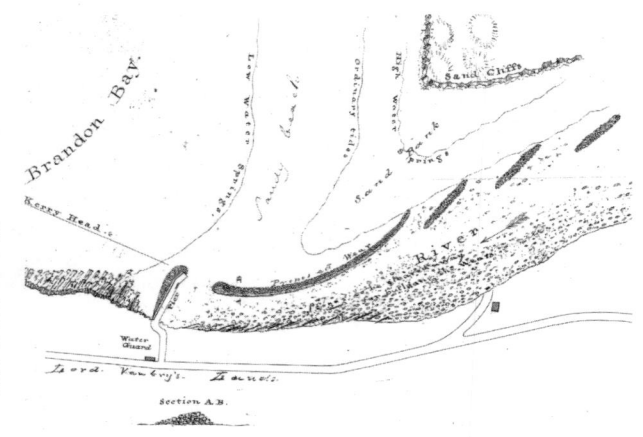

FIGURE APP. 1.34 Nimmo's sketch map of the proposed pier at Brandon. Note the stone weirs he proposed to erect in the river, deflecting the flow towards the pier for scouring. (From Ref 129. Reproduced with permission of the National Archives of Ireland.)

was advanced and a contract signed to finish it. Donnell despaired of it ever being useful, so he recommended that it be abandoned in favour of a new site.[125] The commissioners agreed, giving no more money to it and Barra dropped out of the list of fishery piers after 1827.[126]

Brandon Pier

Location: N 52 16.07, W 010 09.35

This site was in the second part of the 1822 list of piers, receiving £370 from the commissioners, including £100 from the London committee.[127] Ben Meredith started the work immediately on contract and the pier, 143 feet long and sixteen feet high at the head, was completed in 1823.[128] In Donnell's opinion it was badly executed and was severely damaged the following winter. After complaints that the site was a poor one, Nimmo visited in February 1824. The pier head had indeed collapsed because it had been erected upright rather than with a slope, but it could be rectified for the small sum of £60. In Nimmo's opinion, Meredith had already given sufficient value in what he had done so whatever extra money was needed should come from the commissioners. He attached a plan of the site showing a series of weirs that he proposed to erect in the flow of the river and illustrating the existing damage and the expanded sea pavement that needed to be built. The weirs were meant to deflect the flow of the river towards the pier, thereby increasing the scouring effect (figure App. 1.34). When tenders were invited for the work, an offer was put in by two local landowners, but Nimmo rejected it as too costly.[129] Neither had they included the weirs that he wanted. Nothing had happened by the end of the year, so he felt obliged to wait until the following summer.[130] By May 1825 Donnell had been requested to make a new plan for the repairs and he recommended that the pier be rebuilt a little further downstream, to which the commissioners agreed.[131] It was 1827 before the contract to build the new pier was awarded to Joe Burke of Cork, who provided a plan and section of the work in July. By May 1829 the work was almost completed and a plan showing the old pier and the plan of the new is shown in the Tenth Report of the commissioners.[132]

That was not the end of the Brandon pier problems. It needed further repairs in 1834 and in 1843 when it had been 'undermined and fractured and totally carried away by the late tremendous storm'.[133] Nimmo had chosen his original site to take advantage of the shelter given by a ledge of rock immediately outside of it, precisely where Donnell's pier was built.[134] With hindsight, Donnell's chosen location proved not to be any better than Nimmo's, and was in fact far more exposed.

Thomas Rhodes was the next to propose a pier at Brandon together with a quay and village, the plan of which he illustrated in his report.135 This was never built and Brandon pier was largely replaced by a new structure in 1886. The remains of Nimmo's original pier are still evident, about one hundred yards upstream of the existing pier.

Dingle Pier
Location: N 52 08.19, W 10 16.43
The pier constructed here in 1822, at a cost of £450, was an extension and development of an existing small pier.[136] Nimmo examined it on a number of occasions and Donnell reported it still in good order by 1826, having sustained only the usual damage from boats that was easily repaired. The pier has been considerably improved since then.

Cahirsiveen Pier
Location: N 51 56.54, W 10 13.50
The commissioners granted £370 for a pier at Cahirsiveen in 1822, including £92 from the London committee. Called Cahirswine in the commissioners' list of piers, it was built of hewn stone, extending 181 feet from the root.[137] Although visited by Nimmo and by Donnell, both of whom found it in good order or needing only small repairs, Cahirsiveen never featured subsequently in the fishery reports.

Another pier, built on the estate of the Revd Denis Mahony at **Ballinskelligs**[138] not far distant, was also funded by the Commissioners and the London Committee and it, too, rarely featured again in the Commissioners' reports. The two piers were probably overshadowed by their illustrious neighbour, the pier at Knightstown, Valentia.

Knightstown Pier, Valentia
Location: N 51 55.34, W 010 17.12
The pier is at the foot of the main street of the village, which can be reached by road from Portmagee on the mainland, or by a short car ferry crossing from Cahirsiveen.

FIGURE APP. 1.35
The root of Nimmo's pier at Knightstown. To the right of the white bollards is the oldest, first section. From the steps and bollards outwards is the extension added by Meredith. Note the different style of stone work in the sections.

This pier, designed by Nimmo, was funded by the Fisheries Commissioners and the Mansion House Fund. It has been significantly altered since then, but its constructional history is clear in the stonework (figure App. 1.35). The landward section is built of flagstones, mostly laid flat in rough courses and topped with schistus set on its edge. This section was commenced in 1822 and finished in 1825. It was not built exactly where Nimmo had specified, nor to his entire satisfaction.[139] He feared that it would soon become silted by the washing-in of gravel, and refused to sanction payment to the contractor, Ben Meredith, until an extension was added. Meredith added a new twelve-foot extension according to Nimmo's specification, starting from the position of the steps close to the original pier head. This extension was built with cut stone facing, topped with schistus laid on its edge, and was completed in 1825, whereon Nimmo approved the full payment.[140] By 1828 the pier needed repair, maintenance and improvement, which Maurice Fitzgerald agreed to fund in part; so well he might, since it was principally used for the shipment of slate from his works and its extension into deeper water would be beneficial to him and to the steam line to America that he proposed. Donnell designed the new extension in cut stone with a slip on the right-hand side, the new pier head being displaced slightly to the left in his plan, as shown in the Tenth Report of the Commissioners.[141] In Donnell's plan the slate-yard dominates the pier, probably intentionally, with a view either to impress (or maybe embarrass?) Fitzgerald, who was a fisheries commissioner.

The basic form of the pier today is largely unchanged since then (figure App. 1.36). It is the only structure in Knightstown that can be assigned incontrovertibly to Nimmo. The pier to the north of Nimmo's is a newer construction, although it looks much older and did not exist in his time. The car ferry from Cahersiveen lands at the slip to the south of Nimmo's.

Kenmare Pier

Location: N 51 52.21, W 09 35.18

Leave Kenmare by the N71. Watch for a turn to the right just before the end of the town, leading to the pier.

APPENDIX 1 A GAZETTEER OF NIMMO'S PIERS AND HARBOURS

FIGURE APP. 1.36
The pier at Knightstown as it is today. The pier to its left is more modern although it looks older.

The Marquis of Lansdowne wanted a pier at his new town of Kenmare, so he asked Nimmo for assistance. Nimmo made a survey and submitted a plan and estimate for one costing £1,296; the marquis agreed to contribute half of this and the Fisheries Commissioners advanced the remaining £648, somewhat above their normal contribution.

Donnell was ordered to proceed to Kenmare in May 1825 to carry the proposed measure into effect by contract.[142] He was firmly opposed to the site chosen by Nimmo, so he gave his own plan for a harbour in a different place. His objection rested on the shallow depth of water and the difficulty and expense of making a secure pier in the soft mud, which was over eight feet deep at Nimmo's site. Lansdowne referred this to Nimmo, who made a revised plan that was accepted.[143] To overcome the problem of the mud a rough mole of stone was to be laid in the alignment of the pier and the structure built on this only when it had become firm; its strength and durability were to be certified by the commission's officers. The pier, of hammered ashlars in face, would extend along the inside of a ledge of rocks, starting 100 feet from the gravel bank near the land and extending outwards for 300 feet. At ordinary spring tides it was to have twelve feet of water alongside, 200 feet from the head. A channel thirty yards wide was to be dug in the mud to low water along the

FIGURE APP. 1.37
Sea face of the pier at Kenmare with poor stone work that contrasts markedly with the good quality ashlar on the inside face (not shown).

quay out as far as the main channel. Later on, a clay cliff near the gravel bank could be dug down to pier level and the area filled in. Nimmo's specification and plan are given in the Tenth Report of the Commissioners.[144]

The pier today is substantially as Nimmo designed it, although it has suffered, or enjoyed, the usual modifications over time (figure App. 1.37). The gravel bank from which it projected, and which Nimmo had proposed for 'wharf yards', has been reclaimed and built over with homes, the whole area possessing a quiet, genteel ambience.

COUNTY CORK

When requested by the Fisheries Commissioners to examine possible sites for piers in 1822, Nimmo had no time to tour the Cork coast, so he deputed Alexander McGill from Dunmore, who had twenty years' experience of harbour works, to the task.145 Neither had he been able to extend his coast survey to the southern part of the country, so he never produced a consolidated report on the piers and harbours of the south. Understandably perhaps, Cork had no piers in the top twelve of the 1822 list, but had four in the second group (table 6), and there is relatively little information on his contribution to these, the records being confined to references in his letters to the commissioners. Most of the published information comes from Donnell's reports.

Castletown Berehaven

Location: N 51 39.03, W 009 54.28

Travel from Kenmare to Castletown Bere by the R571 through Eyeries, or the R572 over the Healy Pass. The original quay is in the centre of the village.

Being in the second part of the 1822 list, Castletown was granted only £184, including £53 from the London committee in 1822, sufficient to build a landing quay, 190 feet long, against the face of a rock in a sheltered bight opposite Bere Island.[146] Nimmo visited in 1824 and was happy with progress.[147] The quay answered the local needs well and Donnell was happy with it in 1826. By 1827 it was proving useful not only for the fisheries but for general trade also. The principal harbour at Berehaven is now at Laurence Cove on Bere Island. Nimmo does not appear to have made any contribution to it.

Ballycrohane (Coulagh) Pier

Location N 51 42.49, W 009 56.59

Coulagh Bay is on the southern side of the Kenmare River. Leave Kenmare by the N71. Just over the bridge, turn right on to the R571 and travel as far as Ardgroom. About two miles beyond the village watch for a turn to the right leading to the pier.

Nimmo made a tour of the south early in 1824, reporting on piers in progress and making recommendations for new ones to be started. Ballycrohane, called

APPENDIX 1 A GAZETTEER OF NIMMO'S PIERS AND HARBOURS

FIGURE APP.1.38
Quay and pier at Ballycrohane today. The original jetty and quay are now encased in concrete, but the original stone work is evident at the innermost part of the quay. Note the random rubble stones in rough courses with the top layer laid on their sides.

Cullagh in the 1822 list, was granted £369 that included £100 from the London committee.[148] Nimmo regarded the location as an excellent natural harbour, quite unknown to navigators. A contractor was engaged to erect a quay, which he built too far up the creek. Nimmo insisted it be taken down and replaced with one further to seaward. This consisted of a quay, 162 feet long, built along the rocks with a jetty at the head projecting outwards fifty-four feet.[149] When Nimmo visited in 1824 it was completed except for the coping. To improve its accessibility he recommended that the jetty be extended outwards to a sunken rock about forty feet away at a cost of £150, illustrating his plan with a large sketch.[150]

Donnell reported on the harbour in 1826, finding it in a poor state because it had been defectively executed and had not been raised above high spring tide level.[151] He recommended that it be raised eighteen inches, and be properly coped, filled and paved. His published recommendations read exactly as Nimmo's manuscript reports, although he made his recommendations as his own, not Nimmo's. By 1826 the repairs had not yet been undertaken and Donnell provided detailed new specifications for the work, excluding the extension of the jetty.[152] Today, the site is still much as Nimmo designed it, but the jetty has been extensively concreted over and modified by the addition of a slip and other extensions. The original stonework can still be seen at the landward end of the quay (figure App. 1.38).

Clonakilty Pier

Location: N 51 36.55, W 08 52.17

In 1822 the commissioners granted £340 and the London committee £122 for a pier at Clonakilty.[153] Alexander McGill inspected the site on Nimmo's behalf, but no indicative original plan has been found. Sited on the east side of the bay at Ringarundel and built by William England of Dunmore harbour, it consisted of a curved pier, 230 feet in length, faced on both sides. Nimmo examined it in 1824 and reported that it was progressing well, but Donnell found it distressed in 1826.[154] He drew up specifications, accompanied by a plan, for its repair, which were implemented later making the pier useful for fishery and general purposes.[155]

The Fisheries Commission funded a number of other piers in County Cork, but Nimmo had little enough to do with them according to the records. **Courtmacsherry** (misspelt Contmasheny in the 1822 list), the last of the four Cork sites that were funded by the commissioners and the charities, was granted £461 in 1822 (£122 from the London committee) but work was slow in starting. Nothing substantial was done by mid-1825 when Donnell replaced Nimmo. Donnell was instructed to get the work under way in 1826. He failed to see how such a small sum would be useful for a new structure, so he recommended the extension of an existing pier, a course approved by the commissioners.[156]

At Kearn's Port on the island of **Cape Clear**, Nimmo presented a plan for the extension and improvement of a small pier and harbour on the estate of William Beecher, who was to contribute half of the estimated cost. Beecher did not favour Nimmo's plan, possibly on the grounds of cost, and Donnell eventually provided a new one, ending Nimmo's connection with the project.[157]

A Mr Leslie at **Dunworly**, near Barryroe, also had a close fiscal call with Nimmo. Nimmo had advised that it was useless to build a pier at Dunworly unless it was also determined to make a cut or canal from it to communicate with the lake behind (now largely a marsh). In a letter to Townsend in 1824, Leslie pledged to execute the cut as well as the pier 'whatever the expense may turn out to be', a really reckless offer where Nimmo was concerned.[158] J. Redmond Barry, an inspector of fisheries who lived and owned property in the area (and who would go on to become an illustrious chief inspector of Irish Fisheries much later in the century) may have been involved in the plan for a pier, which was never implemented. Nimmo, who was very impressed by the state of agriculture around Clonakilty, had met him and had extensive discussions with him around that time.[159] Today, there is a cut into the marsh behind the estuary, probably made by Leslie, with clear evidence of stone works along the right side of the river. This may have been a roadside quay, like Nimmo made at Cloonisle, but the local residents have no knowledge of a working quay ever at this site.

COUNTY WEXFORD

The distribution of the funds of the Fisheries Commission and the charitable bodies in 1822 was confined exclusively to those regions that suffered the greatest famine, namely the western and south-western counties; the inhabitants of the southern and eastern coasts felt, understandably, that they had not shared a fair proportion of the commission's largesse.[160] For its part, the commission felt obliged to allocate funds in accordance with the desires of the charitable bodies who were liberal contributors to the scheme. All the piers on the south and east coasts were therefore aided by private donors, usually as improvements to their own estates, for example Courtown, Skerries and Killough harbours, and so on. One exception was **Crossfarnogue** (now **Kilmore Quay**) pier in County Wexford.

Crossfarnogue (Kilmore Quay)

Location: N 52 10.20, W 006 35.1

The quay and pier are the centre of the fishing village of Kilmore Quay, at the termination of the R739 road from Wexford town. Nimmo designed a pier for this place in 1822, shown in Plate 15.[161] By opposing a breakwater to a new extension of the main pier, a double harbour was made possible with a sheltered inner part and a more open outer part. The whole was sited well above low water springs and in consequence would be of limited value. It must have been one of the first fishery harbours planned by Nimmo. Donnell made only a brief but informative comment on Crossfarnogue: 'Plan and other documents have not been ordered by the Board, as local contribution was not available.'[162] Nimmo had signed his plan 'Alexander Nimmo, engineer to the Irish Fisheries 1822', which indicates that the plan had been made ostensibly with the commission's approval, if only tentative. The strictness with which the commissioners insisted on funding only those projects that could guarantee local (or charitable) funding is clear in Donnell's comment. Once Crossfarnogue failed to garner local support it was doomed, and it never appeared subsequently in the commissioners' reports. A detached pier with breakwater opposite are shown in the first edition of the six-inch Ordnance Survey map of the area, indicating that funds must have become available before 1842 for what was to become one of Ireland's major fishing ports in the twentieth century.

COUNTY WICKLOW

Greystones Harbour

Location: N 53 09.03, W 06 03.50

A harbour at Greystones was planned for the south shore of Rathdown Bay on the estate of Peter LaTouche. Many different plans had been advanced to improve the cove among which was one by Nimmo costing £3,424.[163] He had surveyed the site in 1822 and drawn up a plan for an outer harbour of two acres with an inner, excavated dock of about one acre. After some years the noblemen, landowners and residents of Wicklow sent a memorial to the chief secretary advocating Nimmo's plan, which they said would benefit the area and be good for the employment of the poor.[164] Goulburn sent it to the commissioners, who asked Nimmo for a detailed plan and specifications. On examining these, they agreed to grant £500 in aid of the harbour 'if done to Nimmo's plan and specifications'.[165] While that was the maximum sum the commission would normally grant, it was much less than the memorialists had hoped for, so nothing was done towards advancing the proposal. By 1826 even Nimmo's plan was missing[166] and Donnell advised that the grant be rescinded and be 're-granted whenever the local fund shall be really subscribed, or when any other circumstance shall occur that will evince a serious intention of going on with the undertaking' – a scathing comment on the whole enterprise.[167] Nothing further was heard of Greystones in the commissioners' reports.

COUNTY DUBLIN

Nimmo had completed his survey of the north-east coast, from Belfast to Dublin, by May 1822, when he was ordered to the west coast and his recommendations were published in the Fourth Report of the Commissioners.[168] Being completed so early in his fisheries career, the plans he drew up for piers and harbours were skimpy in the extreme. Plans for locations in County Dublin include those for **Rush**, **Loughshinny**, **Skerries** and **Balbriggan**; of these only Balbriggan was given any really serious consideration.[169] He claimed to have made out a design for a pier at **Lambay** island, which he thought was a good place for a fishing village.[170] The Commissioners had added this site to the 1822 list, granting it the relatively huge sum of £900, to which Colonel Talbot added £425. Talbot executed the work, but whether to Nimmo's design or that of someone else is not known.

Balbriggan Harbour

Location: N 53 36.41, W 06 10.43

The southern pier already existed when Nimmo examined the area. He proposed to build a breakwater on the north side of the harbour, with a wharf wall (effec-

FIGURE APP. 1.39
Nimmo's plan for a quay and breakwater at Balbriggan, Co. Dublin (top part of plan). (From ref. 171)

tively a jetty) from the end of Grace Lane joining it mid-way along its length. The jetty would be pierced by tunnels with sluices so that the water impounded north of it could be released at low water to scour the harbour. It is shown in his plan, reproduced here as figure App. 1.39.[171] He estimated the likely cost at £3,255, a large sum but relatively low compared, for example, with a breakwater in Loughshinny that would, in his view, have cost £12,000. Although Donnell did not visit the site he reported that the work was in progress in 1826, for which the commissioners had given £1,569 and the inhabitants had subscribed £523. Donnell anticipated that it would be successfully completed in 1829, but no detailed plans appear to have survived.[172] A northern jetty was eventually erected in the harbour but not where Nimmo suggested.

COUNTY LOUTH

As with county Dublin, Nimmo's plans for piers in County Louth were very insubstantial, not more than indicative sketches. They include **Clogher Head**, **Dunany Point** and **Giles's Quay**, plans for which are shown in the Fourth Report of the Commissioners.[173] Only at Clogher Head, where Wallop Brabazon undertook the harbour works, was anything major done at the Louth sites before 1835.[174] Nimmo also claimed that he had done some work at Dundalk, but the entrance to Drogheda port, described earlier, was his principal contribution to the county.

COUNTY DOWN

In no other county did Nimmo propose as many piers as he did in County Down, which he investigated in the course of the first year of his coastal survey. He considered and reported on twenty-five separate sites, giving indicative sketches of his proposals for nineteen of these in the Fourth Report of the Commissioners.[175] These were, as might be predicted, very simple plans covering the following locations: **Greencastle Point, Annalong, Newcastle, Killough, Ardglass, Quentin Bay, Kerney Point, Ballyhalbert, Ballywalter, Copeland Islands, Mill Isle, Derryogue, Bangor, Sand Eel Bay, Tara Bay, Killeef, Groomsport, Ballywilliam** and **Rossglass**. Of the twenty-five sites he discussed, only ten were mentioned in Donnell's list of piers executed or in progress in 1829. Of that ten, only four had piers in progress or in repair and six had had their plans relinquished or rejected by the Fisheries Commissioners. In the period between 1822 and 1829 Nimmo had been actively involved in developments at three of the sites, Bangor, Ardglass and Killough.

Bangor Harbour
Location: N 54 39.55, W 005 40.13

At Bangor, which was a terminal for vessels to and from Portpatrick in

Scotland, there was a small pier operating in 1822 that was often full with sail, much of it involved in the transport of live cattle. Colonel Ward, the proprietor, was endeavouring to improve and extend the port in the hope of its becoming a packet station. On Nimmo's advice, he commenced a mole from the south-east shore to run 400 feet diagonally across the bay in order to increase the shelter and provide eventually for increased quays for the accommodation of larger vessels.[176] However, when Ward had completed the first one hundred feet, Nimmo thought the stones used were too small and the work appeared 'feeble' to him. The proprietor commissioned Nimmo to build a proper harbour, entirely separate from his fisheries brief. The work was never properly completed in Nimmo's time and was never mentioned again by the Commissioners.

Ardglass Harbour and Kimmersport

Location: N 54 15.437, W 005 36.237

Ardglass to the south of Strangford Lough was an important fishery port even before Nimmo, as well as a terminus for two packets from the Isle of Man. The herring fleet made a rendezvous there at the start of each season and the village had nineteen herring smacks, thirty yawls and four sloops permanently stationed there. Yet the harbour was open to the south-east, therefore turbulent and dangerous in winds from that quarter. Nimmo's description of the problem was comprehensive:

> When S.E. winds set in…a heavy ground-swell runs into Ardglass, and being brought-up in a short space by the ledge of the perch-rock, is deflected into the cove forming the present harbour; which being so much bounded by wall and steep rock becomes excessively turbulent; the craft thump much on the bottom, and can with difficulty hold on with all their tackle; nor is there any means of escaping at such times

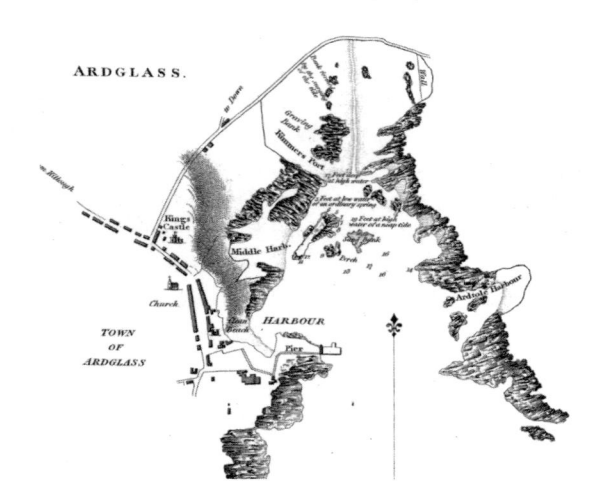

FIGURE APP. 1.40
Indicative sketch by Nimmo of Ardglass and Kimmersport. (From ref. 177.)

APPENDIX 1 A GAZETTEER OF NIMMO'S PIERS AND HARBOURS

into any of the other coves of the bay so that vessels are frequently seriously damaged.[177]

Very expensive works had been carried out at an earlier time to no great advantage. Nimmo proposed to build a breakwater separating the outer and middle harbour and to cut away rock from the latter so as to quell the turbulence and increase the safety of the area within. His indicative plan, reproduced from the Fourth Report of the Commissioners, is shown in figure App. 1.40. This would cost over £2,200; if that amount were not available, his advice was to build a landing quay further in, along the rocks at the entrance to Kimmersport (the innermost part of Ardglass harbour) at a cost of £350. That, at least, would give shelter for the fishing boats as well as providing a place to land their fish. In 1824 the commissioners granted £157 for the proposed Kimmersport works, partly funded by the proprietor, Mr Ogilvie. Work was not completed to Donnell's satisfaction so he (Donnell) made a new plan that was still under consideration when the Fisheries Commission was dissolved.[178] The new Office of Public Works took Ardglass over and expended £6,650 in executing works under the direction of Sir John Rennie.[179] Ogilvie also completed extensive repairs to the main pier. Ardglass harbour today is much as Nimmo had proposed, having an inner harbour at Kimmersport and a harbour breakwater exactly where he suggested. The main pier built by Rennie, which extends well beyond the original, was completely rebuilt in 1838, having been swept away in a great storm, confirming Nimmo's opinion of the turbulence at Ardglass. It has been extensively altered in recent years.

Killough Harbour

Location: N 54 15.07, W 005 38.13

Nimmo accepted private commissions as he proceeded on the first leg of the coastal survey around the north coast, and one was to design harbour works

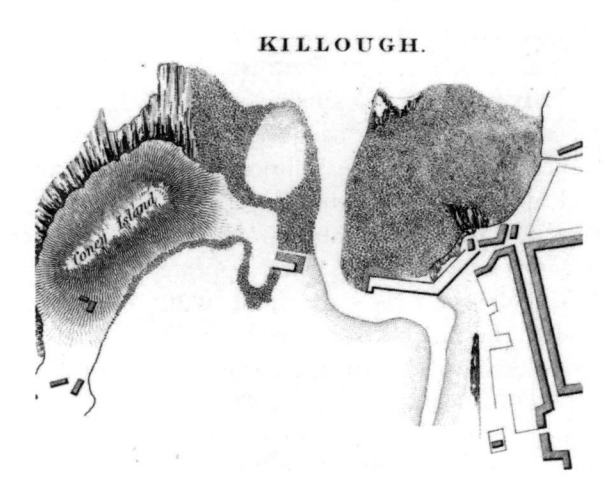

FIGURE APP. 1.41
Indicative sketch of Killough by Nimmo. (From ref. 180)

at Killough, a little to the west of Ardglass, the property of Lord Bangor and his heirs. It already had a small quay that he proposed to extend and which, when finished, would be 500 feet long on the village side. His indicative plan, given in the Fourth Report of the Commissioners, is shown in figure App. 1.41.[180] He planned to build another jetty, one hundred feet long, extending outwards from Coney Island, thereby enclosing an extensive harbour with ten to eighteen feet of water alongside the piers at high tide. Following a detailed survey he prepared a coloured map at eight inches to one mile reproduced here as Plate 16.[181] He later reduced the scale of this and extended it to include Ardglass and some of the surrounding creeks for the use of the commissioners of fisheries, who published it in due course.[182] Interestingly, the soundings were inserted vertically on the Killough part of the latter map (meaning that they were his own observations) and at right angles (meaning that they were taken from existing surveys) on the Ardglass side; Killough had paid for their survey, Ardglass had not!

The family of Lord Bangor now commissioned him to make the necessary improvements at Killough, on which they expended over fourteen thousand pounds.[183] In the course of the work Nimmo employed men such as his assistant O'Hara to do the skilled stone masonry and to investigate the building quality of greenstone from Valentia supplied by Maurice Fitzgerald. Many of those who worked at Killough would be called by Nimmo to work at other pier sites in the west in 1823 and 1824 when the Killough work was completed.

Having received no funds from the Fisheries commissioners Killough never appeared in their reports. In recent years the harbour has been regarded as derelict and useless.

COUNTY ANTRIM

Nimmo proposed harbour works, and gave indicative plans of them, at seven sites in county Antrim: **Portrush, Port Ballintrae, Ballycastle, Cushendun and Red Bay**.[184] In 1829 Donnell mentioned eight Antrim sites in his list of sites of piers; of these, three had been rejected or relinquished (Mill Isle, Mill Bay, Stephen's Port), one was almost complete (Portmuck), and two (Cushendun and Port Ballintrae) were temporarily suspended.[185] Nimmo's contributions to Portrush and Port Ballintrae are discussed in chapter 9. The former was never a fisheries harbour, so it never appeared in the commissioners' later reports. Today, Nimmo's section is part of the outermost pier on the left over the bridge. The much longer western pier was added later to enlarge the harbour considerably and form the present-day port. Nimmo's signed coloured plan for Port Ballintrae is shown in Plate 17.[186] He does not appear to have had anything further to do with the other sites and his plan for Cushendun was later replaced by Donnell's.[187]

APPENDIX 1 A GAZETTEER OF NIMMO'S PIERS AND HARBOURS

COUNTY DONEGAL

Nimmo gave indicative plans of four sites in Donegal: **Arran Roads**, **Burtonport**, **the island of Arran** and **Greencastle**; they were skimpy beyond all reason. They were made in the very earliest days of his coast survey, when he was still unfamiliar with maritime engineering and was still without the use of the *Dunmore*. The only place that he favoured, and for which he gave a few soundings, was Greencastle.

Greencastle Harbour.

Location: N 55 12.50, W 06 57.26

On his first visit to Greencastle he ventured only to suggest that a harbour of four acres could be formed by the construction of two short piers at a cost of £4,000.[188] Aware, perhaps, of the inadequacy of his very first harbour proposals he revisited Raughley, Mullaghmore and Greencastle in July of 1821 and expanded on his earlier plans.[189] At Greencastle, the southern pier should be constructed in a westerly direction along the rocky ledge 700 feet to the 'half-tide rock' and have a return-head jetty projecting one hundred feet to the north-west from there. This would adequately quell the ocean swell and give good shelter for half-deckers that could not otherwise use the anchorage. From the revenue-house on the opposite shore another pier could be extended in a southerly direction to reduce the entrance to 120 feet and make a really sheltered haven. He thought the Customs Board might assist financially with the latter pier, as it would greatly benefit from it. Other improvements could follow later: the beach could be deepened, a scouring basin made near the revenue-house and the river diverted to pass closer to the southern pier. He envisaged a fishing village arising at the site, 'if the proprietor of the land would engage to accommodate the fishermen with building lots upon moderate terms'. In the meantime, he advised that the works be constructed in the cheapest possible form, namely, 'of the hammer-dressed flag of the neighbourhood, laid in large scantlings, but without any cut stone, except for the coping course, which should be of Arran limestone'. His new estimate for the lot was a much more reasonable £2,730.

In January 1824 the commissioners granted £1,597 for Greencastle on condition that a like sum would be raised locally and plans and specifications drawn up.[190] According to Donnell who had not visited the site, work commenced in spring of 1828 and proceeded satisfactorily. The plan at that time was generally along the lines proposed by Nimmo, but the pier from the revenue-house was never built. Donnell's specifications and plan, dated January 1828, are shown in the Tenth Report of the Commissioners.[191]

NOTES

1. J. Donnell, 'Report of James Donnell to the Board of Commissioners of Irish Fisheries', Tenth Report of the Commissioners for the Irish Fisheries, appendix 10 (B.P.P.1829 [329], vol. XIII, mf 31.93– 4), hereafter referred to as Tenth Fisheries Report.
2. A. Nimmo, 'Report on the fishing Stations of the West Coast of Ireland, from Errishead southwards', Eighth Report of the Commissioners for the Irish Fisheries, appendix 10 (B.P.P.1826/7 [487], vol. XI, mf 29.89–90, hereafter referred to as Eighth Fisheries Report.
3. J. Donnell, 'The Report of James Donnell, Civil Engineer, on the State of the Fishery Piers and Landing quays on the Coast of Ireland', Appendix 11 in Eighth Fisheries Report.
4. A. Nimmo, 'Report on the Fishing Stations on the North Coast of Ireland by Alexander Nimmo, engineer to the Board of Fisheries', Fourth Report of the Commissioners for the Irish Fisheries, plate V (B.P.P. 1823 [383], vol. X, mf 25.85–6, hereafter referred to as Fourth Fisheries Report.
5. A. Nimmo, 'Report on the Fishery Stations of the Coast of Connaught.' Sixth Report of the Commissioners for the Irish Fisheries, plate I (B.P.P.1825 [385], vol. XV, mf 27.124–5, hereafter referred to as Sixth Fisheries Report.
6. J. McGowan, In the shadow of Benbulben (Sligo: Aeolus, 1993). Donnell's plan is reproduced as a plate in thispublication.
7. Fourth Fisheries Report, plate V.
8. Sixth Fisheries Report, plate I.
9. Tenth Fisheries Report, plan no. 33.
10. Sixth Fisheries Report, plate II.
11. Sixth Fisheries Report, p. 30.
12. Memorial from the clergy, traders and inhabitants of Ballina, 9 January 1823 (NAI, CSORP 1823 5144 [1]).
13. A. Nimmo to H. Goulburn, 30 January 1823 (NAI, CSORP 1823 5144 [2]).
14. Ballina Impartial (newspaper), 20 October 1823.
15. Plan and Section of the Moy River from the Bridge of Ballina to the Quay at Coolnalecka with the works done under the direction of Alexander Nimmo, 1822 (NAI, CSORP 1823 5144 [3]).
16. Sixth Fisheries Report, plate II, figure 1.
17. Nimmo Sixth Fisheries Report, p. 34.
18. P. Knight, Erris and the Irish Highlands and the Atlantic Railway (Dublin: Keene and Sons, 1836).
19. Sixth Fisheries Report, plate III, figure 2.
20. Eighth Fisheries Report, p. 21.
21. Ibid, p. 20.
22. Ibid., plate II, figure 1.
23. Ibid., plate II, figure 2.
24. Ibid., pp.22–3.
25. Ibid., p. 50.
26. Ibid., pp. 27–8.
27. Ibid., plate IV figure 2.
28. Ibid., p. 24.
29. Ibid., plate III, figure 1.
30. Ibid., plate V, figure 1.
31. Ibid., pp. 52–53.
32. Tenth Fisheries Report, pp 44–5 and plan 19.
33. Marquis of Sligo's offer dated 22 May, 1822 and Commissioners of Fisheries reply dated 10 June 1822. (NAI, OPW8/54[1]).
34. A. Nimmo, Articles of Agreement, dated 12 August 1824, with coloured map of the Harbour of Bundurra and the proposed works, signed by Nimmo (NAI, OPW8/54 [8]).
35. W. Hildebrand to Fisheries Commissioners, 1 January 1825 (NAI, OPW8/54 [15]).
36. A. Nimmo to Fisheries Commissioners, 26 December 1824 and related correspondence (NAI, OPW8/54 [2]).
37. Eighth Fisheries Report, p. 31.
38. Eighth Fisheries Report, p. 35 and Plate VI, figure 1.
39. Eighth Fisheries Report, plate 6, figure 2.
40. A. Nimmo to Fisheries Commissioners, 2 March 1824 (NAI, OPW 8/83 [30]).
41. Eighth Fisheries Report, p. 54.
42. A. Nimmo to Fisheries Commissioners, 16 July 1822 (NAI, OPW8/83 [1]).
43. Tenth Fisheries Report, plan 11.
44. T. Robinson, Connemara. Map and Gazetteer (Roundstone, Co. Galway: Folding Landscapes, 1990).
45. Eighth Fisheries Report, p. 37.

APPENDIX 1 A GAZETTEER OF NIMMO'S PIERS AND HARBOURS

46. A. Nimmo, Report on Clifden Quay, with colour-washed map (NAI, OPW8/84 [9]). This appears to be Nimmo's original plan for Clifden.
47. Eighth Fisheries Report, plate VII.
48. J. Donnell, Specification of work to secure the boat quay and to prolong the ship quay in Clifden by James Donnell, Clifden, 5 November 1825 (NAI, OPW8/84 [2]).
49. Eighth Fisheries Report, plate VIII.
50. Ibid., p. 38.
51. Tenth Fisheries Report, p. 22.
52. Eighth Fisheries Report, p.39.
53. Eighth Fisheries Report, plate IX, figure 1.
54. Ibid., p. 40.
55. Ibid.,, plate IX, figure 2.
56. Eighth Fisheries Report, p. 40.
57. Ibid., pp. 56–7.
58. Ibid., plate 10.
59. Ibid., p. 41.
60. Ibid., p. 57.
61. Ibid., p. 42.
62. Ibid., pp. 57–8.
63. Ibid., plate XI, figure 1.
64. Tenth Fisheries Report, plan 5.
65. Eighth Fisheries Report, p. 43.
66. Eighth Fisheries Report, plate XI, figure 2.
67. J. Donnell, Specifications to rebuild, extend and improve the piers and quays of the Claddagh (NAI, OPW8/79 [3])).
68. Eighth Fisheries Report, pp.42–3.
69. Reproduced in P. O'Dowd, *Down by the Claddagh* (Galway: Kenny's, 1993).
70. Royal Commission on Tidal Harbours. Second Report and Minutes of Evidence, p.58a. (B.P.P. 1846 [692], vol. XVIII, mf 50.169–77).
71. Memorial from the Chamber of Commerce of Galway, 6 July 1822 (NAI, CSORP 1822/1371).
72. 11 Geo IV & 1 Wm IV, cap. cxxii. *An Act for making and maintaining a navigable cut or canal from Lough Corrib to the Bay of Galway, and for the improvement of the harbour of Galway.* 1830.
73. J. Stephens, J. Evidence in Royal Commission on Tidal Harbours, Second Report and Minutes of Evidence, Appendix B, No. 88, p. 167 (B.P.P. 1846 [692], vol. XVIII mf 50.169–77.
74. A. Nimmo, 'The Report of Mr. Nimmo on the Bogs in that Part of the County of Galway to the West of Lough Corrib', in Fourth Report of the Commissioners for the Bogs of Ireland, appendix 12 (B.P.P. 1813/1814 [131], vol.VI, pt 2, mf 15.34–).
75. 1 & 2 Wm IV cap. liv. *An Act to amend and enlarge the powers of the Act passed in the eleventh year of His late Majesty King George the fourth, intitled An Act for making and maintaining a navigable cut or canal from Lough Corrib to the Bay of Galway, and for the improvement of the harbour of Galway.* 1831.
76. First Report of the Commissioners for Public Works Ireland, p. 6 (B.P.P. 1833 [75], vol. XVII, mf 36.127–8.
77. Ibid., sketch of part of the town and harbour of Galway showing improvements.
78. Eighth Fisheries Report, plate XII, figure 1.
79. Ibid., p. 58.
80. Tenth Fisheries Report, p.52
81. Eighth Fisheries Report, plate 12, figure 2.
82. A. Nimmo, Plan of Ballinacourty pier (NAI, OPW8/ 17 [43]).
83. A. Nimmo to H. Townsend, 2 April 1824 (NAI, OPW8/ 17 [2]).
84. Offer to complete piers at Ballinacourty and St Kitts by White and Coen, 17 May 1824 (NAI, OPW8/17 [5]).
85. A. Nimmo to Fisheries Commissioners, 24 September 1824 (NAI, OPW8/17 [11])).
86. Tenth Fisheries Report, p. 32.
87. Eighth Fisheries Report, p. 45.
88. Ibid., plate XIII, figure 1.
89. Offer to complete piers at Ballinacourty and St. Kitts by White and Coen, 17 May 1824 (NAI, OPW8/17 [5]).
90. Eighth Fisheries Report, p. 59.
91. A. Nimmo to Fisheries Commissioners, 5 April 1825 (NAI, OPW8/ 17 [15]).
92. A. Nimmo to Fisheries Commissioners, 2April 1824 (NAI, OPW8/ 17 [15])).
93. Eighth Fisheries Report, plate XIII, figure 2.
94. Ibid., p. 59.
95. Ibid., plate XIV, figure 2.

96. Ibid., p. 47.
97. G. Nimmo to E. Lees, 25 October 1822 (Library of P. O'Dowd).
98. Eighth Fisheries Report, p. 56.
99. Ibid., p. 46.
100. Ibid., p. 59.
101. A. Nimmo to H. Townsend, 16 May 1822 (NAI, OPW8/ 236/1 [3]).
102. A. Nimmo to H. Townsend, 24 May 1822 (NAI, OPW8/ 236/1 [4]).
103. A. Nimmo to H. Townsend, 8 April 1824 (NAI, OPW8/236/1 [6])).
104. Eighth Fisheries Report, p. 60.
105. Acceptance of McDonagh's tender for Liscannor pier (NAI, OPW8/ 236/1 [18]).
106. J. Killaly to H. Townsend, 11 November 1827 (NAI, OPW8/ 263/2 [9])).
107. J. Madden to H. Townsend, 20 January 1828 (NAI, OPW8/236/2 [17])).
108. Plan for a Fishery Harbour at Liscannor by J. Donnell, harbour engineer, etc (NAI, OPW8/236/ 1 [19]). Paper overlays attached to this copy of Donnell's plan show the modifications made to Nimmo's and Donnell's plans as proposed by Killaly.
109. Instructions to Nimmo (NAI, OPW 8/236/1 [1]).
110. A. Nimmo to H. Townsend, 16 May 1822. (NAI, OPW 8 236/1 [3]; A. Nimmo to H. Townsend, 24 May 1822 (NAI, OPW 8/236/1 [4]).
111. Small, coloured sketch map of Seafield (NLI, maps, 15 B [9] 3).
112. Eighth Fisheries Report, p. 61.
113. Report of James Donnell, Tenth Fisheries Report, appendix 10, p. 24.
114. Ibid., p. 61 and plan 35.
115. Report of James Donnell, Eighth Fisheries Report, appendix 11, p. 61.
116. Ibid., pp. 61 – 2.
117. Report of James Donnell, Tenth Fisheries Report, appendix 10, pp. 48–50.
118. Ibid., plan 24.
119. Carrigaholt, County Clare. Coloured plan with proposed extension of the old pier by Alexander Nimmo. (NLI, maps, 15 B 9 (2)).
120. A. Nimmo to H. Townsend, 20 January 1824 (NAI, OPW 8/65 /7).
121. Report of James Donnell, Eighth Fisheries Report, appendix 11, pp. 62–3.
122. Letter of E. Russell, 21 August 1833 (NAI, OPW 8/65 [20]).
123. Coloured map of Barra Harbour. (NLI, maps, 16 H 8 [1]).
124. A. Nimmo to H. Townsend, 29 February 1824 (NAI, OPW 8/48/12).
125. Eighth Fisheries Report, p. 63.
126. Tenth Fisheries Report, p. 24.
127. Eighth Fisheries Report, pp. 63–4.
128. A. Nimmo, Report on the Pier at Brandon (undated, 1823/4) (NAI, OPW 8/48/11).
129. A. Nimmo to H. Townsend, 22 September 1824 (NAI, OPW 8/48/11/5).
130. A. Nimmo to H. Townsend, 15 December 1824 (NAI, OPW 8/48/13).
131. Application for a new pier at Brandon (NAI, OPW 8/48/25).
132. Tenth Fisheries Report, plan 6.
133. Letter from the Coast Guard to the Fisheries Commissioners, 19 January 1843 (NAI, OPW 8/48/40).
134. A. Nimmo, Report on the Pier at Brandon (undated, 1823/4) (NAI, OPW 8/48/11).
135. Report by Thomas Rhodes on Brandon Pier, 15 June 1846 (NAI, OPW 8/48/48).
136. Eighth Fisheries Report, p. 64.
137. Ibid.
138. Tenth Fisheries Report, p. 24; Eighth Fisheries Report, p. 65.
139. A. Nimmo, 'Report on the pier at Valentia' (NAI, OPW8/359 [3]).
140. A. Nimmo to the Fisheries Commissioners, 5 April 1825. (NAI, OPW 8/359 [5]).
141. Tenth Fisheries Report, pp. 61–62 and plan 36.
142. Eighth Fisheries Report, p. 65.
143. Tenth Fisheries Report, p. 24.
144. A. Nimmo, Copy of 'Specifications for the Pier at Kenmare' in Tenth Fisheries Report, p. 47 and plan 22.
145 Eighth Fisheries, p. 36.
146. Ibid., p. 66.
147. A. Nimmo to H. Townsend, 29 February 1824 (NAI, OPW 8/48).
148. Eighth Fisheries Report, p. 66.
149. A. Nimmo, Report on Ballycrohane, 29 February 1824 (NAI, OPW 8/48).
150. A. Nimmo, Further Report on Ballycrohane, 8 April 1824 (NAI, OPW 8/48).
151. Eighth Fisheries Report, p. 66.

APPENDIX 1 A GAZETTEER OF NIMMO'S PIERS AND HARBOURS

152. J. Donnell, Specification for Coulough or Ballycrohane (NAI, OPW 8/95 [3]); J. Donnell, 'Coulough or Ballycrovane', in Tenth Fisheries Report, p. 42 and plan 15.
153. Eighth Fisheries Report, p. 67.
154. A. Nimmo to H. Townsend, 29 February 1824 (NAI, OPW 8/48).
155. Tenth Fisheries Report, p. 38 and plan 10.
156. Eighth Fisheries Report, pp. 67–8.
157. Ibid., pp. 66–7.
158. Taylor to H. Townsend, 1 June 1824; Woodward to H. Townsend, 30 June 1824 (NAI, OPW 8/135).
159. A. Nimmo, Evidence, p. 159, in Minutes of Evidence taken before the Select Committee of the House of Lords to examine into the Nature and Extent of the Disturbances which have prevailed in those districts of Ireland which are now subject to the Provisions of the Insurrection Act and to report to the House (B.P.P. 1825 [200], vol. VII. mf 27.60–63), hereafter referred to as House of Lords Enquiry.
160. Eighth Fisheries Report, p. 71.
161. A. Nimmo, Plan of Pier at Crossfarnogue (NLI, maps, 16 J 13 [2]).
162. Eighth Fisheries Report, p. 69.
163. A. Nimmo to H. Townsend, 24 January 1825 (NAI, CSORP 1825/11543 [4]).
164. Memorial from the noblemen, landowners and residents of Wicklow for a harbour at Greystones. (NAI, CSORP 1825/11543 [3]).
165. H. Townsend to H. Goulburn, 28 April 1825 (NAI, CSORP 1825/11543 [3]).
166. P. LaTouche to W. Gregory, 29 February 1826; J. LaTouche to W. Goulburn, 29 January 1826 (NAI, CSORP 1825/11543 [1]) and [2]).
167. Tenth Fisheries Report, p. 27.
168. A. Nimmo, 'On the Fishing Stations on the North East of Ireland', pp. 33–52 in Fourth Fisheries Report.
169. Fourth Fisheries Report, pp. 49 – 51.
170. Ibid., p. 51.
171. Ibid., plate IV.
172. Eighth Fisheries Report, p. 69.
173. Fourth Fisheries Report, pp. 46–48 and plate VI.
174. Tenth Fisheries Report, p. 69.
175. Fourth Fisheries Report, pp. 34–45.
176. Ibid., pp. 34–5 and plate VI.
177. Ibid., pp. 41–2 and plate IX.
178. Tenth Fisheries Report, p. 27.
179. 'Ardglass Harbour', in First Report from the Commissioners of Public Works in Ireland, p.6 (B.P.P. 1833 [75], vol. XXVII, mf 36.127–8); Statement of Loans made under the Commissioners of Public Works Ireland to 1 January 1835, appendix 2, p. 302, in Report from the Select Committee on Public Works (Ireland) (B.P.P. 1835 [537], vol. XX, mf 38.143–8).
180. Fourth Fisheries Report, p. 42 and plate IX.
181. A. Nimmo, Survey of the Bay and Town of Killough. Coloured map signed by Alexander Nimmo. (PRONI, D 2581/1).
182. Killough and Ardglass. Surveyed by Alexander Nimmo, civil engineer. Published for the use of the Irish Fisheries 1821, plate II, Fourth Fisheries Report.
183. Nimmo, evidence, p. 153, in House of Lords Enquiry.
184. Fourth Fisheries Report, pp. 28–9 and plate IX.
185. Tenth Fisheries Report, pp. 28–9.
186. Port Ballantrae. Coloured plan by Nimmo (NLI, maps, 15 B 1 [1]).
187. Tenth Fisheries Report, plan 16.
188. Fourth Fisheries Report, p. 28.
189. A. Nimmo, Further Report on Greencastle Pier, 5 July 1821, in Fourth Fisheries Report, pp. 31–2.
190. Eighth Fisheries Report, p. 70.
191. Tenth Fisheries Report, pp. 45–6 and plan 16.

Appendix 2
A Listing of Nimmo's Reports, Maps, Plans and Writings and their Locations

MS REPORTS

MS reports that were never published. Each item is unique and its location is indicated in brackets at the end of each entry. Dates in square brackets are inferred.

1806
Nimmo, A., Journal along the North, East and South of Inverness Shire ends at Fort William (NLS, Adv. MS 34.4.20).

1822
Nimmo, A., Report on Clifden Quay by Alexander Nimmo. (NAI, OPW 8/84/9).

Nimmo, A., Extracts from the Report of Alexander Nimmo civil engineer on the present state of and means of improving the Port and Harbour of Sligo, 20 March 1822. (NAI, CSORP 1822/856).

Killaly, J. and Nimmo, A., Report of the Engineers on the proposed Bridge near Youghal, 8 February 1822 (NAI, CSORP 1822/2590).

1823
Telford, T. and Nimmo, A., Report on the Mersey Navigation, 17 June, 1823. (Institution of Civil Engineers London, cat. 627.49427.1).

1823/4
Nimmo, A., Report on the Pier at Brandon (undated, c.1823/4) (NAI, OPW 8/48/11/5).

1824
Nimmo, A., Mr Nimmo's account of the Ross Mines on the Lakes of Killarney, 8 July, 1824. (NLI, MS 657).

APPENDIX 2 A LISTING OF NIMMO'S REPORTS, MAPS, PLANS, WRITINGS AND LOCATIONS

Nimmo, A., Report on the Road proposed for the Colliery District in the West of Lough Allen. Prepared for the Hibernian Mining Company by A. Nimmo (NLI, MS 657).

Nimmo, A., Report on the Quay at Ballycrohane in Cuolagh Bay, 29 February 1824 (NAI, OPW 8/95).

Nimmo, A., Further Report on Ballycrohane Quay, 8 April 1824 (NAI, OPW 8/95).

1827

Telford, T., Brown, N. and Nimmo, A., The Report of Thomas Telford, Nicholas Brown and Alexander Nimmo on the lower part of the Mersey and Irwell Navigation, 19 May, 1827 (ICE London, Library cat. 627.4 (427.2)).

1828

Stevenson, R. and Nimmo, A., Report of Robert Stevenson and Alexander Nimmo to the Mayor of Liverpool on the state of the Wallasey Leasowe, 23 February, 1828 (Liverpool Record Office, 352 CLE/TRA 1/2/15 book 14).

Telford, T., Stevenson R. and Nimmo, A., The Report of Thomas Telford, Robert Stevenson and Alexander Nimmo, civil engineers, recommending two extensive new sea ports etc. etc. in the rivers Dee and Mersey adjacent to Liverpool with a Floating Harbour or Ship Canal to connect them. (NRO, 3410/wat/2/3 pp. 11–39). This is a fair copy of the Report, not in the hand of any of the authors. A printed copy is referenced below.

1830

Vignoles, C. B. and Nimmo, A., Memorandum relative to the experiments made at Mr Laird's works at North Birkenhead (opposite Liverpool) with the new Low Pressure Boiler on the exhausting principle of Messrs Braithwaite and Ericsson, 29 May, 1830 (Institution of Civil Engineers, London. O.C. 64).

Nimmo, A., The Report of Alexander Nimmo, Engineer, on the Progress of the Works of Courtown Harbour, 28 Feb. 1830 (NLS, MS 19971 ff 20–2. Also NAI, OPW8/97 [17]).

[1831]

Alexander Nimmo to the Promoters of the Dublin and Kingstown Railway. [Modified plan for the Dublin and Kingstown Railway] (NAI, OP 974/137).

1831

Nimmo, A., Report of Alexander Nimmo civil engineer on the progress of the public works in the western district in 1830. Dated March 1831 (NAI, OP 973/73).

Nimmo, A., Considerations on the proposed Ship Canal from Dublin to Kingstown as compared with the Railway, 2 February 1831 (NAI, CSORP 1831 2119).

Nimmo, A., Report from Mr Nimmo on the construction of the Dock at Runcorn Gap. Report to the directors of the St Helen's Railway and Canal Company, 8 October 1831 (Nat. Archs. UK, RAIL 593/1).

PUBLISHED REPORTS

1807

Nimmo, A., Historical Statement of the Erection and Boundaries of the Shires of Inverness, Ross, Cromarty, Sutherland and Caithness. Appendix U, pp. 67 – 84, in Third Report of the Commissioners for Highland Roads and Bridges. B.P.P. 1807 (100) vol. III, mf 8.12–13.

1811

Nimmo, A., The Report of Mr Alexander Nimmo on the Bogs in the Barony of Iveragh, in the County of Roscommon [sic]. Report on the practicability of Draining and Cultivating the Bogs in the Barony of Iveragh, 1811. Appendix No. 5 in Fourth Report of the Commissioners on the Bogs of Ireland. B.P.P. 1813/1814 (131), vol. VI, mf 15.33–6.

1812

Nimmo, A., Report on the practicability of Draining and Cultivating the Bogs in various parts of the Counties of Kerry and Cork, 1812. Appendix No. 6 in: Fourth Report of the Commissioners on the Bogs of Ireland. B.P.P. 1813/1814 (131), vol.VI, mf 15.33–6.

Nimmo, A., The Report of Mr Alexander Nimmo on the Bogs in that part of the County of Galway to the West of Lough Corrib, 1812. Appendix No. 12 in Fourth Report of the Commissioners on the Bogs of Ireland. B.P.P. 1813/1814 (131), vol. VI, mf 15.33–6.

1814

Nimmo, A., Estimate of the Expense of Forming a Packet Harbour at Portcullin Cove, near Dunmore, 25 May 1814 (B.P.P. 1813/1814 (200), vol. VII, mf 15.42.

1815

Nimmo, A. Report of Mr Nimmo, engineer to the Commissioners for Improving the River and Harbour of Cork. Dated September 11, 1815. Cork, Edwards and Savage, 1815. Reprinted in 1973 by Fercor Press of Cork.

APPENDIX 2 A LISTING OF NIMMO'S REPORTS, MAPS, PLANS, WRITINGS AND LOCATIONS

1819

Nimmo, A., Some Account of the Drainages of Holland etc., in Second Report of the Select Committee on the State of Disease in Ireland (B.P.P. 1819 [409], vol. VIII, mf 20.69).

Nimmo, A., Short Account of the Drainage of the Fens in England, in Second Report of the Select Committee on the State of Disease in Ireland (B.P.P. 1819 [409], vol. VIII, mf 20.69).

1821

Nimmo, A., Mr Nimmo's Report, dated 20th February 1821, upon the works of Dunmore Harbour (B.P.P. 1821 [492], vol. XI, mf 23.64).

1822

Nimmo, A., Report of Alexander Nimmo, Civil Engineer, on the present state and means of improving the Port and Harbour of Sligo, in Royal Commission on Tidal Harbours, Second Report, Appendix B, No. 92, pp. 155–159. B.P.P. 1846 (692), vol. XVIII, mf 50.169–77.

1823

Nimmo, A., Public Works and the Employment of the Poor. Report dated 16 October 1822 on the Western District by Alexander Nimmo, Civil Engineer. B.P.P. 1823 (249) vol. X, mf 25.86–7.

Nimmo, A., Report on the Fishing Stations on the North Coast of Ireland by Alexander Nimmo, Engineer to the Board of Fisheries, in Fourth Report of the Commissioners of the Irish Fisheries. B.P.P. 1823 (383) vol. X, mf 25.85 – 6.

Nimmo, A., Report No. 2. On the Fishing Stations on the North East of Ireland, in Fourth Report of the Commissioners of the Irish Fisheries. B.P.P. 1823 (383) vol. X, mf 25.85–6.

1824

Nimmo, A., Report of Mr Nimmo to Mr Galloway dated 1 March 1824 on the subject of the road through Glenbegh and [sic] County of Kerry. B.P.P. 1824 (282) vol. XXI, mf 26.142.

1825

Nimmo, A., Table of the comparative numbers of Persons relieved by Poor Rate, in Minutes of Evidence taken before the Select Committee of the House of Lords to examine into the Nature and Extent of the Disturbances which have prevailed in those districts of Ireland which are now subject to the Provisions of the Insurrection Act and to report to the House. B.P.P. 1825 (200), vol. VII, Mf 27.60–3, hereafter referred to as House of Lords Enquiry.

Nimmo, A., A Report as to the Roads wanted to give Access to the Harbours and Fishing Stations on the Coast of Ireland in House of Lords Enquiry. B.P.P. 1825 (200), vol. VII, mf 27.60–3.

Nimmo, A., Report on the Progress of the Public Works of the Western District in 1824, Appendix 1, Estimate of the sum that will be required for carrying on certain Public Works in Ireland during the year ending 5 January 1826. B.P.P.1825 (279), vol. XXII, mf 27.174.

Nimmo, A. Report on the Fishery Stations of the Coast of Connaught, in Sixth Report of the Commissioners for the Irish Fisheries. B.P.P.1825 (385), vol. XV, mf 27.124–5.

1826

Nimmo, A., *The Report of Alexander Nimmo, Engineer, on the Improvement of the Harbour of Drogheda. Report dated May 10th 1826.* Drogheda, 1826. Printed by P. Kelly (NLS, 6.1241(50)).

1827

Nimmo, A., Report on the Fishing Stations of the West Coast of Ireland, from Errishead southwards. Appendix 10 in Eighth Report of the Commissioners of the Irish Fisheries. B.P.P. 1826/7 (487), vol. XI, mf 29.89–90.

Nimmo, A. Mr Nimmo's description of the system of road making in the counties of Galway, Mayo, Sligo, Leitrim and Roscommon, in Report of the Select Committee on Grand Jury Presentments. B.P.P. 1826/1827 (555), vol. III, mf 29. 23–4.

1828

Stevenson, R. and Nimmo, A., Preliminary Report of Robert Stevenson and Alexander Nimmo, civil engineers on the proposed improvements at Wallasey Pool, Feb. 23 1828. Printed in D. Stevenson, *Life of Robert Stevenson* (Edinburgh: A. and C. Black, Publisher, 1878).

Telford, T., Stevenson, R. and Nimmo, A., The Report of Thomas Telford, Robert Stevenson and Alexander Nimmo, civil engineers recommending two extensive new sea ports etc. etc. in the rivers Dee and Mersey adjacent to Liverpool with a Floating Harbour or Ship Canal to connect them. In Stevenson, *Life of Robert Stevenson.*

Telford, T., Stevenson, R. and Nimmo, A., 'Further Report respecting the proposed two new ports etc., on the rivers Dee and Mersey, adjacent to Liverpool', in Stevenson, *Life of Robert Stevenson.*

Nimmo, A., Report of Alexander Nimmo Esq. to the Parliamentary Commissioners of the Holyhead Road, dated London, 11 June 1828 in Select Committee on the State of the Roads under the care of the Whetstone and St Albans Trustees, appendix 1. B.P.P. 1828 (546), vol. IV, 30, 25–26.

APPENDIX 2 A LISTING OF NIMMO'S REPORTS, MAPS, PLANS, WRITINGS AND LOCATIONS

1829.

Nimmo, A., Public Works (Ireland). Report of Mr Nimmo dated 17 June 1829 on the Western District 1828. B.P.P. 1829 (348), vol. XXII, mf 31.136.

1830.

Nimmo, A., Public Works Ireland. Report of Mr Nimmo dated March 1830 on the Progress of the Public Works in the Western District of Ireland in the year 1829. B.P.P. 1830 (199), vol. XXVII, mf 32.194.

Nimmo, A., Public Works Ireland. A Return dated 7 July 1830 of all the Public Money expended under the direction of Mr Nimmo on Public Works in Ireland; which of those Works have been completed, and which left in an unfinished state; and of all Piers built on the Coast, which of them now remaining in a state of perfection, and which in a dilapidated state. B.P.P. 1830 (624), vol. XXVII, mf 32.194.

1831

Nimmo, A., Dublin and Kingstown Railway. Mr Nimmo's letter to the Marquis of Anglesey dated 25 March 1831, together with his observations on the proposed ship canal (RIA, 3B53.56/411).

1838

Vignoles, C. B. and Nimmo, A. Memoranda relative to the experiments made at Mr Laird's works at North Birkenhead with the new Low Pressure Boiler on the exhausting principle of Messrs Braithwaite and Ericsson. Note E, pp. 733–4 in appendix to *A Practical Treatise on Railroads* by Nicholas Wood, third edition, London, 1838.

PUBLISHED BOOKS AND PAPERS

1811

Nimmo, A., 'Boscovich's Theory of Natural Philosophy', in *Edinburgh Encyclopaedia*, vol. 3, parts 3 and 4, pp. 749–68 (Edinburgh: Blackwell, 1811).

Nimmo, A., 'Theory of Bridges', part of section 'Bridges', in *Edinburgh Encyclopaedia*, vol. 4, part 2, pp. 489–512. (Edinburgh: Blackwell, 1811).

1812

Nicholson, P. and Nimmo, A., 'Theory of Carpentry', in *Edinburgh Encyclopaedia*, vol. 5, pp. 749–68. (Edinburgh: Blackwell, 1812).

1821

Nimmo, A., [and others], 'Inland Navigation', in *Edinburgh Encyclopaedia*, vol. 15, pp. 209–315 (Edinburgh: Blackwell, 1821).

1824

Nimmo, A., 'Draining', in *Edinburgh Encyclopaedia*, vol. 8, pp. 63–81 (Edinburgh: Blackwell, 1824).

1825

Nimmo, A., 'On the Application of the Science of Geology to the purpose of Practical Navigation', *Transactions of the Royal Irish Academy* vol. XIV (1825), part 1, pp.39–50 (Paper read October 1823).

Nimmo, A., 'On the Application of the Science of Geology to the purpose of Practical Navigation', *Dublin Philosophical Journal and Scientific Review* vol. 1 (1825), part 1, pp. 152–61.

Nimmo, A., 'On Railways', *Dublin Philosophical Journal and Scientific Review* vol. 1 (1825), part 1, pp. 221–6.

1826

Nimmo, A., 'Report of Alexander Nimmo, Civil Engineer, on the proposed Railway between Limerick and Waterford', *Dublin Philosophical Journal and Scientific Review* vol. 2, (1826), part 3, pp. 1–19. (February).

1832

Nimmo, A. *New Piloting Directions for St George's Channel and the Coast of Ireland. Written to accompany the Chart of St George's Channel and the Coast of Ireland, drawn for the Corporation for Improving the Port of Dublin etc.* (Dublin: A. Thom, 1832).

MAPS, CHARTS AND PLANS IN MANUSCRIPT

These maps and plans are entirely in manuscript, or contain important manuscript annotations. Dates given in brackets are estimated from later publications or from other evidence. In some cases, MS maps and plans were subsequently published, often in the papers of the House of Commons, in which case they also appear under the heading 'Published Maps and Plans' below. The originals often contain MS annotations and insertions that distinguish them from the published copies. For this reason the location of each item is indicated.

1805

Plan of Fort George. Small pencil sketch (Highland Folk Museum, Kingussie, Scotland. C.C.17).

1812

Bogs in the South-western part of the District of Kerry surveyed by order of the Commissioners for enquiring into the nature and extent of the several Bogs in Ireland and the Practicability of Draining and Cultivating them. By Alexander Nimmo, F.R.S Edinburgh, etc., 1812.

A large (size approx 120 by 90 inches) coloured manuscript map of the Iveragh peninsula, divided horizontally into three separate sections: (1) the northern part, approx. 120 x 30 inches; (2) the middle part, approx. 120 x 30 inches; (3) the southern part including Kenmare, approx. 120 x 30 inches.

Accompanying these items are the following separate items by Nimmo:
- Coloured manuscript chart entitled 'Longitudinal section of the navigable level between Valentia and the River Eeny.' Scale about 6 inches to one mile. (This shows sections of a proposed canal)
- Coloured manuscript chart entitled 'Profile of the River Maine from the Castlemaine Haven and mouth of the Laune to its source near Castle Island.'
- Coloured manuscript chart entitled 'Section across the summit between Tralee and the Maine below Castle Island.' (NLI, maps, 16 E. 3).

Bogs on the river Cashen etc., in the north of Kerry surveyed by order of the Commissioners for enquiring into the nature and extent of the several Bogs in Ireland and the Practicability of Draining and Cultivating them. By Alexander Nimmo, F.R.S Edinburgh, etc., 1812.

A large, coloured manuscript map, divided into two parts.

Also accompanying this item are:
- Coloured MS chart entitled 'Section of an extension of the summit level north side of Dirreen River'
- Coloured MS chart entitled 'Section from Ballyheigue Bay to the junction of the Brick and Feale' (NLI, maps, 16 E. 4)

Bogs in the Eastern part of the District of Kerry levelled and surveyed by order of the Commissioners for enquiring into the nature and extent of the several Bogs in Ireland and the Practicability of Draining and Cultivating them. By Alexander Nimmo, F.R.S Edinburgh, etc., 1812. Three large coloured manuscript maps.

Also accompanying is a separate, coloured MS chart entitled 'Sections on Gurrane Bog' (NLI, maps, 16 E. 2).

[1813?]

Geological Map of Connemara, Ireland. Colour-washed geological sketch map. (Geol. Soc. Lond., LDGSL 999).

A new map of Ireland ... by Alexander Taylor. A copy coloured geologically that has MS notes by A. Nimmo (NLI, Maps,16 B. 3 (15)).

[1814]

Manuscript plan, coloured: ground plan of two leaseholds, the property of Sir Charles Coote, in Lower Baggot Street, Dublin. With notes in the handwriting of A. Nimmo (NLI, maps, 16 G. 9 (5)).

[1815?]
Map, road from Ballinrave to Carrick-on-Shannon and Battlebridge to Drumsna, Co. Leitrim (NAI, OPW5HC/6/0066).

1815
Map, road from Boyle to Ballina by Foxford. Barony of Coolavin, Co. Mayo. (NAI, OPW5HC/6/0105).
Map, road from Boyle to Tobbercurry (NAI, OPW5 HC/6/0108).
Map, road from Tobbercurry to Ballina (NAI, OPW5 HC/6/0425).

1816
Plan for Ballingaule Pier and Fishing Village (PRONI, Villiers-Stuart papers, T/3131/H/7. Also MIC 464/22).

1818
Map, road from Cahir[siveen] to Killarney, with proposed improvements. Lissbawn, Coolnaharrigal and Cappaghmore, Barony of Iveragh. (NAI, OPW5HC/6/0306).

1819
Design for a Harbour for small craft at Breenogue Head in the County of Wexford by Alexander Nimmo, 1819 (NLI, maps, 16 J 13 (3)).

[1821]
Manuscript plan: Breakwater pier at Port Ballintrae (Co. Antrim). Signed Alexander Nimmo (NLI, maps, 15 B. 1(1)).
Manuscript map: Map of Belfast and Larne compiled from various authorities for the use of the Irish Fisheries by A. Nimmo (coast from Ballygally to Ballyhalbert with soundings). (NLI, maps, 16 B. 22 (21)).
Manuscript map: Map of Belfast and Larne compiled from various authorities for the use of the Irish Fisheries by A. Nimmo (coast from Ballygally to Ballyhalbert with soundings). Draft copy of the previous item, this copy lacking place names and other details (NLI, maps, 16 B. 22 (22)).
Chapel Bay, Co. Down. One colour-washed and one uncoloured MS map of Chapel Bay, with estimates for piers. Probably by Nimmo, 1822. (NLI, maps, 15 B. 19 (21)).
The Coast of Down from the Lee Stone to St John's Point. Surveyed for the Commissioners of Irish Fisheries by Alexander Nimmo. Undated, engraved map with MS corrections and emendations addressed to engraver by Nimmo shown in red (NLI, maps, 15 B 19 (19)).
Plan and sections of the Lower Newry Navigation with a design for its improvement by Alexander Nimmo, C.E. etc (NLI, maps, 16 B 22 (4)).

1821

Map, road from Killarney to Cahirciveen. Keelnabrack and Ballymakilly, Barony of Dunkerron. (with Ben Meredith) (Nat. Archs., OPW5 HC/6/0307).

Map, road from Killala to Broadhaven by William Bald. (with MS annotations by A. Nimmo) (NAI, OPW5 HC/6/0301).

Manuscript map, coloured: Sligo Harbour (Nat. Archs. UK, MPD 1/32/2).

Strangford River or Entrance into Lough Cone. Surveyed for the Commissioners of Irish Fisheries by Alexander Nimmo civil engineer. Engraved chart dated 1821. This copy incorporates the Butter Plady MS correction with bearings shown in red (NLI, maps, 15 B 19 (2)).

Survey of Bay and Town of Killough. Coloured map signed by Alexander Nimmo. (PRONI, D 2581/1).

Carlingford Lough, surveyed for the Commissioners of Irish Fisheries by Alexander Nimmo, FRSE, MRIA, civil engineer 1821. Engraved charts. Various drafts in course of preparation for engraving (NLI, maps, 16 B 22 (5 to 12)).

[1822]

Coloured plan entitled 'Proposed Bridge over the Liffy at the Fall of Poulafouca.' No name, no date but most likely by Nimmo. (NAI, OPW 5HC/6/0386).

Coloured plan entitled 'Proposed Bridge over the Hollow next the Liffey.' No name, no date but most likely by Nimmo. (NAI, OPW 5HC/6/0387).

Coloured MS map entitled: Cleggan pier, Co. Galway by Alexander Nimmo (NLI, maps, 16 H. 5 (6)).

Colour-washed map of Clifden accompanying 'Report on Clifden Quay.' (NAI, OPW 8/84/9).

Coloured MS map entitled: Clifton (Clifden) harbour by Alexander Nimmo (NLI, maps, 16 H. 5 (7)).

Coloured MS map entitled: Costello Harbour (Co. Galway), showing present and proposed piers, by A. Nimmo (NLI, maps, 16 H. 5 (9)).

Coloured MS map entitled: Bay of Balahaline, (Co. Clare), and design of a proposed Breakwater by A. Nimmo. With MS annotation by Nimmo (NLI, maps, 15 B 9 (1)).

Coloured MS map entitled: Carrigaholt harbour (Co. Clare) by Alexander Nimmo, civil engineer. With proposed extension of old pier (NLI, maps, 15 B 9 (2)).

Small, coloured sketch plan of Seafield, Co. Clare with proposal for pier. Unsigned, but most likely by Nimmo (NLI, maps, 15 B 9 (3)).

Coloured MS map entitled: A map of the Port and Harbour of Kilrush Co. Clare, with observations on the necessity and utility of building a pier at Point of Scough. Probably by Nimmo (NLI, maps, 16 B. 22 (19)).

1822

Design for a Pier at Crossfarnogue by Alexander Nimmo Engineer to the Irish Fisheries, 1822. Colour-washed plan (NLI, maps, 16 J 13 (2)).

Plan and section of the Moy River from the Bridge of Ballina to the Quay at Coolnalecka with the Works done under the direction of Alexander Nimmo. Watercolour map and plan (NAI, CSORP 1823 5144).

[1823]

Manuscript plan: Elevation of the new bridge of Limerick over the River Shannon. By Alexander Nimmo (NLI, maps, 16 H. 22 (3)).

Coloured MS map entitled: The shore at Bray, Co. Wicklow, showing proposed pier and soundings. Unsigned, undated, but likely to be by Nimmo (NLI, maps, 16 J 17 (2)).

Coloured MS map entitled: The Harbour in Clare Island (Co. Mayo) with a design for a pier and inner dock by A. Nimmo (NLI, maps, 16 I. 3 (1)).

1823

Manuscript map: Dublin Bay surveyed for the Commissioners of Irish Fisheries by A. Nimmo, 1823 (NLI, maps, 16 G. 33).

Plan and Elevation of a Proposed Bridge over the Shannon at Limerick (NAI, OPW5HC/6/0340A).

[1824]

Coloured MS map entitled: Plan and section of the Jetties of the Claddagh of Galway. By Alexander Nimmo (NLI, maps, 16 H. 5 (3)).

Coloured plan of Ballinacourty Pier. (NAI, OPW 8/17).

Coloured MS map entitled: Barra Harbour (Co. Kerry) with sketch of proposed pier by A. Nimmo (NLI, maps, 16 H. 8(1)).

1824

Colour-washed map of the proposed 'Harbour of Bundorra' (NAI, OPW 8/54).

[1826?]

Plan of Pier at New Harbour (Co. Galway) by Alexander Nimmo, C.E. Engraved, colour-washed plan (NAI OPW, 8/267 (1)).

1826

Map and Section of an intended Railway or Tram Road between the Cities of Limerick and Waterford with Branches to Thurles, Killenaule and Carrick. Designed for the Hibernian Railway Company by Alexander Nimmo, C.E. and Benjamin Meredith, surveyor, 1826 (NLS, SIGNET.s.145).

Coloured MS map entitled: Cleggan Harbour, Co. Galway. By Alexander Nimmo (NLI, maps, 16 H. 5 (5)).

APPENDIX 2 A LISTING OF NIMMO'S REPORTS, MAPS, PLANS, WRITINGS AND LOCATIONS

1827

Coloured map entitled 'Map of a proposed turnpike road from Armagh to Monaghan within the County of Monaghan' signed by Alexander Nimmo. (NAI, OPW5 HC/6/ 0016).

[1828]

Plan, Elevation and Sections of a proposed timber Bridge over the Blackwater at Rincrew near Youghal by George Nimmo. Undated MS plan. This is the second plan for a bridge at Youghal by Nimmo and is probably by Alexander, although it is attributed to George. The bridge was not built to this plan, but to that given below for [1828] (PRONI, Villiers-Stuart Papers, T3131/H/10/13/1. Also on microfilm MIC/464/22).

1828

Map of the Barony of Ballinahinch, showing works underway in Connemara in 1828 (NAI, CSORP 1828 1468).

1834

Map of road from Clifden to Bunowen Bay, Co. Galway (NAI, OPW5 HC/6/0163). The date indicates that this is a later copy.

ENGRAVED AND/OR PUBLISHED MAPS AND CHARTS

Some of these are published copies of maps and charts the MS versions of which are listed above.

1814

Engraved, coloured map, Iveragh in the county of Kerry levelled and surveyed by order of the Commissioners for enquiring into the Nature and Extent of the Bogs of Ireland by Alexander Nimmo, C.E. etc., 1811. Plate 1 in Appendix No. 5 to the Fourth Report of the Commissioners on the Bogs in Ireland. B.P.P. 1813/1814 (131), vol.VI, mf 15.34–5. For a fine, modern reproduction of this map, see Horner, *Iveragh. Nimmo's Map*.

Engraved, coloured map, bogs on the River Kenmare in Kerry surveyed by order of the Commissioners for enquiring into the Nature and Extent of the Bogs of Ireland by Alexander Nimmo, C. E. etc., 1812. Plate II in Appendix No. 6 to the Fourth Report of the Commissioners on the Bogs in Ireland. B.P.P. 1813/1814 (131), vol.VI, mf 15.34–5. For a fine, modern reproduction of this map, see Horner, *Kenmare River. Nimmo's Map*.

Engraved, coloured map, Bogs on the Rivers Laune and Lower Maine in Kerry surveyed by order of the Commissioners for enquiring into the Nature and Extent of the Bogs of Ireland by Alexander Nimmo, C. E. etc., 1812. Plate IV in Appendix No. 6 to the Fourth Report of the Commissioners on the Bogs in Ireland. B.P.P. 1813/1814 (131), vol.VI, mf 15.34–5.

Engraved, coloured map, Bogs of the Upper Maine etc. surveyed by order of the Commissioners for enquiring into the Nature and Extent of the Bogs of Ireland by Alexander Nimmo, C.E. etc., 1812. Plate V in Appendix No. 6 to the Fourth Report of the Commissioners on the Bogs in Ireland. B.P.P. 1813/1814 (131), vol. VI, mf 15.34–5.

Engraved, coloured map, Bogs of Slieveluaghar, Counties Kerry and Cork surveyed by order of the Commissioners for enquiring into the Nature and Extent of the Bogs of Ireland by Alexander Nimmo, C.E. etc., 1812. Plate VI in Appendix No. 6 to the Fourth Report of the Commissioners on the Bogs in Ireland. B.P.P. 1813/1814 (131), vol. VI, mf 15.34–5.

Engraved, coloured map, Bogs of the River Cashen etc., north of Kerry surveyed by order of the Commissioners for enquiring into the Nature and Extent of the Bogs of Ireland by Alexander Nimmo, C. E. etc., 1812. Plate VII in Appendix No. 6 to the Fourth Report of the Commissioners on the Bogs in Ireland. B.P.P. 1813/1814 (131) vol. VI, mf 15.34–5.

Engraved coloured maps, General Profile of the Middle district of Kerry. Plate VIII in Appendix No. 6 to the Fourth Report of the Commissioners on the Bogs in Ireland. B.P.P. 1813/1814 (131),vol. VI, mf 15.34–5.

1818

Plan of the Works of the Harbour of Dunmore by Alexander Nimmo FRSE, MRIA, Civil Engineer, May 1818. Engraved plan (BL., Maps, 13748 (6)).

1820

Map of Raughley. (Nat. Archs. UK, M2579 M 2579).

Plan of the proposed Addition to the Pier of Killough by Alexander Nimmo Civil Engineer FRSE, MRIA, etc. etc. 1820. Engraved plan (Down County Museum 1990–5)

1821

Bay and Harbour of Sligo surveyed for the Commissioners of the Port by Alexander Nimmo FRSE, MRIA civil engineer 1821. Published for the use of the Irish Fisheries. Engraved chart. (BL, maps, 13450 (1)).

The Harbours of Newport and Westport from the Survey of the County of Mayo by William Bald with Soundings etc. taken for the Commissioners of Irish fisheries under the direction of Alexander Nimmo (Royal Geographic Society).

Carlingford Lough, surveyed for the Commissioners of Irish Fisheries by Alexander Nimmo, FRSE, MRIA, civil engineer 1821. Engraved chart (BL, maps, 12875 (1)).

[1822?]

Newry Navigation. Plan and sections of Lower Newry Navigation with a design for its Improvement by Alexander Nimmo. Lithograph by Allen's, Dame St Dublin (NLI, maps, 16 B 22 (4)).

1822

Part of the coast of Dublin, from Howth to Balbriggan, by Alexander Nimmo, C. E. 1822 surveyed for the Commissioners of Irish Fisheries. Engraved chart (BL, maps, 11795 (1)).

1823

Strangford River or Entrance into Lough Cone. Surveyed for the Commissioners of Irish fisheries by Alexander Nimmo civil engineer. Engraved map in Fourth Report of the Commissioners of Irish Fisheries. B.P.P. 1823 (383), vol. X, plate 1, mf 25.85–6.

Killough and Ardglass. Surveyed by Alexander Nimmo F.R.S.E., M.R.I.A. civil engineer, published for the use of the Irish Fisheries 1821. Engraved map in Fourth Report of the Commissioners of Irish Fisheries. B.P.P. 1823 (383), vol. X, Plate 2, mf 25.85–6.

The Coast of Down from the Lee Stone to St John's Point. Surveyed for the Commissioners of Irish Fisheries by Alexander Nimmo. Engraved map in Fourth Report of the Commissioners of Irish Fisheries. B.P.P. 1823 (383), vol. X, plate 3, mf 25.85–6.

Piers in the County of Antrim (Portrush, Port Ballantrae, Ballycastle, Cushendun and Red Bay). Engraved plans in Fourth Report of the Commissioners of the Irish Fisheries. B.P.P. 1823 (383), vol. X, plate 4, mf 25.85–6.

Piers, County of Sligo (Raughley, Mullaghmore and Rosses) and County of Donegal (Arran Roads, Island of Arran, Burtonport and Greencastle). Engraved plans in Fourth Report of the Commissioners of the Irish Fisheries. B.P.P. 1823 (383), vol. X, plate 5, mf 25.85–6.

Bangor. Engraved plan in Fourth Report of the Commissioners of the Irish Fisheries. B.P.P. 1823 (383), vol. X, plate 7, mf 25.85–6.

Untitled. Engraved plans of the ports of Annalong, Giles's Quay, Skerries, Derryogue, Clogher Head, Loughshinny, Greencastle Point, Dunany Point, Balbriggen and Rush, in Fourth Report of the Commissioners of the Irish Fisheries. B.P.P. 1823 (383), vol. X, plate 6, mf 25.85–6.

Untitled. Engraved plans of the ports of Groomsport, Chapel Bay, Ballywalter, Sand Eel Bay (Orlog), Mill Isle, Ballywilliam, The Boat Hole (Woburn), Ballyhemlan near Ballyhalbert and Sandyland near Ballywalter, in Fourth Report of the Commissioners of the Irish Fisheries. B.P.P. 1823 (383), vol. X, plate 8, mf 25.85–6.

Untitled. Engraved plans of the ports of Newcastle in Ards, Ardglass, Killough, Cove at Kerney Point, Quinton Bay, Tara Bay, Rossglass and Newcastle, in Fourth Report of the Commissioners of the Irish Fisheries. B.P.P. 1823 (383), vol. X, plate 9, mf 25.85–6.

1824

Map of a proposed new road and bridge between the City of Limerick and Parteen in the County of Clare, surveyed under the direction of Alexander Nimmo by Ben Meredith. Lithograph, road shown in red with elevation and section of proposed chain suspension bridge at Athlunkard (NLI, maps, 16 H 22 (4)).

1825

Raughley Pier and Mullaghmore. Engraved plans in Sixth Report of the Commissioners of the Irish Fisheries. B.P.P. 1825 (385), vol. XV, mf 27.124 –5. Killala and Pullagheeny. Engraved plans in Sixth Report of the Commissioners of the Irish Fisheries. B.P.P. 1825 (385), vol. XV, plate 2, mf 27.124–5.

Buntrahur and Saleen. Engraved plans in Sixth Report of the Commissioners of the Irish Fisheries. B.P.P. 1825 (385), vol. XV, plate 3, mf 27.124–5.

Geology rendered practicable for the purpose of navigation. Engraved, coloured map accompanying the paper 'On the application of the Science of Geology to the purposes of Practical Navigation' in *Transactions of the Royal Irish Academy*, vol. XIV, (1825), part 1, 39–50. (paper read October 1823). Not all archived copies of this map are coloured. The R.I.A. copy is uncoloured, the TCD copy, and a copy in the author's possession, are hand-coloured.

1826

Isles of Iniskea and Annagh in the Mullet. Engraved plans in Eighth Report of the Commissioners of the Irish Fisheries, appendix 10, plate 1. B.P.P. 1826/7 (487), vol. XI, mf 29.89–90.

Tarmon Pier and Achill Sound. Engraved plans in Eighth Report of the Commissioners of the Irish Fisheries, Appendix 10, plate 2. B.P.P. 1826/7 (487), vol. XI, mf 29.89–90.

Oldhead Pier and Clare Island Harbour. Engraved plans in Eighth Report of the Commissioners of the Irish Fisheries, Appendix 10, plate 3. B.P.P. 1826/7 (487), vol. XI, mf 29.89–90.

Clare Island Harbour and Port-a-chuilla proposed Boat Pier. Engraved plans in Eighth Report of the Commissioners of the Irish Fisheries, Appendix 10, Plate 4. B.P.P. 1826/7 (487), vol. XI, mf 29.89–90.

Ennisturk and Bundurra. Engraved plans in Eighth Report of the Commissioners of the Irish Fisheries, appendix 10, Plate 5. B.P.P. 1826/7 (487), vol. XI, mf 29.89–90.

Derryinver Pier and Cleggan Bay and Pier. Engraved plans in Eighth Report of the Commissioners of the Irish Fisheries, Appendix 10, plate 6. B.P.P. 1826/7 (487), vol. XI, mf 29.89– 90.

Ardbear or Clifden Harbour. Engraved plan in Eighth Report of the

Commissioners of the Irish Fisheries, Appendix 10, Plate 7. B.P.P. 1826/7 (487), vol. XI, mf 29.89–90.

Roundstone Pier. Engraved plan in Eighth Report of the Commissioners of the Irish Fisheries, Appendix 10, plate 8. B.P.P. 1826/7 (487), vol. XI, mf 29.89–90.

Greatman's Bay and Costello Bay. Engraved plans in Eighth Report of the Commissioners of the Irish Fisheries, Appendix 10, plate 9. B.P.P. 1826/7 (487), vol. XI, mf 29.89–90.

Spiddle, Galway Bay. Engraved plan in Eighth Report of the Commissioners of the Irish Fisheries, Appendix 10, Plate 10. B.P.P. 1826/7 (487), vol. XI, mf 29.89–90.

Barna, Galway Bay and Claddagh, Galway. Engraved plans in Eighth Report of the Commissioners of the Irish Fisheries, Appendix 10, plate 11. B.P.P. 1826/7 (487), vol. XI, mf 29.89–90.

New Harbour, Galway Bay and Ballinacourty, Galway Bay. Engraved plans in Eighth Report of the Commissioners of the Irish Fisheries, Appendix 10, plate 12. B.P.P. 1826/7 (487) vol. XI, mf 29.89–90.

St Kitts Galway Bay and Duras, Galway Bay. Engraved plans in Eighth Report of the Commissioners of the Irish Fisheries, Appendix 10, plate 13. B.P.P. 1826/7 (487), vol. XI, mf 29.89–90.

Burrin New Quay, Galway Bay and Killeeny Harbour. Engraved plans in Eighth Report of the Commissioners of the Irish Fisheries, Appendix 10, plate 14. B.P.P. 1826/7 (487), vol. XI, mf 29.89–90.

1827

Plan of the proposed improvements of the New Passage Ferry on the Monmouth side by Alexander Nimmo, C. E. etc. Engraved plan in Select Committee on the Milford Haven Communication, Second Report and Minutes of Evidence, plan 6. B.P.P. 1827 (472), vol. III, mf 29.22–3.

Plan of the proposed improvements of the New Passage Ferry on the Gloucestershire side by Alexander Nimmo, C.E. etc. Engraved plan in Select Committee on the Milford Haven Communication, Second Report and Minutes of Evidence, plan 7. B.P.P. 1827 (472), vol. III, mf 29.22–3.

Plan of pier at Hobbs Point. Engraved plan in Select Committee on the Milford Haven Communication, Second Report and Minutes of Evidence, plan 5. B.P.P. 1827 (472), vol. III, mf 29.22–3.

[1828].

Bay and Harbour of Youghal surveyed under the direction of Alexander Nimmo C.E. etc. exhibiting the proposed bridge. Engraved map and section, undated. The section shows the timber bridge that was actually built at this site between 1829 and 1832 (BL, maps, 11500 (3)).

1831

The Harbour of Valentia by Alexander Nimmo civil engineer, dated 1831. London, published by Act of Parliament at the Hydrographical Office of the Admiralty, June 15th 1832. Engraved chart of Valentia harbour with enlarged insets of the north-western entrance and the western entrance (BL, maps, SEC 1. (49)).

1832

Chart of the Coasts of Ireland and St George's Channel exhibiting the sea and harbour lights under their management drawn chiefly from original surveys, respectfully dedicated to the Corporation for improving the Port of Dublin by Alexander Nimmo, civil engineer and hydrographer. (Irish Architectural Archive, IEI 534).

1835

Bald, W. Chart of the River Boyne from Drogheda to the Sea at the Bar, showing the various works now in progress to improve the channel under the direction of William Bald, Civil Engineer, 1834. Plan 21 in B.P.P. 1835 (76), vol. XXXVI, mf 38. 296–8. According to Bald, this chart is a copy of Nimmo's original chart, based on a survey done for Nimmo by John MacNeill.

PUBLISHED COPIES OF NIMMO'S ORAL EVIDENCE TO PARLIAMENTARY COMMITTEES

1817

Report of the Select Committee on means of preventing explosions on steamboats, 24 June 1817. B.P.P. 1817 (422), vol. VI, mf 18.30.

1819

Second Report of the Select Committee to enquire into the State of Ireland, as to Disease ... and the Condition of the Labouring Poor ... with a view to facilitate the application of the funds of private individuals and Associations for their employment in useful and productive labour. Minutes of Evidence, 8–14 May. B.P.P. 1819 (409), vol. VIII, mf 20.68–9.

1824

Report of the Select Committee on the Survey and Valuation of Ireland. Minutes of Evidence (Nimmo's evidence on 14, 17, 18, 27 May 1824). B.P.P. 1824 (445), vol. VIII, mf 26.47–51.

1825

Select Committee of the House of Lords appointed to examine into the nature and extent of the disturbances which have prevailed in those districts of

APPENDIX 2 A LISTING OF NIMMO'S REPORTS, MAPS, PLANS, WRITINGS AND LOCATIONS

Ireland which are not subject to the provisions of the Insurrection Act. Minutes of Evidence. B.P.P. 1825 (200), vol. VII, mf 27.60–63.

Evidence to Committee on Grand Jury Presentments. B.P.P. 1826/7 (555) vol. III, mf 29.23–24.

1826

[First] Report from the Select Committee on Emigration from the United Kingdom (Nimmo's evidence on 4 May 1826). B.P.P. 1826 (404), vol. IV, mf 28.20–24

Committee on a Bill for making and maintaining navigable communication for Ships between Norwich and the Sea at Lowestoft. Minutes of Evidence (Nimmo's evidence on 11, 12 and 18 April 1826). B.P.P. 1826 (369), vol. IV, mf 28.24–26.

1827

Select Committee on the Milford Haven Communication. Second Report and Minutes of Evidence (Nimmo's evidence on 14 and 22 May and 7 June 1827). B.P.P. 1827 (472), vol. III, mf 29.22

Second and Third reports of the Committee on Emigration from the United Kingdom (Nimmo's evidence on 8 May 1827). B.P.P. 1826/7 (237), vol. V, mf 20.39–41.

WORKS NOT YET FOUND

These are charts, plans and maps for whose existence there is contemporary evidence but which have not been seen or encountered in the extensive research for this book. The references give indicative evidence for the existence of each item.

Survey charts of Galway, Clew and Killala Bays (1822–26)

'Some charts appertaining to this survey have been published for the use of the fisheries, and sought after with much earnestness. A few more are in the hands of the engraver, comprehending some of the most interesting Bays of the western and northwestern coasts; namely, *Galway, Clew* and *Killala* Bays.' Seventh Report of the Commissioners for Irish Fisheries: Dated 19 May 1826.

Survey chart of the River of Cork [1815]

A chart, drawn to the scale of 20 inches to one mile, of the River Lee from the city of Cork to Passage. Report of Mr Nimmo, engineer to the Commissioners for Improving the River and Harbour of Cork. Dated September 11, 1815.

Survey chart of Greatman's Bay, Co. Galway

'Greatman's Bay has been carefully surveyed and the soundings marked in the neighbourhood of the new pier, engraved on the same sheet with Costello and a sketch of the neighbouring isles is inserted from my bog surveys.' A. Nimmo in Eighth Report of the Commissioners of Irish Fisheries.

Survey chart of the entrance to the River Boyne
'I have delineated the course of this proposed channel upon the plan', p. 8, and 'I have drawn it upon the plan', p. 10, in *The Report of Alexander Nimmo, Engineer, on the Improvement of the Harbour of Drogheda*. Drogheda, 1826. According to Bald (B.P.P. 1835 [76], vol. XXXVI) the survey was carried out by John MacNeill for Nimmo.

Plan for a bridge at the lower end of Swansea
'...accordingly we proposed, and both he [Telford] and I have given plans to that effect to obtain the same thing by crossing at the lower end of Swansea', p. 663, in Minutes of Evidence to the Select Committee on the Milford Communication. (B.P.P. 1826–1827 [472], vol. III mf 29.22).

Appendix 3
Account of all the structural works, and their state of completion, carried on by Nimmo in the western district as listed by him in June 1830.

(Reprinted from the British Parliamentary Paper 1830 [624], vol. XXVII)

County of the Town of Galway	State of Works
Road from Galway to Claregalway.	Completed.
Woodstock Road, near Galway.	Ditto.
Clonliffe road.	Ditto.
Road from Galway to Menlough.	Ditto.
Slate Pier, at Galway.	Ditto.
Claddach Pier, at same.	Ditto.
Galway market-house.	Ditto.

County of Galway	
Conamara Coast road, from Galway to Killary Harbour.	In progress.
Kilkerran road, from Kilkerran Bay to join Killary road.	Ditto.
Road from Spiddle to Moycullen.	Completed.
From Killeries to Cong.	Ditto.
From Headfort to Galway.	Ditto.
From Oughterard to Clifden	In progress.
Central Conamara road.	Ditto.
Road from Kinvara to Kilcolgan.	Completed.
Crushua road.	Ditto.
Road from Creggs to Galway.	Unfinished.
From Killough to Kilconnell.	Completed.
From Keelogues to New Forest.	Unfinished.
Castlekelly Road.	Completed.
Road from Tullynadaly to Beagh.	Ditto.
From Tuam to Galway.	Ditto.

From Tuam to Clontoo.	Ditto.
From coast road through Roundstone to Bunowen.	In progress.
To Costello pier.	Unfinished.
To Greatman's Bay pier.	Ditto.
To Derryinver pier.	Completed.
Blindwell Hill.	Ditto.
Ironpool Turlough road.	Ditto.
Ballinderry Bridge.	Ditto.
Corofin Quay and cut.	Unfinished.
Sewer at Tuam.	Completed.
Annadown and Oldbury piers in Lough Corrib.	Ditto.
Ferry of Knock, piers and road.	Ditto.
Leenane Quay.	Ditto.
Derryinver quay.	Ditto.
Cloonisle quay.	Ditto.
Greatman's Bay quay.	Unfinished.
Clifden Doobeg quay.	Completed.
Clifden fishery pier.	Unfinished.
Cleggan fishery pier.	Completed.
Roundstone pier.	Ditto.
Costello pier.	Ditto.
Barna pier.	Unfinished.
Burrin pier.	Completed.

County of Mayo

Road from Westport to Killary Harbour.	In progress.
Ballintobber road.	Completed.
Bundurra Bridge.	Ditto.
Road from Hollymount to Claremorris and Ballyhaunis.	Unfinished.
From Ballinrobe to Hollymount.	Completed.
From Westport to Claremorris (Ballyglass).	Ditto.
From Newport to Achill Head.	Ditto.
Bridge of Annies.	
	Ditto.
Erris road and its branches	
	Ditto.
River Moy navigation works and roads.	Unfinished.
Road from Killala to Ballina, Foxford and Swineford.	Completed.
From Ballina to Moyne.	Ditto.
Castle Connor ferry road.	Ditto.
Road from Bonnyconlon to Swineford	Unfinished.
Killala pier and road to same.	Completed.
Kilnacarra Bridge.	Ditto.

Road from Killala to Broadhaven. | In progress.
    From Bangor to Tulloghan Bay | Unfinished.
    From Bangor to Ballyveeny and Newport. | Ditto.

County of Sligo

Sligo Committee roads.	Completed.
Sligo Committee Harbour.	Unfinished.
Cartron road and wall.	Completed.
Raughly pier.	Ditto.
Ballyfarnon road.	Ditto.
Drumcliffe road.	Ditto.
Gortlownane road.	Ditto.
Circuit or Templehouse road.	Ditto.
Road from Bonnyconlon to Pullagheeny pier.	Ditto.
Geevah colliery road.	Ditto.

County of Roscommon

Road from Castlerea to Ballymoe.	Completed.
Boyle new entrance.	Ditto.
Road from Boyle to Tarmonbarry and Slievebawn.	Ditto.
From Carrick-on-Shannon to French Park.	Ditto.
From the Arigna iron works to Mount Allen.	Ditto.
Mount Allen road to Keadue.	Ditto.

County of Leitrim

Dromahair colliery road.	In progress.
Road from Cloone to Mohil.	Completed.
From Ballinamore to Belturbet.	Ditto.

Signed: Alexander Nimmo. June 18th, 1830

Appendix 4
Statement of the Commissioners for Auditing the Public Accounts in Ireland regarding the Accounts of Alexander Nimmo.

(Reprinted from the Twenty-First Report of the Commissioners,
B.P. P. 1833 [102], vol. XVII mf 36.124–127)

In our seventeenth report we have given a detail of the peculiar circumstances connected with the expenditure of the late Alexander Nimmo, as civil engineer, on account of public works in the western district, pursuant to the 3 Geo 4, c. 34, in the years 1822, 1823, 1824 and 1825, and on which account disallowances were made by us, to the amount of £19,368-14s for the reasons then specified.

In the course of the accounts subsequently furnished between the year 1825 and the present final account of the representatives of the late Mr Nimmo, these disallowances have been gradually reduced to the amount of £8972-10s-4¾d by the production of vouchers.

In the present account, a credit has been claimed by the representatives of the late Mr Nimmo, for the sums still remaining as charges against their late brother, and from which, with the exception of one credit, we have thought it our duty to relieve them, though these credits have not been supported by such perfect vouchers as are rigorously extracted from other accountants, where the expenditure has been incurred under ordinary circumstances.

It has been most clearly shown to us, that every exertion had been made by the late Mr Nimmo previous to his death, and subsequently by the officers connected with the establishment, and by the representatives, to remedy, in every way practicable, the errors which had originally existed in the inaccuracy of the accounts, and to supply the deficiencies in vouchers which had been the consequence of an acknowledged want of system in the early stages of the late Mr Nimmo's mode of payment of the expenditure on the public works in the western district, and of which irregularities the most fruitful sources have been stated at large in our seventeenth report above alluded to, and to which we would respectfully beg leave to refer. These exertions, earnest as they seem to have been, have in many, indeed in most instances proved unavailing, to the full accomplishment of the object. In the absence therefore of those proofs, which we had hoped might have been obtained in the interval between our examination of the first series of accounts and that of the one now under consideration, we have given the most severe and searching scrutiny into the value of the next best secondary evidence that the accountants have been able to produce, and the result has been, that we feel fully justified in the declaration of our opinion, that no moral doubt exists of the faithful appropriation of the moneys stated to have been advanced for the execution of the public works; these advances have, in many instances, been made under the express orders of the Irish Government to country gentlemen, residing in the neighbourhood of those districts where the pressure of distress was heaviest, and whence the loudest calls were heard for immediate relief, which was as instantly as benevolently yielded by the Government.

In all cases, unquestionable evidence has been given to us, either by the affidavit of the late Mr Nimmo, or by the oaths of the superintending engineers, of the actual execution of the works taken credit for in the account: and where any money has been advanced by Mr Nimmo, from the grants entrusted to him, to private individuals, for the declared purpose of giving employment to the poor, in the construction or repairing of public works, the advances appear to have been given under the orders of the Irish Government, originating on an application for such assistance by the parties resident amongst and witnessing the scenes of the distress: the receipts of these parties for the money entrusted to them, with their affidavit of its appropriation, have been in most cases produced; in the absence of the latter, the testimony of the superintending engineer has supplied the deficiency, and in those cases where the actual acknowledgement of any person said to have been entrusted with the public money has not been sent, the cancelled notes of the Bank of Ireland, drawn in favour of that party, and carrying his endorsement, have been exhibited; and thus the chain of evidence, connecting the party receiving the money with the execution of the public works under the late Mr Nimmo's superintendence, is perfect in all its links.

Sources and Bibliography

MANUSCRIPT SOURCES

Items are referenced in the chapter notes and references by the archive location codes as abbreviated below, followed by the series and/or call number. Thus, letters from 1823 in the National Archives of Ireland, Chief Secretary's Office Registered Papers are referenced:
NAI, CSORP 1823/No.

National Library of Ireland (NLI)
    Maps
    Hibernian Mining Company Records
    William Smith O'Brien Papers

British Library (BL)
    Peel Papers
    Wellesley Papers
    Maps

National Library of Scotland (NLS)
    Robert Stevenson and Co. Records
    Nimmo's Inverness Survey
    Royal Society of Edinburgh Minute Book
    Maps, Signet Library

National Archives of Ireland (NAI)
    Chief Secretary's Office Registered Papers (CSORP)
    Official Papers (OP)
    OPW engineering and architectural drawings, roads and bridges (OPW 5HC)
    OPW engineering and architectural drawings, piers and harbour structures (OPW 8)

SOURCES AND SELECT BIBLIOGRAPHY

National Archives of the UK (Nat. Archs. UK)
RAIL 370/1; RAIL 593/1; BT 41/906/5326
Works of A. Nimmo, AO17
Maps

National Archives of Scotland (NAS)
Chanonry Kirk Session Minutes

Highland Council Archive, Inverness Library
Inverness Royal Academy, Minute Book, Board of Directors

Public Records Office of Northern Ireland (PRONI)
Fitzgerald Papers
Villiers-Stuart Papers
Downshire Papers

University of Durham Library
Grey of Howick papers

University of Cambridge Library
Greenough Papers

University College London
Greenough Papers

Library, Institute of Civil Engineers, London (ICE)

Geological Society of London Library (GSL)

Princeton University, Firestone Library

PARLIAMENTARY PAPERS

Parliamentary papers are quoted as follows: British Parliamentary Papers (B.P.P.), year of publication, paper number (in square brackets), Chadwick-Healy volume number and the Chadwick-Healy microfiche number. Thus, the Third Report of the Commissioners of the Irish Fisheries is referenced: B.P.P. 1822 [428], vol. XIV, mf 24.109.

The parliamentary papers quoted include:
Commissioners for the Highland Roads and Bridges
Commissioners for Auditing the Public Accounts of Ireland
Commissioners for the Bogs of Ireland
Commissioners for the Irish Fisheries
Commissioners for Repairing, Maintaining and keeping in repair certain Roads and Bridges in Ireland
Commissioners of Inquiry into the Collection and Management of the Revenue in Ireland
Commission of Inquiry into Excise Establishments
House of Lords Enquiry into the State of Ireland
Public Works Ireland
Royal Commission on Tidal Harbours

Select Committee on the Survey and Valuation of Ireland;
Select Committee on Post Office Communication with Ireland;
Select Committee on Grand Jury Presentments in Ireland;
Select Committee on the Employment of the Poor in Ireland;
Select Committee on the State of the Poor;
Select Committee on Irish Estimates;
Select Committee on the Milford Communication;
Select Committee on the Irish Estimates;
Select Committee on the Dublin and Kingstown Ship Canal;
Select Committee on the Means of Preventing Explosions on Steamboats.

SECONDARY PRINTED SOURCES, SELECT BIBLIOGRAPHY

Andrews, J. H., *A Paper Landscape*. (Dublin: Four Courts Press, 2002).

Ashton, T. S., *The Industrial Revolution, 1760–1830*. (Oxford: Oxford University Press, 1969).

Berry, H. F., *A History of the Royal Dublin Society* (London: Longman, Green and Co., 1915).

Billington, D. P., *The Tower and the Bridge*. (New York: Basic Books, 1983).

Chart, D. A., *An Economic History of Ireland*. (Dublin: Talbot Press, 1920).

Cox, R. C. and Gould, M. H., *Civil Engineering Heritage, Ireland* (London: Thomas Telford, 1998).

De Courcy, S., 'Alexander Nimmo: Engineer extraordinary', *Connemara. Journal of the Clifden and Connemara Heritage Group*, 2 (1995), pp. 47–56.

Delaney, R., *Ireland's Inland Waterways. Celebrating 300 years*. (Belfast: Appletree Press, 2004).

Dunlop, J., *The British Fisheries Society 1786–1893*. (Edinburgh: John Donald, 1978).

Foster, J.W. and Chesney, H. C. G. (eds), *Nature in Ireland: A Scientific and Cultural History* (Montreal: McGill-Queen's University Press, 1997).

Fryer, C. E. J., *The Waterford and Limerick Railway* (Usk, Monmouthshire: Oakwood Press, 2000).

Goring, R., *Chambers Scottish Biographical Dictionary* (Edinburgh: Chambers, 1992).

Griffith, R., *A Practical Guide to the Valuation of Rents in Ireland*, ed. J. V. Fitzgerald (Dublin: E. Ponsonby, 1881).

Haldane, A. R. B., *New Ways through the Glens* (Edinburgh: Thomas Nelson and Sons, 1962).

Herries-Davies, G. L., *Sheets of Many Colours* (Dublin: Royal Dublin Society, 1983).

Herries-Davies, G. L., and Mollan, R. C., (eds), *Richard Griffith 1784–1878* (Dublin: Royal Dublin Society, 1980).

Horner, A., *Iveragh Co. Kerry in 1811. Alexander Nimmo's Map for the Bogs Commissioners* (Dublin: Glen Maps, 2002).

Horner, A., *Kenmare River Co. Kerry in 1812. Alexander Nimmo's Map for the Bogs Commissioners* (Dublin: Glen Maps, 2003).

Jenkins, B., *Era of Emancipation* (Montreal: McGill-Queen's University Press, 1988).

Keegan, B. F. and O'Connor, R. (eds), *Irish Marine Science 1995* (Galway: Galway University Press, 1966).

Kinsella, A. *The Windswept Shore. A History of the Courtown District* (Dublin: Graphic Services, 1982).

Kirwan, R., *Elements of Mineralogy* (London: McKinley, 1810).

Lee, A. J., and Ramster, J. W. (eds), *Atlas of the Seas around the British Isles* (London: Ministry of Agriculture and Food, 1981).

Lee, S. (ed.), *Dictionary of National Biography* (London: Smith, Elder and Co., 1906).

Lewis, S., *A Topographical Dictionary of Ireland* (London: S. Lewis, 1837).

—— *A Topographical Dictionary of Wales*, 4th edition (London: S. Lewis, 1844).

Marmion, A., *The Ancient and Modern History of the Maritime Ports of Ireland* (London: V. H. Banks, 1855).

Matthew, H. G. and Harrison, B. (eds), *Oxford Dictionary of National Biography* (Oxford: Oxford University Press, 2004).

McCutcheon, W. A., *The Canals of the North of Ireland* (London: David and Charles, 1965).

Mitchell, J., *Reminiscences of my Life in the Highlands*, 2 vols (London: privately printed, 1883, 1884).

Mohr, P., *Wind, Rain and Rocks* (Galway: the author, n.d. [c 2001]).

Mudie, R. (ed.), *The Surveyor, Engineer and Architect for the Year 1842* (London: Bell and Wood, 1842).

Murray, K. A., *Ireland's First Railway* (Dublin: Irish Railway Record Society, 1981).

O'Brien, W., *Ross Island. Mining, Metal and Society in Early Ireland* (Galway: National University of Ireland Press, 2006).

O'Flaherty, R., *A Chorographical Description of West or H-Iar Connaught*, ed. J. Hardiman (Dublin: Irish Archaeological Society, 1846).

Rickman, J. (ed.), *Life of Thomas Telford, Civil Engineer, Written by Himself* (London: Hansard and Sons, 1838).

Ryland, R. H., *The History, Topography and Antiquities of the County and City of Waterford* (London: John Murray, 1824).

Sirr, J. D., *A Memoir of the Honourable and Most Reverend Power Le Poer Trench, Last Archbishop of Tuam* (Dublin: William Curry, 1845).

Skempton, A. W., *Biographical Dictionary of Civil Engineers. Vol. 1, 1500–1830* (London: Thomas Telford, 2002).

Smart, R. N., *Biographical Register of the University of St Andrews 1747–1897* (Fife: University of St Andrews Library, 2004).

Steers, J. A., *The Coastline of Scotland* (Cambridge: Cambridge University Press, 1973).

Stevenson, D., *Life of Robert Stevenson* (Edinburgh: A. and C. Black, 1878).

Trant, K., *The Blessington Estate 1667–1908*, (Dublin: Anvil Books, 2004).

Villiers-Tuthill, K., *Alexander Nimmo and the Western District*. (Clifden: Connemara Girl Publications, 2006).
―― History of Clifden 1810–1860 (Clifden, the author, 1992).
Ward, A. W., Prothero, G. W. and Stanley, L. (eds), *The Cambridge Modern History*. Vol. 12, *The Restoration* (Cambridge: Cambridge University Press, 1907).
Webster, T., *Minutes of Evidence and Proceedings of the Liverpool and Birkenhead Dock bills for the sessions 1848, 1850, 1851 and 1852* (London: John Weale, 1853).
Wilkinson, G. H. and Hogg, T. J., *A Report of the Proceedings of a Court of Inquiry into the Existing State of the Corporation of Liverpool* (Liverpool: J. and J. Mawdsley, 1833).

Index

Abbot, Charles, 21
Aberdeen, 4, 13, 26, 28, 106
Achill Sound Pier (Co. Mayo), 190, 348, 351, 351
Act of Union (1801), ix, 43, 163
Admiralty, 143, 144, 145, 172–9, 414
agriculture, 43, 44–5, 54, 61–3, 79, 90, 148–9, 191
   Connemara, 87–8, 89, 90, 91, 92, 93
   in Iveragh, 58, 59, 60, 61–4
   in Scotland, 19, 20, 34–5
'Agriculture' article in *Edinburgh Encyclopaedia*, 39
Ainslie, John, 2, 14, 23, 24
Allen, Lough, 263, 264
American and Colonial Steam Navigation Company, 268–71
American War of Independence, 18
Anglesey, Marquis of, 209–10, 211
Annios Bridge (Co. Mayo), 245, 247
Antrim, Co, P17, 165, 168, 257, 392, 406, 411
Aran Islands, 89, 182, 198, 373, 413
Ardbear (Salt Lake) New Bridge (Co. Galway), 220, 245, 248
Ardglass Harbour and Kimmersport (Co. Down), 160, 167, 168, 390, 390–1, 411
Arigna (Co. Leitrim), 263, 264, 267
Arrowsmith, Aaron, 23, 24, 25, 25, 33, 34
Athlunkard Bridge (proposed in Limerick), 217, 229, 229, 412
Atlantic trade routes, 259, 261, 266–73
Audit Commissioners (Public Accounts Office), 160, 205, 207, 420–1

Balbriggan Harbour (Co. Dublin), 388, 388–9
Bald, William, 2–3, 14, 15, 135, 136, 143, 160, 168, 189
   Bogs Commission and, 15, 48–9, 50, 53, 55, 128
   harbours and piers, 189, 260–1, 349, 350–1, 352, 354, 410
   maps and charts, 407, 410, 414
Ballina (Co. Mayo), P6, 168–9, 192, 200, 345–6
Ballinacourty Pier (Co. Galway), 369–70, 408, 413
Ballinahinch (Co. Galway), 87, 92, 93, 94, 191, 206, 241, 264, 328, 357, 360
Benabola (Canal) Bridge, 246, 248, 248, 249
Ballinderry Bridge (Co. Galway), 247
Ballingaule Pier and Village, 253–4, 254, 406
Ballycrohane (Coulagh) Pier (Co. Cork), 384–5, 385, 399
Bangor Harbour (Co. Down), 168, 258, 389–90, 411
Barna Pier (Co. Galway), 365, 365, 413
Barren Land Act, 62–3
Barrow Harbour (Co. Kerry), 77, 379
Barry, J. Redmond, 386
Bealaclugga Bridge (Co. Clare), 220, 221
Beaufort, Francis, 271, 319, 320
Beaufort, W., 116
Belfast Harbour, 168
Belmullet Canal (Broadhaven, Co. Mayo), 197, 347–8, 348
Benabola (Canal) Bridge (Co. Galway), 246, 248, 248, 249
Bianconi Coach Inn (Maam), 197, 197
Billington, D.P., 38
Bingham, Colonel, 205, 325

Blackrock (Co. Cork), 115, 117, 118, 123
Blacksod (Tarmon) Pier (Co. Mayo), 189, 190, 350, 350–1, 412
Blackwater River (Co. Kerry), P1, 67, 68, 72, 157, 217, 230
Blake, Manus, 198–9, 325
Blake family of Renvyle, 92, 94, 198, 206, 357
Bland family of Derryquin, 67, 68
Bligh, Captain, 172, 174
Board of Civil Engineers, 134–6
Board of Inland Navigation, 121, 122–3, 129, 153, 159–60, 184, 259
   John Killaly and, 122–3, 202, 207–8, 258, 305
   roads and, 153, 193–4, 202, 207, 247
Bog of Allen, 47–8, 176
bog reclamation, 44–5, 47, 48, 54, 59–60, 64, 72–3, 76, 138, 140
Bogs Commission, ix, 12, 13, 15, 39, 43–55, 57–77, 83, 128, 140, 158, 176
   Connemara and, 79–85, 87–99, 190–1, 194, 262
   maps and charts, P1, P2, P3, 46, 51–4, 79, 83, 94, 95, 98–9, 103–5
   maps and charts (Co. Kerry), P1, 58, 61, 65, 66, 67–8, 404–5, 409–10
   AN and, 15, 49–51, 53, 55, 57–66, 103, 158, 329, 333
   written reports to, 51–4, 58–9, 66, 68, 74, 81, 82–4, 87, 103, 400
Bolton, Archbishop Theophilus (of Cashel), 44
'Book of Splittings', the, 27
'Boscovich' article in *Edinburgh Encyclopaedia* (AN), 37, 39, 40, 103, 403
Boundary Survey *see* Survey and Valuation of Ireland (Boundary Survey)
Boyne, River, 259, 414, 416
Brandon Pier (Co. Kerry), 380, 380–1, 398
Brewster, David (1781-1858), 3–5, 37–8, 39
bridges, 66, 71, 114, 153, 157, 158, 160, 214–52, 230, 236–49
   Dunmore Harbour, 103, 113–14, 220
   Edinburgh Encyclopaedia article (AN and Thomas Telford), 37–9, 103, 332, 403
   Leenane to Westport road, 220, 222, 236–9, 238, 239
   maps and plans, 407, 408, 409, 412, 413
   pointed-arch, 217–24, 236, 239
   Ring of Kerry, 66, 71, 214
   in Scotland, 20, 21, 27, 139
   timber, 217, 230–6
   urban, 224–30
   utilitarian nature of, 214, 224
Britain, AN's works in, 276–97
   Haddiscoe cut (Yare Navigation, Norfolk), 276–7, 331
   Mersey and Irwell Navigation Company (M&I), 203, 280–1
   see also Scotland
British Fisheries Society, 19–20, 21, 137, 141
Brown, Nicholas, 280, 399
Browne, Denis, 206, 325, 328

Brownrigg, John, 258
Bundorragha Bridge (Co. Mayo), 244
Bundurra Quay and Pier (Co. Mayo), P13, 354–5, 355, 408, 412
Burrin New Quay (Co. Clare), 373–4, 374, 413
butter trade, 61–2, 115, 266, 271

Cahir Bridge (Co. Waterford), 158
Cahirsiveen Pier (Co. Kerry), 381
Caledonian Canal, 12, 20, 21, 23, 40, 139
Canada, 267–8
Canal (Benabola) Bridge (Co. Galway), 246, 248, 248, 249
canals, 45, 51, 74–6, 98, 285, 337, 400, 406
    Belmullet Canal (Broadhaven, Co. Mayo), 197, 347–8, 348
    Caledonian Canal, 12, 20, 21, 23, 40, 139
    Corrib Canal (Co. Galway), 367–9
    Dublin and, 314, 315–16, 400, 403
    Haddiscoe cut (Yare Navigation, Norfolk), 276–7, 331
    MBB canal, 308, 309–11, 315
    Newry Navigation, 258–9, 406, 410
    proposed, 64–5, 118, 119, 122, 124
    railways and, 300, 301, 302, 305, 306, 309–10, 313, 314–16
    Shannon navigation, 12, 193–4, 225, 263, 264, 267
    Wallasey Pool Plan, 281, 282, 283–97, 288
Carlingford, Lough, 167, 258, 407, 410
'Carpentry' article in Edinburgh Encyclopaedia (AN and Peter Nicholson), 37, 39, 103, 403
Carrigaholt Pier (Co. Clare), P14, 378–9, 379, 407
Carter, Samson, 319, 320, 330, 334, 336
Cashen, River, 57, 76, 410
Casla Bridge (Co. Galway), 244
Castlebar (Co. Mayo), 200, 205, 208
Castlemaine (Co. Kerry), 72, 73
Castletown Berehaven (Co. Cork), 384
cattle pastures, 62
Chapman, William, 280
Chester (Cheshire), 286, 288, 295–6, 297
Chrimes, M., 107
Church, Irish, 63
City of Dublin Steam Packet Company, 264
Clare, Co., P14, 89, 196, 202, 220, 221, 373–9, 407, 412, 413
Clare Island (Co. Mayo), P12, 352, 408, 412
Cleggan Pier (Co. Galway), P7, 196, 198, 201, 357–8, 407, 408, 412
Clifden Quay (Co. Galway), 182, 197, 198, 358–60, 359, 398, 407, 412–13
Clonakilty Pier (Co. Cork), 385
Cloongullane Bridge (Co. Mayo), 247
Cloonisle Quay (Co. Galway), 356, 361, 361–2
coal industry, 43, 54, 263, 264, 300, 309, 323
coastal survey, Fisheries Commission, 144, 163–9, 172–82, 203, 220, 265–6, 388, 389, 391
    termination of (1824), 145, 178, 183–4, 203
coastal surveys, Admiralty, 172, 173, 174, 175, 176–7, 178
Colby, Major Thomas, 145
Colthurst, Sir Nicholas, 121
Commissioners for Highland Roads and Bridges, 13, 15, 21, 23–37, 49, 50, 139, 140, 141, 195–6, 400
Commissioners for the Relief of the Distressed Poor (Dublin Castle) see Dublin Castle Fund
Connaught, 168, 172, 187–211, 333, 379
    famine (1822), 151, 181–4, 187–8, 197–8, 362, 373
    piers and harbours, 168, 187, 191–2, 192, 193, 196, 198, 205
    see also western district: piers and harbours
    roads, 146, 187, 189–91, 192, 192–5, 196, 198, 202, 203, 206, 207–11
Connemara (Co. Galway), 79–99
    Bogs Commission and, 79–85, 87–99, 190–1, 194, 262
coast of, 91, 96, 169
    famine (1822), 181–4, 197–8, 362
    geology of, P2, P3, 79–85, 87, 103–5
    maps and charts, x, 13, 79, 95
    maps and charts, geological, P2, P3, 14, 85–7, 94, 98–9, 105, 405

roads, P11, 89, 92–3, 94–6, 97–8, 101, 190–1, 206, 239, 356, 361, 363
Consolidated Fund, 129, 152, 153, 156, 159, 190, 206, 225, 233, 305
Conyngham, William Burton, 164
Coote, Sir Charles, 101, 102, 104, 220, 405
Cork, Co., 13, 49, 57, 61, 75, 106, 124, 196, 202, 384–6, 399, 410
    see also Youghal Bridge (Co. Cork)
Cork River and Harbour, P4, 115–24, 120, 171, 182, 184, 253, 400, 415
    River Lee, 107, 115, 116–18, 121–2, 123, 124, 415
Cornamona Bridge (Co. Mayo), 239, 239–41
Corrib, Lough, 55, 81, 82, 87, 88–9, 90, 96–8, 367–9
Corrib, River, 80, 87, 92, 94
County Assessment Acts, 23
County Surveyors, 134, 135–6, 141
Courtmacsherry Pier (Co. Cork), 386
Courtown (Brenogue) Harbour (Co. Wexford), 157, 160, 183, 204, 254–7, 255, 308, 329, 335, 399
Cox, Lemuel, 231, 235
Cox, R.C., 214
Croker, John Wilson, 143, 145, 176–7, 178, 179, 325
Croker, Thomas Crofton, 319
Crossfarnogue (Kilmore Quay, Co. Wexford), P15, 387, 408
Cubitt, William, 276, 319, 332
Currane, Lough, 58, 59

D'Arcy family of Clifden, 92, 94, 198, 206, 328, 359
Dargan, William, 316
de Courcy, Sean, 58, 82, 218, 319
de Valera Bridge (Cork), 124
Deane, Thomas, 122, 123, 124
Dee, River (England), 281, 283–4, 285–7, 286, 291, 293, 294–6
Derryinver Pier (Co. Galway), 356–7, 357, 412
Derryquin (Co. Kerry), P1, 67, 68
Dingle Pier (Co. Kerry), 381
diving bells, x, 107–8, 110, 111, 228, 279, 307, 322
docks and quays, 118–19, 122, 123–4, 225, 228–9, 278–9, 286, 304
    Wallasey Pool Plan, 283, 284, 287, 288, 288, 289
    see also piers and harbours
Donegal, Co., 164–5, 166–7, 393, 411
Donnell, James (Harbour Engineer, Fisheries Commission), 342, 347, 350, 352, 364, 371, 372, 374, 381, 384, 386, 389
    abandonments and, 165, 340, 345, 365, 379, 388
    appointment of (June 1825), 183–4, 340
    Ennisturk pier and, 205, 354
    funding issues and, 165, 184, 353, 358, 361, 365, 387, 388
    list of piers, 389, 392
    modifications/alterations and, 340, 342, 343–4, 361, 375–6, 377–8, 380–1, 382, 383, 385
    'relief' works and, 184, 358
    repairs and, 351, 358, 360, 363, 369, 370, 377, 379
    specifications/plans, 343–4, 353–5, 358, 359, 365, 373, 375–7, 382, 385, 391, 393
Dooras (Duras) Pier (Co. Galway), 182, 370, 372, 372–3
Doughbeg Wall (Co. Galway), 358–60
Down, Co., 196, 406
    piers and harbours, P16, 160, 167, 168, 183, 257, 258, 389–92, 390, 391, 407, 411
Downshire, Lord, 218
drainage of land, 44, 51, 54, 60–1, 92, 138, 140, 400, 401
    irrigation and, 59–60, 62
'Draining' article in Edinburgh Encyclopaedia (AN), 39–40, 404
Drogheda Harbour (Co. Louth), 160, 183, 259–61, 329, 389
Dublin, Co., 47, 83, 177, 196, 388–9, 411
Dublin and Kingstown Railway Company, 314–15
Dublin Bay, 168, 170, 171, 172, 174, 207, 408
Dublin Castle Fund, 182, 183, 188–9, 194
    piers and harbours, 343, 349, 350, 351, 354, 374, 377
    piers and harbours (Co. Galway), 357, 360, 363, 364, 365, 369, 370, 373

INDEX

Dublin Philosophical Journal, 300, 404
Dublin Society, 43–4, 46–7, 48, 54, 83, 85, 86, 101
Dublin Steam Packet Company, 270, 271
Dunkerron (Co. Kerry), 66, 67–72
Dunmore (survey cutter), 164, 165, 167, 181
Dunmore Harbour, 102–14, 113, 114, 115, 138, 159, 181, 183, 184, 253, 265, 332
   bridges, 103, 113–14, 220
   financial issues, 103, 109–11, 112, 113, 182, 329
   mail boat route to Wales, 102, 230, 277
   maps and plans, 104, 105, 106–7, 410
   Parliamentary bill, 105–6, 108, 109, 115, 308
   reports on, 400, 401

economics, 34–5, 43, 44, 45, 55, 61–4, 77, 88–94, 264–5
   local taxation, 27, 128, 129, 130, 131–4, 140, 141–5, 148, 150, 152, 156
   Scotland, 18–21
   trade routes, 259, 261, 264, 266–73
   see also famine; financial issues; poverty and distress
Edgeworth, Maria, 89, 243, 321, 328, 329, 335
Edgeworth, Richard Lovell, 48
Edgeworth, William, 135, 143, 265, 269, 272, 321, 328
Edinburgh, 1, 2, 3–5, 4, 7, 14
Edinburgh Encyclopaedia, 5, 37–41, 103, 332, 403–4
Edinburgh Magazine (later *Edinburgh Philosophical Journal*), 3–5, 37
Ellis, Rev. Philip Brabazon, 330, 336
emigration, 18, 20–1, 129, 415
Eminent Men of Fife (M.F. Connolly, 1866), 319
Employment of the Poor Act (1822), 151–2, 153, 156–7, 158, 187–9, 190, 206
   see also western district (under Employment of the Poor Act, 1822)
engineering, civil, 48–54, 128, 131, 134–7, 149–50, 151–2, 153, 155–9
   training and, 41, 136–7, 332
England, William, 385
Ennisturk (Inishturk) pier (Co. Mayo), 200, 205, 353–4, 412
The Entrance to Strangford Lough (chart by AN), 167, 172–5, 173, 175, 177, 196, 407, 411
Erne, Lough, 263–4, 304, 342
Erriff Bridge (Co. Mayo), 221, 239, 239, 241, 248, 332

famine, ix, 54, 106, 119, 128, 137, 148–50
   1822 (western/south-western counties), 151, 181–4, 187–9, 197–8, 362, 373, 386
Fennelly, June, 220
Ferguson, William, 1–2, 320
financial affairs of AN, 13, 49, 51, 160, 205, 265, 329–30, 420–1
   expenses, 108, 109, 184, 311, 312
   fees, 13, 36, 108, 112, 116, 121, 124, 158, 166, 312–13
   problems with accounting, 51, 160, 205, 210, 211, 311–13, 329, 420–1
   salary, 5, 13, 48, 101, 112, 157–8, 184, 210, 211, 293, 313, 329
financial issues, 206–11, 225, 257, 260
   criticism of AN, 160, 199, 202–3, 205–6, 207, 209–10, 211
   speculation, 264–5, 300
   see also economics
fisheries, Irish, 43, 89, 91, 114, 129, 138–9, 140, 163–5, 166–8
   see also Fisheries Commission
fisheries, Scottish, 19–20, 21, 138–9, 163
Fisheries Act, Irish (1819), 139, 141, 163
Fisheries Commission, x, 128, 141, 157, 159, 160, 163–9, 256, 313, 327, 329, 340
   appointment of James Donnell (June 1825), 183–4, 340
   coastal survey *see* coastal survey, Fisheries Commission
   Maurice Fitzgerald and, 164, 265
   maps and charts, 167–8, 169, 172–4, 173, 175, 203, 406–8, 410–12, 415
   piers and harbours, 164–9, 178–9, 181–4, 190, 191, 206, 340–51, 357–61, 374–93, 407, 408

   see also Donnell, James (Harbour Engineer, Fisheries Commission)
   poverty relief works and, 129, 181–4, 341, 346, 358, 362
   written reports to, 165, 172, 401, 402
Fitzgerald, Sir Maurice (1772-1849), 159, 164, 265, 302, 333, 336
   AN and, 63, 65, 160, 164, 179, 183, 203, 205–6, 260, 265–73, 290, 328–9
   Select Committees and, 131, 143, 159
   Valentia (Co. Kerry) and, 65, 265–73, 382, 392
Fitzgerald, William Vesey, 121, 130, 131, 136, 143, 303
Fitzwilliam, Lord, 255, 306
Fort George, 4, 15–16, 27, 404
Fort William, 4, 21, 27, 32, 33
Fortrose Academy, 4, 5–6, 7, 10, 11, 14, 15, 24, 141
Foster, John, 44, 47, 49, 143
Foyle, Lough, 164
France, 103, 106, 136, 159, 224, 225
Fraser, Robert, 137, 139, 141
Fraser of Lovat, 25–6
funding of public works, 139–41, 149–61, 192, 194, 201–3
   Consolidated Fund, 129, 152, 153, 156, 159, 190, 206, 225, 233, 305
   Grand Juries and *see* Grand Juries: public works and presentments legislation *see* legislation, Parliamentary
   see also Dublin Castle Fund; London Committee for the Relief of Irish Distress (London Tavern Committee); Mansion House Committee

Gaelic place names, 34, 324
Galway, Co., 13, 47, 55, 156, 160, 184, 189, 190–1, 198, 202, 417–18
   bridges, 220, 240, 242, 243, 244–7, 245, 246, 248, 248, 249
   Joyce's Country, 81, 87, 88, 89, 91, 191, 196–7, 197, 220
   piers and harbours, P7, 200, 324, 356–73, 407, 408, 412–13
   piers and harbours (Killary), 87, 91, 94, 95, 191, 209, 355, 356
   see also Ballinahinch (Co. Galway); Connemara (Co. Galway)
Galway Bay, 171, 177, 195
Galway town, 82, 87, 88, 94, 198, 199–201, 279, 417
   Floating Harbour, 367–9, 368
Gentleman's Magazine, 319
Geological Society of London, 14, 25, 37, 84, 85, 103
Geological Survey of Ireland, x, 55, 84, 86–7, 171
geology, 3, 13–14, 58–9, 67, 72, 79–88, 404
   Connemara, P2, P3, 14, 79–85, 87, 103–5
   sea-bed survey, P5, 169–71, 172, 173, 203
Glenanean Bridge (Co. Mayo), 221, 223, 239, 332
Glencoe, 23, 31–2, 35
Glengall, Earl of, 301, 302, 303
Glenocally Bridge (on N59 road, Co. Mayo), 220, 222, 236
Gorumna Pier (Greatman's Bay, Co. Galway), 362, 362
Goulburn, H., 157, 172, 187, 188, 189, 199, 201, 219, 345, 387
   criticism of AN, 194, 195, 196
Gould, M.H., 214
Gowlaun Bridge (Co. Mayo), 236, 237
Grand Canal, 47, 48, 128, 193, 194, 264, 305, 313, 314, 315
Grand Juries, 15, 128–9, 130–4, 142–6, 221
   local taxation and, 128, 130, 140, 141–5
   Parliamentary Acts, 134, 136, 141, 150–2, 153–7
   public works and presentments, 105, 128–9, 130–7, 140, 141–6, 148–57, 159, 189, 201, 219
   public works and presentments (roads), 134, 153, 189, 190, 195, 218, 265
Grant, Charles, 40, 109, 137, 140, 141, 156
Grant, Duncan, 26, 141
Grant, William, 181
Greatman's Bay (Co. Galway), 362, 362, 413, 415–16
Greencastle Harbour (Co. Donegal), 164, 165, 166–7, 393
Greenough, George Bellas, P2, 14, 84–5, 86, 87, 135
Gregory, St J.G., 267, 268, 269–70
Gregory, William, 110, 148, 149, 152, 156, 179, 187, 188, 194, 232

429

Grey, Earl, 210, 211
Greystones Harbour (Co. Wicklow), 387–8
Griffith, Richard, 14–15, 47, 55, 135, 136, 143, 145, 160, 263
    Bogs Commission and, 14, 48, 50, 53, 55, 83–4, 98–9, 128, 158
    geological map of Ireland, x, 55, 83, 84, 86–7, 171
    maps and charts and, 84, 85–7, 98–9, 141–2, 171
    southern district engineer, 151–2, 187, 197, 200, 201, 202, 206–7, 266
Griffith, Richard (Snr), 46–7, 48, 99
Groome, F., 28
Grubb, Robert, 110, 112
Guinness, Arthur, 303

Haddiscoe cut (Yare Navigation, Norfolk), 276–7, 331
Haldane, A.R.B., 36
Hall, Sir James, 3, 14
Halpin, George, 113, 256, 334
Hamilton, William Rowan, 111, 227, 321, 322–4
harbours and seaports *see* piers and harbours
Hawkshaw, John, 311, 312
Hay, Alexander, 357, 360, 370
Heather, William, 178
Helbrè Island (Dee estuary), 283–4, 285, 286, 287–8, 289, 290, 292, 294, 295
Herries-Davies, G.L., 50, 82–3, 85–6
Hibernian General Railway Company, 301–6, 313, 408
Hibernian Mining Company, 262–5, 399
Highland clearances, 18, 34
Highland Society of London, 19, 20, 21
Hill, Sir George, 109
Holland, 103, 106, 138, 278, 401
Horner, A., 58
House of Lords, 40, 146, 193, 263, 308, 322, 326–7, 331
Hutton, James, 2, 3
Hutton, Robert, 84, 85, 86, 103, 135
hydrography, ix, x, 16–17, 67–8, 165–79, 183, 333
    charts, 116, 166, 167–75, 173, 175, 177, 196, 272–3, 407, 411
    sea-bed survey, x, P5, 169–71, 172, 173, 203

Iar-Connacht (Co. Galway), 80, 87, 88, 89, 92–3, 94–5
industry, 43, 54, 58, 65, 91, 263, 264, 300, 309, 323
    mining, 261–5, 323
'Inland Navigation' article in *Edinburgh Encyclopaedia*, 39–40, 403
Institution of Civil Engineers, 307, 311
Inverness, 4, 8, 14, 17–18, 21, 28
Inverness Academy, 4, 6–13, 14, 16, 17–18, 33, 49, 141
'Inverness Perambulation' (survey), 15, 23–37
Irish Sea, 170
irrigation, 59–60, 62
Iveragh peninsula (Co. Kerry), 55, 57–66, 68, 72, 181, 265, 271, 405, 409

Jameson, Robert, 3, 5, 14
Jessop, William, 40, 106, 314
Johnston, Francis, 135
Jones, John E., iii, 233–4, 235, 235, 236
Jones, Samuel, 206, 249, 328, 359
Joyce's Country (Co. Galway), 81, 87, 88, 89, 91, 191, 196–7, 197, 220

Kearn's Port (Cape Clear, Co. Cork), 386
kelp-making, 89–90
Kenmare Pier (Co. Kerry), 262, 329, 382–4, 383
Kerry, Co., 49, 196
    Bogs Commission and, 13, 55, 57–77, 404, 405, 409, 410
    Iveragh peninsula, 55, 57–66, 68, 72, 265, 271, 405, 409
    piers and harbours in, 77, 262, 266–73, 329, 379–84, 398, 408, 414
    see also Valentia (Co. Kerry)
    roads, ix, 61, 65, 66, 67, 68, 72, 101, 156, 157, 181, 262
    roads (proposed), P1, 65–6, 68–72, 77
Kilbaha Pier (Co. Clare), 377, 377–8, 378
Kilhale Bridge (Co. Mayo), 243, 244

Killala Harbour (Co. Mayo), 168–9, 172, 181, 190, 344, 345, 346–7, 347, 415
Killaly, Hamilton, 209, 368
Killaly, John, 136–7, 203, 205, 209, 219–20, 221, 305, 314, 376
    Board of Inland Navigation and, 122–3, 202, 207–8, 258, 305
    bridges and, 219–20, 221, 230, 232, 233, 239, 241, 244, 398
    directing engineer for Mayo and Galway (1831-2), 207–11
    Loan Commissioners and, 122, 149–50, 156–7, 203, 368
    middle district engineer, 151–2, 156–7, 187, 197, 201, 206, 207
    Moiety Act (1820) and, 151, 156–7, 158, 219, 255
    Office of Public Works and, 368
    roads and, 157, 195, 202, 203, 207–11, 220, 221
Killany (Killeany) Pier (Aran Islands), 373, 413
Killarney (Co. Kerry), 57, 61, 62, 65, 68, 72, 74, 75, 204
Killary Harbour (Co. Galway), 87, 91, 94, 95, 191, 209, 355, 356
Killorglin (Co. Kerry), 61, 65, 67, 68, 72, 73, 74
Killough Harbour (Co. Down), P16, 167, 168, 183, 257, 391, 391–2, 407, 411
King's House (near Ardfeadh), 4, 30, 31, 35–6, 37
Kingussie (Scotland), 4, 28–9, 34
Kirkcaldy (Fife), 1, 4, 15, 103, 320
Kirwan, Richard (1733-1812), 79–80
Knight, Patrick, 181, 189–90, 205, 348
Knightstown Pier (Valentia, Co. Kerry), 381–2, 382
Knightstown Village, 272–3
Knox, John, 19, 23

labour issues, 199–201, 327–8
Laird, William, 282–6, 289–95, 297, 307
Lamb, William, 205, 207
land ownership and management, 54, 55, 61–4, 67, 90–1
Lansdowne, Lord (Henry Petty-Fitzmaurice), 265–6, 269, 329, 333, 382–3
Larkin, William, 79, 135
Larne Harbour (Co. Antrim), 168
LaTouche, Peter, 160, 188, 303, 330, 387
Laune, River, 57, 60, 74, 75
Lee, River (River of Cork) *see* Cork River and Harbour
Leenane Bridge (Co. Galway), 243
Leenane Pier (Co. Galway), 196, 197, 356, 356
Lees, Edward, 107, 108, 109, 110, 111–12, 218, 230–1, 232, 268
legislation, Parliamentary, ix, 21, 23, 62–3, 128–9, 149–60, 202, 232–3
    bog reclamation, 44–5, 48, 55, 63
    drafting and promotion of bills, 105, 108, 225–6, 308, 310, 311, 315
    Dunmore Harbour bill, 105–6, 108, 109
    Employment of the Poor Act (1822), 151–2, 153, 156–7, 158, 187–9, 190, 206
    Fisheries Act, Irish (1819), 139, 141, 163
    Grand Juries and, 134, 136, 141, 150–2, 153–7
    Loan Act (1817), 122, 129, 140, 149–51, 153, 156, 157, 190, 221, 225–7, 230, 256
    local/private acts, ix, 115, 122, 232–3, 255, 264–5, 268–9, 270, 308, 309–10, 367–8
    Moiety Act (1820), 151, 153, 155, 156, 157–8, 190, 206, 218, 219, 221, 255, 262
    Railway Act, Irish (1826), 305, 306, 316
    Road Acts, 101, 134, 136, 153, 194
Leitrim, Co., 156, 164, 202, 263, 264, 267, 419
Leveson Gower, Lord, 205, 206
Lewis, Samuel, 218, 219
lighthouses, 102, 107, 109, 112, 114
lime-making, 91
Limerick, Co., 13, 49, 57, 75, 111
Limerick City, 12, 204, 224–30, 304
Limerick to Waterford Railway, ix, 80, 81, 225, 264, 300–6, 313, 316, 404, 408

INDEX

limestone, 74–5, 82, 87–8, 89, 91, 92, 93
Liscannor Pier (Co. Clare), 181, 183, 327, 375, 375–6
Liverpool, 259, 266, 269, 271, 279–97, 307, 310, 311, 313, 399, 402
Liverpool and Leeds Railway Company, 211, 307–13, 329
Loan Act (1817), 122, 129, 140, 149–51, 153, 156, 157, 190, 221, 225–7, 230, 256
Loan Commissioners, 138, 149–50, 151, 155, 157, 203, 224, 225–7, 305, 368
    John Killaly and, 122, 149–50, 156–7, 203, 368
    AN and, 149–50, 157, 232, 329
Loch Ness, 15, 16–17
London Committee for the Relief of Irish Distress (London Tavern Committee), 182, 183, 187–8
    piers and harbours, 343, 350, 354, 374, 378, 380, 381, 384, 385, 386
    piers and harbours (Co. Galway), 364, 365, 369, 370, 373
Longford, Co., 47, 48
Lords-Lieutenant, 153, 163, 172, 179, 202–3, 209–10, 218, 221, 345
    public works legislation and, 134, 149–57, 187, 188–9, 206, 218, 219, 221, 255, 262, 305
Louth, Co., 160, 177, 183, 259–61, 329, 389
Lowestoft (Suffolk), 276, 277
Lynch, Patrick, 370, 372

Maam (Joyce's Country), 34, 196–7, 197, 200, 220, 240, 241, 241, 248, 335
Mackenzie, Murdoch, 167–8, 172, 174, 177, 178
Mackenzie, Sir George Steuart, 25, 41, 65, 68
MacNeill, John (1793-1880), 261
Mahon, Lough, 118, 124, 166
mail boats, 102, 106, 108, 111, 112–13, 230–6, 304, 307
    Dunmore to Wales, 102, 230, 277
mail coach roads, 101, 105, 189, 194–5, 218, 265, 277–9
Maine, River, 57, 60, 73, 410
Manchester, 279–80, 281, 308, 309, 311
Manchester and Liverpool Railway Company, 307–8
Mansion House Committee, 182, 183, 349, 351, 352, 357, 360, 363, 377, 379, 381
maps and charts, x, 13–14, 172–9, 333, 404–9, 415–16
    alleged errors in, 172–4, 173, 175, 196
    Bogs Commission and, P1, P2, P3, 46, 51–4, 79, 83, 94, 95, 98–9, 103–5
    Bogs Commission (Co. Kerry) and, P1, 58, 61, 65, 66, 67–8, 404–5, 409–10
    of Co. Kerry, 57, 58, 61, 65, 66, 67–8, 404–5, 409–10
    of Co. Mayo, P12, 48
    of Connemara see Connemara (Co. Galway): maps and charts of England and Wales, 84, 87
    engraved and/or published, 167, 169, 172, 179, 409–14
    Richard Griffith and, 84, 86–7, 98–9, 141–2, 171
    of Ireland, x, 47, 51–4, 84, 85–7, 144–5
    Knightstown Village and, 272–3
    local taxation and, 141–2
    maritime charts, 166, 167–8, 169, 172–4, 173, 175, 177–8, 203
    piers and harbours, P12–P17, 179, 261, 272–3, 347, 368, 377, 380, 388, 390, 391, 392
    piers and harbours (Cork), 116, 119–21, 120, 415
    piers and harbours (Dunmore), 104, 105, 106–7, 410
    piers and harbours (listing and locations), 407–8, 410, 411, 412–13, 414
    piers and harbours (Valentia), 179, 272–3, 414
    River Moy, P6, 346
    roads, P11, 35, 95, 101, 105, 406, 407, 409, 412
    of Scotland, 23, 25, 25, 26, 35
    sea-bed survey, P5, 169, 170, 171
Marchant, T.R., 83
market house (Galway town), 198, 279
Marmion, A., 123
Martin, Richard, 80, 198, 241, 301, 303
Martin family of Ballinahinch, 92, 93, 94, 206, 321, 328, 357, 360
Mask, Lough, 80, 81, 82, 92, 95
mathematics, 38–9, 41, 127, 227, 320, 332

Mayo, Co., 15, 48–9, 50, 72, 83, 98, 101, 105, 143, 156, 168, 177
    bridges, 220, 221, 222, 223, 226, 236–48, 237, 239, 243, 245, 332
    listing of all works, 418–19
    piers and harbours, P12, P13, 189, 190, 192, 200, 205, 345–56, 408, 410, 412
    see also Killala Harbour (Co. Mayo)
    roads, 105, 189–90, 194–5, 202, 221
MBB canal, 308, 309–11, 315
McDonagh, Patrick, 244–7
McGill, Alex (resident engineer at Dunmore), 108, 110, 181, 200, 256, 354, 358, 361, 384, 385
mechanical engineering, 307
Menzies, Sir Robert, 29, 30
Meredith, Ben, 66, 181, 229, 273, 303, 311, 380, 382, 408, 412
Mersey and Irwell Navigation Company (M&I), 280–1
Mersey navigation, 203, 279–97, 398, 399
Milford Haven, 102, 230–1, 277–9, 304, 413
mineralogy, 43, 58–9, 79–88
    see also geology
mining, 72, 261–5, 323
Mitchell, James, 8–9, 11, 12
M'Manus, Henry, 234, 335
Mohr, Professor P., P3, 80
Moiety Act (1820), 151, 153, 155, 156, 157–8, 190, 206, 218, 219, 221, 255, 262
Monaghan, Co., 409
Moray Firth, 15–16, 26
mountains, x, 24, 57, 66, 72, 74, 75, 80, 82, 93
    Grampian, 27, 28, 29
Moy, River, P6, 168, 171, 192, 247, 344, 345–6, 408
Mudie, Robert, 10–11, 265, 319, 325, 330
Mullaghmore Harbour (Co. Sligo), 164, 166, 183, 197, 341–2, 342, 393, 412
Munster, 63, 79

Napoleonic Wars, 9, 54, 148
National Archives of Ireland, 101, 189, 219, 345, 370, 376
National Library of Ireland, 54, 175, 220, 378, 379
Navigation wall (Cork), 116–8, P4, 120, 121, 123
Neagh, Lough, 258, 259, 264
Neptunists and Plutonists, 3, 14
Neuilly, Pont (Paris), 224, 225
New Harbour (Rinville, Co. Galway), 369, 369, 413
New Piloting Directions for St George's Channel and the Coast of Ireland (AN, 1832), 171, 207, 334, 404
Newcastle, 300
Newport, Sir John, 137–8, 143, 333
Newry Navigation, 258–9, 406, 410
Nicholson, Peter (1765-1844), 37, 39, 403
Nimmo, Alexander (nephew of AN) 356–7
Nimmo, Alexander (AN, 1783-1832)
    articles published, 37–40, 103, 300, 332, 403, 404
    biographical information, ix, x, 1–5, 103, 112, 160, 196, 319–37
    biographies of, 82, 319
    books and papers published, 403–4
    bust of by J.E. Jones, iii, 234
    contemporary anecdotes about, 11, 319–20, 321, 322–4, 330
    criticism of, 160, 194, 195, 196, 199, 202–3, 204–6, 207, 209–10, 211, 325
    family of, 1, 13, 196, 319, 320, 330, 334–7
    see also Nimmo, George (brother of AN)
    health of, 187, 204, 207, 211, 310, 334
    integrated approach of, 193–4, 196, 197–8, 334
    personal qualities, ix, 321, 322, 323–4, 325, 326–8, 330–1
    political arena and, 14, 127, 145, 296, 329
    reports (MS, unpublished), 398–400
    reports (published), 82, 400–3
    as schoolmaster, 5–13, 141, 321, 324
    staff and assistants of see staff and assistants of AN

structural legacies, 331, 332–3
Nimmo, George (brother of AN), 13, 200, 220, 231–2, 233, 234, 255, 257, 334–6, 409
  Wallasey Pool Plan and, 284, 285, 290, 294, 297
Nimmo, John (brother of AN), 196, 319, 320, 334, 336
Nimmo's Bridge (Ring of Kerry), 66, 71

O'Brien, W., 263
Observations on the distillery Laws of the United Kingdom (George Nimmo, 1838), 335
oceanography, 171, 172
O'Connell, Daniel, 269, 302–3, 328–9
O'Connell's Bridge (Ring of Kerry), 66
Office of Public Works (OPW), x, 112, 128, 228, 333, 334, 335, 349
  establishment of (1831), 129, 159–61, 184, 202
  harbours and, 112–13, 160, 257, 261, 341, 368, 391
O'Flaherty, Roderick, 79
O'Hara, William, 200, 256, 354, 355
O'Keeffe, P., 214, 217–18, 221
Old Head Pier (Co. Mayo), 352–3, 353, 355, 412
O'Malley, Sir Samuel, 352
'On Railways' (AN article in *Dublin Philosophical Journal*), 300
Onandoo Bridge (Co. Mayo), 221, 226
Ordnance Survey of England, 176–7
Ordnance Survey of Ireland, x, 53, 55, 128–9, 137, 142–5, 146, 176–9, 333
Otway, Rev. Caesar, 323–4, 335
Oughterard (Co. Galway), 80, 81, 84, 87, 94, 190, 191
Owen, Jacob, 257

P&O Steam Navigation Company, 270
packet boats *see* mail boats
Paine, Henry, 112
Palmerston, Lord, 166, 183, 342
Parliament, 44–5, 47, 48, 268–9, 270, 401–2
  fisheries and, 19, 43, 129, 139, 140, 141, 163
  Grand Juries and, 134, 136, 141, 150–2
  House of Lords, 40, 146, 193, 263, 308, 322, 326–7, 331
  railways and, 302–3, 305, 306, 308, 309, 310, 314–15
  see also legislation, Parliamentary; Select Committees
Paxton, R., 82, 319
Peel, Robert, 121, 127, 131, 134–5, 136–7, 148, 149, 265, 330
  Dunmore Harbour and, 102, 105–6, 107, 108, 109
Perronet, Jean, 224, 225, 228
Perth (Scotland), 4, 26, 27, 28, 30, 35–6
piers and harbours, ix, 96, 103, 166–7, 179, 181–4, 402
  in Co. Antrim, P17, 165, 168, 257, 392, 406
  in Co. Clare, P14, 181, 183, 373–9, 407, 413
  in Co. Cork, 384–6, 399
  see also Cork River and Harbour
  in Co. Donegal, 164, 165, 166–7, 393
  in Co. Down, P16, 160, 167, 168, 183, 257, 258, 389–92, 390, 391, 407, 411
  in Co. Dublin, 388–9
  in Co. Kerry, 77, 262, 266–73, 329, 379–84, 398, 408, 414
  see also Valentia (Co. Kerry)
  in Co. Louth, 160, 183, 259–61, 329, 389
  in Co. Mayo, P6, P13, 189, 190, 192, 200, 205, 345–56, 408, 410, 412
  see also Killala Harbour (Co. Mayo)
  in Co. Wexford, 386–7P15, 408
  see also Courtown (Brenogue) Harbour
  in Co. Wicklow, 387–8
  in Connaught, 168, 187, 191–2, 192, 193, 196, 198, 205
  see also western district: piers and harbours
  Fisheries Commission and, 164–9, 178–9, 181–4, 190, 191, 206, 340–51, 357–61, 374–93, 407, 408
  see also Donnell, James (Harbour Engineer, Fisheries Commission)
  later modifications to, 114, 340, 361
  maps and charts *see* maps and charts: piers and harbours
  Moy River (Co. Mayo), P6, 345, 345–6, 408
  'Nimmo's Pier' (Galway), 324, 366–7, 368

site selection, 340–1, 342
  see also Galway, Co.: piers and harbours; Sligo, Co.: harbours
Pim, James, 303, 313, 314
Playfair, John, 1–2, 3, 5, 7–8, 14, 15, 16, 37, 320, 325
Plutonists and Neptunists, 3, 14
political and social issues, 44, 63–4, 90, 92, 106, 107, 122, 128–35, 137–8, 209, 326–9
  tithes, 62–3, 90
  see also poverty and distress; public works: as poverty relief strategy
Pollagheeny Harbour (Co. Sligo), 344, 344–5
Ponts et Chaussées (Corp and Ingeneurs) 136, 161
Poor Laws and Rates, 326–7, 328, 401
Port Ballintrae (Co. Antrim), P17, 165, 257, 406
Porter, Robert, 121, 122, 123, 124
Post Office, 101, 159, 189, 194, 195, 206, 218, 268, 271, 329
  Dunmore Harbour and, 102, 109, 110
  Youghal Bridge and, 230–1, 232–3
potato cultivation, 62–3, 89, 148, 188
Poulaphuca Bridge (Co. Wicklow), P8, P9, 114, 157, 214, 217–20, 221
poverty and distress, 61, 92, 106, 119, 137–41, 148–61, 187–203, 326–8, 401
  Fisheries Commission and, 129, 181–4, 341, 346, 358, 362
  see also famine; public works: as poverty relief strategy
Presbyterian Church, Scottish, 5
Preston and Wigan Railway Company, 313
Price, Henry H. (1794-1839), 107, 278
Princeton University (USA), 335
Prior, Thomas, 44
private commissions, 253–73
  Ballingaule Pier and Village, 253–4, 254, 406
  Courtown (Brenogue) Harbour, 157, 160, 183, 204, 254–7, 255, 308, 329, 335, 399
  Drogheda Harbour (Co. Louth), 160, 183, 259–61, 329, 389
  Killough Harbour (Co. Down), P16, 167, 168, 183, 391, 391–2, 407, 411
  Liverpool and Leeds Railway Company, 211, 307–13, 329
  Newry Navigation, 258–9, 406, 410
  Valentia (Co. Kerry), 266–73, 414
  see also Cork River and Harbour
public works, 128–41, 149–61, 199–201, 401, 403
  funding of *see* funding of public works
  legislation, 128–9, 134, 136, 149–60, 194, 202, 232–3, 255, 305
  see also Employment of the Poor Act (1822); Loan Act (1817); Moiety Act (1820)
  Lords-Lieutenant and, 134, 149–57, 187, 188–9, 206, 218, 219, 221, 255, 262, 305
  as poverty relief strategy, 20, 106, 107, 119, 121, 129, 137–41, 148–61, 187–203, 206–11, 401
  see also western district (under Employment of the Poor Act, 1822)
  as poverty relief strategy (Fisheries Commission), 181–4, 341, 346, 358, 362
public bodies *see* Bogs Commission; Fisheries Commission; Office of Public Works (OPW)
Select Committees and, 129, 137–41, 158, 221–4, 231, 232
  see also bridges; canals; piers and harbours; railways; roads
Pulteney, William, 19–20

Railway Act, Irish (1826), 305, 306, 316
railways, 65, 75–6, 157, 263, 297, 300–16, 329, 404, 408
  Dublin to Kingstown, ix–x, 160, 207, 297, 306, 313–16
  Limerick to Waterford, ix, 225, 264, 300–6, 316, 404, 408
  Liverpool and Leeds Railway Company, 211, 307–13, 329
  Parliament and, 302–3, 305, 306, 308, 309, 310, 314–15
Rannoch Moor, 27, 29, 30–1, 33, 35–7
Raughley Harbour (Co. Sligo), 164, 166, 200, 332, 342–4, 343, 349–50, 356, 393, 410, 412
Rennie, John, 1, 50, 51, 106, 258, 280, 314, 331, 332, 391

INDEX

Dunmore Harbour and, 102, 105, 106, 107
Rhodes, Thomas, 12, 15, 225, 228–9, 264, 381
Rickman, John, 21, 33, 34, 35, 281–2, 283, 284, 292, 294, 297
Ring of Kerry (N70 road), ix, 66, 67, 68, 71, 214
Ritson, James, 284, 285, 294
roads
  Bealagh Beama Pass road, 72
  Board of Inland Navigation and, 153, 193–4, 202, 207, 247
  Bogs Commission and, 45
  Co. Mayo, 105, 189–90, 194–5, 202, 221
  Connaught, 146, 187, 189–91, 192, 192–5, 196, 198, 202, 203, 206, 207–11
  Connemara, P11, 89, 92–3, 94–6, 97–8, 101, 190–1, 206, 239, 356, 361, 363
  Drung Hill road, P1, 61, 65, 66, 181
  fisheries and, 138, 140, 191
  funding of, 150, 151–2, 153, 156, 157
  General Wade's, 18, 27
  Grand Juries and, 134, 153, 189, 190, 195, 218, 265
  Holyhead Road, 261
  John Killaly and, 157, 202, 203, 207–11, 220, 221
  in Kerry, ix, 61, 65, 66, 67, 68, 72, 101, 156, 157, 181, 262
  Killarney to Kenmare, 68, 72, 157, 262, 332
  mail coaches and, 101, 105, 189, 194–5, 218, 265, 277–9
  maps and plans, P11, 35, 95, 101, 105, 406, 407, 409, 412
  Office of Public Works and, 160
  Parliamentary Acts, 101, 134, 136, 153, 194
  proposed, P1, 65–6, 68–72, 77, 94–6, 229
reports on, 399, 401, 402
Ring of Kerry (N70), ix, 66, 67, 68, 71, 214
  in Scotland, 18, 19–20, 21, 27, 35–6, 139
Robe, River, bridge over (Co. Mayo), 244, 247
Roberts, Thomas Soutell, P4, 117
Robinson, Dr (astronomer), 321, 323, 324
Robinson, Tom (cartographer), 361
Robison, John, 3
Rocket, Stevenson's, 306
Roscommon, Co., 57, 101, 105, 143, 156, 194, 202, 419
Ross Island (near Killarney), 72, 261–5, 322, 329, 398
Rossaveal (Co. Galway), 363–4
Roundstone Harbour (Co. Galway), 196, 200, 201, 360, 360–1, 413
Roundstone village (Co. Galway), 196, 200, 201, 335, 336
Royal Canal, 47, 128, 193, 194, 315
Royal Dublin Society, iii, 234
Royal Hibernian Academy, 234
Royal Irish Academy, 68, 135, 169, 203, 412
Royal Society of Edinburgh, 7, 16, 38, 41, 49, 59, 103
Ruddock, Ted, 40, 82, 103, 227–8, 319
Runcorn Gap, 313, 400
Rusheen Bay Bridge (failed, Co. Galway), 244–7

Saleen (Ely Erris) Harbour (Co. Mayo), 189, 190, 349, 349–50, 412
Salt Lake (Ardbear) Bridge (Co. Galway), 220, 245, 248
sand, 62, 64, 68, 72–3, 88, 89, 91, 93
Scotland, 1–21, 4, 37–41, 63, 139–40
  commissions (est. 1803), 13, 15, 21, 23–37, 49, 50, 139, 140, 141, 195–6, 400
  'Inverness Perambulation', 15, 23–37
  roads, 18, 19–20, 21, 27, 35–6, 139
  sheep farming in, 18, 20, 34–5
  surveys and mapping, 14, 15–18, 20–1, 23–37, 25, 177
sea-bed survey, x, P5, 169–71, 172, 173, 203
Seafield Pier (Co. Clare), 376–7, 407
seaports and harbours *see* piers and harbours
Sefton, Earl, 309–10
Select Committees, x, 19, 92, 105, 128–9, 130–4, 195–6, 337, 415
  on Grand Jury system, 128, 130–4, 142–6
  Milford Haven Communication (1827), 230–1, 277–9, 413, 415

AN and, 137–8, 143–5, 158–9, 203, 231, 307, 401, 413, 414–15
public works and, 129, 137–41, 158, 221–4, 231, 232
on the State of Ireland as to Disease and the Labouring Poor, 40, 129, 137–41, 146, 158, 159, 163, 203, 328, 414
on Survey and Valuation of Ireland, 128–9, 143–5, 176–8, 203, 414
Thomas Telford and, 137, 139–40, 195–6
Sevastopulo, G.D., 83
Severn, River, 277–9
Shannon, River, 47, 48, 57, 76, 101, 191, 193, 224, 263, 264, 304
  crossings of, 192, 195, 229, 229–30, 408
Shannon navigation, 12, 193–4, 225, 263, 264, 267
Shaughnessy's Bridge (Connemara), 114, 220, 222, 242, 243
Shindela Bridge (Galway), 242, 243
Simington, T., 214, 217–18, 221
Slate breakwater (Galway), 198, 200
Sliabh Luachra, 72, 74, 266
Sligo, Co., 47, 156, 164, 166, 419
  harbours, 165–7, 168–9, 183, 192, 200, 341–5, 398, 401, 407, 410, 411, 412
  see also Mullaghmore harbour (Co. Sligo); Raughley Harbour (Co. Sligo)
  Port of Sligo, 166, 168, 182, 191–2, 329, 398, 401, 407, 410
Sligo, Marquis of, 169, 183, 353, 355
Smith, Adam, 328
Smith, William, 87
social and political issues *see* political and social issues
Southey, Robert, 324, 330
Spiddle (An Spidéal) Pier (Co. Galway), 364, 364, 413
Spring Rice, Thomas, 143, 158, 159, 160, 196, 224, 225, 329, 333, 335
Srahlea Bridge (Co. Mayo), 239
St. Andrews University, 1, 2, 3, 4, 5, 320
St Helens Railway and Canal Company, 313, 400
St Kitt's (Killeenaran) Pier (Co. Galway), 370–2, 371, 413
St Leger, Noblett, 311, 312, 319, 322, 334, 336
staff and assistants of AN, 51, 136–7, 143, 179–81, 199–201, 261, 332, 334–5, 340, 370
  expenses, 311–13, 329
  offices, 34, 196–7, 200, 240, 311–13, 322, 330, 335, 360
  see also Carter, Samson; England, William; Knight, Patrick; McGill, Alex (resident engineer at Dunmore); O'Hara, William; St Leger, Noblett
Staigue Fort (Co. Kerry), 68
Stanley, Edward, 207–9, 210, 211
steam boats, 112, 259, 260, 264, 266–73, 307
steam engines, x, 262, 300, 303, 306–7
Stevenson, David, 292, 331–2
Stevenson, Robert, 1, 17, 204, 296–7, 312, 331–2, 334
  Wallasey Pool Plan and, 283, 284–5, 287, 289–90, 292–3, 294, 399, 402
Strangford River and Lough, 167, 172–5, 173, 175, 177, 196, 407, 411
Subrine, Mons., 80
Suir, River, 102, 107, 231, 304, 305
Sunfish Bank, 169, 172
Survey and Valuation of Ireland (Boundary Survey), 55, 128–9, 142–6
Select Committee (Spring Rice Committee), 128–9, 143–5, 176–8, 203, 414
surveyors, 26, 48, 130, 131, 134–6, 144, 177–8
  training of, 136–7
  see also engineering, civil
Swansea, 231, 262, 277, 278, 416

Talbot, Lord, 135, 136
taxation, local, 27, 128, 129, 130, 131–4, 140, 148, 150, 152, 156
  boundaries and, 130, 131–4, 141–5
Taylor, Major Alexander, 85–6, 134, 177

Telford, Thomas, 12, 106, 110, 111, 181, 261, 276, 277, 305, 331, 332
   Board of Civil Engineers and, 134–5, 136
   Edinburgh Encyclopaedia and, 37–9
   AN and, ix, 24, 36–9, 49–50, 110, 166, 256, 276–82, 285, 294, 297, 325
   Scotland and, 1, 17, 19–21, 24, 36–7, 139–40, 141, 195–6
   Select Committees and, 137, 139–40, 195–6
   Wallasey Pool Plan and, 281–2, 283, 284–5, 286, 290, 292–3, 294, 297, 398, 399, 402
Tiernakill Bridge (Connemara), 241, 241–3
Tipperary, Co., 47, 48, 75
tithes, 62–3, 90, 130
Tobermory (Mull), 19, 253
Tobin, Sir John, 307, 308, 310–11
   Wallasey Pool Plan and, 282–3, 284, 285, 287–8, 290–1, 293, 294–5, 297
Toombeola Bridge (Co. Galway), 240
A Topographical Dictionary of Ireland (Samuel Lewis, 1835), 218, 219
Trench, Archbishop, 149, 187–8, 198, 336
Trigonometrical Survey of Ireland (Ordnance Survey), x, 53, 55, 128–9, 137, 142–5, 146, 176–9, 333
An tSean Céibh Pier (Rossaveal, Co. Galway), 363, 363–4
Tully Mountain (Co. Galway), 80, 82
Tumneenaun Bridge (Co. Mayo), 220, 223, 241
turf industry, 65, 75, 89, 263

United States of America, 266–73, 335

Valentia (Co. Kerry), 58, 59, 61, 64, 65, 66, 67, 171, 178–9, 207, 264, 414
   Atlantic trade route scheme, 65, 266–73
Vallancey, General Charles, 46, 47
Vignoles, Charles, 228, 307, 310, 313, 316, 337, 399, 403
Villiers-Tuthill, Kathleen, 193

Wade, General (1673-1748), 18, 27
Wallasey Pool Plan, 281–97, 286, 398, 399
   proposal documents, 284, 285–7, 286, 288, 295, 296
warping (controlled flooding), 73, 76
Waterford, Co., 102, 196

Waterford city and port, 102, 108, 112, 115, 121, 122–3, 171, 231, 235, 304
Waterford to Limerick Railway, ix, 80, 81, 225, 264, 300–6, 313, 316, 404, 408
waterways, 51, 64–5, 73–6, 88–9, 96–8, 102
warping (controlled flooding), 73, 76
   see also canals; and under entries for individual loughs and rivers
Watt, James, 3
Weir Bridge (Recess, Co. Galway), 246, 248, 248, 249
Wellesley, Richard, 189, 194, 207, 208, 225, 301
Wellesley Bridge (Limerick City), P10, 111, 160, 203, 214–17, 224–9, 226, 227, 304, 322, 329, 332
western district (under Employment of the Poor Act, 1822), 151, 157, 187–8, 189–203, 204–7, 313, 325, 399, 402, 403
   AN's neglect of, 203–4, 211
   AN's removal from, 207–11
   financial issues, 312, 329, 334
   listing of all works, 417–19
   piers and harbours, 340, 341, 343, 345, 346, 354, 379
   piers and harbours (Co. Galway), 356, 357, 358, 359, 361, 362, 363, 364, 365
   roads, 146, 193, 196, 202
   see also Connaught; Connemara (Co. Galway); Galway, Co.; Mayo, Co.
Westmeath, Co., 47, 48
Westport (Co. Galway), 92, 94, 95, 191, 209, 211
Westport (Co. Mayo), 169, 410
Wexford, Co., P15, 196, 386–7, 406, 408
   Courtown (Brenogue) Harbour, 157, 160, 183, 204, 254–7, 255, 308, 329, 335, 399
Wick (Pulteneytown), 19, 253
Wicklow, Co., 47, 83, 98, 196, 387–8, 408
Poulaphuca Bridge, P8, P9, 114, 157, 214, 217–20, 221
Williams, Charles Wye, 264, 307, 319, 324, 329–30, 331
Woods, Joseph, 80
Wordsworth, William, 324

Yganavan, Lough, 72–3
Youghal Bridge (Co. Cork), 217, 230–6, 234, 235, 237, 329, 335, 398, 409, 413